18th International Symposium on Magnetic Bearings (ISMB18)

Lyon, France
18-21 July 2023

ISBN: 978-1-7138-9516-9

Printed from e-media with permission by:

Curran Associates, Inc.
57 Morehouse Lane
Red Hook, NY 12571

Some format issues inherent in the e-media version may also appear in this print version.

Printed by Curran Associates, Inc. (2024)

For permission requests, please contact Linz Center of Mechatronics GmbH
at the address below.

Linz Center of Mechatronics GmbH
Altenberger Straße 69
4040 Linz
Osterreich

Phone: +43 732 2468-6002
Fax: +43 732 2468-6005

office@lcm.at

Additional copies of this publication are available from:

Curran Associates, Inc.
57 Morehouse Lane
Red Hook, NY 12571 USA
Phone: 845-758-0400
Fax: 845-758-2633
Email: curran@proceedings.com
Web: www.proceedings.com

FOREWORD

The International Symposium on Magnetic Bearings (ISMB) was held for the first time in 1988. Since then, the community of universities, research institutions and industrial companies working in the field of magnetic bearings has organized the meeting every two years. The conference is usually held alternately in the Americas, China, Europe, and Japan and brings together global experts and leaders in the progressing field of magnetic bearing technology.

All scientific papers undergo a peer-review process and are available digitally on the web-portal www.magneticbearings.org after the conferences. The Linz Center of Mechatronics launched this portal in 2014 and operates it in coordination with the international community.

Special thanks go to all the authors of the scientific contributions, the organizers of ISMB 18, assoc. Prof. J. Mafoud and his team, the members of the Advisory Committee, the sponsors of the web portal and the involved funding agency: The publication work has been supported by the COMET-K2 Center of the Linz Center of Mechatronics (LCM) funded by the Austrian federal government and the federal state of Upper Austria.

TABLE OF CONTENTS

Limitations of Reluctance Networks to Model the Frequency-Dependent Leakage and Fringing Fluxes in Active Magnetic Thrust Bearings

Robert SEIFERT

VEM Sachsenwerk GmbH, Pirnaer Landstraße 176, 01257 Dresden, Germany, robert.seifert@vem-group.com

Abstract

Previous works have demonstrated that analytical high–fidelity models of nonlaminated actuators and magnetic thrust bearings cannot just describe the magnetic skin effect inside the solid core, but also be applied directly within the control circuit. By an appropriate rational approximation a digital implementation on a microcontroller becomes possible. These approximated models generally do not consider frequency–dependent fringing and leakage fluxes, which may account for more than 7 % of the total flux. Reluctance networks are a popular choice to address this discrepancy. When calculated carefully, they can highly improve the accuracy of static models. However, their limitations in real–world scenarios are usually not discussed, even though the magnetic skin effect significantly changes the flux distribution in the nonlaminated core at already very low frequencies. In this article we review the practicability of reluctance networks and their possible simplifications in the context of real–time control systems. Depending on the control's objective we find they may be even discouraged, while simple correction factors allow for consistent results over the entire frequency range.

Keywords: Leakage Fluxes, Reluctance Networks, Eddy Currents, Active Magnetic Thrust Bearings, FO–Systems

1. Introduction

In recent years active magnetic bearings (AMBs) have seen an increase in their industrial applications due to a reduction in costs of the required power electronics and controllers but also in maintenance and operations, due to the establishment of remote service. This opens the market for new applications and prompts an increasing number of OEMs of traditional products, such as compressors, pumps and turbines to consider AMBs. As a consequence, there emerges a desire for high–stiffness AMBs with a disturbance suppression $\Delta K_{\mathrm{DS}} > 120\,\mathrm{dB}\frac{\mathrm{m}}{\mathrm{N}}$, which will match the stiffness of their mechanical counterparts ($\Delta K_{\mathrm{DS}} \approx 160\,\mathrm{dB}\frac{\mathrm{m}}{\mathrm{N}}$) to some degree. This is especially challenging for active magnetic thrust bearings (AMTB) as well as other solid–core actuators, as eddy current flow inside the core and the resultant magnetic skin effect is usually not avoidable (Seifert et al., 2021a).

The magnetic skin effect causes a significant lag between the force–generating magnetic flux and the measurable coil current, which results into a substantial damping of the actuator force in the often current controlled systems (Zhu et al., 2010). Alternatives, like a direct voltage control (Vischer, 1988) as well as hybrid forms (Ferreira et al., 2017; Keith, 1993) have not become established widely, considered only as a compromise. In previous works, we proposed a non–compromising control approach (Seifert et al., 2021a): instead of the measurable coil current, we control the actual force–generating flux as shown in Fig. 1. In the feedback branch we introduced a fractional–order flux estimator, which is able to determine the air gap flux from the measurable coil current in real–time. By use of the diffusion equation, the flux estimator mainly compensates the consequences of the magnetic skin effect caused by the eddy currents, but other nonlinearities may be considered as well. While the hysteresis is deemed neglectable, the core saturation can only be taken in account by a linearizing the core's permeability at a carefully determined operating point. Lastly, the inclusion of leakage and fringing fluxes is discussed in this paper. We review the practicability of reluctance networks as obvious choice often proposed in literature (e g. Le et al. (2016) and Sun et al. (2009) amongst others).

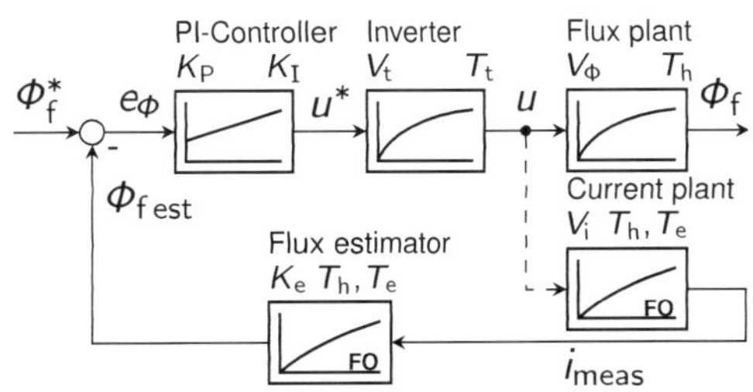

Figure 1: Closed–loop flux control with fractional–order flux estimator based on the measured coil current i_{meas}

Fringing and leakage fluxes are commonly subject of investigation in actuator modeling. However, in case of nonlaminated actuators like magnetic thrust bearings their accurate computation is challenging and the authors find that – more often than not – approaches found in the literature are based on oversimplified assumptions. Usually, reluctance networks calculated for the magnetostatic case are also applied to the frequency domain, without discussing the impact of a changing flux distribution due to the magnetic skin effect. That is why we review the common practice of reluctance networks for the dynamic case of nonlaminated actuators. We use of a revised set of analytical expressions we proposed in Seifert et al., 2021b to model the reluctances of the fringing and leakage flux paths more accurately. In addition, possible simplifications are suggested to include leakage and fringing fluxes in real–time control models and distinguish them depending on the control's objective.

This paper is organized as follows. Section 2 gives a brief introduction into the existing eddy–current model, on which the flux model and estimator is based on. For a deeper understanding of the physics and the mathematical modeling based on fractional–order systems, we like to refer to our previous work Seifert et al., 2019. In the main part of the article we use a FE–analysis (FEA) to identify all flux paths and form a reluctance network for the static case in section 3. The inverted model can then be used in section 4 to calculate all frequency–dependent fringing/leakage fluxes for a given magnetomotive force. In the final section 5 we discuss the accuracy of the reluctance model and its actual practicability in comparison with simple constant correction factors.

2. Eddy current effects and their modeling

Eddy currents in nonlaminated actuators have already been discussed thoroughly in the literature in recent years. The foundations of any high–fidelity model in the frequency–domain were laid down by Stoll, 1974, further refined by Feeley, 1996 to establish the so–called "*eddy–inductance*", which we used for the flux estimator. Cylindrical actuators and magnetic bearings were first considered by Rabinovici et al., 1992 and Kucera et al., 1996, respectively, leading to the almost complete model by Zhu et al., 2010.

In our previous work (Seifert et al., 2019) we presented a full historical review of eddy current models, a study on their physical impact as well as final model refinements. Due to this extensive groundwork we will introduce only the most essential equations here.

In a first step, we divide the actuator geometry (Fig. 2) into core parts, which are only permeated by a one–dimensional magnetic flux $\underline{\Phi}_i(\mathrm{j}\omega)$. This we calculate by solving the diffusion equation $\mathrm{curl}(\mathrm{curl}\,\vec{\underline{B}}) = -\underline{\alpha}^2\vec{\underline{B}}$ in its complex form for every core element, where $\underline{\alpha} = \sqrt{\mathrm{j}\omega\kappa\mu}$ denotes the complex wave propagation constant with the constant permeability $\mu = \mu_0\mu_r$ and electrical conductivity κ (for assumed isotropic and homogeneous materials). By help of the magnetomotive force (mmf) $\underline{\Theta}(\mathrm{j}\omega)$, we obtain the overall *effective reluctance* $\underline{\mathcal{R}}_{\mathrm{eff}}(\mathrm{j}\omega)$ as the sum of the part reluctances $\underline{\mathcal{R}}_i(\mathrm{j}\omega)$ of every core element $\underline{\mathcal{R}}_{\mathrm{eff}}(\mathrm{j}\omega)$, which translates into the *effective inductance* $\underline{L}_{\mathrm{eff}}(\mathrm{j}\omega)$ and the definition of the desired flux estimator:

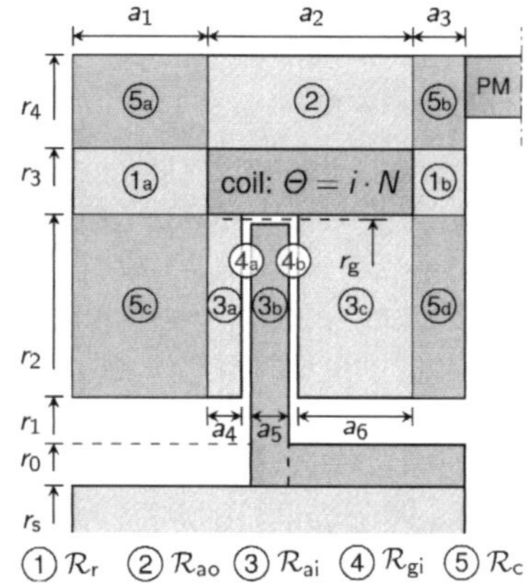

Figure 2: Magnetic circuit of AMTB divided into five classes of effective reluctances (cf. Seifert et al., 2019)

$$\underline{\mathcal{R}}_{\mathrm{eff}}(\mathrm{j}\omega) = \frac{N^2}{\underline{L}_{\mathrm{eff}}(\mathrm{j}\omega)} = \sum_i \frac{\underline{\Theta}(\mathrm{j}\omega)}{\underline{\Phi}_i(\mathrm{j}\omega)} = \sum_i \underline{\mathcal{R}}_i(\mathrm{j}\omega) \quad \Rightarrow \quad \underline{G}_{\mathrm{FE}}(\mathrm{j}\omega) = \frac{\underline{\Phi}_{\mathrm{f\,est}}(\mathrm{j}\omega)}{\underline{i}_{\mathrm{meas}}(\mathrm{j}\omega)} = \frac{L_{\mathrm{h}}}{N}\frac{R_{\mathrm{Cu}} + s\underline{L}_{\mathrm{eff}}(\mathrm{j}\omega)}{R_{\mathrm{Cu}} + sL_{\mathrm{h}}}, \tag{1}$$

with the number of coil turns N. The main inductance $L_{\mathrm{h}} = L_{\mathrm{eff}}(\omega = 0)$ corresponds to the force–generating component of current and flux, which is independent from the eddy currents.

3. Static Fringing and Leakage Flux Model

The basis for every dynamic model of fringing Φ^{υ} and leakage fluxes Φ^{σ} is a magnetostatic model. For the sake of readability we speak of *non–core fluxes* $\Phi^{\upsilon|\sigma}$ covering both flux types as opposed to the core or *main path fluxes* exclusively permeating through the iron core and the geometrical air gap. Air gap fluxes Φ_{gi}, directly crossing the gap without bulging, are attributed to the core fluxes, as they share a comparable frequency behavior. Early adaptations of reluctance networks considering non–core fluxes in radial magnetic bearings go back to Meeker et al. (1996). Later they have been applied to thrust bearings by Sun et al., 2009 and Wang et al., 2014 as well as combined bearings by Le et al., 2016 and Zhong et al., 2017, like in our case, but with significantly differing leakage paths and sometimes only for the magnetostatic case. Non–core fluxes in laminated radial bearings and conventional rotating machines are usually subject to only minor fluctuations and can be considered by a constant leakage k_σ and fringing factor k_υ, respectively. In nonlaminated AMTBs the magnetic skin effect alters the flux paths and the overall impact of the non–core fluxes depends on the frequency. For this reasons, we study the potential and practicability of a reluctance network and check whether it can be implemented within the flux estimator.

As a first step we assume a symmetrical thrust bearing (Fig. 3, bias flux omitted), to simplify the identification of the non–core flux paths in the field results of a magnetostatic FEA. Otherwise, negligible but misleading asymmetrical compensation fluxes occur. After identifying the reluctance network we calculate quantitative values for the non–core reluctances $\mathcal{R}^{\upsilon|\sigma}$. To achieve this, we use the fluxes determined for every network branch by the FEA for a given magnetomotive force (mmf) Θ. In section 4, the same network can then be used backwards to calculate the dynamic flux distribution for various frequencies and the real asymmetrical thrust bearing (cf. Fig. 2).

Generally, in thrust bearings we can differ between fringing and leakage fluxes as well as passive non–core fluxes. The latter have no influence on the force, as they do not bypass any air gaps. On the other hand, fringing fluxes bypass a single air gap in the direction of the force and therefore contribute to it. Leakage fluxes permeate perpendicular to the force or bypass both air gaps and are not part of the force generating main flux Φ_h. The following core Φ, fringing Φ^υ and leakage fluxes Φ^σ can be identified unambiguously from the FEA in Fig. 3:

a) Core and Air Gap Fluxes

- Φ_ao closes main flux path outside of coil in the outer stator and defines the total flux Φ_t quantitatively.
- $\Phi_{\mathrm{giL|R}}$ cross the air gaps in the main flux path. They are the major force–generating fluxes.
- Φ_di crosses the disk and rotor and equals the force–generating main flux, after subtraction of Φ^σ_dc.

b) Fringing Fluxes Φ^υ

- $\Phi^\upsilon_{\mathrm{dioL|R}}$ bypass single coil–sided air gap.
- $\Phi^\upsilon_{\mathrm{aiiL|R}}$ bypass single shaft–sided air gap.

c) Leakage Fluxes Φ^σ

- $\Phi^\sigma_{\mathrm{ciiL|R}}$ bypass between single air gap and shaft, perpendicular to thrust force. They vanish for high frequencies.
- Φ^σ_dc bypasses whole disk and therefore both air gaps.
- $\Phi^\sigma_{\mathrm{riL|R}}$ leak from inner side of the radial core elements $\mathcal{R}_{\mathrm{rL|R}}$. They form negligible flux swirls, which alter their shape even for small changes of frequency, load or symmetry. To simplify the network calculation, they are neglected.

d) Passive non–core Fluxes

- $\Phi^\sigma_{\mathrm{rcL|R}}$ cross the coil through its center. Depending on the symmetry, they form swirls only for low frequencies. For high frequencies they change direction and exclusively bypass core elements with no direct impact on the force.

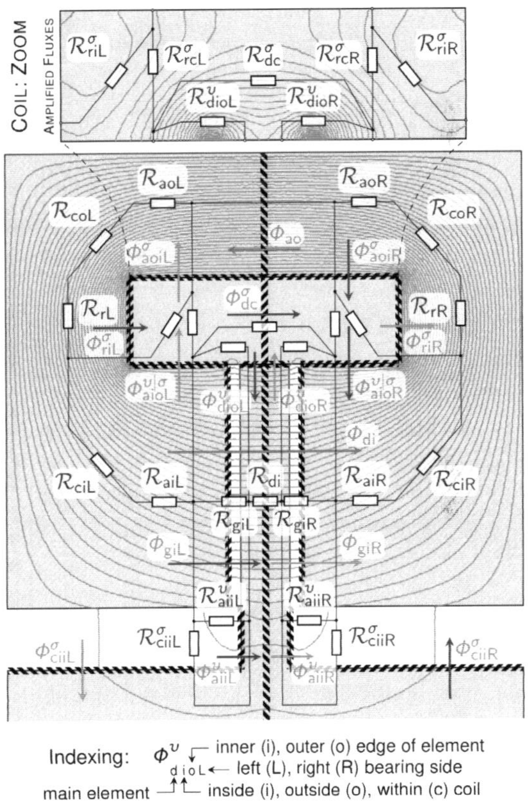

Figure 3: Magnetostatic flux distribution of symmetrical AMTB with omitted bias flux: Identification of fringing Φ^υ and leakage fluxes Φ^σ and reluctances from FEA

Every one of these non–core fluxes $\Phi^{\upsilon|\sigma}$ is assigned to a non–core reluctance $\mathcal{R}^{\upsilon|\sigma}$. In combination with the core reluctances $\mathcal{R}$ in Fig. 2 (calculated in Seifert et al., 2019) we obtain the complete reluctance network in Fig. 4.

Physically, the mmf Θ is a distributed quantity and should be considered by a multitude of sources in the network. However, since the exact distribution is not known, it would increase the number of unknowns and the network could not be solved. Rather we place the mmf where the total magnetic flux Φ_t is at its maximum: in the outer stator yoke. To be able to model the swirls $\Phi^\sigma_{\mathrm{rcL|R}} = \Phi^\sigma_{\mathrm{aoiL|R}}$ (with $\Phi^\sigma_{\mathrm{riL|R}}$ neglected), we distribute Θ to the outer corners and the center of the stator (Fig. 4). However, to consider the changing directions of the fluxes $\Phi^\sigma_{\mathrm{aoiL|R}}$ and to keep the network solvable, the manually chosen weighting has to be adapted for the static and dynamic as well as symmetric and asymmetric case. We remind that in magnetic circuits a mesh is defined as $\sum \Theta_i = \sum \mathcal{R}_i \Phi_i$, so the sum of mmf sources equals the sum of magnetic voltage drops $\sum \mathcal{R}_i \Phi_i$.

Hence, the reluctance network is described by the system of linear equations $\mathcal{R} \cdot \boldsymbol{\Psi} = \boldsymbol{\Theta}$ with the reluctance matrix $\mathcal{R}$ as well as the vectors $\boldsymbol{\Psi}$ and $\boldsymbol{\Theta}$ describing the flux linkages of every mesh and the mmf sources, respectively. Using Kirchhoff's circuit laws, the flux linkages

$$\boldsymbol{\Psi} = \left[\Psi_{\mathrm{rL}}\ \Psi_{\mathrm{rR}}\ \Psi_{\mathrm{cil}}\ \Psi_{\mathrm{ciR}}\ \Psi_{\mathrm{giL}}\ \Psi_{\mathrm{giR}}\ \Psi_{\mathrm{goL}}\ \Psi_{\mathrm{goR}}\ \Psi_{\mathrm{c}}\ \Psi_{\mathrm{g}}\right]^{\mathrm{T}} \quad (2)$$

are calculated step–wise and the fluxes determined from the magnetostatic FE–analysis (as indicated in Fig. 3):

$$\begin{aligned} \Psi_{\mathrm{ciL}} &= \Phi^{\sigma}_{\mathrm{ciiL}} & \Psi_{\mathrm{giL}} &= \Phi^{\upsilon}_{\mathrm{aiiL}} + \Phi^{\sigma}_{\mathrm{ciiL}} \\ \Psi_{\mathrm{ciR}} &= \Phi^{\sigma}_{\mathrm{ciiR}} & \Psi_{\mathrm{giR}} &= \Phi^{\upsilon}_{\mathrm{aiiR}} + \Phi^{\sigma}_{\mathrm{ciiR}} \\ \Psi_{\mathrm{c}} &= \Phi_{\mathrm{ao}} & \Psi_{\mathrm{goL}} &= \Phi_{\mathrm{ao}} - \Phi^{\sigma}_{\mathrm{dc}} - \Phi^{\upsilon}_{\mathrm{dioL}} \\ \Psi_{\mathrm{rL}} &= \Phi_{\mathrm{ao}} + \Phi^{\sigma}_{\mathrm{aoiL}} & \Psi_{\mathrm{goR}} &= \Phi_{\mathrm{ao}} - \Phi^{\sigma}_{\mathrm{dc}} - \Phi^{\upsilon}_{\mathrm{dioR}} \\ \Psi_{\mathrm{rR}} &= \Phi_{\mathrm{ao}} + \Phi^{\sigma}_{\mathrm{aoiR}} & \Psi_{\mathrm{g}} &= \Phi_{\mathrm{ao}} - \Phi^{\sigma}_{\mathrm{dc}} \quad . \end{aligned} \quad (3)$$

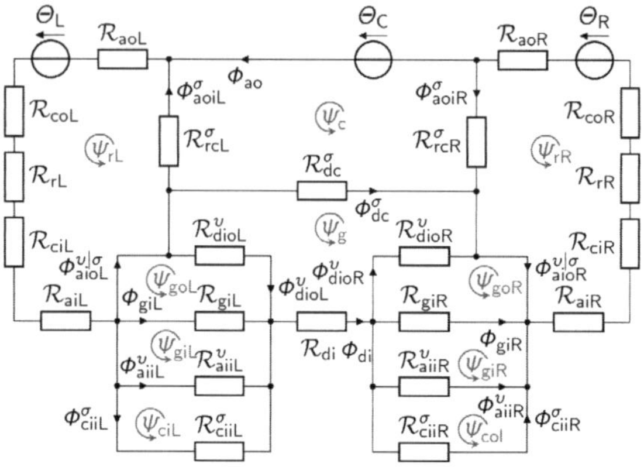

Figure 4: Complete reluctance network including fringing $\mathcal{R}^{\upsilon}_{i}$ and leakage reluctances $\mathcal{R}^{\sigma}_{i}$

With more equations the system would be overdetermined, so the focus lies on the total flux $\Phi_{\mathrm{ao}} = \Phi_{\mathrm{t}}$ which can be quantified exactly and defines the reference flux. For the static and symmetric case the mmf sources are distributed equally to the stator corners, to maintain the swirl character of $\Phi^{\sigma}_{\mathrm{aoiL|R}}$:

$$\Theta_{\mathrm{L}} = \Theta_{\mathrm{R}} = \Theta/2 \quad \text{and} \quad \Theta_{\mathrm{C}} = 0 \quad \text{leading to the vector} \quad \boldsymbol{\Theta} = \left[\Theta_{\mathrm{L}}\ \Theta_{\mathrm{R}}\ 0\ \ 0\ \ 0\ \ 0\ \ 0\ \ 0\ \ \Theta_{\mathrm{C}}\ \ 0\right]^{\mathrm{T}} . \quad (4)$$

The reluctance matrix $\mathcal{R}$ describing the network is composed of the sought–after non–core reluctances $\mathcal{R}^{\upsilon|\sigma}_{i}$ as well as the known analytical solutions for the static part reluctances $\mathcal{R}_{0\,i}$ from Fig. 2. It is depicted in detail in Seifert et al., 2021b. To solve the equation system for all $\mathcal{R}^{\upsilon|\sigma}_{i}$ e. g. the Gaussian elimination algorithm or a computer algebra system (CAS) can be used. At this point we omit the last mesh equation for Ψ_{g}, as there are only nine unknown reluctances, but it is later needed for the backward calculation of the leakage fluxes from the network. The quantitative values of the determined non–core reluctances $\mathcal{R}^{\upsilon|\sigma}_{i}$ are summarized in Table 1 and compared with the parallel core reluctances $\mathcal{R}_{||}$ they bypass. It becomes apparent, that in the operational bandwidth of the thrust bearing (up to 1 kHz) the central leakage reluctances $\mathcal{R}^{\sigma}_{\mathrm{rcL|R}}$ have the least impact and can be neglected. On the other hand, a comparison of Table 1 with Fig. 5 shows that the high impact of the reluctances $\mathcal{R}^{\upsilon}_{\mathrm{aiiL|R}}$ and $\mathcal{R}^{\sigma}_{\mathrm{ciiL|R}}$ on the side of the shaft, with 7.8 % and 14 %, respectively, is misleading. The actual fluxes $\Phi^{\upsilon}_{\mathrm{aiiL|R}}$ and $\Phi^{\sigma}_{\mathrm{ciiL|R}}$ fall below 1 % for more than 1 kHz, due to the magnetic skin effect. This discrepancy can not be compensated by the reluctance network, as we will show later (cf. Fig. 8).

4. Dynamic Fringing and Leakage Flux Model

Already for frequencies below 1 Hz the magnetic skin effect causes the core fluxes to permeate closer to the coil. As a consequence, the leakage flux $\Phi^{\sigma}_{\mathrm{rcL|R}}$ change their direction, so the mmf–source has to be arranged centrally in the outer branch of the reluctance network in Fig. 4, such that $\Theta_{\mathrm{L}} = \Theta_{\mathrm{R}} = 0$ and $\Theta_{\mathrm{C}} = \Theta$. In the fully computed reluctance network, we can now substitute the static part reluctances $\mathcal{R}_{0\,i}$ by the frequency–dependent, and therefore complex, effective part reluctances $\underline{\mathcal{R}}_{i}$ (Seifert et al., 2019). For a given mmf Θ it is then possible to calculate the mesh flux linkages $\underline{\Psi}$ and hence every core flux $\underline{\Phi}_{i}$ and non–core flux $\underline{\Phi}^{\upsilon|\sigma}$ for any frequency.

Table 1: Fringing $\mathcal{R}^{\upsilon}_{i}$ and leakage reluctances $\mathcal{R}^{\sigma}_{i}$ determined by FEA compared to bypassed core reluctances $\mathcal{R}_{||}$ (symmetric case)

$\mathcal{R}^{\upsilon\vert\sigma}_{i}$	in A/Vs	$\mathcal{R}_{\vert\vert}$	in A/Vs	$\mathcal{R}_{\vert\vert}/\mathcal{R}^{\upsilon\vert\sigma}_{i}$	
SYMMETRIC CASE:				0 Hz	1 kHz
$\mathcal{R}^{\sigma}_{\mathrm{rcL\vert R}}$	$2.36 \cdot 10^{7}$	$\mathcal{R}_{\mathrm{soL\vert R}}$	$1.08 \cdot 10^{4}$	0.05 %	2.06 %
$\mathcal{R}^{\upsilon}_{\mathrm{dioL\vert R}}$	$5.94 \cdot 10^{6}$	$\mathcal{R}_{\mathrm{giL\vert R}}$	$1.01 \cdot 10^{5}$	2.37 %	6.22 %
$\mathcal{R}^{\upsilon}_{\mathrm{aiiL\vert R}}$	$3.39 \cdot 10^{6}$	$\mathcal{R}_{\mathrm{giL\vert R}}$	$1.01 \cdot 10^{5}$	2.99 %	7.84 %
$\mathcal{R}^{\sigma}_{\mathrm{ciiL\vert R}}$	$1.89 \cdot 10^{6}$	$\mathcal{R}_{\mathrm{giL\vert R}}$	$1.01 \cdot 10^{5}$	5.36 %	14.06 %
$\mathcal{R}^{\sigma}_{\mathrm{dc}}$	$1.51 \cdot 10^{7}$	$\mathcal{R}_{\mathrm{di+giL+giR}}$	$2.03 \cdot 10^{5}$	1.34 %	4.18 %

*$\mathcal{R}_{\mathrm{soL|R}} = \mathcal{R}_{\mathrm{aoL|R}} + \mathcal{R}_{\mathrm{coL|R}} + \mathcal{R}_{\mathrm{rL|R}} + \mathcal{R}_{\mathrm{ciL|R}} + \mathcal{R}_{\mathrm{aiL|R}}$

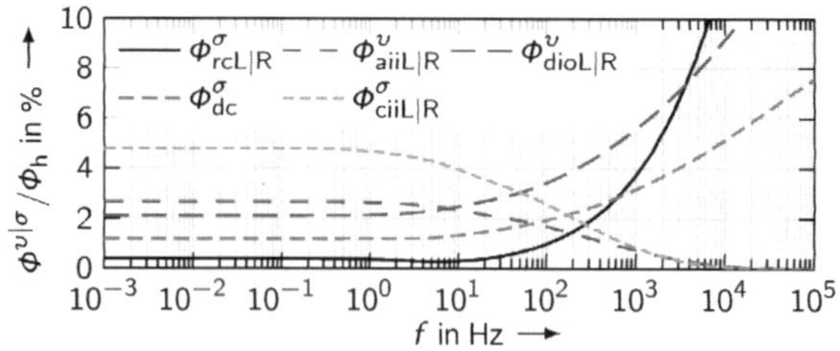

Figure 5: Share of fringing Φ^{υ} and leakage fluxes Φ^{σ} on the total flux $\Phi_{\mathrm{t}} = \Phi_{\mathrm{ao}}$ for the symmetric thrust bearing

This way, the sole analytical overall effective reluctance $\underline{\mathcal{R}}_{\mathrm{eff}}$ can be corrected to the leakage- and fringing–accounting effective reluctance $\underline{\mathcal{R}}'_{\mathrm{eff\,t}} = \Theta/\underline{\Phi}_{\mathrm{ao}} = \Theta/\underline{\Phi}_{\mathrm{t}}$, which we compare with a dynamic FEA. Fig. 6 shows, that the amplitude error of $\underline{\mathcal{R}}_{\mathrm{eff}}$ (ca. 1 dB) towards the FEA results can be eliminated almost entirely by the reluctance network. The leakage and fringing fluxes are also responsible for a considerable decay of the phase from its theoretical limit of 45° for high frequencies, which is describable by the network as well. Only the modeling error of the original analytical solution around the eddy current edge frequency f_{e} (cf. Seifert et al., 2019) persists.

4.1 Simplified dynamic reluctance network

The complexity of the complete network (Fig. 4) leads to a significant increase in the system order m_{S} of the rational form of $\underline{\mathcal{R}}_{\mathrm{eff}}$ (Seifert et al., 2019) and is therefore not applicable for the estimator G_{FE}. Hence, we propose the following simplifications, based on Table 1, Fig. 5 and the qualitative evaluation of the FEA field distribution.

1. The outer axial core reluctances $\underline{\mathcal{R}}_{\mathrm{aoL|R}}$ are summed up to $\underline{\mathcal{R}}_{\mathrm{ao}}$ by addition of their characteristic lengths.

2. The fringing fluxes $\underline{\Phi}^{\upsilon}_{\mathrm{dioL|R}}$, $\underline{\Phi}^{\upsilon}_{\mathrm{aiiL|R}}$ (and corresponding reluctances) are combined and assigned to the air gap fringing reluctances $\mathcal{R}^{\upsilon}_{\mathrm{giL|R}}$:

$$\mathcal{R}^{\upsilon}_{\mathrm{giL|R}} = \frac{\mathcal{R}^{\upsilon}_{\mathrm{dioL|R}}\mathcal{R}^{\upsilon}_{\mathrm{aiiL|R}}}{\mathcal{R}^{\upsilon}_{\mathrm{dioL|R}}+\mathcal{R}^{\upsilon}_{\mathrm{aiiL|R}}}\,. \tag{5}$$

3. The central leakage reluctance $\mathcal{R}^{\sigma}_{\mathrm{dc}}$, bypassing both air gaps, is divided into two equal parts, assigned to a single air gap each. Together with the leakage reluctance $\mathcal{R}^{\sigma}_{\mathrm{ciiL|R}}$ on the shaft side, they form the air gap leakage reluctances $\mathcal{R}^{\sigma}_{\mathrm{giL|R}}$:

$$\mathcal{R}^{\sigma}_{\mathrm{giL|R}} = \frac{\frac{1}{2}\mathcal{R}^{\sigma}_{\mathrm{dc}}\mathcal{R}^{\sigma}_{\mathrm{ciiL|R}}}{\frac{1}{2}\mathcal{R}^{\sigma}_{\mathrm{dc}}+\mathcal{R}^{\sigma}_{\mathrm{ciiL|R}}}\,. \tag{6}$$

4. To determine the corrected total effective reluctance $\underline{\mathcal{R}}'_{\mathrm{eff\,t}}$, relating to the total flux $\underline{\Phi}_{\mathrm{ao}} = \underline{\Phi}_{\mathrm{t}}$ and not considering the force, we merge $\mathcal{R}^{\upsilon}_{\mathrm{giL|R}}$ and $\mathcal{R}^{\sigma}_{\mathrm{giL|R}}$ to the combined air gap fringing and leakage reluctance $\mathcal{R}^{\upsilon\sigma}_{\mathrm{giL|R}}$:

$$\mathcal{R}^{\upsilon\sigma}_{\mathrm{giL|R}} = \frac{\mathcal{R}^{\upsilon}_{\mathrm{giL|R}}\mathcal{R}^{\sigma}_{\mathrm{giL|R}}}{\mathcal{R}^{\upsilon}_{\mathrm{giL|R}}+\mathcal{R}^{\sigma}_{\mathrm{giL|R}}}\,. \tag{7}$$

5. The respective "geometrical" air gap reluctances in the main flux path are equal for both bearing sides, independently from the symmetry, so that $\underline{\mathcal{R}}_{\mathrm{giL|R}} = \underline{\mathcal{R}}_{\mathrm{gi}}$. Furthermore we calculate the arithmetic mean of $\mathcal{R}^{\upsilon\sigma}_{\mathrm{giL|R}}$ to obtain the single corrected air gap reluctance:

$$\underline{\mathcal{R}}'_{\mathrm{gi}} = 2\left(\overline{\mathcal{R}}^{\upsilon\sigma}_{\mathrm{gi}} \,||\, \underline{\mathcal{R}}_{\mathrm{gi}}\right) \quad \text{with} \quad \overline{\mathcal{R}}^{\upsilon\sigma}_{\mathrm{gi}} = \frac{1}{2}\left(\mathcal{R}^{\upsilon\sigma}_{\mathrm{giL}}+\mathcal{R}^{\upsilon\sigma}_{\mathrm{giR}}\right) \tag{8}$$

The introduced error of $\underline{\mathcal{R}}'_{\mathrm{gi}}$ compared to $\underline{\mathcal{R}}_{\mathrm{giL|R}} \,||\, \mathcal{R}^{\upsilon\sigma}_{\mathrm{giL|R}}$ is only 0.03 %, which is why we also adopt this step for

$$\overline{\mathcal{R}}^{\upsilon}_{\mathrm{gi}} = \frac{1}{2}\left(\mathcal{R}^{\upsilon}_{\mathrm{giL}}+\mathcal{R}^{\upsilon}_{\mathrm{giR}}\right),\quad \overline{\mathcal{R}}^{\sigma}_{\mathrm{gi}} = \frac{1}{2}\left(\mathcal{R}^{\sigma}_{\mathrm{giL}}+\mathcal{R}^{\sigma}_{\mathrm{giR}}\right). \tag{9}$$

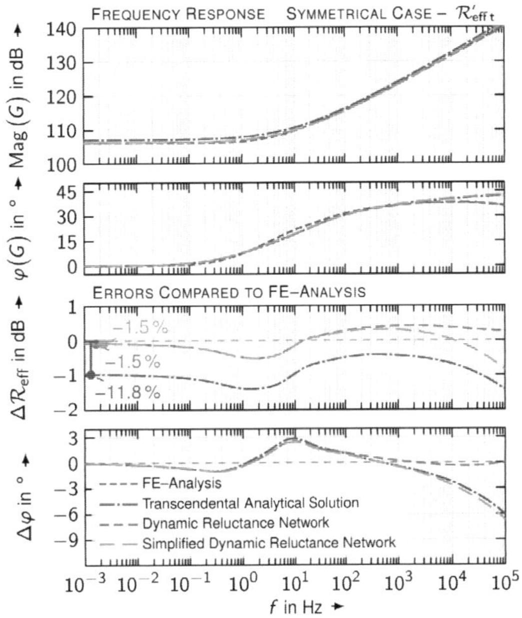

Figure 6: Frequency responses for the corrected total eff. reluctance $\mathcal{R}'_{\mathrm{eff\,t}}$ relating to the total flux Φ_{t} and its absol. magnitude- and phase errors in respect to reference FEA

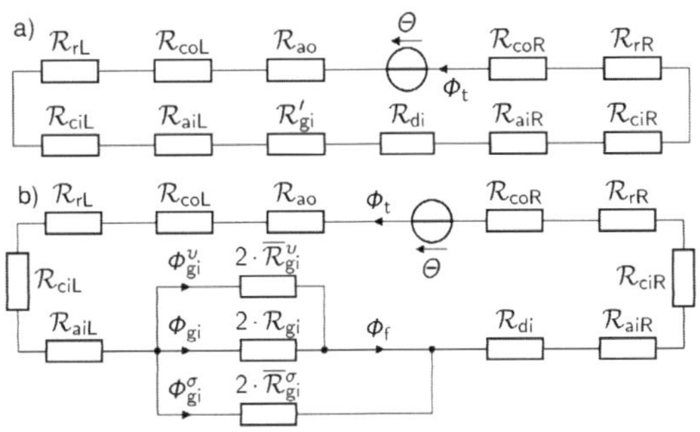

Figure 7: Simplified dynamic reluctance network with a) corrected air gap reluctance or b) with total flux path split into the force–generating fringing Φ^{υ} and leakage part Φ^{σ}

6. As a last step we neglect the transversal reluctances $\mathcal{R}^{\sigma}_{\mathrm{rcL|R}}$, since their impact for low frequencies is negligible (cf. Fig. 5). Although in the dynamic case for high frequencies above 1 kHz the respective fluxes $\underline{\Phi}^{\sigma}_{\mathrm{rcL|R}}$ are dominant, the impact of $\mathcal{R}^{\sigma}_{\mathrm{rcL|R}}$ on the magnitude response of $\underline{\mathcal{R}}'_{\mathrm{eff\,t}}$ is limited due its passive characteristic. However, the previously observed phase decay for frequencies above 1 kHz cannot be modeled without $\mathcal{R}^{\sigma}_{\mathrm{rcL|R}}$, which is deemed uncritical, as it lies outside of the bearings operating bandwidth. If higher bandwidths require the consideration of $\mathcal{R}^{\sigma}_{\mathrm{rcL|R}}$, one should be aware that the system order m_{S} almost doubles.

As result we obtain the simplified dynamic reluctance network in Fig. 7, which forms the leakage- and fringing–accounting total effective reluctance $\underline{\mathcal{R}}'_{\mathrm{eff\,t}} = \Theta/\underline{\Phi}_{\mathrm{t}}$ by summing up of its part reluctances. Up to ca. 100 Hz both, the full and the simplified network, lead to identical results. In the further relevant bandwidth up to 1 kHz the magnitude and phase error remains low with < 0.4 dB and $< 1°$, respectively (cf. Fig. 6).

However, in this form the network cannot just yet be applied to the flux estimator, as the magnetic bearing control relies on the force–generating main flux $\underline{\Phi}_{\mathrm{h}} = \underline{\Phi}_{\mathrm{f}} = \underline{\Phi}_{\mathrm{gi}} + \underline{\Phi}^{\upsilon}$ and not the total flux $\underline{\Phi}_{\mathrm{t}}$ as illustrated in Fig. 7. Consequentially, the force–related total effective reluctance $\underline{\mathcal{R}}'_{\mathrm{eff\,f}} = \Theta/\underline{\Phi}_{\mathrm{f}}$ only refers to the force–generating flux $\underline{\Phi}_{\mathrm{f}}$. The results for the accordingly adapted $\underline{\mathcal{R}}'_{\mathrm{eff\,f}}$ will be shown in section 4.3

4.2 Transition to Asymmetric Bearing

In the asymmetric real thrust bearing the position of the coil was shifted to the right as depicted in Fig. 2. We took this measure to equalize the static reluctances of the bias flux paths and therefore enable equal forces in both air gaps. In terms of fringing and leakage fluxes, we observe that they are facilitated for high frequencies $f > 1$ kHz especially on the left side of the bearing (Seifert et al., 2021b). Furthermore, the right–sided central leakage flux $\underline{\Phi}^{\sigma}_{\mathrm{aoiR}}$ remains a swirl, as in Fig. 4, over the entire frequency range, while its left–sided counterpart $\underline{\Phi}^{\sigma}_{\mathrm{aoiL}}$ is always passive (it changes its direction compared to Fig. 4). To account for this asymmetry, the magnetic voltage source has to be placed in the outer right corner of the network, such that $\Theta_{\mathrm{L}} = \Theta_{\mathrm{C}} = 0$ and $\Theta_{\mathrm{R}} = \Theta$. By this slight adjustment, the full and the simplified reluctance network enables comparable results for the corrected total effective reluctance $\underline{\mathcal{R}}'_{\mathrm{eff\,t}}$ like in the symmetric case, at least in the bandwidth below 1 kHz. Quantitative error data is disclosed in Seifert et al., 2021b.

4.3 Implementation and Conclusion

Previously we indicated, that for the flux estimator not only the corrected total effective reluctance $\underline{\mathcal{R}}'_{\mathrm{eff\,t}} = \Theta/\underline{\Phi}_{\mathrm{t}}$ is decisive, but also the force–related effective reluctance $\underline{\mathcal{R}}'_{\mathrm{eff\,f}} = \Theta/\underline{\Phi}_{\mathrm{f}}$. The latter relates solely to the actual force–generating part $\underline{\Phi}_{\mathrm{f}}$ of the total flux $\underline{\Phi}_{\mathrm{t}}$, which we calculate with the flux divider following Fig. 7. By dividing both fluxes, $\underline{\Phi}_{\mathrm{f}}$ and $\underline{\Phi}_{\mathrm{t}}$, through the mmf $\Theta = i \cdot N$ and calculating the reciprocal, we obtain the desired form:

$$\underline{\Phi}_{\mathrm{f}} = \frac{\mathcal{R}^{\sigma}_{\mathrm{gi}}}{\mathcal{R}^{\sigma}_{\mathrm{gi}} + \underline{\mathcal{R}}_{\mathrm{gf}}}\,\underline{\Phi}_{\mathrm{t}} \quad \text{with} \quad \underline{\mathcal{R}}_{\mathrm{gf}} = 2\frac{\overline{\mathcal{R}}^{\upsilon}_{\mathrm{gi}}\,\underline{\mathcal{R}}_{\mathrm{gi}}}{\overline{\mathcal{R}}^{\upsilon}_{\mathrm{gi}} + \underline{\mathcal{R}}_{\mathrm{gi}}} \quad \text{and} \quad \mathcal{R}^{\sigma}_{\mathrm{gi}} = 2\overline{\mathcal{R}}^{\sigma}_{\mathrm{gi}} \quad \overset{:\,\Theta}{\Longrightarrow} \quad \underline{\mathcal{R}}'_{\mathrm{eff\,f}} = \left(1 + \frac{\underline{\mathcal{R}}_{\mathrm{gf}}}{\mathcal{R}^{\sigma}_{\mathrm{gi}}}\right)\underline{\mathcal{R}}'_{\mathrm{eff\,t}}\,. \tag{10}$$

The implementation of the combined air gap leakage reluctances $\mathcal{R}^{\upsilon\sigma}_{\mathrm{giL|R}}$, leading to $\underline{\mathcal{R}}'_{\mathrm{eff\,t}}$, is carried out together with the calculation of the air gap reluctance $\underline{\mathcal{R}}_{\mathrm{gi}}$ or the air gap element (fractional order $\gamma = 1/4$) of the equivalent implicit system $\underline{\mathcal{R}}_{\mathrm{EIS}}$ (cf. Seifert et al., 2019, eq. (58)) according to (8). In the next step, we derive the force–related reluctance $\underline{\mathcal{R}}'_{\mathrm{eff\,f}}$ from (10). All subsequent calculations affect neither the approximations nor the discretization. Only the system orders $m_{\mathrm{S}}, n_{\mathrm{S}}$ of the approximated overall system (sum of all part reluctances in Fig. 7 or equivalent implicit system Seifert et al., 2019) increase – in case of $\underline{\mathcal{R}}'_{\mathrm{eff\,t}}$ only by 1, regarding $\mathcal{R}'_{\mathrm{eff\,f}}$, the flux divider implicates a further increase of the total order by the order $m_{\mathcal{R}_{\mathrm{g}}}$ of the reluctance $\underline{\mathcal{R}}_{\mathrm{gi}}$:

$$m_{\mathrm{S\,t}} = m_{\mathrm{S}} + 1, \quad n_{\mathrm{S\,t}} = m_{\mathrm{S\,t}} - 1 \quad \text{vs.} \quad m_{\mathrm{S\,f}} = m_{\mathrm{S}} + m_{\mathcal{R}_{\mathrm{g}}} + 1 = m_{\mathrm{S\,t}} + m_{\mathcal{R}_{\mathrm{g}}}, \quad n_{\mathrm{S\,f}} = m_{\mathrm{S\,f}} - 1\,. \tag{11}$$

However, in Seifert et al., 2019 we proposed to undertake an additional Padé–approximation (PASR), which overrides the increase in order of $m_{\mathcal{R}_{\mathrm{g}}} = 8 \ldots 29$ (for $f < 21$ kHz). Therefore, the order of the PASR–solution $m_{\mathcal{P}}$ remains equal to the fringing/leakage–free model. This is a considerable advantage compared to the other proposed approximation method MAEIS, where considering the leakage within $\underline{\mathcal{R}}'_{\mathrm{eff\,f}}$ increases the total system order by $m_{\mathcal{R}_{\mathrm{g}}} = (m_{\mathrm{S}} + 1)/2 = 5 \ldots 25$, limiting the applicability of the approach.

5. Alternative: Constant Correction Factor

The simplest way to take fringing and leakage fluxes into account, is the introduction of constant correction factors $k_{\upsilon\sigma\mathrm{t}}$ and $k_{\upsilon\sigma\mathrm{f}}$ (relating to the current i). Without an increase to the system order, we can correct the total effective reluctance $\underline{\mathcal{R}}_{\mathrm{eff}}$ or inductance $\underline{L}_{\mathrm{eff}}$ with the correction factor $k_{\upsilon\sigma\mathrm{t}}$, by forming the ratio of the static corrected overall reluctance $\mathcal{R}'_{0\mathrm{t}}$ with the static fringing/leakage–free reluctance $\mathcal{R}_0 = \mathcal{R}_{\mathrm{eff}}(f = 0)$:

$$\underline{L}_{\mathrm{eff\,t}} = k_{\upsilon\sigma\mathrm{t}} N \frac{\underline{\Phi}_{\mathrm{t}}}{i} = \frac{k_{\upsilon\sigma\mathrm{t}} N^2}{\underline{\mathcal{R}}_{\mathrm{eff}}} = \frac{N^2}{\underline{\mathcal{R}}'_{\mathrm{eff\,t}}} \Rightarrow \underline{\mathcal{R}}'_{\mathrm{eff\,t}} = \frac{\underline{\mathcal{R}}_{\mathrm{eff}}}{k_{\upsilon\sigma\mathrm{t}}} \quad \Rightarrow \quad k_{\upsilon\sigma\mathrm{t}} = \frac{\mathcal{R}_0}{\mathcal{R}'_{0\mathrm{t}}} = \mathcal{R}_0 \cdot \frac{\Phi_{0\mathrm{t}}}{\Theta}\,, \tag{12}$$

The determination of $\mathcal{R}_{0\mathrm{t}}$ in (10) or Φ_t for that matter, can be conducted by analytical calculation or a FEA. By using the flux divider from (10) we determine analogously the force–related effective inductance and the corresponding correction factor $k_{\upsilon\sigma\mathrm{f}}$:

$$\underline{L}_{\mathrm{eff\,f}} = k_{\upsilon\sigma\mathrm{f}} N \frac{\Phi_\mathrm{f}}{i} = \frac{k_{\upsilon\sigma\mathrm{f}} N^2}{\underline{\mathcal{R}}_\mathrm{eff}} = \frac{N^2}{\underline{\mathcal{R}}'_{\mathrm{eff\,f}}} \Rightarrow \underline{\mathcal{R}}'_{\mathrm{eff\,f}} = \frac{\underline{\mathcal{R}}_\mathrm{eff}}{k_{\upsilon\sigma\mathrm{f}}} \quad \Rightarrow \quad k_{\upsilon\sigma\mathrm{f}} = \frac{\mathcal{R}_0}{\mathcal{R}'_{0\mathrm{f}}} = \frac{\mathcal{R}_0}{\Theta/\Phi_{0\mathrm{f}}} = \frac{\mathcal{R}^\sigma_\mathrm{gi}}{\mathcal{R}^\sigma_\mathrm{gi} + \mathcal{R}_{0\,\mathrm{gf}}} \frac{\mathcal{R}_0}{\mathcal{R}'_{0\mathrm{t}}}. \tag{13}$$

5.1 Comparison

Up to a frequency of 1 Hz the implementation of the force–related reluctance $\underline{\mathcal{R}}'_{\mathrm{eff\,f}}$ by means of the flux divider (10) or the correction factor $k_{\upsilon\sigma\mathrm{f}}$ (13) leads to equal results, as Fig. 8 illustrates. The absolute error in respect to the FEA can be reduced by 0.35 dB to < 0.1 dB (stationary) compared to the fringing/leakage–free $\underline{\mathcal{R}}_\mathrm{eff}$, at which point the simulatively determined reluctance is more accurate.

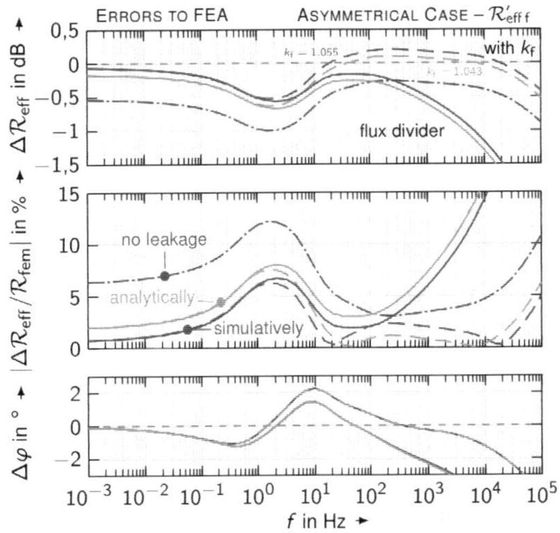

Figure 8: Absolute and relative magnitude and phase error for the force–generating reluctance $\mathcal{R}'_{\mathrm{eff\,f}}$ based on the flux divider (simplified reluctance network) or correction factor $k_{\upsilon\sigma\mathrm{f}}$ in respect to reference FEA

However, as opposed to the corrected total effective reluctance $\underline{\mathcal{R}}'_{\mathrm{eff\,t}}$ (cf. Fig. 6), the force–related reluctance $\mathcal{R}'_{\mathrm{eff\,f}}$, which is based on the flux divider, does not provide satisfying results in the relevant bandwidth up to 1 kHz. The reason for this derives from the the shift in the leakage flux distribution from the shaft to the coil for high frequencies, which cannot be modeled by the reluctance network. Although occurring errors cancel each other out for $\underline{\mathcal{R}}'_{\mathrm{eff\,t}}$, this is not the case for $\underline{\mathcal{R}}'_{\mathrm{eff\,f}}$ resulting in an inaccurate representation of the force–generating flux $\underline{\Phi}_\mathrm{f}$. On the other hand, compared to the original analytical solution the correction factor $k_{\upsilon\sigma\mathrm{f}}$ allows only a slight improvement in accuracy for $1\,\mathrm{Hz} < f < 1\,\mathrm{kHz}$ (Fig. 8), but more importantly does not degrade it. Hence we conclude, if the force–generating flux $\underline{\Phi}_\mathrm{f}$ is to be the focus, the correction factor is the preferred method.

Furthermore, Fig. 8 reveals a possibly unexpected relation. The errors of $|\underline{\mathcal{R}}'_{\mathrm{eff\,f}}|$ in respect to $|\underline{\mathcal{R}}_\mathrm{eff}|$ are smaller than in the case of $\underline{\mathcal{R}}'_{\mathrm{eff\,t}}$ and remain negative. This in turn means that $|\underline{\mathcal{R}}'_{\mathrm{eff\,f}}|$ is smaller than $|\underline{\mathcal{R}}_\mathrm{eff}|$ and the force f is actually increased by the presence of fringing and leakage fluxes. Although, the force–generating flux $\underline{\Phi}_\mathrm{f}$ is indeed smaller than the total flux $\underline{\Phi}_\mathrm{t}$ (reduced by $\underline{\Phi}^\sigma$), the fringing and leakage fluxes cause a general reduction of the total reluctance $|\underline{\mathcal{R}}'_{\mathrm{eff\,t}}| \ll |\underline{\mathcal{R}}_\mathrm{eff}|$ and thus an increase of the actuator force. In our case the latter effect is dominant, resulting in correction factors $k_{\upsilon\sigma\mathrm{t|f}} > 1$. But it is important to note, that this observation cannot be generalized for all actuators or magnetic bearings and is only probable in case the force–generating fringing fluxes dominate over the leakage fluxes.

5.2 Correction of Flux Estimator

By recalling the definition of the flux estimator from (1), we see that there are multiple occurrences of the effective inductance $\underline{L}_\mathrm{eff}$ and its stationary counterpart $L_\mathrm{h} = \lim_{\omega\to 0} \underline{L}_\mathrm{eff}(\mathrm{j}\omega)$, the main inductance. While the first term is the actual force–related field building component, which we correct with the factor $k_{\upsilon\sigma\mathrm{f}}$, the second term describes the magnetizing currents, that relate to the total flux $\underline{\Phi}_\mathrm{t}$. Hence, the latter has to be corrected with the factor $k_{\upsilon\sigma\mathrm{t}}$ finally leading to:

$$\underline{G}'_\mathrm{FE}(\mathrm{j}\omega) = \underbrace{\frac{\underline{\Phi}_{\mathrm{f\,est}}(\mathrm{j}\omega)}{\underline{i}_\mathrm{meas}(\mathrm{j}\omega)}}_{\text{measurable: / force–related:}} = \underbrace{\frac{L_\mathrm{hf}}{N}}_{\text{field}} \cdot \frac{\overbrace{R_\mathrm{Cu} + s\underline{L}_{\mathrm{eff\,t}}(\mathrm{j}\omega)}^{\text{magnetizing current}}}{\underbrace{R_\mathrm{Cu} + sL_\mathrm{h\,t}}_{\text{magnetizing current}}}, \tag{14}$$

Whether or not this differentiation is appropriate depends on the actuator geometry and the balance between fringing and leakage fluxes. In our case $k_{\upsilon\sigma\mathrm{f}}$ and $k_{\upsilon\sigma\mathrm{t}}$ differ by considerable 5.2 % and its inclusion comes at no cost. We note, that in fact the main inductances $L_\mathrm{h\,t}$ and $L_\mathrm{h\,f}$ are not directly affected by the magnetic skin effect, but by the redistribution of the fringing/leakage fluxes. We still assume them to be constant. Theoretically, it would be possible to isolate the frequency–dependent influence of these non–core fluxes and project them on the main inductances as well. However, the improvement would be very low compared to the additional computing effort.

6. Conclusion and outlook

The control of electromagnetic actuators and magnetic bearings usually relies on simplified models and perturbations like eddy currents, hysteresis, saturation, fringing and leakage fluxes are omitted. Especially in case of nonlaminated cores, it is widely accepted that eddy currents have a significant impact on the actuators performance and should be considered within the design of the control. In previous works we laid the groundwork for the first practical digital implementation of a flux estimator to fully compensate eddy currents in the inner actuator control loop. The considerably minor perturbations fringing and leakage remain solely subjects of academic research. Although various studies exist, they are usually limited to the analytical modeling and FE–analyses of the actuator. For the first time, we discussed the actual applicability of the known analytical models to control design and further improved them.

Our study regarding the fringing and leakage fluxes in a magnetic thrust bearing shows, that the practicality of a reluctance network highly depends on whether the total flux $\underline{\Phi}_{\mathrm{t}}$ and its respective corrected total reluctance $\underline{\mathcal{R}}'_{\mathrm{eff\,t}}$ is of interest or the actual force–generating flux $\underline{\Phi}_{\mathrm{f}}$ relating to $\underline{\mathcal{R}}'_{\mathrm{eff\,f}}$. In the former case, the proposed networks can significantly reduce model errors (static: 11.8 %) by 10 percentage points over a wide range of frequencies. The full reluctance network is only beneficial over the simplified network for less relevant frequencies above 1 kHz, where the characteristic phase drop is reproduced. In the bandwidth of interest below 1 kHz both networks are equally accurate. In the latter case, due to the magnetic skin effect, neither the full nor the simplified reluctance network can model the changing flux distribution for frequencies above 1 Hz, leading to an incorrect representation of the force-generating flux $\underline{\Phi}_{\mathrm{f}}$ and its according reluctance $\underline{\mathcal{R}}_{\mathrm{eff\,f}}$. However, constant correction factors, possibly greater than 1, reduce the model errors (static: 6.4 %) by satisfying 5.6 percentage points. Since they do not increase the system order and are most simple to implement we deem them to be the preferred variant for the flux estimator and control models in general.

We conclude, that our findings in this article only slightly improve the previously proposed flux control based on a fractional–order flux estimator. However, for the commonly omitted and supposedly minor perturbations fringing and leakage fluxes, we provided an extensive review about modeling approaches and discussed their impact as well as practical applicability as reference for future studies.

References

Feeley, J. J. (1996). "A simple dynamic model for eddy currents in a magnetic actuator". *IEEE Transactions on Magnetics* 32.2, pp. 453–458.

Ferreira, J., E. Maslen, and R. Fittro (2017). "Transpermeance Amplifier Applied to Magnetic Bearings". *Actuators* 6.9, pp. 1–20.

Keith, F. J. (1993). "Implicit flux feedback control for magnetic bearings". English. Dissertation. University of Virginia.

Kucera, L. and M. Ahrens (1996). "A Model for Axial Magnetic Bearings Including Eddy Currents". *3rd International Symposium on Magnetic Suspension Technology*. .2. Tallahassee, United States: NASA, pp. 421–437.

Le, Y. and K. Wang (2016). "Design and Optimization Method of Magnetic Bearing for High-Speed Motor Considering Eddy Current Effects". *IEEE/ASME Transactions on Mechatronics* 21.4, pp. 2061–2072.

Meeker, D. C., E. H. Maslen, and M. D. Noh (1996). "An augmented circuit model for magnetic bearings including eddy currents, fringing, and leakage". *IEEE Transactions on Magnetics* 32.4, pp. 3219–3227.

Rabinovici, R. and B. Kaplan (1992). "Effective magnetization and forces due to eddy currents". *IEEE Transactions on Magnetics* 28.3, pp. 1863–1869.

Seifert, R. and W. Hofmann (2021a). "Highly Dynamic Thrust Bearing Control Based on a Fractional-Order Flux Estimator". *IEEE Transactions on Industry Applications* 57.6, pp. 6988–6999.

Seifert, R., J. Porstmann, and W. Hofmann (2021b). "Consideration of Hysteresis, Saturation, Fringing and Leakage Fluxes in the Frequency–Dependent Analytical Model of Nonlaminated Cylindrical Actuators". *IEEE TechRxiv*. Preprint, pp. 1–17.

Seifert, R., K. Röbenack, and W. Hofmann (2019). "Rational Approximation of the Analytical Model of Nonlaminated Cylindrical Magnetic Actuators for Flux Estimation and Control". *IEEE Transactions on Magnetics* 55.12, pp. 1–16.

Stoll, R. (1974). *The analysis of eddy currents*. Monographs in electrical and electronic engineering. Oxford: Clarendon Press.

Sun, Y., Y.-S. Ho, and L. Yu (2009). "Dynamic Stiffnesses of Active Magnetic Thrust Bearing Including Eddy-Current Effects". *IEEE Transactions on Magnetics* 45.1, pp. 139–149.

Vischer, D. (1988). "Sensorlose und spannungsgesteuerte Magnetlager". Dissertation. ETH Zürich.

Wang, K., D. Wang, H. Lin, Y. Shen, X. Zhang, and H. Yang (2014). "Analytical Modeling of Permanent Magnet Biased Axial Magnetic Bearing With Multiple Air Gaps". *IEEE Transactions on Magnetics* 50.11, pp. 1–4.

Zhong, Y., L. Wu, X. Huang, Y. Fang, and J. Zhang (2017). "An Improved Magnetic Circuit Model of a 3-DOF Magnetic Bearing Considering Leakage and Cross-Coupling Effects". *IEEE Transactions on Magnetics* 53.11, pp. 1–6.

Zhu, L. and C. Knospe (2010). "Modeling of Nonlaminated Electromagnetic Suspension Systems". *IEEE Transactions on Mechatronics* 15.1, pp. 59–69.

Flywheel Energy Storage System with Thermal Insulation

Chin-Hsiang LIN [a], Shyh-Leh CHEN [a]

a Department of Mechanical Engineering and Advanced Institute of Manufacturing with High-tech Innovations, National Chung Cheng University, Chiayi 621, Taiwan, imeslc@ccu.edu.tw

Abstract

This paper proposes a novel design of a magnetically supported flywheel energy storage system with thermal insulation. It utilizes a magnetic coupler to directly transmit the power. The proposed design can induce almost no energy loss. If the power is transmitted indirectly by electromagnetic induction, the energy transmission efficiency will be lower. The motor and the flywheel share the same rotor shaft, and the shaft is supported by two sets of five-degree-of-freedom magnetic bearings. Because the bearings are non-contact, the speed of the flywheel can be increased, thereby the stored energy can be increased. The driving heat source and the heat source of the magnetic bearings are both outside the vacuum chamber of the flywheel, which is easy to dissipate and will not affect the flywheel.

Keywords: *Flywheel Storage Energy System, Magnetic Bearing, Magnetic Coupler*

1. Introduction

Flywheel energy storage system (FESS) with magnetic bearings can realize high speed rotation and store the kinetic energy with high efficiency. Due to its great potential, a large number of research results have been reported in recent years. One critical issue of FESS is the heat dissipation. Since it is in general contained in a vacuum chamber, the heat generated by the motor/generator is difficult to be dissipated. A flywheel with cooling vents was designed in (Song, et al., 2020), where the airflow can be generated when rotating the flywheel. However, the heat from the motor cannot be efficiently removed. In (Qian, et al., 2017), a heat conduction ring and a cooling support are designed to export the heat out of the flywheel chamber. This design will increase the cost, and due to the use of mechanical contact bearings, high speed can hardly be achieved. Some designs separate the drive motor from the flywheel, and use the concept of electromagnetic coupler to drive the flywheel for rotation (Ben, et al., 2017) (Hu, et al., 2019). However, the motor and flywheel are still supported by mechanical bearings. In (Dubois, et al., 2012), the motor is separated from the magnetically supported flywheel, but the motor still supported by the mechanical bearings.

As one can see, to keep the heat source out of the flywheel chamber, the motor is designed outside and the driving torque and power are transmitted through electromagnetic induction. In this way, the motor and flywheel do not share the same rotor shaft and they need to be supported by separate bearings. Contact bearing is the most common choice, but it suffers from the friction problems. An alternative is the magnetic bearing, but the heat dissipation problem in a vacuum environment needs to be resolved.

In this study, the motor and the flywheel are designed to share the same rotor shaft, and the power is transmitted by the magnetic coupler, which is relatively direct and has almost no energy loss problem. If the power is transmitted by electromagnetic induction, there is a problem of energy transmission efficiency. The integrated rotor shaft is supported by a set of five-degree-of-freedom magnetic bearings. Because the bearings are all non-contact, the flywheel speed can be increased, thereby increasing the stored energy. Both the driving heat source and the heat source of the magnetic bearing are outside the vacuum chamber of the flywheel, which is easy to dissipate heat and will not affect the flywheel.

2. System design and modeling

The system under study is shown in Figure 1. The FESS is designed to be 3 KW with the rotational speed of 60,000 rpm. The motor is located on top of the lower flywheel with the same rotor shaft. The rotor of the motor is the outer rotor of the magnetic coupler. The inner rotor of the magnetic coupler is the rotor shaft. Therefore, the motor will drive the magnetic coupler that will drive the rotor shaft and the flywheel. The flywheel (and the whole rotor shaft) is located inside a vacuum chamber so that higher speed can be achieved without the friction loss caused by air. Since the motor and the stator of the magnetic bearings are all placed outside the isolation cover and the vacuum chamber, the heat will not be accumulated within the flywheel system.

The weight of the shaft and flywheel is supported by two sets of passive axial magnetic bearings (upper and lower PMB), where the lower one is placed inside the vacuum chamber. The weight of outer rotor of the magnetic coupler is supported by another two sets of the passive axial magnetic bearings (PMB). All of the passive axial magnetic bearings are designed with Halbach arrays, as shown in Figure 2. In addition, the radial displacement of the shaft and the flywheel is controlled by the upper and lower active magnetic bearings (AMB, as shown in Figure 3) so that the position of the shaft and the flywheel is kept at the center position, as shown in Figure 3. These two sets of AMBs are located outside of the vacuum chamber that are easier for heat dissipation. Finally, the radial motion of the outer rotor of the magnetic coupler is constrained by the ball bearings. Although the friction of the ball bearing will generate heat at high speed, it is outside the vacuum chamber and can be easily removed. It is also possible to replace the ball bearings with AMB, but the dynamics will be very complicated that makes the levitation controller more challenging.

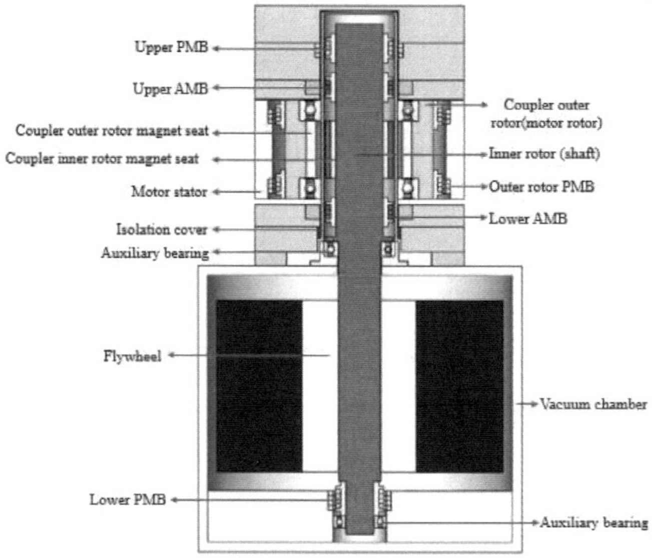

Figure 1 Flywheel system

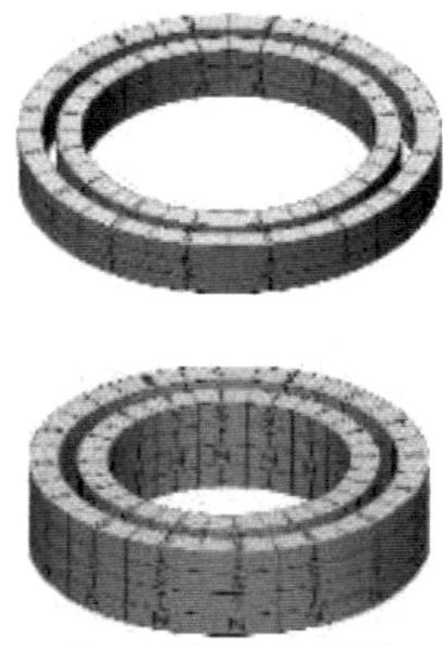

Figure 2. The axial PMB

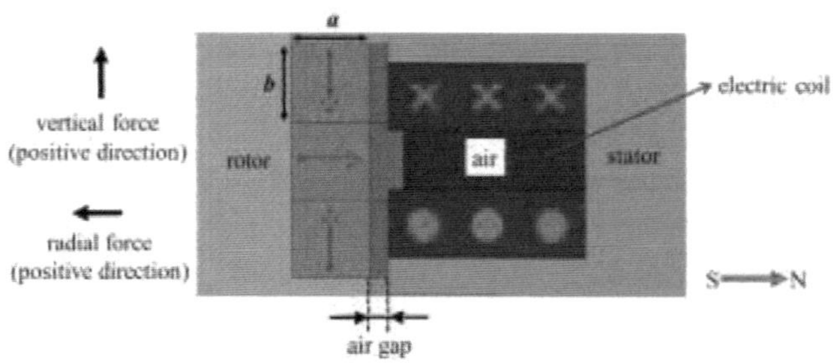

Figure 3. The AMB

Based on the above design, the flywheel system is considered to be 5 degrees of freedom (DOF). The weight of the rotor shaft will be supported by axial passive magnetic bearings and no levitation controller is needed. Also, the axial dynamics of the shaft is decoupled from the radial dynamics and can be neglected. Therefore, only the 4-DOF levitation dynamics in the radial directions will be considered in the mathematical model. When representing the mathematical model of the system, the bearing coordinates of the system are first transformed into the body coordinates of the system. The mathematical model of the system is expressed as

$$M\ddot{q}+G\dot{q}=f \tag{1}$$

where *M* is the inertia matrix consisting of the mass and rotational inertia of the system, *G* is the matrix caused by the gyroscopic effect, $q \in R^4$ is the generalized displacement and $f \in R^4$ is the generalized force from the magnetic force. The generalized force *f* contains several contributions from the PMBs and AMBs, which will be discussed below.

From the experimental identification, the magnetic forces generated by AMBs can be expressed as

$$f_{atx} \approx k_1 i_{tx} \tag{2}$$

$$f_{aty} \approx k_2 i_{ty} \tag{3}$$

$$f_{aby} \approx k_4 i_{by} \tag{4}$$

$$f_{aby} \approx k_4 i_{by} \tag{5}$$

where the subscript *a* represents the active magnetic bearing, *t* is the upper end, *b* is the lower end, $i_{tx,}i_{ty,}i_{bx,}i_{by}$ are the coil currents, and $k_1 \sim k_4$ are constants. It is interesting to note that the magnetic forces generated by the AMB do not depend on the rotor displacements. It depends only on the control currents linearly. This is due to the symmetric coil winding scheme and there is no iron core. As a result, there is no magnetic flux between rotor and stator as long as the coil current is zero, even the rotor position is not zero. Such design makes the levitation controller design easier.

Although the axial dynamics will be neglected and the PMBs are used to provide the supporting axial force direction, the PMBs also generate the forces in the other directions. These forces are assumed to be cancelled out in the radial directions. However, the forces generated by the PMBs can create non-negligible moments for the radial DOFs since the rotor position is not always at the center. In addition to the radial force of the passive magnetic bearing, the axial force also has an effect. The axial force relative to the shaft center also generates a moment that affects the whole dynamics. The forces caused by the passive magnetic bearing are then expressed as

$$f_{ptx} \approx k_5 x_t + k_6 x_b \tag{6}$$

$$f_{pty} \approx k_7 y_t + k_8 y_b \tag{7}$$

$$f_{pbx} \approx k_9 x_t + k_{10} x_b \tag{8}$$

$$f_{pby} \approx k_{11} y_t + k_{12} y_b \tag{9}$$

where *p* denotes the passive magnetic bearing, $x_{t,}y_{t,}x_{b,}y_b$ are the radial rotor displacements, and $k_5 \sim k_{12}$ are constants. According to Newton's law, the equations of motion can be expressed as

$$\sum F_x = m\frac{l_t\ddot{x}_b + l_b\ddot{x}_t}{l_t + l_b} \tag{10}$$

$$\sum F_y = m\frac{l_t\ddot{y}_b + l_b\ddot{y}_t}{l_t + l_b} \tag{11}$$

$$\sum M_x = J_t\frac{\ddot{y}_b - \ddot{y}_t}{l_b + l_t} \tag{12}$$

$$\sum M_y = J_t \frac{\ddot{x}_t - \ddot{x}_b}{l_b + l_t} \tag{13}$$

and

$$\sum F_x = k_1 i_{tx} + k_3 i_{bx} + (k_5 + k_9) x_t + (k_6 + k_{10}) x_b \tag{14}$$

$$\sum F_y = k_2 i_{ty} + k_4 i_{by} + (k_7 + k_{11}) y_t + (k_8 + k_{12}) y_b \tag{15}$$

$$\sum M_x = k_4 i_{by} l_b - k_2 i_{ty} l_t + k_{13} y_t + k_{14} y_b \tag{16}$$

$$\sum M_y = k_1 i_{tx} l_t - k_3 i_{bx} l_b + k_{15} x_t + k_{16} x_b \tag{17}$$

where l_b and l_t are the lengths from the mass center to the lower and upper ends of the active magnetic bearing, m is the mass of the rotor, and J_t is the rotational inertia. With these equations of motion, one can then design the levitation controller for the system.

3. Numerical simulation

The generated magnetic torques and forces are simulated using COMSOL. Figure 5 shows that the torque of the magnetic coupler (shown in Figure 4) if its outer and inner rotors are displaced by some angles. It can be seen that the maximum torque is 33.137Nm. The PMB is used to support the force in the axial direction (weight of the shaft and flywheel), which is around 347.4N. Since there are two sets of PMBs, each set of PMBs needs to support about 170N of force. The magnetic force of PMB is not only related to the magnet, but also related to the air gap and offset. The larger the air gap, the smaller the magnetic force; and the amount of offset is also related to the magnetic force. When the magnet moves up and down vertically, the magnetic force varies.

As shown in Figure 6, when the residual magnetic flux density (B_r) of the magnet is 0.4T, the maximum axial force is below 170N (17kg), so the residual magnetic flux density must be increased. When the magnet residual flux density (B_r) is increased to 1.3T, the magnetic force of 170N corresponds to offset value of about 3.6mm, and when the magnet residual magnetic flux density (B_r) is adjusted to 1.0T, the magnetic force of 170N corresponds to offset of 1.8mm. Therefore, the magnet with a residual magnetic flux density of 1.3 will be selected in the end, because there is a larger space for the rotor to go down, and the operation flexibility is greater.

In Figure 7, when the residual magnetic flux density is 0.45T, the maximum axial force is below 170N (17kg), so the residual magnetic flux density must be increased. Since the lower passive magnetic bearing is a combination of four sets of magnets, although the size is smaller than the upper passive magnetic bearing, the magnetic force corresponding to the increased residual magnetic flux density will be greatly improved, so the residual magnetic flux density is slightly increased to 0.75T, the magnetic force of 170N corresponds to offset of about 2.3mm, which is a value with considerable range for operation as an operating point.

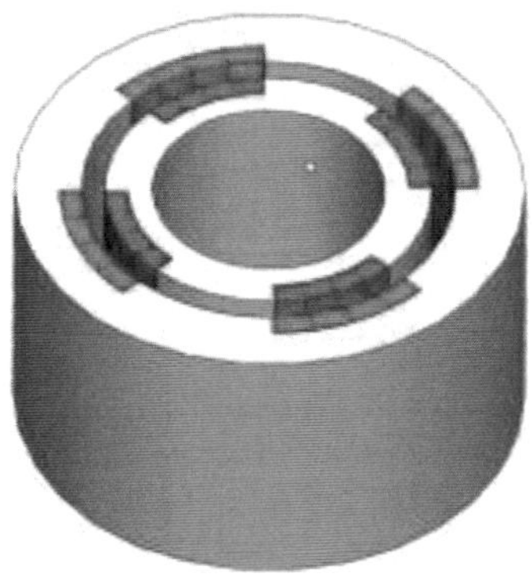

Figure 4. The magnetic coupler

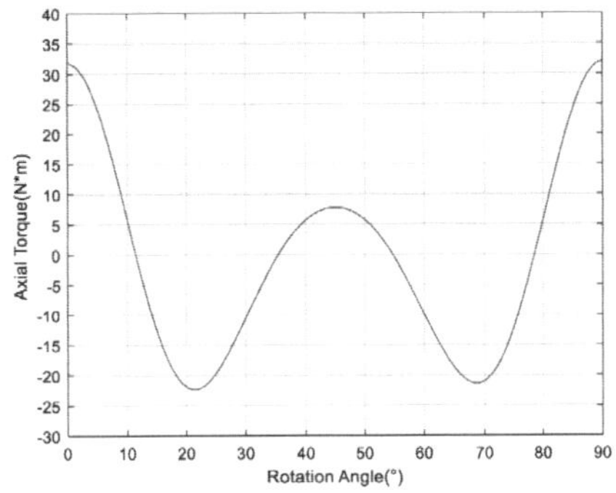

Figure 5. The torque of the magnetic coupler

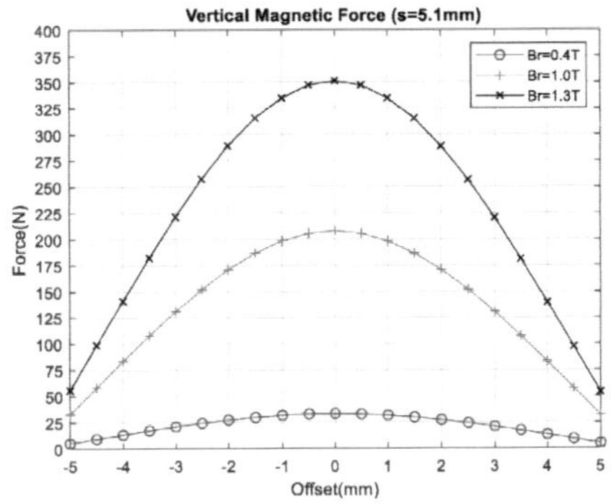

Figure 6 Axial force of upper PMB

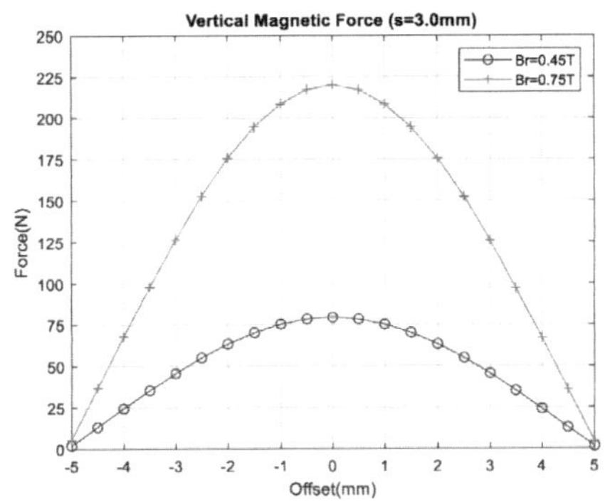

Figure 7 Axial force of lower PMB

4. Conclusions

According to the above simulation, it can be seen that the magnetic coupler can provide large enough torque and the passive magnetic bearing can support the weight of the whole rotor system. In addition, the magnetic coupler method of power transmission is more direct, and there is almost no energy loss problem, while if the electromagnetic induction method of power transmission, there is a problem of energy transmission efficiency.

References

[1]H. W. Song, D. L. Ma, Z. Y. Cui, T. Jia and C. Song, "Heat dissipation flywheel and engine assembly," China Patent 210240443U, Apr. 3, 2020.

[2]Z. Qian, R. Q. Wu and L. Zheng, "Whole heat abstractor of high energy storage flywheel system," China Patent 206807215U, Dec. 26, 2017.

[3]D. I. Ben, N. Zohar and D. Pincu, "Magnetically coupled flywheel," U. S. Patent 20170244300A1, Aug. 24, 2017.

[4]Q. C. Hu, Q. L. Mai and J. P. Xie, "Flywheel-motor magnetic coupling transmission hybrid power system," China Patent 110365158A, Oct. 22, 2019.

[5]M. Dubois, M. Desjardins and L. Tremblay, "System for decoupling a rotor from a stator of a permanent motor and flywheel storage system using the same," U.S. Patent 20120176074A1, Jul. 12, 2012.

Implementation of the DSP-based Fractional Order PID Controller for a Centrifugal Compressor by AMBs

Chia-Juei, Fan and Chao-Yun, Chen

Green Energy and Environment Research Laboratories, Industrial Technology Research Institute, ROC, CJFan@itri.org.tw

Abstract

Active magnetic bearings (AMBs) have many advantages compared with the traditional bearings, especially in the non-contact characteristic that improves applicability of AMBs supported rotor to high rotation speed applications such as the centrifugal compressor that has played a key role in high-end air-conditioning systems. However, precise positioning and high stability are indispensable in controller design for AMBs. Unavoidably, there are many harmful elements which affect the AMBs supported rotor in actual operations. The conventional PID controller has gained popularity in AMBs systems because of easily implementation and tuning intuitive. However, the PID controller may not be sufficient to stabilize complex dynamics of AMBs supported rotor. In order to enhance the stability of the AMBs supported rotor, this study proposed to apply the fractional order PID (FOPID) controller to the centrifugal compressor, in which the centrifugal compressor equipped with five-degree of freedom AMBs and the FOPID has been implemented by digital signal processor (DSP). The discretization implementations are described in detail, including the difficulties encountered and the corresponding solutions. Experimental results show the performance of FOPID compared with the traditional integer order PID (IOPID), and the FOPID can exhibit satisfactory performance, that included levitation and rotation.

Keywords: *Magnetic bearings, Centrifugal compressor, Fractional order PID, DSP*

1. Introduction

Centrifugal compressor is the key hardware components of the high efficiency chiller, which is used to the air conditioning systems for large buildings. The most critical element of centrifugal compressor is the bearings that support the shaft at rotation motion. However, the high coefficient of performance (COP) of the chiller is the indispensable for the energy saving strategy. Therefore, there are many technologies in improving COP of the chiller, in which the AMBs supported rotor that allows the centrifugal compressor to the higher rotation speed, and the compact designed compressor will be obtained. Active magnetic bearings have many advantages compared with the traditional bearings, especially in the non-contact characteristic. However, the mechanical stiffness of AMBs is negative and an open loop AMB is an unstable dynamic system, the task of stabilizing this unstable system by a suitable controller comes down to finding an appropriate current command signal [1]. Consequently, precise positioning and high stability are indispensable in controller design for AMBs.

In pace with advancement of semiconductor process technology, the advanced control theory can be more easily implemented in DSP. Chen et al. applied integral sliding mode control to the three-pole AMB system with assembly error and non-uniform flux distribution [2] Yoon et al. proposed to applied output regulation approach to deal with the rotor unbalance problem of AMB systems [3]. Yoon et al. implemented the LQG control and μ-synthesis to the maglev centrifugal compressor [4]. Although the modern control theory has the ability to deal with the nonlinear phenomenon of maglev centrifugal compressor, however these advanced control approach requires the accurate system modeling technology, and their mathematical method is too complex, even if the advanced control approach has the strongly performance.

In actual industrial applications, the conventional PID controller has gained popularity in AMBs systems because of easily implementation and tuning intuitive. However, the PID controller may not be sufficient to stabilize complex dynamics of AMBs supported rotor. Recently, PID controller have been extended to another expressed form using the all-coefficient adaptive control approach (ACAC). Di et al. applied ACAC to a High-speed desorption pump supported by AMBs, the ACAC can guarantee the close-loop stability and minimize the vibration during the speed increase as well as at the operational speed [5]. Furthermore, fractional order PID Controller has been developed to the control engineering, FOPID is based on the PID generalized form using fractional calculus [6], Matignon proved the stability results as well as the theory of the controllability and observability of finite-dimensional linear fractional differential systems [7]. Oustaloup was the first to propose the concept of fractional PID controller and successfully used in the CRONE control [8]. Padula proposed the tuning rules for FOPID and the tuning rules allow to minimize the integrated absolute error subject to a constraint on the maximum sensitivity [9]. In actually, a fractional-order system which includes a fractional-order controller, a fractional-order controlled object, and both in practical applications.

According the aforementioned literature reviews, the advanced control approach and real time adaptive control method are not suitable for AMB systems in industrial applications, the reasons included that the actual industrial control products require the strongly reliability and stability. Moreover, the easily maintenance is a critical point for the service engineers. In view of this, FOPID has the advantages that included easily implementation and tuning intuitive, and FOPID can be more closely fit the actual system dynamics. This study has been applied the FOPID to the centrifugal compressor by AMBs, and experimental results show that the FOPID exhibits satisfactory performance compared with the traditional PID.

The rest of this paper is organized as follows. Section 2 gives an introduction of the AMB system modeling. Section 3 introduces the fractional order PID Controller and its implementation. Experimental setup and results are included in Section 4, while conclusions are drawn in Section 5.

2. AMB system description and dynamics equation

The rotor-bearing model diagram of centrifugal compressor can be shown in Fig. 1 [1].

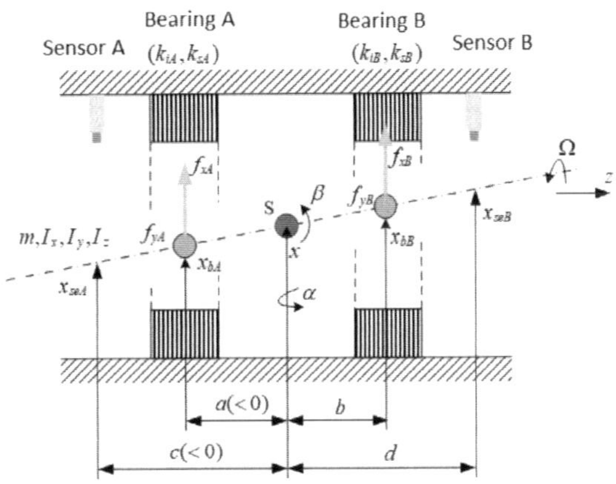

Figure 1 Rotor-bearing model diagram [1].

According to Fig. 1, the center of mass displacements *x* and *y* as well as the Euler angles *α* and *β*, all combined into the vector *q*, the measured rotor displacements x_{seA}, x_{seB}, y_{seA}, and y_{seB}, are comprised in the output vector *Y*. the rotor-bearing dynamics equation can be written as below [1]

$$M\ddot{q} + G\dot{q} = Bu_f \tag{1}$$

$$Y = Cq \tag{2}$$

where $M = \begin{bmatrix} I_y & 0 & 0 & 0 \\ 0 & m & 0 & 0 \\ 0 & 0 & I_x & 0 \\ 0 & 0 & 0 & m \end{bmatrix}$, $B = \begin{bmatrix} a & b & 0 & 0 \\ 1 & 1 & 0 & 0 \\ 0 & 0 & a & b \\ 0 & 0 & 1 & 1 \end{bmatrix}$, $G = \begin{bmatrix} 0 & 0 & I_z\Omega & 0 \\ 0 & 0 & 0 & 0 \\ -I_z\Omega & 0 & 0 & 0 \\ 0 & 0 & 0 & 0 \end{bmatrix}$, $C = \begin{bmatrix} c & 1 & 0 & 0 \\ d & 1 & 0 & 0 \\ 0 & 0 & c & 1 \\ 0 & 0 & d & 1 \end{bmatrix}$,

$q = (\beta, x, -\alpha, y)^T$, $u_f = (f_{xA}, f_{xB}, f_{yA}, f_{yB})^T$ $Y = (x_{seA}, x_{seB}, y_{seA}, y_{seB})^T$. the magnetic bearing force u_f can be described as

$$u_f = \begin{bmatrix} f_{xA} \\ f_{xB} \\ f_{yA} \\ f_{yB} \end{bmatrix} = -\begin{bmatrix} k_{sA} & 0 & 0 & 0 \\ 0 & k_{sB} & 0 & 0 \\ 0 & 0 & k_{sA} & 0 \\ 0 & 0 & 0 & k_{sB} \end{bmatrix} \begin{bmatrix} x_{bA} \\ x_{bB} \\ y_{bA} \\ y_{bB} \end{bmatrix} + \begin{bmatrix} k_{iA} & 0 & 0 & 0 \\ 0 & k_{iB} & 0 & 0 \\ 0 & 0 & k_{iA} & 0 \\ 0 & 0 & 0 & k_{iB} \end{bmatrix} \begin{bmatrix} i_{xA} \\ i_{xB} \\ i_{yA} \\ i_{yB} \end{bmatrix}$$

$$= -K_s q_b + K_i i \tag{3}$$

where k_s is the displacement stiffness and k_i is the current stiffness.

3. Fractional Order PID Controller and its implementation

3.1 A brief introduction to fractional order calculus

Fractional, or non-integer order fundamental operator of differentiation and integration is denoted by

$$ {}_aD_t^\alpha = \begin{cases} \frac{d^\alpha}{dt^\alpha}, & \alpha > 0, \\ 1, & \alpha = 0, \\ \int_a^t (d\tau)^{-\alpha}, & \alpha < 0, \end{cases} \tag{4} $$

Where α is the fractional order of the differentiation or integration, and typically $\alpha \in \Re$ but it also can be complex number [10]. This article discusses the case where the fractional order is a real number. And the Riemann-Liouville definition is used to define fractional calculus as follows.

$$ {}_aD_t^\alpha f(t) = \lim_{h\to 0} \frac{1}{\Gamma(m-\alpha)} \left(\frac{d}{dt}\right)^m \int_0^t \frac{f(\tau)}{(t-\tau)^{1-m+\alpha}} d\tau \tag{5} $$

where $m\text{-}1 < \alpha < m$ and $\Gamma(m-\alpha) = \int_0^\infty e^{-t} t^{(m-\alpha)-1} dt$ is Euler's gamma function. After establishing the definitions of fractional order calculus, the formula for the Laplace transform of the derivative of order α has been discussed. It is shown in [11] that the Laplace transform of an α , the derivative of signal $x(t)$ is given by

$$ L\{D^\alpha x(t)\} = \int_0^\infty e^{-st} {}_0D_t^\alpha x(t) dt = s^\alpha X(s) - \sum_{k=0}^{m-1} s^k {}_0D_t^{\alpha-k-1} x(t)\big|_{t=0} \tag{6} $$

A fractional order differential equation can be expressed in a transfer function form which it provided both the input and output signal as fallow

$$ G(s) = \frac{a_1 s^{\alpha_1} + a_2 s^{\alpha_2} + \cdots + a_m s^{\alpha_m}}{b_1 s^{\beta_1} + b_2 s^{\beta_2} + \cdots + b_m s^{\beta_m}} \tag{7} $$

3.2 Fractional order transfer function approximation algorithm and its implementation

Oustaloup [8] presented an approximation algorithm which can fit the fractional order operator to a bank of integer order filters in the specified frequency band. The fractional order α in the specified frequency band [ω_l, ω_h] is given as follows

$$ s^\alpha = (\omega_h)^\alpha \prod_{k=1}^{N} \frac{s+\omega_k'}{s+\omega_k}, \quad 0 < \alpha < 1 \tag{8} $$

Where N is the number of poles and zeros which are evaluated as $\omega_k' = \omega_l (\omega_h / \omega_l)^{2k-1-\alpha/2N}$ and $\omega_k' = \omega_l (\omega_h / \omega_l)^{2k-1+\alpha/2N}$. For the case α<0, the Eq. (8) would be inverted. For the case $|\alpha| > 1$, s^α should be rearranged to $s^\alpha = s^n s^\sigma$ .Where n is an integer number and $\sigma \in [0,1]$, s^σ term needs to be approximated by Eq. (8).

Bilinear transformation is the IIR filter design technology used in this article. The s-domain transfer function, H(s), is mapped to the z-domain through Eq. (9) to become the transfer function, H(z). For implementation, The H(z) needs to be sorted into Eq. (10). There will be no spectrum aliasing problem regardless of the original s-domain transfer function and bandwidth. After bilinear transformation, the amplitude frequency response range will not exceed half of the sampling frequency, fs. Even if there is no spectral aliasing problem in bilinear transformation, the input data must satisfy the sampling theorem. If the original input data is under sampled due to a low sampling frequency and violates the sampling theorem, then any filter cannot eliminate the errors caused by under sampling. However, it must be noted that there will be nonlinear distortion with the original s-domain frequency after bilinear transformation conversion, and the frequency distortion effect will be more apparent as it approaches half of the sampling frequency.

$$ s = \frac{2}{T_s}\left(\frac{1-z^{-1}}{1+z^{-1}}\right) \tag{9} $$

where Ts is the sampling period and z is a complex number.

$$H(z) = \frac{\sum_{k=0}^{N} b(k) z^{-k}}{1 - \sum_{k=1}^{M} a(k) z^{-k}} \tag{10}$$

In order to implement the differential equation in digital signal processor, the digital filter uses Direct-Form I [12]. While there are many variants of this type of filter structure, the Direct-Form I method is considered to have the best performance against filter coefficient quantization errors and stability problems, and is also the method used in this article. To balance the performance of the fractional order PID with the difficulty of DSP implementation, the number of poles and zeros N is set to 5, and the frequency limits ω_l and ω_h to 0.1 and 50,000 respectively. However, since the system order is still high, the filter poles and zeros in the direct structure are very sensitive to various errors in the filter coefficients. For instance, quantization errors in the coefficients can cause the filter poles and zeros to move on the z plane, leading to amplitude frequency response that fails to meet the requirements. Additionally, rounding errors may accumulate with the number of operations, leading to filter instability, while overflow errors can cause the filter to oscillate. Fortunately, JTAG(Joint Test Action Group) can confirm the DSP variables of the embedded system in real-time, allowing for the detection of overflow errors quickly.

During the practice, it was found that the simulated filter coefficients were stable during computer analysis but unstable during DSP implementation. The instability of the control system caused by such filter coefficients was resolved by changing single-precision floating-point numbers to double-precision floating-point numbers. At the same time, the possibility of overflow errors in numerical calculations is reduced. In addition to consuming more memory space, changing from single-precision floating-point numbers to double-precision floating-point numbers posed a main difficulty, which was that the DSP floating-point arithmetic unit in used was FPU32, which could only calculate 32-bit floating-point numbers. Therefore, the operation of 64-bit double-precision floating-point numbers could not be effectively accelerated. The AMB control involves the control of five degrees of freedom. With the decoupling architecture, it includes the translational, inclination displacements of the radial X-Y axis, and axial position control. Each degree of freedom has a corresponding fractional order PID controller. In addition to the AMB controller, the control system also includes several filters for suppressing position and current signal noise and AMB system resonance. Besides, a serial communication interface program is in operation to monitor the AMB trajectory on a personal computer. Due to limited DSP resources, the AMB position control loop in this article is choose as 3kHz. To fully utilize the performance of fractional order PID controller in five degrees of freedom AMB control system, a DSP with a 64-bit double-precision floating-point arithmetic unit is necessary and crucial.

4. Experimental setup and results

4.1 Experimental Setup

The hardware of the experimental system, which is shown in Fig. 2, consists of a centrifugal compressor equipped with five degrees of freedom AMBs. Its composition includes an impeller, an induction motor, a rotor, two radial AMBs, one axial AMB, inductive-type sensors, a variable-frequency drive, and a magnetic bearing drive controller. The magnetic bearing drive controller, which includes a drive circuit, a DSP-based control circuit, and a sensor circuit, is embedded on the side of the centrifugal compressor. The sensor circuit provides five-degree of freedom position feedback and pulse signal of rotor rotation. The magnetic centrifugal compressor is used to evaluate the performance of FOPID controller, which had implemented by digital signal processor, TI-TMS320F28335. The fractional order setting of FOPID derivative term is 0.97 and N=5.

Figure 2 Maglev centrifugal compressor equipped with five-degree of freedom AMBs.

4.2 Experimental results

Experiment #1: Rotor levitation for evaluating the performance of FOPID

Figure 3 shows the rotor levitation result using FOPID, the black circle is the backup bearing clearance, the red dashed circle is the 30% of the backup bearing clearance from international standard (ISO-14839-2) [13]. According to the results of Fig. 3, both the front radial orbit and the rear radial orbit can be precisely controlled in the center of the backup bearing clearance, the maximum vibration percentages are 0.339% and 0.579%, the RMS percentages of orbit are 0.13% and 0.199%. From the vibration of the trajectory, FOPID demonstrates sufficient stability.

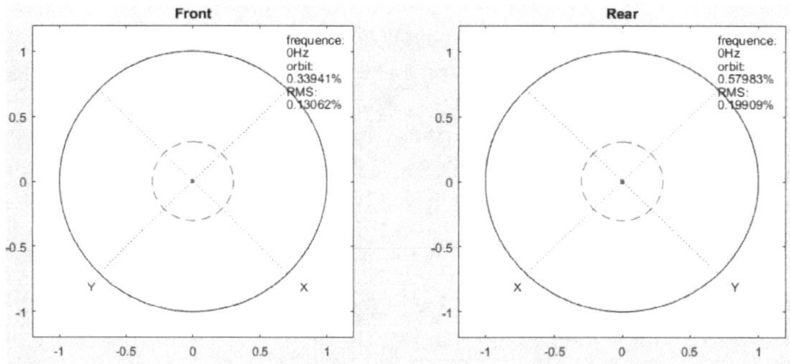

Figure 3 The measured rotor orbit of the levitation verification.

Experiment #2: Rotor rotated motion for evaluating the performance compared with the FOPID and IOPID

Figures 4 and 5 show the measured rotor orbit of the centrifugal compressor at 9,000rpm (150Hz), in which the IOPID represents the traditional integer order PID, the black circle is the backup bearing clearance, the red dashed circle is the 30% of the backup bearing clearance (ISO-14839-2). In the experimental result of using IOPID, the maximum vibration percentages are 6.915% and 6.236%. In the experimental result of using FOPID, the maximum vibration percentages are 5.034% and 4.85%. Clearly, The FOPID exhibits satisfactory performance compared with the IOPID. Which means FOPID can be more closely fit the actual system dynamics.

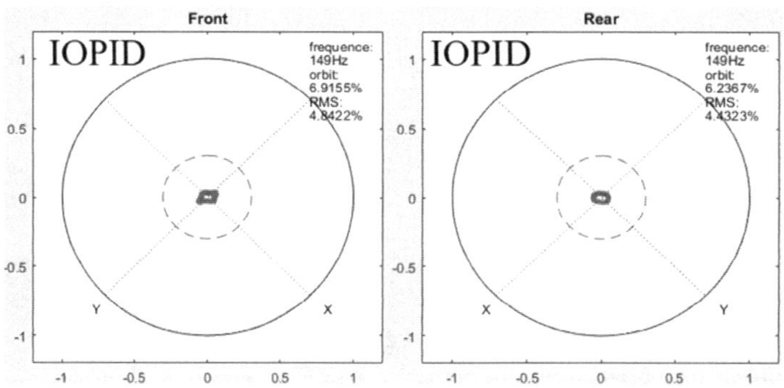

Figure 4 The measured rotor orbit of the centrifugal compressor at 9,000rpm using IOPID.

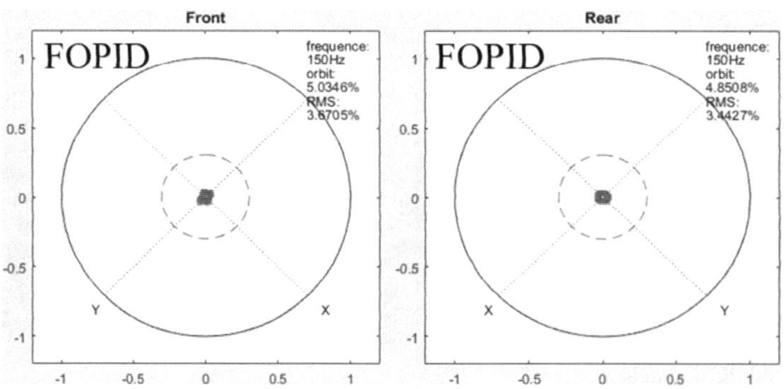

Figure 5 The measured rotor orbit of the centrifugal compressor at 9,000rpm using FOPID.

5. Conclusions

In the actual industrial applications of the AMBs system, the requirements of controller design include the strongly reliability, stability, and easily maintenance. However, the advanced control approach requires the accurate system modeling technology, and the maintenance is extremely difficult for the service engineers. Therefore, this article applied FOPID to the magnetic centrifugal compressor, the advantages of FOPID are that the controller can be more closely fit the actual system dynamics, easily implementation and tuning intuitive. The fractional order transfer function approximation algorithm and its discretization implementation are described in detail, including the difficulties encountered and the corresponding solutions. The experimental results show that the rotor can be precisely positioned at the center of the backup bearing. In addition, the FOPID controller exhibits superior performance compared to the IOPID controller at high rotation speeds.

Acknowledgements

The authors would like to thank the Bureau of Energy, Ministry of Economic Affairs, Taiwan, R.O.C., for supporting this research.

References

[1] Schweitzer G and Maslen EH (2009) Magnetic bearing. Springer.

[2] Chen SL, Li YT, Lin CH, and Chen CY (2023) Effects of Imperfect Assembly and Magnetic Properties on the Three-Pole AMB System. Applied Sciences-Basel, 13(1): 347.

[3] Yoon SY, Di L, and Lin Z (2016) Unbalance compensation for AMB systems with input delay: An output regulation approach. Control Engineering Practice 46: 166-175.

[4] Yoon SY, Lin Z, and Allaire PE (2012) Control of surge in centrifugal compressors by active magnetic bearings: Theory and implementation. Springer Science and Business Media.

[5]Di L, Chen CY, Lin CH, and Lin Z (2018) Characteristic Model Based All-Coefficient Adaptive Control of a High-Speed Desorption Pump Supported by AMBs. 15th international Symposium on Magnetic Bearings, Japan 3-6 Aug. 2016, T2B4 1-8.

[6] Podlubny I (1999) Fractional Differential Equations. New York, NY, USA: Academic Press.

[7] Matignon D and D'Andrea-Novel B (2002) Observer-based controllers for fractional differential systans. In: 36th IEEE conference on Decision and Control, San Diego,California, 10-12 December 1997, pp.4967–4972. California: IEEE control systems society.

[8] Oustaloup A and Mathieu B (1999) La commande CRONE: Du scalaire au multivariate. Paris, France: Hermes.

[9] Padula F and Visioli A (2011) Tuning rules for optimal PID and fractional-order PID controllers. Journal of Process Control, 21(1): 69-81.

[10] Oustaloup A, Levron F, Mathieu B and Nanot FM (2000) Frequencyband complex noninteger differentiator: characterization and synthesis. IEEE Transactions on Circuits and Systems I 47(1): 25–39.

[11] Podlubny I (1997) The Laplace Transform Method for Linear Differential Equations of the Fractional Order. PhD Thesis, Technical University of Kosice, Slovak.

[12] A. V. Oppenheim and R. W. Schafer (1975) Digital Signal Processing. Englewood Cliffs, NJ: Prentice-Hall.

[13] ISO 14839-2 (2004) Mechanical vibration - Vibration of rotating machinery equipped with active magnetic bearings - Part 2: Evaluation of vibration.

Experimental Results of Unbiased Control of 9-Pole Radial Magnetic Bearing

Myounggyu D. NOH[a], Wonjin JEONG[a], Miseon SONG[a]

a Dept. Mechatronics Engineering, Chungnam National University, 99 Daehak-ro, Yuseong-gu, Daejeon, Korea, mnoh@cnu.ac.kr

Abstract

Previously, a generalized unbiased control strategy has been proposed for unsymmetrical radial magnetic bearings, which reduces the risk of slew rate limiting. In this paper, nine-pole radial bearings are designed and applied them to an industry-scale compressor prototype. The bearing is configured so that a single inverter can drive one radial bearing. In order to implement the control laws, phase selection algorithm is examined. When the force inversion rule is included in the plant dynamics, it is found that the plant seen by the controller is linear. Lead-filter type control laws are designed and implemented. Measurements of the sensitivities confirm the appropriateness of the control.

Keywords: Unsymmetrical bearing, Unbiased control, Force inversion, Phase selection

1. Introduction

Typical radial active magnetic bearings are structurally symmetric. For example, an eight-pole bearing uses two opposing pairs to control one axis by winding two adjacent poles in series. Each pole pair is driven by a power amplifier. Bias linearization is widely used to overcome the quadratic relationship of coil currents to the magnetic force. However, the bias currents increase the ohmic losses and require larger capacity of power amplifiers that are needed to supply control currents.

Unbiased control of symmetric bearings has the critical issue of slew rate limiting (Tsiotras & Wilson 2003). Unsymmetrical three-pole bearings have been studied as well (Chen & Hsu 2002, Hemenway & Severson 2020). However, they are applicable to very small rotors and the control axes are cross-coupled.

A generalized unbiased control strategy for unsymmetrical bearings has been proposed by Meeker (Meeker 2017). The strategy reduces the risk of slew rate limiting. Also, it is possible to use linear control law, if the nonlinearities are handled inside the plant dynamics.

In this research, we have applied the unbiased control strategy to active magnetic bearings supporting an industry-scale compressor. The performance of the control is investigated through experiments.

2. Unbiased Control Strategy

2.1 Current to Force Relationship

The 9-pole radial magnetic bearing considered in this paper is illustrated in Figure 1. The reference axes and pole numbering are also shown. As discussed in Meeker (2017), all nine poles can be independently driven by three inverters with the possibility of fault tolerance. In this paper, however, three adjacent poles are grouped so that a single inverter can drive the bearing.

Using magnetic circuit analysis, the forces that are generated by coil currents can be expressed as

$$f_x = \frac{1}{2}\mathbf{i}^T \mathbf{M}_x \mathbf{i}, \quad f_y = \frac{1}{2}\mathbf{i}^T \mathbf{M}_y \mathbf{i} \tag{1}$$

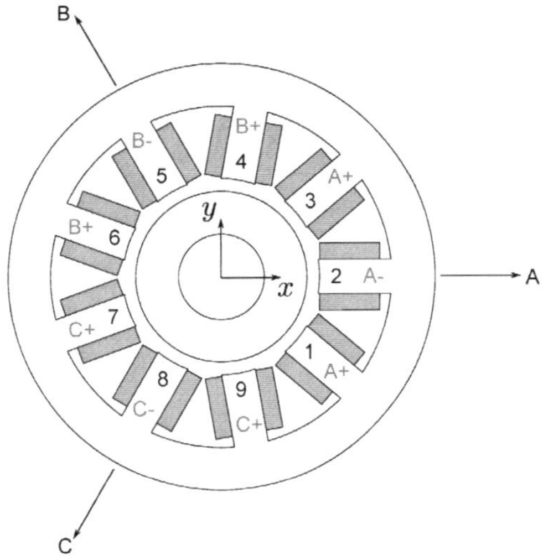

Figure 1: Schematic of 9-pole radial bearing

where $\mathbf{i}$ is the vector of coil currents and defined as $[i_A\ i_B\ i_C]^T$. In (1), $\mathbf{M}_x$ and $\mathbf{M}_y$ are the matrices determined by bearing geometry. Since two-dimensional forces are determined by three currents, the force-current relationship in (1) is not unique. Suppose that two independent currents, i_r and i_i, arranged in a complex form satisfy

$$f = f_x + jf_y = (i_r + ji_i)^2 \tag{2}$$

Then, the forces can be determined by

$$f_x = i_r^2 - i_i^2, \quad f_y = 2i_r i_i \tag{3}$$

Since the mapping from the control currents to coil currents is not unique, a constrained optimization can be set up to find the mapping matrix, where the elements of the matrix and the total flux in the bearing are minimized (Meeker 2017). The mapping matrix for a nine-pole bearing with three adjacent poles wired in series is given as

$$\mathbf{W} = \begin{bmatrix} 1.02623 & 0 \\ -0.51312 & -0.88874 \\ -0.51312 & 0.88874 \end{bmatrix} \tag{4}$$

with respect to the reference axes in Figure 1.

The force-current relationship (1) does not account for the dependence of the force with respect to the position of rotor. The control currents must be adjusted in order to include the position dependence as follows (Meeker 2017):

$$i_r^* = i_r - xi_r - yi_i, \quad i_i^* = i_i + xi_i - yii_r \tag{5}$$

where x and y are the displacements of the rotor.

2.2 Force Inversion and Phase Selection

Figure 2 shows the block diagram of unbiased control. Control law generates the required force of f_x^* and f_y^*. The required control currents can be obtained by inverting the force-current relationship.

$$i_r = \frac{1}{\sqrt{2}}\sqrt{|f| + f_x}, \quad i_i = \frac{\text{sgn}(f_y)}{\sqrt{2}}\sqrt{|f| - f_x} \tag{6}$$

where

$$|f| = \sqrt{f_x^2 + f_y^2}$$

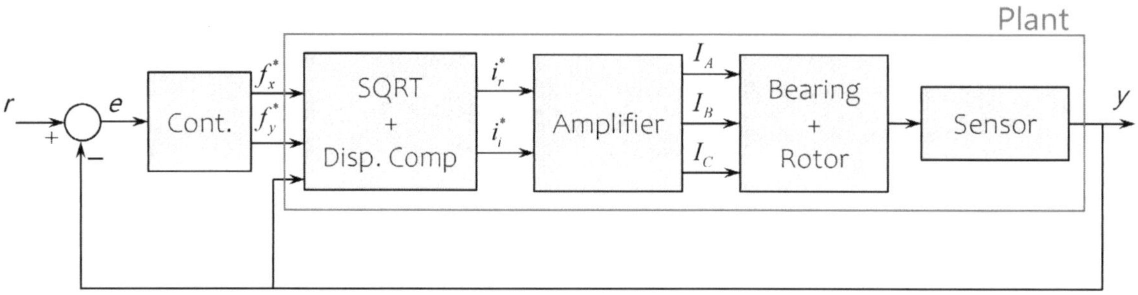

Figure 2: Block diagram of unbiased control

When inverting the force-current equation, it is important to avoid discontinuity in current outputs. As shown in Figure 3, there can be a large jump in i_r when the force command crosses the negative real axis (from F_1 to F_2. If the phase of the current is adjusted by π as in the right diagram, the continuity of the control current can be maintained.

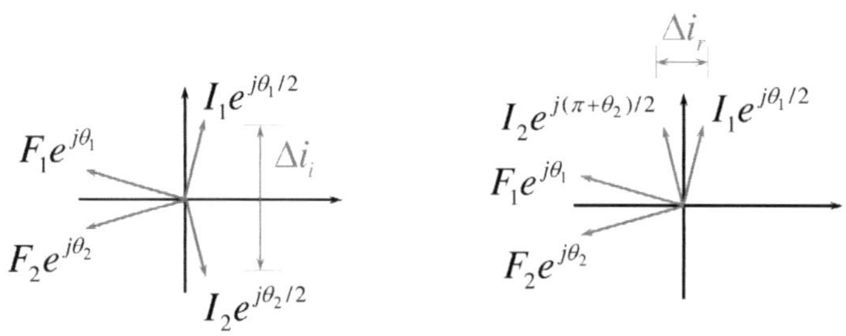

Figure 3: Phase selection

If we treat the force inversion and the displacement compensation as a part of the plant, this plant seen by the force command is linear. Therefore, typical control laws such as lead filters can be used to levitate the bearing.

3. Test Rig for Unbiased Control

Two Nine-pole radial AMBs (motor side and impeller side) are designed to support a chiller compressor (capacity of 500RT and the operating speed of 18,000 rpm). The load capacity of each bearing is 1120 N. Journal diameter is 93.2 mm. The pictures of the compressor and the bearing are shown in Figure 4.

Figure 4: Prototype of chiller compressor and 9-pole radial AMB

Lead filter type control laws are used. Notch filters are employed to suppress the vibration due to flexible modes. Integrators are added to decrease the steady-state error. Figure 5 shows the bode plots of the controllers for motor and impeller bearings.

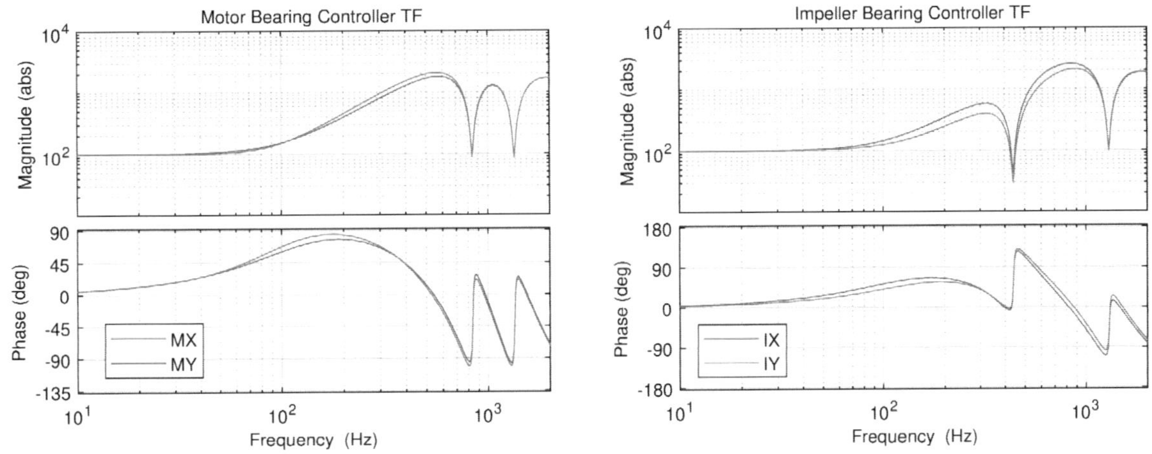

Figure 5: Bode plots of control transfer functions for motor and impeller bearings

4. Test Results

Levitation control is implemented using MATLAB/SimulinkRT at the sampling rate of 10 kHz. Plant transfer fuctions and output sensitivities are measured through sine sweep tests. The plant transfer functions are shown in Figure 6. The output sensitivities are are shown in Figure 7. Measured sensitivities satisfy ISO14839-3 requirements.

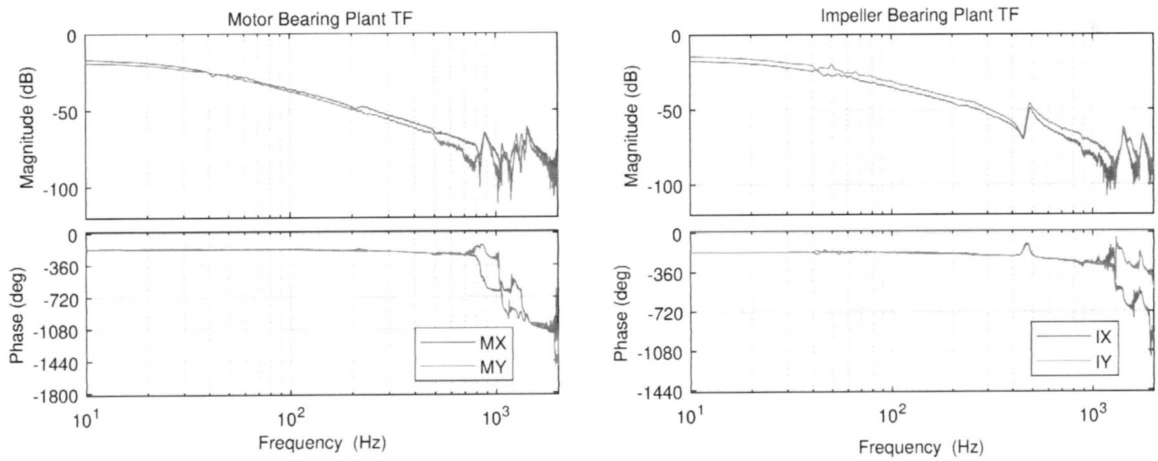

Figure 6: Plant transfer functions of motor and impeller bearings

References

Chen, S.-L. & Hsu, C.-T. (2002), 'Optimal design of a three-pole active magnetic bearing', *IEEE Transactions on Magnetics* **38**(5), 3458–3466.

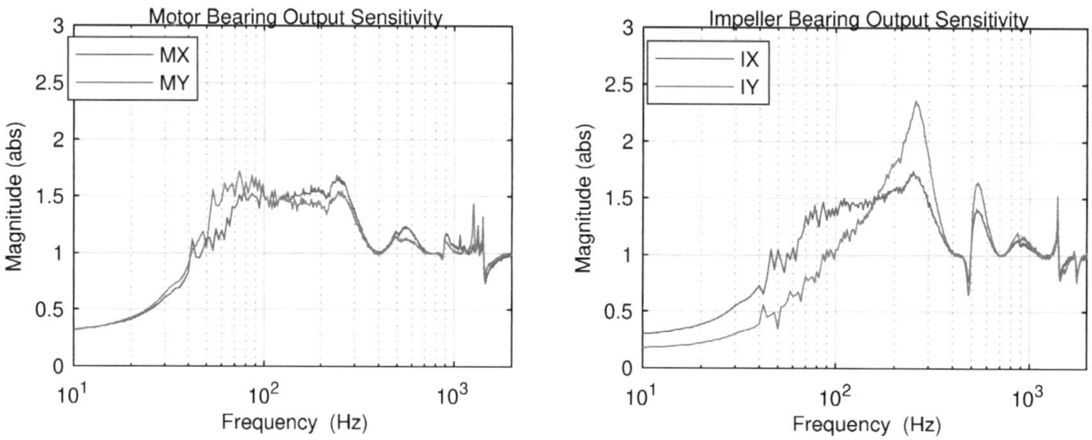

Figure 7: Sensitivity measurements for motor and impeller bearings

Hemenway, N. R. & Severson, E. L. (2020), 'Three-pole magnetic bearing design and actuation', *IEEE Transactions on Industry Applications* **56**(6), 6348–6359.

Meeker, D. (2017), A generalized unbiased control strategy for radial magnetic bearings, *in* 'Actuators', Vol. 6, MDPI, p. 1.

Tsiotras, P. & Wilson, B. C. (2003), 'Zero-and low-bias control designs for active magnetic bearings', *IEEE Transactions on Control Systems Technology* **11**(6), 889–904.

Unbalance Compensation for a 5-DOF Magnetic Rotor-Bearing System

Suo-Chih Wei[a], Shyh-Leh CHEN[a]

a Department of Mechanical Engineering and Advanced Institute of Manufacturing with High-tech Innovations, National Chung Cheng University Chiayi 621, Taiwan, imeslc@ccu.edu.tw

1. Abstract

The objective of this study is to design an unbalance force rejection controller by using the decoupling control law as a levitation controller. The system under study is a 5-DOF rotor magnetic bearing system. It is equipped with two sets of heteropolar radial active magnetic bearings and one axial active magnetic bearing. In this study, an unbalance force rejection control (UFRC) structure is proposed, which is inspired by the coordinate transformation of the decoupling control. By combining the proposed UFRC controller with decoupling control, the system can be divided into two independent parts, parallel mode and conical mode. That is, the system has been decoupled. By decoupling the system, the design and analysis of the UFRC controller can be more intuitive and easier. In addition, the advantage of decoupling control is maintained. The design method and stability analysis are given in the paper. After adding UFRC controller to the control loop, the stability becomes speed dependent. Thus, the UFRC controller parameter varies for different speed range. At the end of this paper, the performance of the proposed concept is verified by numerical simulation. In summary, the proposed method works and makes the rotor turn around its center of mass so that the imbalance force is rejected. Although only the static unbalance is considered in this study, the same concept can be extended to the case where both the static unbalance and the dynamic unbalance are present.

Keywords: *Magnetic bearing, Unbalance compensation, UFRC, Decoupling control*

2. Introduction

In addition, since the magnitude of the unbalance force is proportional to the square of the speed in a magnetic bearing system, other problems such as amplifier saturation and flexible mode excitation can occur at high speeds. This is where unbalance compensation comes into play. Based on the objective, most unbalance compensation algorithms can be divided into two groups. One is unbalance vibration cancellation control (Chen, et al., 2019), which forces the rotor to rotate around its geometric axis. The other is unbalance force rejection control (UFRC) (Zheng, et al., 2016), which forces the rotor to rotate about its principal axis. In this study, a decoupling controller is used to levitate the rotor, and an UFRC controller is designed to cancel the unbalance force. The mathematical model that considers the eccentricity is introduced, and the design of the UFRC controller is investigated.

3. System description and dynamic modeling

The system under study is a 5-DOF rotor-magnetic bearing system shown in Figure 1. The rotor weight is supported by two 8-pole heteropolar active magnetic bearings at each end.

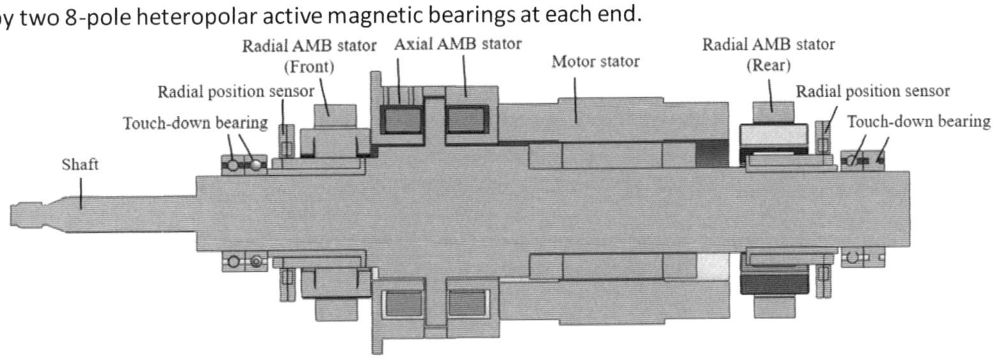

Figure 1 5-DOF rotor-magnetic bearing system

From the book by Maslen and Schweitzer (2009), if we consider only the static unbalance the rotor dynamics can be described by (1).

$$\begin{aligned} &M\ddot{q}+G\dot{q}+K_{sS}q=BK_i i_c+Us \\ &q_{se}=Cq \\ &Us=m\Omega^2\left[0 \quad -e_y\sin\Omega t+e_x\cos\Omega t \quad 0 \quad e_x\sin\Omega t+e_y\cos\Omega t\right]^T \end{aligned} \tag{1}$$

where $q=[\beta, x, -\alpha, y]^T$, $q_{se}=\left[x_f, x_r, y_f, y_r\right]^T$. q is the body coordinate displacement of the rotor, in which x, y are displacements of the rotor and β, $-\alpha$ are the rotation angle of the rotor. q_{se} is the sensor coordinate displacement of the rotor, in other words, rotor displacement measured by position sensors. C is the measurement matrix. Ω is the rotation speed and e_x, e_y are eccentricities of the rotor in x, y directions respectively. M is the inertial matrix and is diagonal. To make the discussion later on easier, let us call x, y the parallel modes of the rotor and β, $-\alpha$ are the conical modes. Next, the decoupling PID control law we utilize to levitated the rotor is

$$i_c=-(BK_i)^{-1}\left[PC^{-1}q_{se}+IC^{-1}\int q_{se}dt+DC^{-1}\dot{q}_{se}\right]-K_{sS,comp}q_{se} \tag{2}$$

where P, I, D are diagonal and $K_{sS,comp}=-\left(BK_i\right)^{-1}K_{sS}C^{-1}$. With the control law, closed loop dynamics become (3)

$$M\ddot{q}+D\dot{q}+Pq+I\int qdt=Us \tag{3}$$

From the left side of (3), it is easy to see that the parallel and conical modes are independent. In this study, only static unbalance is considered. That is, eccentricity exists only in the radial direction of the rotor, while the axial direction of the rotor is in the absence of eccentricity. Thus, only parallel modes are affected by the unbalance force (Us), while conical modes are not. This is a nice property that we will take advantage of when designing the UFRC controller.

4. Design of UFRC controller

4.1 Block diagram of the system and UFRC controller structure

Figure 2 shows the control loop of the system, where block N represents the UFRC controller. Also, in (2), the term $K_{sS,comp}q_{se}$ is used to decouple the parallel and conical modes and must not be changed to achieve the best decoupling effect. Thus, the feedback signal is injected directly into the decoupling PID controller without passing through the UFRC controller. Figure 3 shows the structure of the UFRC controller. The error is transformed into body coordinates when it enters the UFRC controller. Before it leaves, it is transformed back to sensor coordinate. It can be seen that the concept is similar to that of the decoupling controller. Most importantly, for body coordinate errors, only parallel errors pass through the generalized notch filter (Zheng, et al., 2016), where ε, θ are parameters to be designed. This reduces the complexity of designing the notch filter parameters, since the stability of conical modes (whose pole locations are speed-dependent) is not affected by the notch filter.

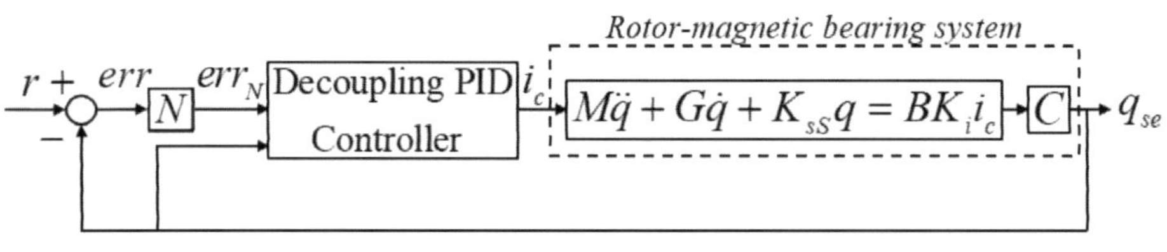

Figure 2 Block diagram of the system with controllers

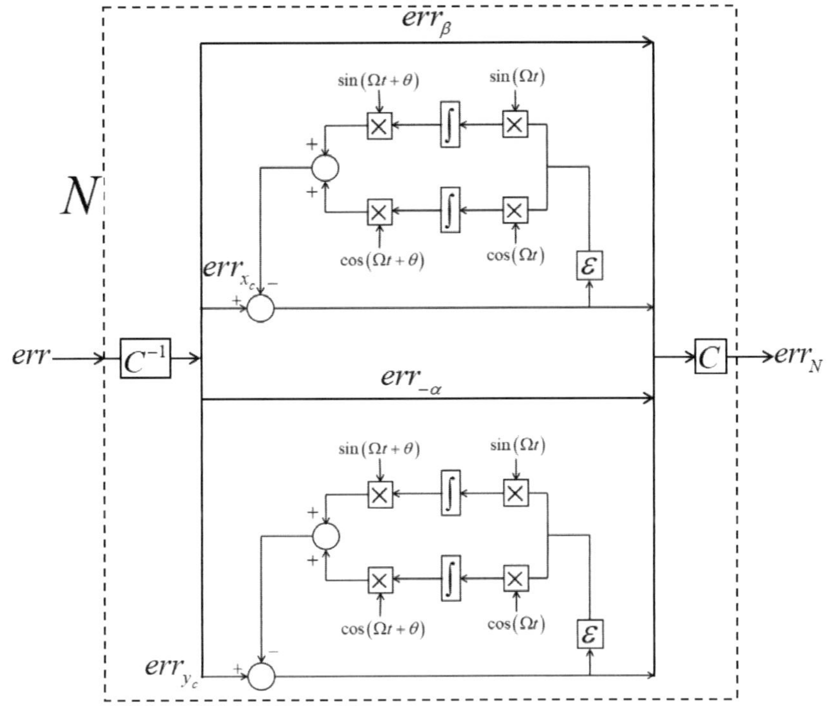

Figure 3 UFRC controller structure

4.2 Simplification of the overall system

After substituting the control law (2) and the UFRC controller structure into the block diagram shown in Figure 2, the system can be represented by the block diagram shown in Figure 4 when the rotor is not rotating. As you can see, a 4-DOF system where each DOF is coupled now becomes four 1-DOF systems where each DOF is independent. Each block in Figure 4 is as follows

$$N_{para} = \frac{s^2+\Omega^2}{s^2+\varepsilon_c s+\Omega^2-\Omega\varepsilon_s}, G_{c,para} = \frac{k_{D,para}s^2+k_{P,para}s+k_{I,para}}{s}, G_{c,coni} = \frac{k_{D,coni}s^2+k_{P,coni}s+k_{I,coni}}{s}$$
$$G_{P,para} = \frac{1}{ms^2}, G_{P,coni} = \frac{1}{I_y s^2}, \varepsilon_c = \varepsilon\cos\theta, \varepsilon_s = \varepsilon\sin\theta \tag{4}$$

N_{para} is the transfer function of the parallel mode part in UFRC controller and is of the form of generalized notch filter. $G_{c,para}$ and $G_{c,coni}$ are PID controller transfer functions correspond to parallel and conical mode respectively. Similarly, $G_{P,para}$ and $G_{P,coni}$ are plant transfer functions for parallel and conical mode respectively. $F_{unbalance}$ is the unbalance force acting on the system and can be treated as external disturbance.

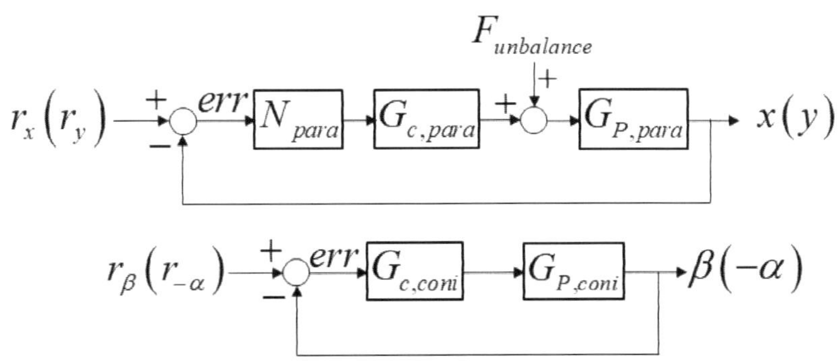

Figure 4 Block diagram of decoupled system (not rotating)

4.3 Design of UFRC controller parameter

After simplifying the system, the UFRC controller parameter design is much easier. From Figure 4, we can easily see that the conical mode block diagram has nothing to do with the notch filter design. Therefore, only the parallel mode characteristics such as stability, performance, etc. need to be investigated. Furthermore, if the dynamic unbalance is taken into account, the notch filter for each mode can be designed independently. From (4), we know that not only the UFRC controller parameter changes the system characteristics, but also the rotational speed, since N_{para} is also a function of the rotational speed Ω. Thus, the closed-loop stability of the system should be checked after introducing the designed notch filter under a certain rotational speed. In addition, since the gyroscopic effect becomes more significant and the unbalance force is stronger at high speed, the system should better preserve its characteristics at high speed after adding the notch filter. By preserving the characteristics, we can make sure that the system characteristics (sensitivity, phase margin, etc.) are similar to what we desired when designing the decoupling controller.

Let us set the operating speed as 20000 rpm, by following the above two baselines, that is, to maintain the stability and characteristics when the rotator rotates at 20000 rpm, the UFRC controller parameters are chosen as ε = 1007.1, θ = -0.1188. Figure 5 shows the dominant pole locations of the parallel mode at different speeds with this set of UFRC controller parameters. From the figure, we can see that the system loses its stability when the speed is below 3800 rpm, also the system retains its characteristics when the speed is above 6300 rpm. Note that the result in Figure 5 was obtained using the linearized model, i.e. the parallel mode block diagram in Figure 4. In the linearized model, the magnetic forces are linearized and the gyroscopic effect is not considered. To obtain a more accurate result, a nonlinear model with nonlinear magnetic force and gyroscopic effect is used for numerical simulation. The simulation result is shown in Figure 6, where UFRC is turned on when t = 1 sec. When the rotational speed is 4500 rpm and UFRC is turned on, the system reaches marginally stable, which means that the system may lose its stability when the rotational speed is below 4500 rpm. As you can see, the result is slightly different from the linear model. Thus, only when the speed is higher than 4500 rpm, the parameter set ε = 1007.1, θ = -0.1188 can be used. For speed below 4500 rpm, we need to design another set of UFRC controller parameter.

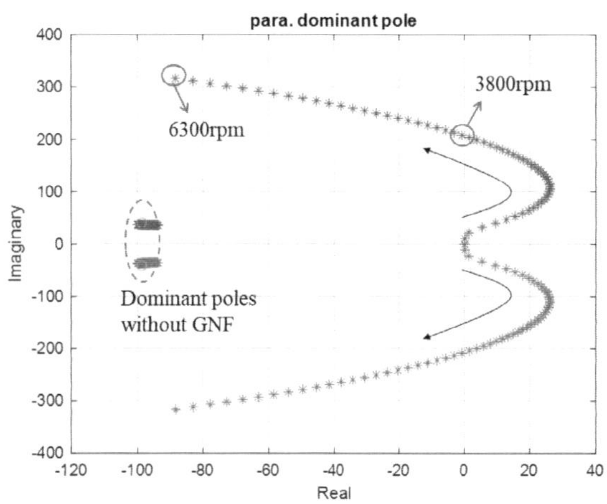

Figure 5 Parallel mode dominant pole locations under different rotation speed (UFRC on, ε=1007.1, θ=-0.1188)

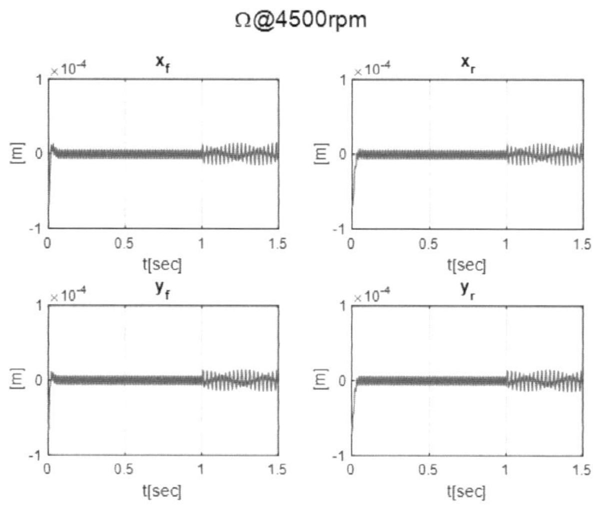

Figure 6 Rotor displacement while rotating at 4500 rpm (ε=1007.1, θ=-0.1188)

Following the same baseline, the UFRC controller parameters are chosen as ε = 200, θ = -0.0477. Figure 7 shows the dominant pole locations of the parallel mode under different rotation speed with ε = 200, θ = -0.0477. Similarly, the system loses its stability when the rotation speed falls below a certain number, which in this case is 2700 rpm. However, the dominant poles are very different from those without notch filter even at high speed, which means that the system characteristics change a lot after turning on UFRC. The simulation result with the nonlinear model is shown in Figure 8, where the UFRC is turned on when t = 1 sec. The system reaches marginally stable when the rotational speed is 2950 rpm, which is also slightly higher than the result obtained from the linear model. For the sake of safety, the first set of UFRC controller parameter is used when $\Omega > 5000$ rpm (higher than 4500 rpm) and the second set is used when 5000RPM $> \Omega \geq 3100$ rpm (higher than 2950 rpm). The URFC controller parameters for different speed ranges are shown in Table1.

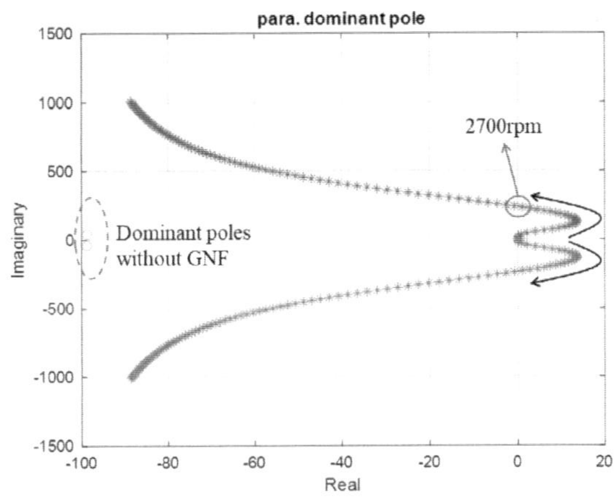

Figure7 Parallel mode dominant pole locations under different rotation speed (UFRC on, ε=200, θ=-0.0477)

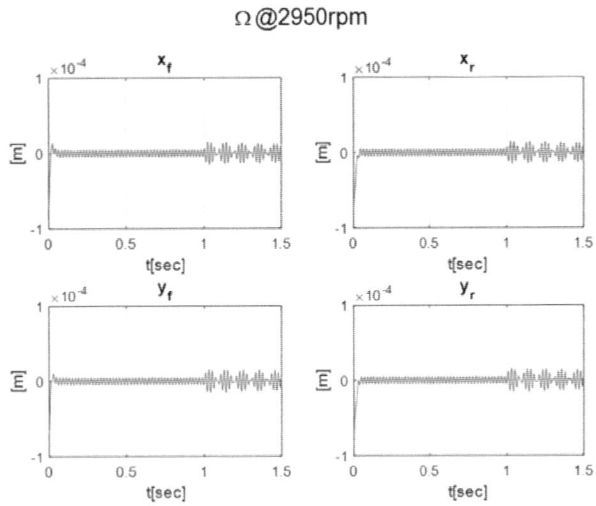

Figure 8 Rotor displacement while rotating at 4500 rpm (ε=200, θ=-0.0477)

Table 1 UFRC controller parameters for different rotation speed range

Rotation Speed [rpm]	ε	θ
5000↑	1007.1	-0.1188
3100 ~ 5000	200	-0.0477
3100↓	0	0

5. Numerical simulation

The rotation speed during simulation is shown in Figure 9. According to ISO1940-1, under the speed of 20000 rpm the maximum admissible eccentricity value is 3x10^{-3}mm. Therefore, the eccentricity of the rotor is set to e_x=3x10^{-3}mm, e_y=0. The simulation result is shown in Figure 10, where the rotor response is shown in sensor coordinates. Initially, the rotor floats from its initial position to the center without rotating. Then the rotor rotates at the speed shown in Figure 9. As can be seen in the zoomed view of Figure 10, the maximum and minimum displacement values of the one with UFRC are almost the same and are around $\sqrt{e_x^2+e_y^2}$. The same observation can be found in all other three plots, which means that the rotor is rotating around its principal axis (center of mass), causing the centroid of the rotor to oscillate.

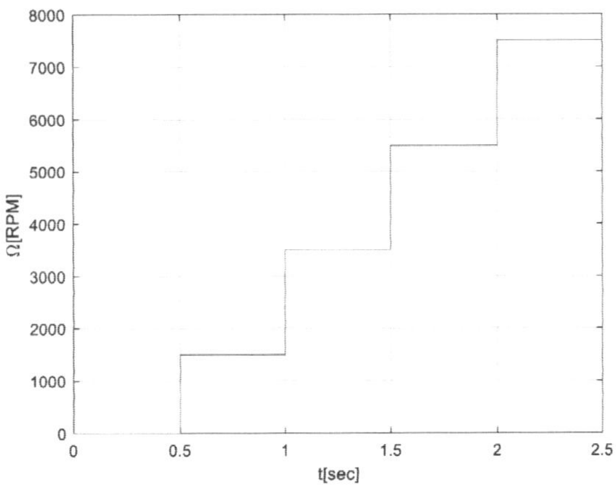

Figure 9 Rotation speed [RPM] versus time

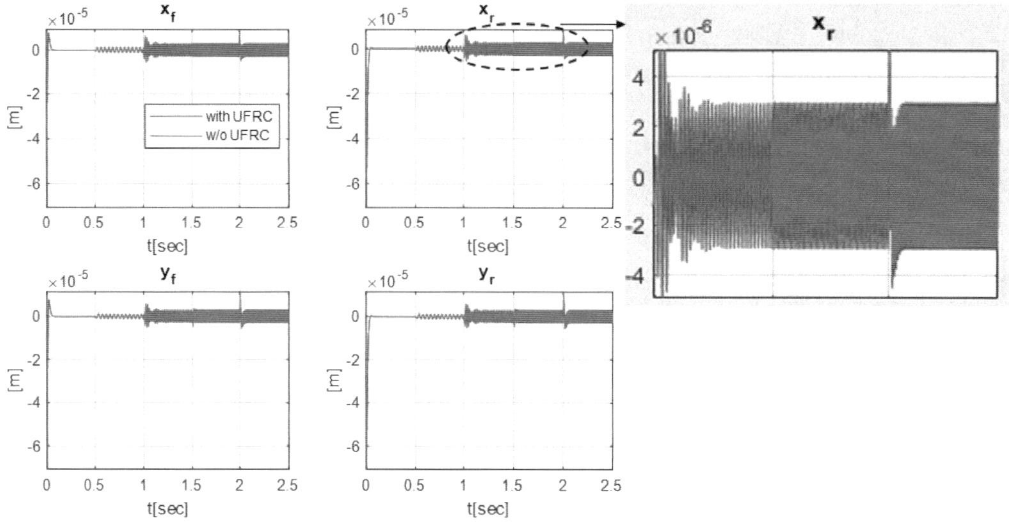

Figure 10 Sensor coordinate response versus time

6.Conclusion

In this study, a UFRC controller structure has been proposed. While using the decoupling control law as the levitation control, the UFRC controller can be easily designed. In addition, the stability of the system can also be easily verified. The UFRC controller design was shown in the paper, and its performance was verified through numerical simulation. From the simulation, the designed UFRC controller was proved to be effective.

References

[1] Chen, S.-L., S.-Y. Lin, and C.-S. Toh (2019) Adaptive unbalance compensation for a three-pole active magnetic bearing system. IEEE Transactions on Industrial Electronics 67(3): p. 2097-2106.

[2] Zheng, S., Q. Chen, and H. Ren (2016) Active balancing control of AMB-rotor systems using a phase-shift notch filter connected in parallel mode. IEEE Transactions on Industrial Electronics 63(6): p. 3777-3785.

[3] Maslen, E.H. and G. Schweitzer (2009) Magnetic bearings: theory, design, and application to rotating machinery. Berlin, Heidelberg:Springer, pp.184-185, 191-194.

Realization of Lateral Flux-path Control Magnetic Suspension Using Rotational Mechanism

Takeshi MIZUNO [a], Shigenori KUROSAWA [b], Yuji ISHINO [a], Masaya TAKASAKI [a]
a Saitama University, Shimo-okubo 255, Sakura-ku, 338-8570 Saitama, Japan, mizar@mech.saitama-u.ac.jp
b Zeon Corporation, Yako 1-2-1, Kawasaki-ku, 210-9507 Kawasaki, Japan

Abstract

A magnetic suspension system using rotational flux-path control mechanism is developed. In the developed flux path control mechanism, a pair of gear-shaped flux control plates are attached to the top and bottom of a permanent magnet (magnetic source); they are surrounded by a ferromagnetic outer yoke with teeth facing the control plates. The gear-shaped control plate is rotated to change the relative angle to the teeth of the outer yoke. The attractive force acting on the ferromagnetic floator is adjusted by rotating the control plates. Stable suspension is achieved in the developed apparatus.

Keywords: *Magnetic suspension, Magnetic bearing ,Flux-path control, Variable flux path, Rotational mechanism*

1. Introduction

Magnetic suspension is one of the most promising methods of achieving non-contact suspension. It has various applications taking advantage of unique characteristics (Jayawant, 1981; Schweitzer and Maslen, 2009). In some applications, a large gap between a magnetic source and a suspended object (floator) is required. Flux-path control magnetic suspension has been proposed to achieve such large-gap suspension with small energy (Mizuno et al., 2007). In the original flux-path control mechanism, ferromagnetic control plates are inserted into the gap between the magnetic source and the ferromagnetic floator.

The authors proposed a novel configuration of flux-path control magnetic suspension named as lateral control type (Ishibashi et al., 2017). In this system, control plates were located around the magnetic source and moved vertically. To study its characteristics experimentally, an apparatus was developed which has three flux-path control mechanisms (Mizuno et al., 2019). Noncontact suspension with three-dimensional motion control was achieved in the apparatus.

However, the size of the apparatus was rather large mainly because each flux-path control mechanism enlarged the motion of the control plate by using see-saw mechanism. Such enlargement was necessary to vary the control force sufficiently to achieve stable suspension. In this work, a novel rotational flux-path control mechanism is developed to make the suspension system smaller. The developed flux-path control mechanism needs no mechanical magnification to obtain sufficient control force for stable noncontact suspension.

2. Flux-path control magnetic suspension systems

2.1 Flux-interrupted type

Figure 1 shows a basic configuration of the original flux-path control suspension system (Mizuno et al., 2007). A pair of ferromagnetic control plates are inserted into the gap between the magnetic source and the floator. When the distance between the control plates becomes wider, more flux reaches the floator from the magnetic source, and the attractive force acting on the floator becomes stronger as shown by Fig.1(b), and vice versa. The suspension system with such configuration is called as flux-interrupted type because the control plates interrupt the flux path from the magnetic source to the floator.

This type has an advantage over the conventional active magnetic suspension with an electromagnet in achieving large-gap suspension mainly because the actuator drives the control plate instead of the floator itself. However, the attractive force action on the floator decreases by inserting the control plates.

2.2 Flux-concentrated Type

Figure 2 shows a basic configuration of the flux-concentrated control suspension system (Mizuno et al., 2013). The control plates are made of permanent magnet instead of ferromagnetic material. This type is superior in producing intensive force to the flux-interrupted type. In addition, the relation of the suspension force to the distance between the

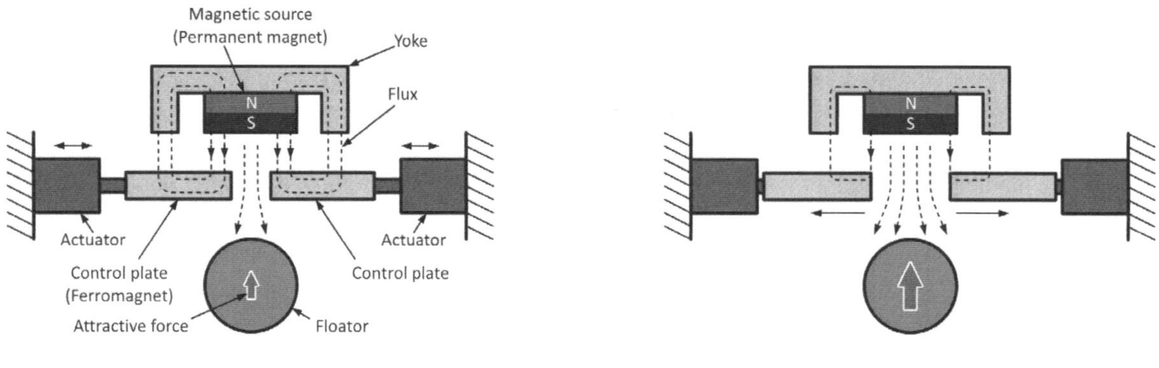

(a) Basic structure *(b) Control plates opened*

Figure 1 Flux-interrupted flux-path control magnetic suspension system

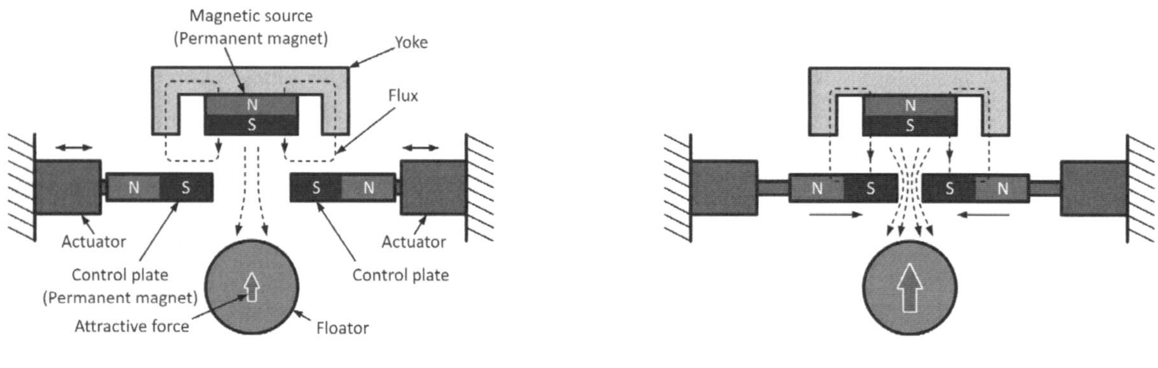

(a) Basic structure *(b) Control plates closed*

Figure 2 Flux-interrupted flux-path control magnetic suspension system

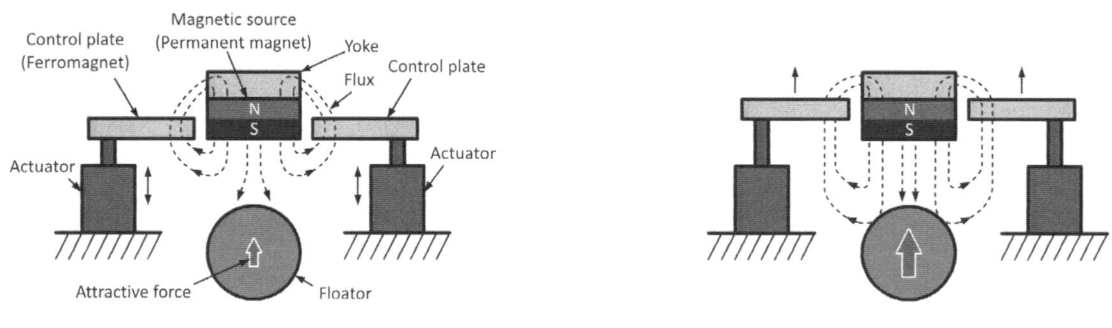

(a) Basic structure *(b) Control plates raised*

Figure 3 Laterally controlled flux-path magnetic suspension system

control plates is contrasting. When the distance between the control plates becomes smaller, more flux is concentrated in the gap so that the attractive force acting on the floator becomes stronger, and vice versa.

2.3 Lateral control type

In both former types, the control plates are inserted between the floator and the magnetic source (permanent magnet). It decreases the effective gap. In the lateral control type, control plates are placed in the lateral of the magnetic source (Ishibashi et al., 2017). The attractive force is adjusted by moving the control plates in the vertical direction beside the permanent magnet. Thus, the effective gap equals that between the floator and the magnetic source. Figure 3 shows

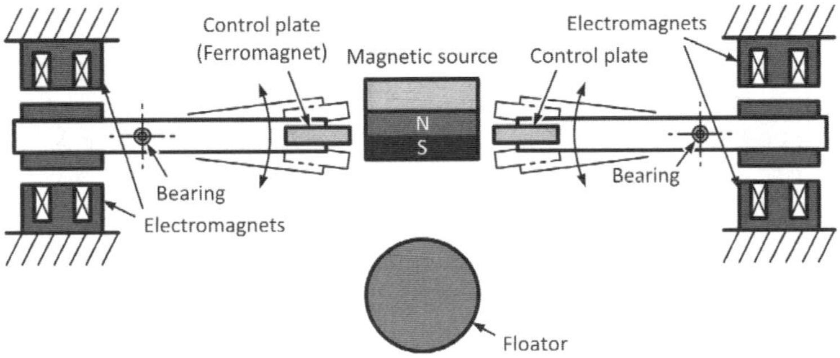

Figure 4 Schematic drawing of magnetic suspension system with fabricated flux-path control mechanisms

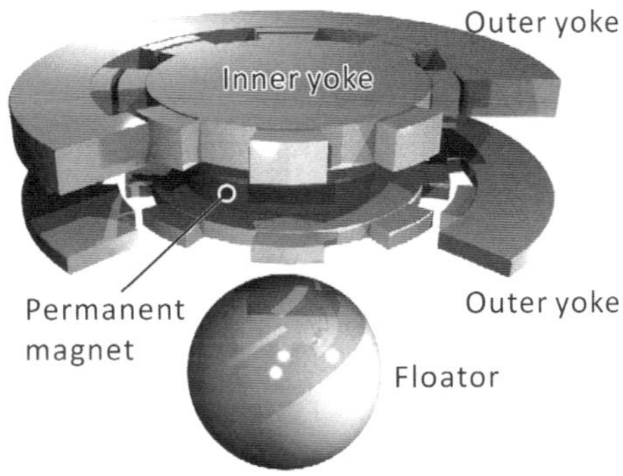

Figure 5 Magnetic suspension with a rotational flux-path control mechanism

the principle of adjustment of the attractive force. When the control plates are placed at the lowest position where the bottom of the control plates is aligned with that of the permanent magnet, more flux flows into the control plates, and the attractive force becomes smaller. When the control plates are placed at the highest position, where the top of the control plates is aligned with that of the permanent magnet, the attractive force becomes larger because the control plates work as yoke.

Figure 4 shows a schematic drawing of the fabricated flux-path control mechanism (Mizuno et al., 2019). In this mechanism, a ferromagnetic control plate is fixed on one end of a lever made of non-magnetic material. The lever is supported by a ball bearing to rotate about a horizontal axis. A pair of electromagnets are placed at an end of the lever. The motion at the electromagnets is mechanically amplified approximately five times at the control plates. Such a structure inevitably made the apparatus rather large (Mizuno et al., 2019).

2.4 Rotational flux-path control type

To reduce the size of suspension system, rotational flux-path control mechanism has been proposed (Kurosawa et al., 2019; Mizuno et al., 2020). Figure 5 shows a conceptual diagram of the proposed magnetic suspension system with a rotational flux-path control mechanism. A gear-shaped ferromagnetic yoke (inner yoke) is attached on the top of a disk-shaped permanent magnet (magnetic source). It is surrounded by a gear-shaped plate (upper outer yoke). The inner yoke is rotated by an actuator to vary the relative angle between a magnetic tooth of the inner yoke and that of the outer yoke. A ferromagnetic sphere (floator) is located under the magnetic source. The attractive force acting on the floator can be adjusted by the angle, which will be explained by Fig.6 where the number of teeth is assumed to be eight. When the teeth face each other (relative angle is 0 deg.) as shown by Fig.6a, flux flows in the outer yoke most and the attractive force is maximum as a result. In contrast, when the tooth of the inner yoke is at the middle of the teeth of the outer yoke as shown by Fig.6b, flux flows in the outer yoke least and the attractive force is minimum. When the number of the teeth is *n*, the corresponding relative angle θ_r is 180/*n* deg

Another gear shaped inner yoke and the corresponding outer yoke are placed below the permanent magnet as shown by Fig.5 and Fig.7. In this case, when the teeth face each other (relative angle is 0 deg.) as shown by Fig.7a, flux flows in the outer yoke most, which leads to minimum attractive force instead. In contrast, when the tooth of the inner yoke is at the middle of the teeth of the outer yoke as shown by Fig.7b, flux flows in the outer yoke least and leads to maximum attractive force.

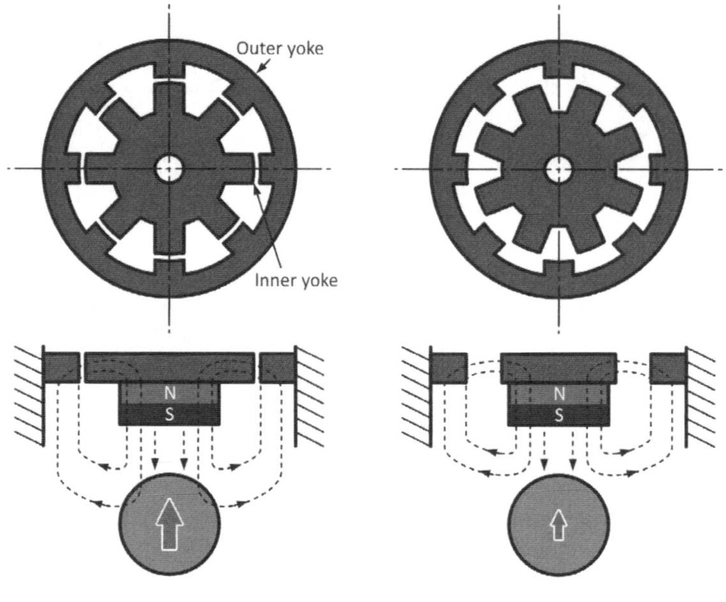

(a) Relative angle: 0 deg *(b) Relative angle: 22.5 deg*

Figure 6 Control plate on the top of magnetic source

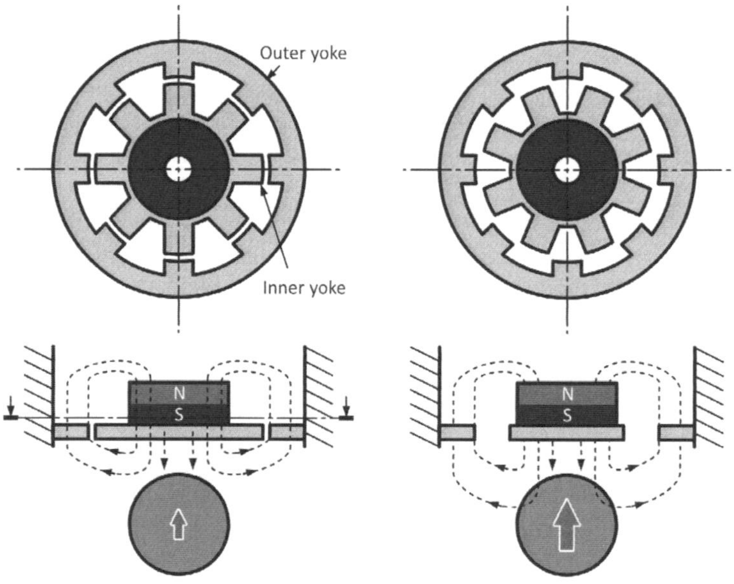

(a) Relative angle: 0 deg *(b) Relative angle: 22.5 deg*

Figure 6 Control plate at the bottom of magnetic source

In the developed apparatus, both upper and lower yokes are combined to maximize the variation of the attractive force (control force). The relative angle between the upper and lower is set to be 180/*n* deg.

3. Experiment Apparatus

Figure 8 shows a picture of the developed apparatus. In this apparatus, the inner yokes sandwiching the permanent magnet are rotated. The number of the teeth is eight so that $\theta_r = 22.5\,\text{deg}$. Figure 9 shows schematic views of the

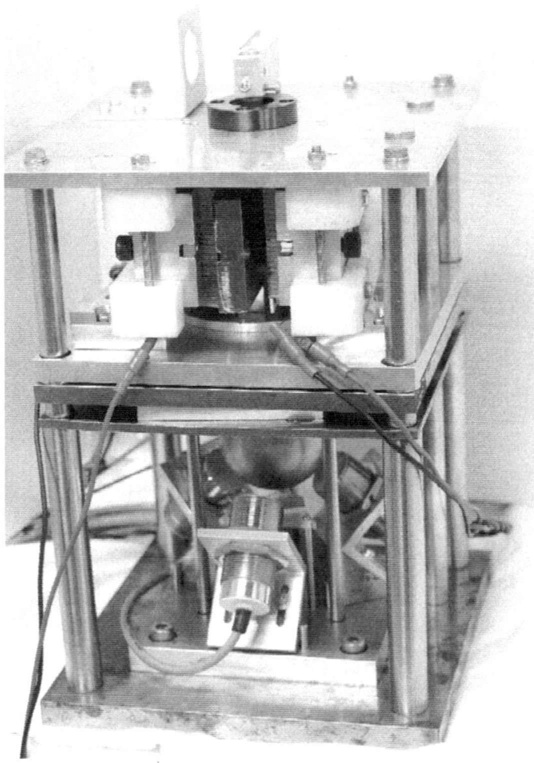
Figure 8 Picture of apparatus

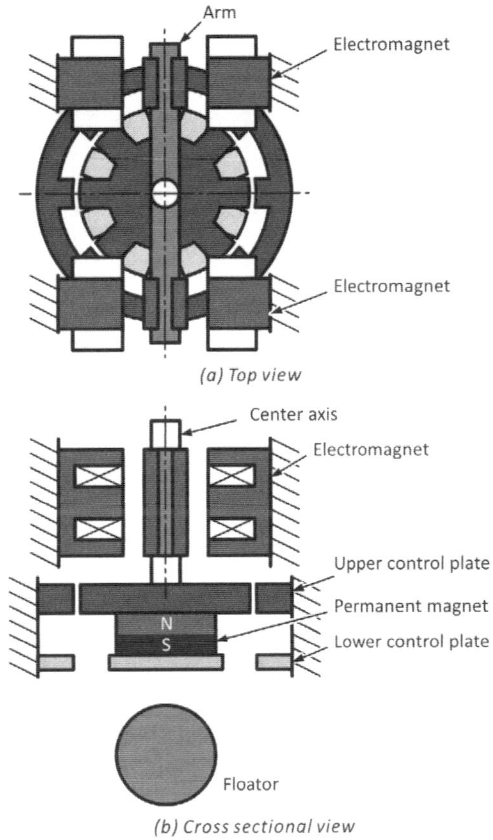

Figure 9 Schematic drawing of rotational flux-path control mechanism

rotational mechanism. An arm is connected to the rotational axis of them. The arm is driven by two pairs of differentially operated electromagnets. They are located above the inner and outer yokes.

4. Experimental Results

The designed controller has a double-loop structure that is similar to that used in the previous work (Mizuno et al., 2018). In the inner loop, the motion of the lever is locally fed back in the flux-control mechanism. In the outer loop, the displacement of floator in the vertical direction is also fed back. In the inner loop, PD control is applied to provide the flux-path control mechanism sufficient stiffness and damping property to suspend the floator. In the outer loop, PD control is also applied to stabilize the suspension system.

Figure 10 shows a measured frequency response of the rotational mechanism (inner loop). In the measurement, a sinusoidal signal was added to the control signal and the output was the angular displacement of the arm. It is found that the resonant frequency is approximately 35 Hz.

Stable suspension was achieved with the designed controller. Figure 11 shows the motions of the floator and the arm during levitation. The floator levitates at the position of 14.05 mm below the magnetic source. It was experimentally certified that the suspension was maintained in the presence of small disturbances.

5. Conclusions

A rotational flux-path control mechanism was introduced to reduce the size of laterally controlled magnetic suspension system. In the developed flux path control mechanism, two gear-shaped flux control plates were attached to the top and bottom of a magnetic source. Each of them was surrounded by a ferromagnetic outer yoke with teeth. The attractive force acting on the ferromagnetic floator was adjusted by changing the relative angle between the inner and outer yokes. Stable suspension was achieved in the developed apparatus.

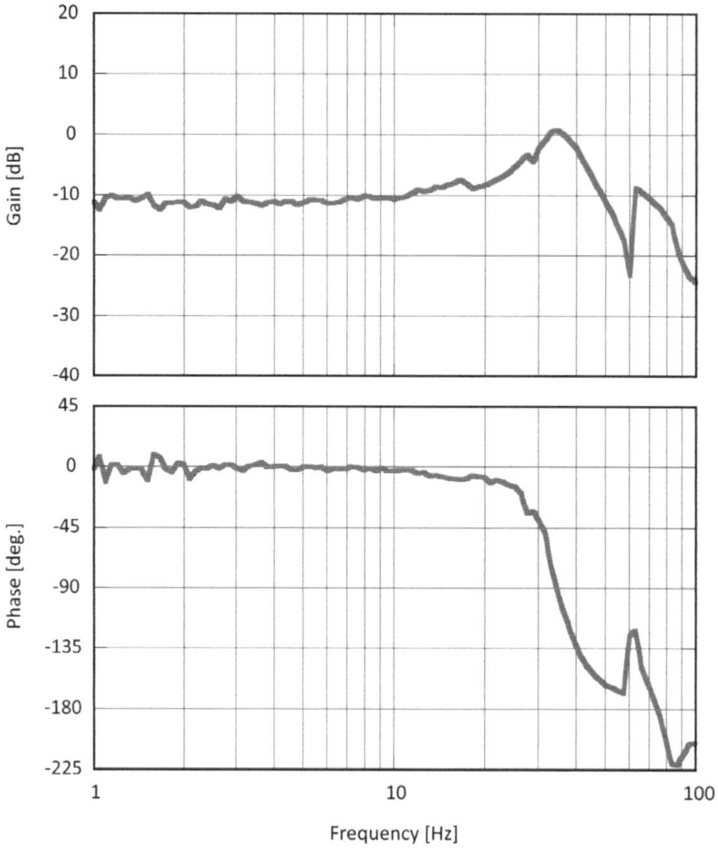

Figure 10 Frequency response of rotational mechanims

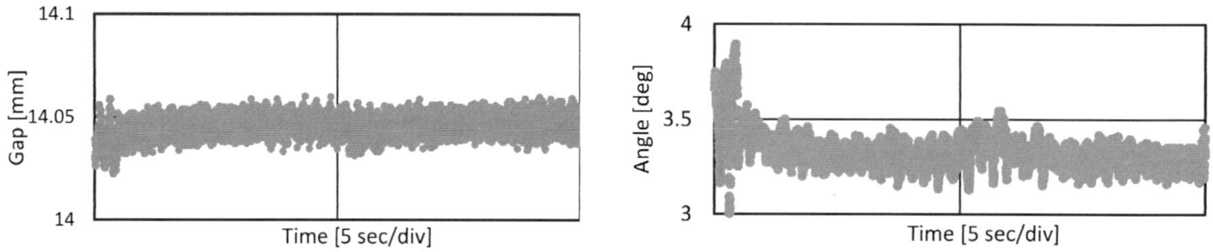

(a) Floator *(b) Angular displacement of arm*

Figure 11 Motions during levitation

References

Ishibashi N, Mizuno T, Ishino Y, Yamaguchi D, Hara M, Takasaki M and Yamada K (2017) The proposal of magnetic suspension using laterally control flux-path mechanism. Actuators 6(11).

Jayawant BV (1981) Electromagnetic Levitation and Suspension Techniques. London: Edward Arnold.

Kurosawa S, Mizuno T, Ishino Y, Yamaguchi D and Takasaki M (2019) Miniaturization of laterally control flux-path mechanism (1st report: Proposal of rotational flux-path control mechanism) (in Japanese). In Symposium on motion and vibration control, Kochi, Japan, 4-6 December 2019, B308. Tokyo: The Japan Society of Mechanical Engineers.

Mizuno T, Hirai Y, Ishino Y and Takasaki M (2007) Flux-path control magnetic suspension system using voice coil motors. Journal of System Design and Dynamics 1(2): 147-158.

Mizuno T, Yoneno Y, Ishino Y and Takasaki M (2013) Proposal of flux-path control magnetic suspension using flux concentration. In: 1st Brazilian workshop on magnetic bearings, Rio de Janeiro, Brazil, 25-26 October 2013.

Mizuno T, Ishibashi N, Ishino Y, Yamaguchi D and Takasaki M (2019): Three-dimensional control of an iron ball by flux-path control mechanisms located around magnetic source. In: 12th International symposium on linear drives for industry applications, Neuchatel, Switzerland, 1-3 July 2019.

Mizuno T, Kurosawa S, Ishino Y, Yamaguchi D and Takasaki M (2020): Miniaturization of laterally control flux-path mechanism (2nd report: Realization of suspension) (in Japanese). In Dynamics and design conference 2020, Online, 1-4 September 2020, 530. Tokyo: The Japan Society of Mechanical Engineers.

Schweitzer G and Maslen EH (eds) (2009) Magnetic Bearings. Berlin Heidelberg: Springer.

Operation Test of Magnetic Bearing by Regenerated Power from High-Speed IPM Motor at Unexpected Power Stop

Yohji OKADA [a], Fumiya KITAYAMA [a], Ken-ichi MATSUDA [a], Ryou KONDO [a]

a Ibaraki University, 3-14-6, Nakanarusawa, Hitachi, Japan, y.okada@vc.ibaraki.ac.jp

Abstract

Two pole Internal Permanent Magnet (IPM) type high speed motor has been developed. Hybrid (HB) magnetic bearing is designed and tested by regenerated energy from high speed motor after sudden power stop. The motor should have wide air gap for levitation and the rotor speed is better to be decreased before touch down. The regenerative brake is useful to reduce the rotor speed and to operate the magnetic bearing by the regenerated energy. The motor is designed with a commercial FEM code MagNet. The previously developed magnetic bearing is modified and used to support one end of the rotor shaft and another end is supported by a ball bearing. The motor-bearing system is fabricated and the sudden power stop is tested.

Keywords: *Active magnetic bearing, High-speed motor, Regenerate power, Soft touch-down*

1. Background and Scheme of high speed IPM motor

Scheme of turbo refrigerator is shown in Fig. 1. Turbo compressor increases the pressure and the temperature of coolant efficiently. Pressure is a kind of energy and the standard nozzle expander changes this pressure to increase coolant temperature. Turbo expander cools the coolant efficiently (Mayekawa Mfg. Co. Ltd., Pascal_Air (2022)). This shows one of good application of Active Magnetic Bearing (AMB). However unexpected power stop damages high-speed rotor (Schweitzer, et al., 2009). This paper proposes regenerative brake to IPM type motor and the regenerated power is used to operate the AMB until the rotor speed slows down (Okada, et al., 2019).

Analytical model of the proposed IPM motor and the photo of closed slot stator are shown in Fig. 2. The rotor has 6 internal permanent magnets (Shin-Etsu, N32EZ) to polarize 2 pole N-S rotor, while the stator has 12 slots with 200 sheets of laminated steel (15HX1000). Each stator stems have two layered winding coils. For smooth rotating flux, 24 coils are connected to three phase current amplifiers. The top two coils are connected both to U phase. Then clockwise direction, the next two coils are connected to U and –W. The third two coils are connected both to -W, and the fourth two coils are to -W and V, and so on. The analytical torque versus speed curves are analyzed and compared with 2 pole 6 slot motor. The results are shown in Fig. 3. The 2 pole 6 slot motor shows high torque (Torque2-6), but the top speed is lower (Okada, et al., 2019). The closed slot of 12 slot stator shows high speed rotation (Torqe2-12C), but the produced torque is lower. For this application high speed rotation is necessary, we decided to use the closed slot 12 stator motor, as shown in the right of Fig. 2. Each 24 winding coils are 42 turns with Ø 0.6 copper wire. The inner ends of stator stem are connected with soft magnetic iron with the thickness of 0.7 mm (SUY-1).

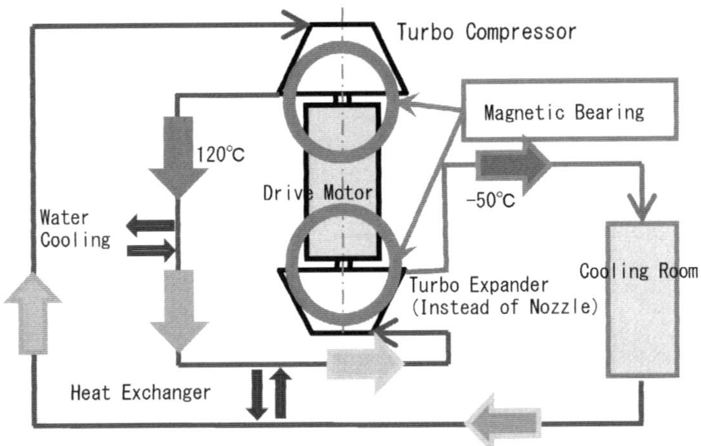

Figure 1 Typical example of using active magnetic bearing to turbo refrigerator.

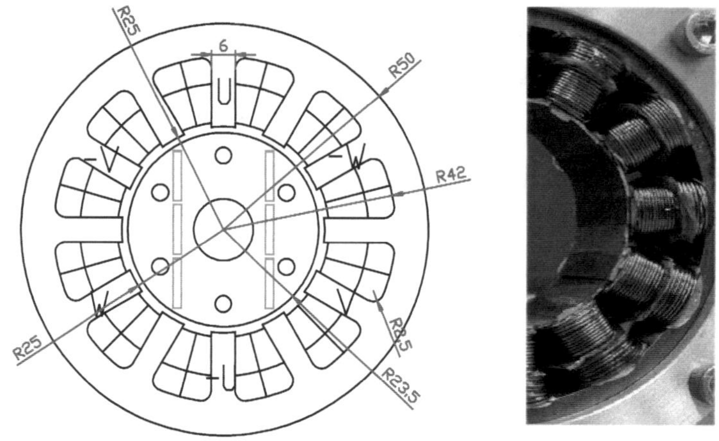

Figure 2 Analytical model of the 2 pole 12 slot IPM motor (left) and the photo of closed slot motor stator (right).

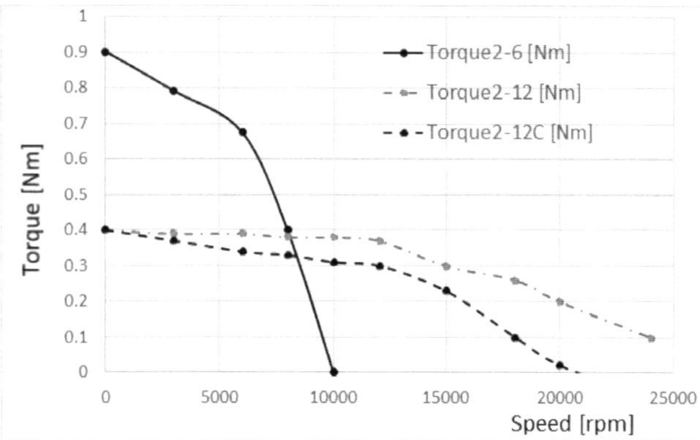

Figure 3 Torque versus speed curves (supply voltage: 72 V).

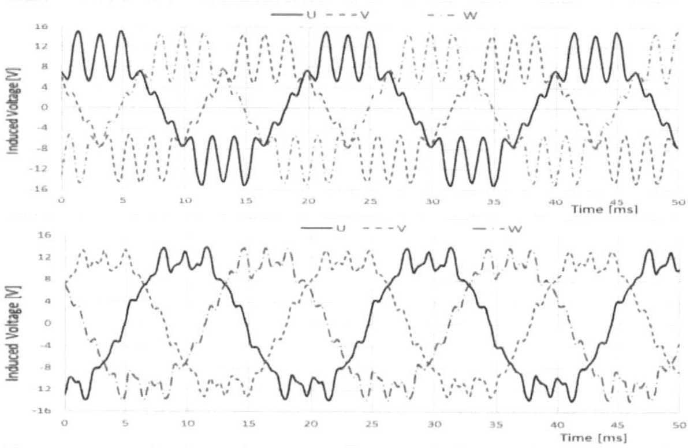

Figure 4 Back electro-motive test results of the motor (upper; open slot, and lower; closed slot ones).

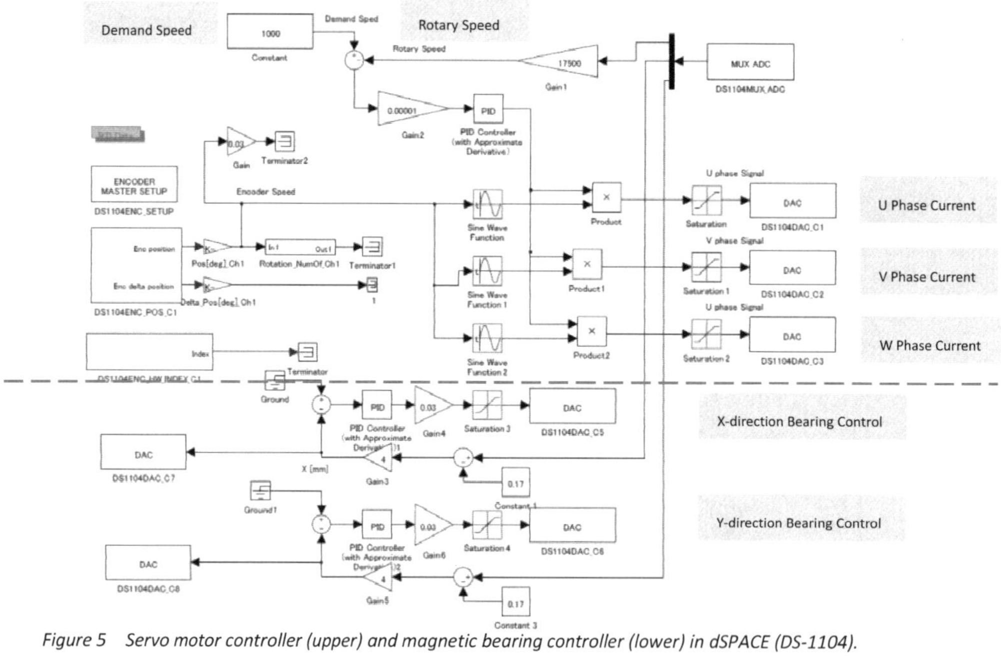

Figure 5 Servo motor controller (upper) and magnetic bearing controller (lower) in dSPACE (DS-1104).

The motor is fabricated and back electro-motive test is carried out at 3,000 rpm with a DC motor (maxon, 148877). The results are shown in Fig. 4. The upper is the result of open slot while the lower is that of closed slot. One can recognize the fundamental rotational wave form. However the closed slot result reduces higher harmonics drastically.

Figure 6 Motor is tested using motor analyzer.

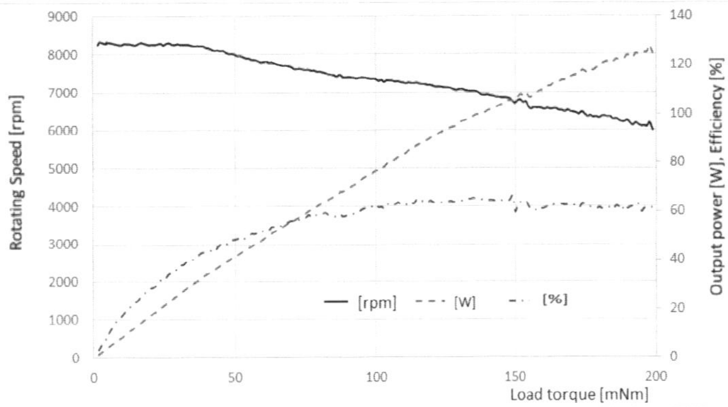

Figure 7 Torque versus speed curves (supply voltage: 72 V).

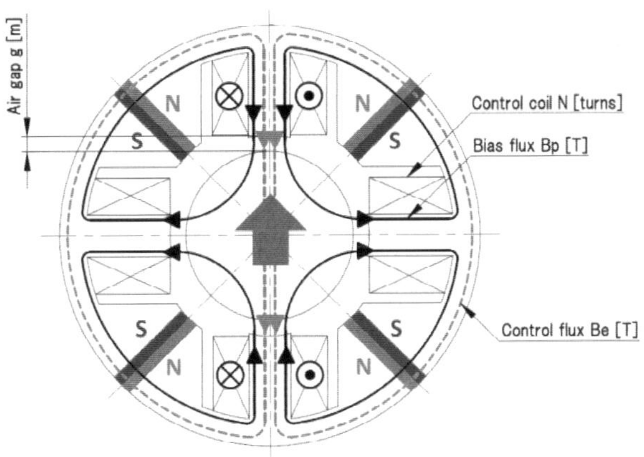

Figure 8 Operating principle of HB type magnetic bearing.

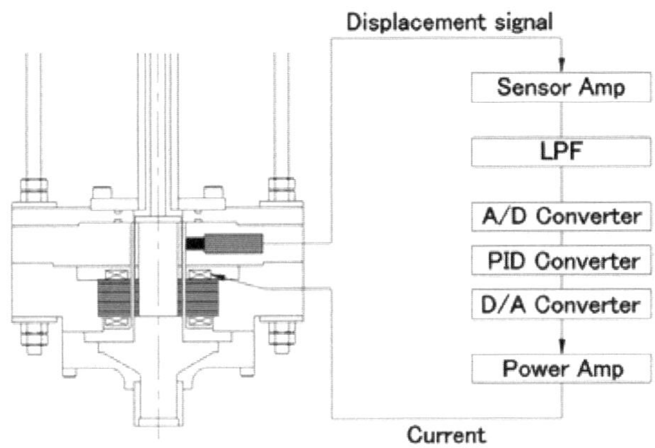

Figure 9 Magnetically supported pump and AMB controller.

2. Experimental setup and control system

Figure 5 shows dSPACE based controller for motor and magnetic bearing. Above the dashed red line is the motor controller, while the lower is the magnetic bearing controller. The motor is controlled by a PID servo motor control using the rotary encoder (E6D-CWZ 2C). The motor can rotate up to 12,000 rpm (Okada, et al., 2021). The turbo refrigerator in Fig. 1 rotates about 18,000 rpm, but in the laboratory test this top speed is considered almost enough.

2.1 Motor load test results

Motor load test is carried out using motor analyzer (Sugawara, PC-SAA2, TB-2KS). Photo of the motor test is shown in Fig. 6. The stable rotating speed with load is little over 8,000 rpm. The motor load test results are shown in Fig. 7. The PID control gains are K_P=2.0, K_I=20, and K_D=0, respectively. The sampling interval used is τ=0.1 ms. The resulting maximum efficiency is about 65 %. The lower efficiency may be due to the wide airgap and shaft is not straight enough.

2.2 Active magnetic bearing and test system

Due to our budget limitation, previously developed magnetic bearing is used to support one end of the rotor shaft and another end is supported by a ball bearing. This magnetic bearing was originally developed to support the liquid nitrogen pump (Okada, et al., 2015). The operating principle is shown in Fig. 8. The scheme of liquid nitrogen pump and the magnetic bearing with controller is shown in Fig. 9. The original designed model of HB magnetic bearing is shown in Fig. 10. The stator has 2 mm thickness cylindrical FRP pipe for separate the liquid nitrogen, hence the magnetic airgap is 3 mm. The bearing part is made by a silicon steel sheet (35A300) with the stack thickness of 30.1 mm. The analyzed flux densities of the control current at 0 A and 2 A are shown in Figs. 11 and 12, respectively. One can recognize the flux of upper stem is increased while the flux of lower stem is decreased by giving control current. The levitation force factor is about 21 N/A as shown in Fig. 13, but the force is not enough. The power amplifier used is analog power amplifier (APEX, PA12A). Hence we modified the rotor part outer radius of magnetic bearing as 1.5 mm larger. The modified design of HB type AMB is shown in Fig. 14. The force factor is about twice bigger and the rotor is levitated stably.

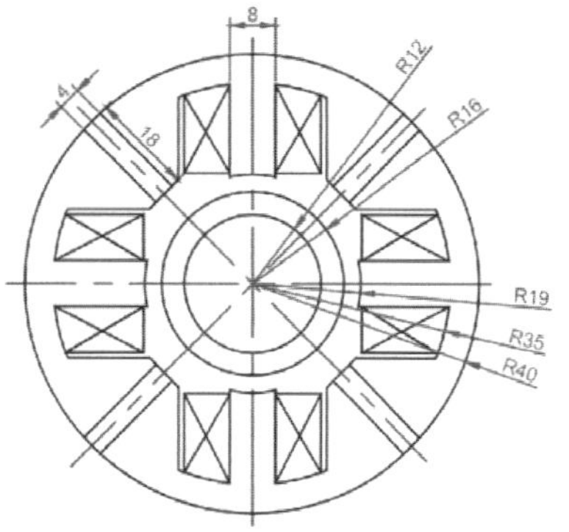

Figure 10 Original design model of HB type magnetic bearing.

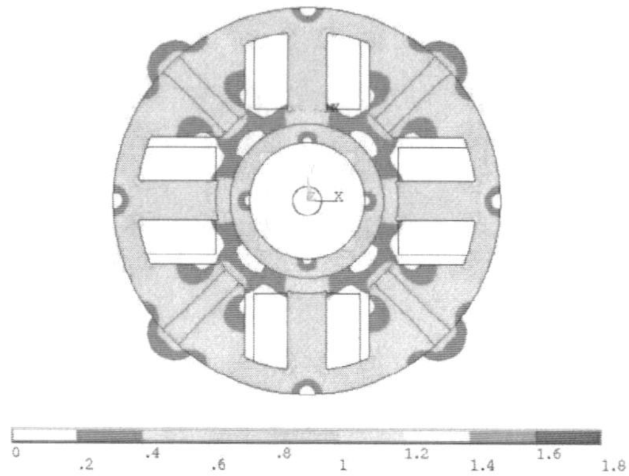

Figure 11 Flux density (Control current: 0 A).

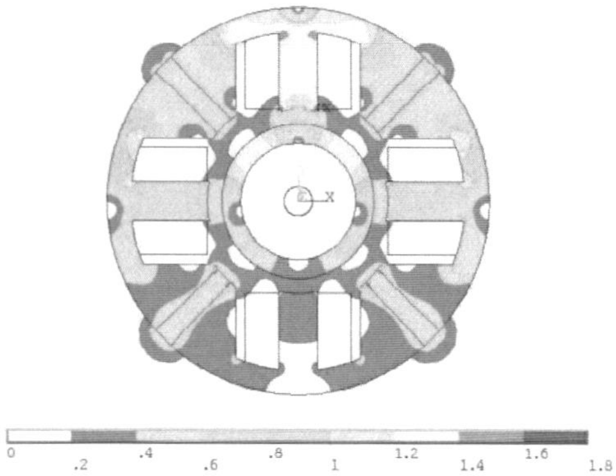

Figure 12 Flux density (Control current: 2 A).

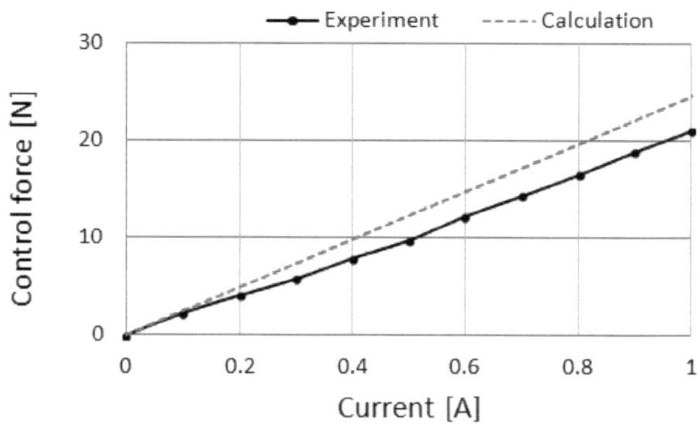

Figure 13 Force property of original magnetic bearing.

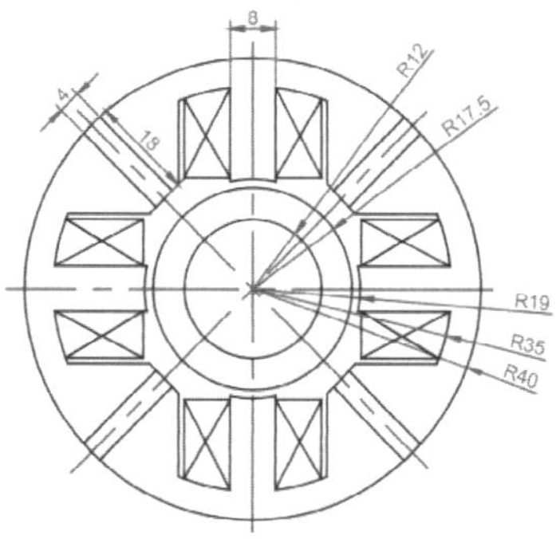

Figure 14 Modified design model of HB type magnetic bearing.

Table 1 Control gains of HB type magnetic bearing.

	Radial Value	
	x-direction	y-direction
K_P [A/m]	300	600
K_I [A · sec/m]	60	60
K_D [A/sec · m]	0.48	12
T_d [ms]	1.667	1.667

2.3 Levitation test of HB type magnetic bearing

The photo of assembled motor with magnetic bearing is shown in Fig. 15. The control parameters of magnetic bearing are shown in Table 1. Where K_p, K_I and K_D are the PID control gains and T_d is the derivative time constant. The step response test is carried out by giving the step signal to x-direction, and then to y-direction. The results are shown in Fig. 16 and Fig.17, respectively. The levitation is stable and mutual interferences are relatively small.

3. Operation test of rotor system at unexpected power stop

During the rotor rotates over 8,000 rpm, unexpected power stop is tested using the rotor system shown in Fig. 15. First, the stable rotation is tested at 12,000 rpm. The driving currents of U, V and W phases are shown in Fig. 18. The motor drivers are PWM amplifiers (Copley Co., model 4212Z), and the current signals are recorded. At the same time, the shaft displacements are measured using eddy current displacement sensors (Keyence, EX-502, EX-008) as shown in Fig. 19. The shaft displacements are under 0.04 mm. The shaft vibration is recognized, but it is relatively stable.

Figure 15 Photo of assembled motor with HB type magnetic bearing.

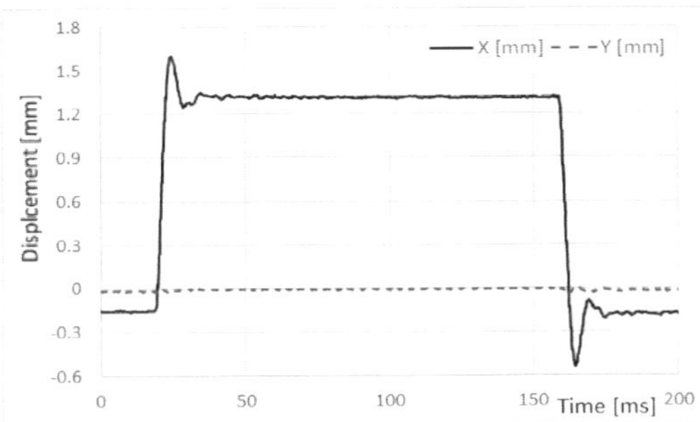

Figure 16 Step response (x-direction excitation).

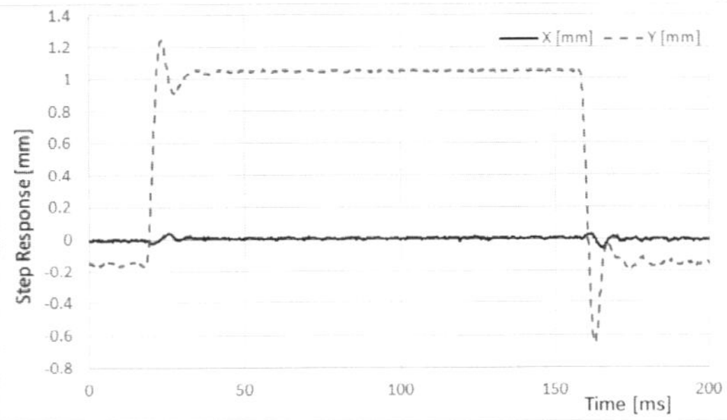

Figure 17 Step response (y-direction excitation).

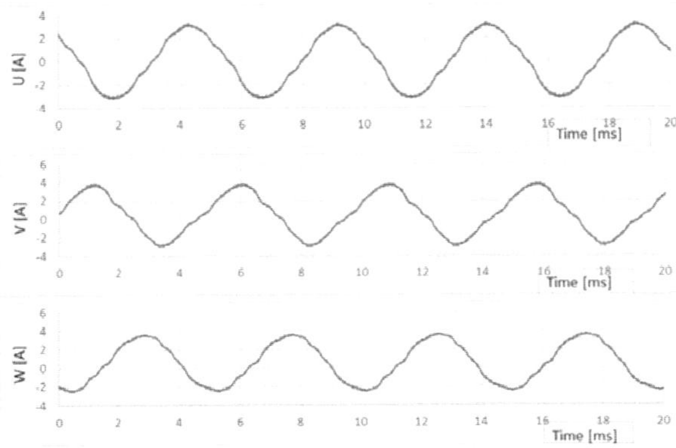

Figure 18 Driving currents of U, V and W phases [A].

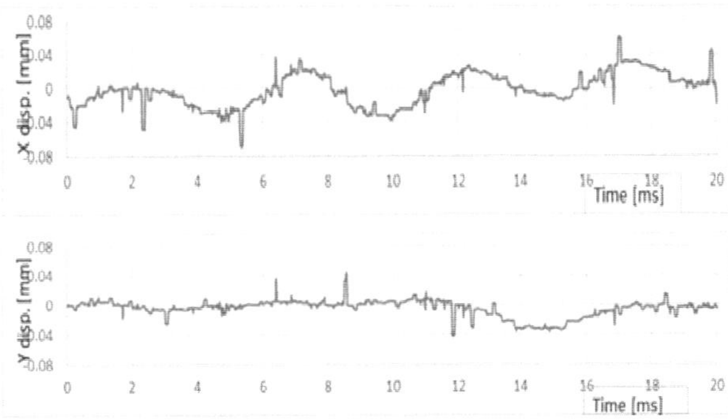

Figure 19 Shaft displacements of x and y directions [mm].

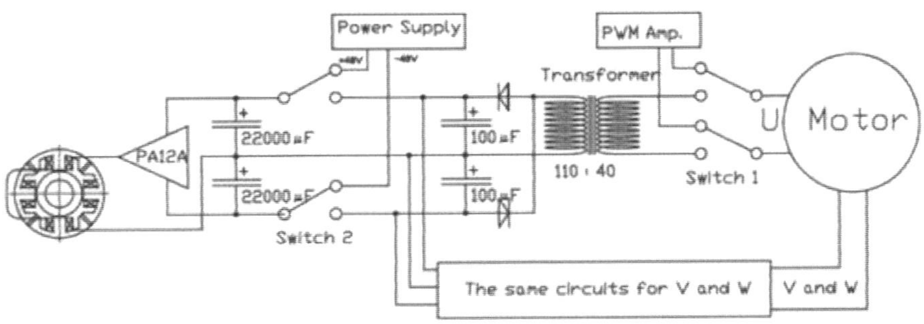

Figure 20 Regenerative circuit from motor to AMB.

3.1 Unexpected power stop

During the motor rotates about 8,000 rpm, unexpected power stop is tested using the regenerative circuit as shown

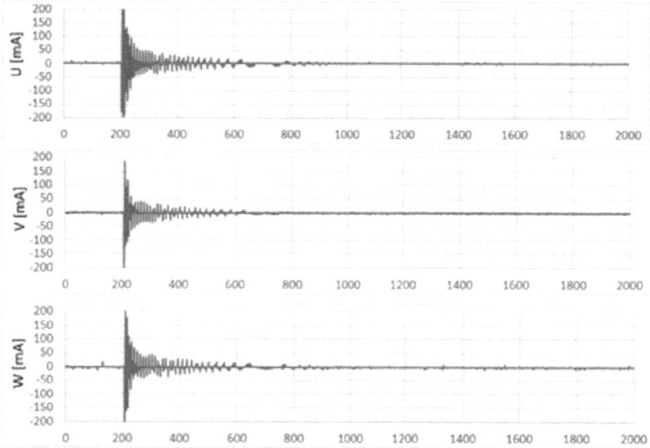

Figure 21 *Regenerated currents of U, V, W phases [mA] versus time [ms].*

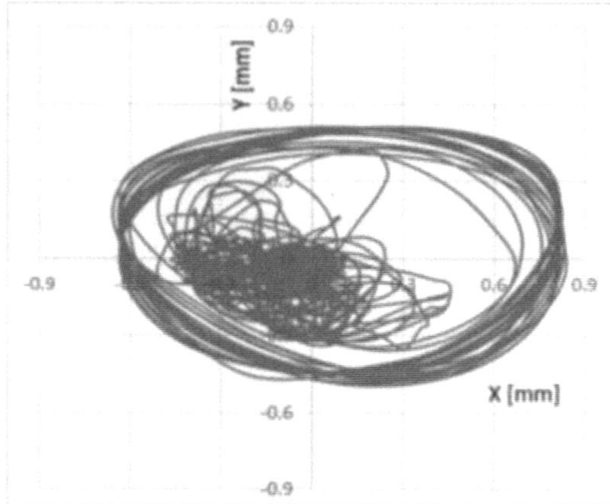

Figure 22 *Orbital x-y trajectory after power shut down.*

in Fig. 20. The supply power is switched by hand (switch 1) and the motor speed is decreased which produces the regenerated power. This regenerated power is supplied to AMB, using the switch 2. The regenerated currents of U, V and W phases are shown in Fig. 21. These currents are used to support the shaft by driving the HB type AMB. Before power shut down, the motor rotates stably by the PWM power amplifiers in the right upper of Fig. 20. The regenerated currents are zero until 200 ms in Fig. 21. After the supply power is shut down the regenerated currents are suddenly increased up to 200 mA and the rotating speed is decreased. But the regeneration time is short (about 300 ms) as shown in Fig. 21. This is considered that the shaft is thin and does not have enough inertial power. The orbital trajectory is shown in Fig. 22. Before power shut down the orbital trajectory remains within the magnitude of 0.05 mm around the center. After power shut down the trajectory increases quickly and the shaft touches down to the emergency bearing safely.

4. Concluding remarks

Two pole IPM type motor is developed to test the active magnetic bearing support for unexpected power shut down. The motor can run up to 12,000 rpm. Small hybrid magnetic bearing is used to support the one end of the rotor and another end is supported by a ball bearing. The stable rotating levitation is recognized up to 12,000 rpm. At the rotating speed of 8,000 rpm this system is tested when the power is unexpectedly shut down. The test results show possibility of supporting the rotor shaft for short time and the shaft speed is reduced by the regenerative brake quickly. This regenerated power can drive the magnetic bearing until the shaft speed decreases and touch down to emergency bearing.

However the regenerated power is not enough due to the low rotary inertia of the shaft. Also the regenerative circuit should be improved to decrease the shaft speed quickly and to produce the regenerated power more effectively.

Acknowledgment

This research is supported by Tsugawa foundation. The authors would like to express sincere appreciation.

References

Mayekawa Mfg. Co. Ltd. (2022), Pascal_Air: https://mayekawa.com/americas/mna/downloads/pdf/Refrigeration%20Systems/Pascal_Air.pdf

Okada Y, Kitayama F, and Kondo R (2019), Design of High Speed Interior Permanent Magnet type Motor for Turbo-Machinery, Proc. of The 11th Japanese-Mediterranean Workshop Applied Electromagnetic Engineering for Superconducting, Multifunctional and Nanomaterials.

Okada Y, et al. (2015), Development of Highly Efficient Magnetic Bearing and Application to Ultra-Low Temperature Fluid Pump, Bulletin of the JSME, Mechanical Engineering Journal, 2(4): 1-10.

Okada Y, Kitayama F, and Kondo R (2021), Test operation of High Speed Interior Permanent Magnet type Motor supported by Hybrid type magnetic bearing, Journal of the Japan Society of Applied Electromagnetics and Mechanics, 29(2): 327-333, (in Japanese).

Schweitzer G., et al. (2009) Magnetic Bearings: Springer, 27-434.

Design & Control of Bearingless drive for Rim Driven Thruster

Wouter MANENSCHIJN[a], Niels KUIPERS, Bauke KALMA

a Insubmo BV, Anne Wadmanwei 4a, 8914BD Leeuwarden, Netherlands, wouter@insumo.nl

Abstract

A motor design and control strategy for a bearingless rim-driven thruster is proposed that obviates the need for a conventional radial bearing, greatly reducing mechanical complexity and wear. The modelling of the system is explained along with the conrol strategy, including a cascaded control scheme. The physical implementation in a working prototype with combined windings is presented and test results are reveiled. These display the functioning of the bearingless drive at various rotational speeds. The results are discussed and future research directions are identified.

Keywords: Bearingless drive, Active Magnetic Bearing, Hub-less Rim-driven thruster

1. Introduction

Hub-less rim-driven thrusters (see figure 1) are a novel type of electric ship propulsion units with numerous advantages. Due to the open propeller configuration organic debris and hazardous materials, like ropes and plastics, can easily find their way through without getting entangled. When designed well, they are also extremely silent, resulting in a very low acoustic and magnetic signature. The open propeller is the only moving part within the thruster, therefore wear of mechanical parts is minimized. However, a good and robust rotor bearing system is needed and this directly designates the main challenge. This paper focuses on the integrated electromagnetic bearing, also known as bearingless drive. Together with a water-lubricated thrust bearing it is part of a uniquely designed hybrid-bearing system. Applying this frictionless hybrid bearing system reduces losses and improves the machines overall efficiency. Furthermore, removing the need for oil as lubricant removes any pollution risks. The novelty of the dual purpose electric rim-drive does ask for a thorough motor design and control strategy, which will be elaborated on in the next sections. First, existing literature will be considered. Then, the motor design and control strategy will be presented, along with the physical implementation in a working prototype. Lastly, results are presented and discussed.

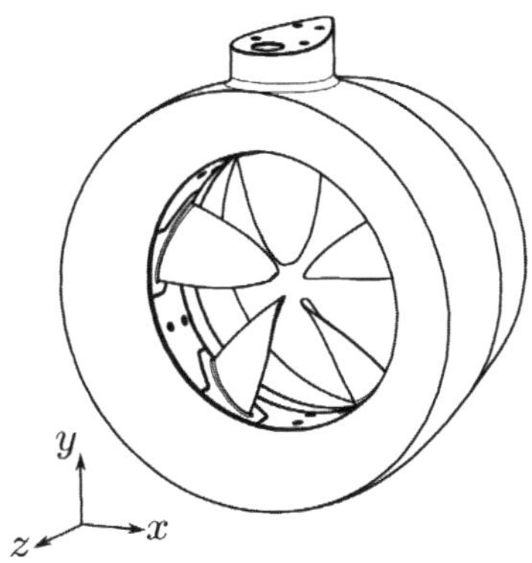

Figure 1: Hub-less rim-driven thruster (Insumo B.V.)

2. Prior Art

Magnetic levitation has been studied for many years, as it provides frictionless rotation with minimal energy dissipation. Furthermore, it has many applications in precise positioning equipment. Regular Active Magnetic Bearings (AMB) generally require a relatively large amount of space, leading to the need for a longer shaft (Chen et al. 2019). Considering the application of a rim-driven thruster, this is highly detrimental. Hence the bearingless drive comes to mind. Bearingless drives have been studied for several different applications, including medical pumps and mixers (Wang et al. 2019). They are an ideal bearing solution for slice drives, as they require minimal space in the axial direction. So far, most bearingless drives are relatively low in power, i.e. less than 30kW (Chen et al. 2019). Furthermore, existing bearingless drives most often apply separate windings to generate the levitating forces. Combined windings have been researched only recently and few industrial applications are known (Sokolov et al. 2021). Another challange for the application of the rim-driven thruster is the fact that the entire machine is submerged during operation. Active magnetic bearings have been applied before in such conditions (Ahad & Ahmad 2021), however a seperate motor is used for the driving torque. What is proposed in this work is, unlike prior art, a bearingless drive using combined windings, applicable for high power rim-driven thrusters that operate in underwater conditions.

3. Design & Modeling

In this paper we will assume that the translations along the z-axis (see figure 1) and rotations around the y and x axes are restrained by the water-lubricated axial bearings. These are the subject of another study, but will be taken for granted here. What remains are translations along x and y, and rotation around z. In order to control these degrees of freedom, the permanent magnet (PM) motor, inrunner design, is electrically split in three equally spaced stator segments. Each segment or motor having a mechanical angle of 120 degrees compared to the next, therefore denoted by subscript 0, 120 and 240. Each segment then is able to provide a torque to the rotor, as well as a radial force on the rotor, as depicted in figure 2.

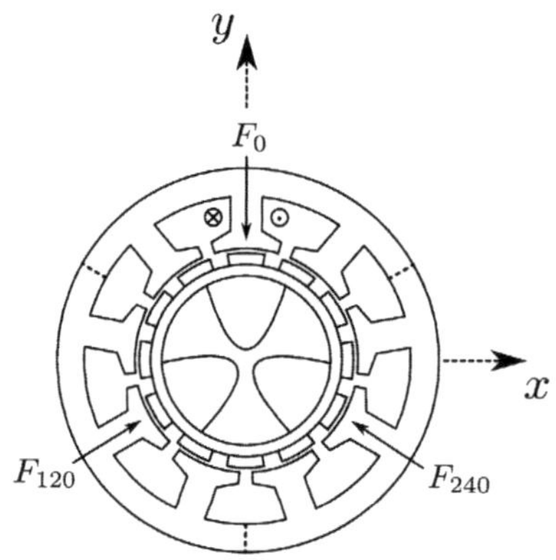

Figure 2: Schematic view of the PM-motor

Now that the motor has been split into three parts, each separate motor remains a three phase machine, resulting in a total of 9 phases. These can be transformed using the Park and Clarke transforms to the dq-reference frame. Overall, we thus need to control six currents: three d-currents that result in forces on the rotor and three q-currents that produce torques. The torques can simply be added as a scalar, while the forces can be added as vectors, resulting in a resultant force on the rotor. Taking into account the negative stiffness of the rotor relative to the stator, we can maintain the rotor in the centre of the stator. Considering the mechanical domain, we can derive the following set of equations:

$$\begin{bmatrix}\dot{x}\\ \ddot{x}\\ \dot{y}\\ \ddot{y}\\ \dot{\theta}\\ \ddot{\theta}\end{bmatrix} = \begin{bmatrix}0 & 1 & 0 & 0 & 0 & 0\\ \frac{c_x}{m} & 0 & 0 & 0 & 0 & 0\\ 0 & 0 & 0 & 1 & 0 & 0\\ 0 & 0 & \frac{c_y}{m} & 0 & 0 & 0\\ 0 & 0 & 0 & 0 & 0 & 1\\ 0 & 0 & 0 & 0 & 0 & p_{prop}\end{bmatrix} \cdot \begin{bmatrix}x\\ \dot{x}\\ y\\ \dot{y}\\ \theta\\ \dot{\theta}\end{bmatrix} + \begin{bmatrix}0 & 0 & 0\\ \frac{1}{m} & 0 & 0\\ 0 & 0 & 0\\ 0 & \frac{1}{m} & 0\\ 0 & 0 & 0\\ 0 & 0 & \frac{1}{J_{zz}}\end{bmatrix} \cdot \begin{bmatrix}F_x\\ F_y\\ T\end{bmatrix} \tag{1}$$

where x denotes the position in x-direction, y the position in y-direction. θ the angular position, c the magnetic stiffness between rotor and stator, m the rotor mass, p_{prop} the rotor frictional losses, J_{zz} the rotor inertia, F_x the input force in x direction, F_y the input force in y direction and T the driving torque on the rotor. Taking into account the geometry of the machine, the input forces and torques are related to the motor currents as follows:

$$\begin{bmatrix}F_x\\ F_y\\ T\end{bmatrix} = \begin{bmatrix}0 & -\frac{1}{2}\sqrt{3}K_F & \frac{1}{2}\sqrt{3}K_F & 0 & 0 & 0\\ K_F & -\frac{1}{2}K_F & -\frac{1}{2}K_F & 0 & 0 & 0\\ 0 & 0 & 0 & K_T & K_T & K_T\end{bmatrix} \cdot \begin{bmatrix}i_{d_0}\\ i_{d_{120}}\\ i_{d_{240}}\\ i_{q_0}\\ i_{q_{120}}\\ i_{q_{240}}\end{bmatrix} \tag{2}$$

where K_F is the current-to-force factor which is assumed equal and constant for each motor segment, K_T the current-to-torque factor, also equal and constant for each motor, i_{d0} the current on the d-axis for motor 0, up to $i_{q_{120}}$, the current on the q-axis for motor 240. Equation 2 is a simplification assuming that the d and q currents are completely separate and do not interfere. Furthermore, in practice, the factor from current to force and torque will depend on the x,y position of the rotor. However, as deviations in x and y are small, this simplification seems justified.

Using equations (1) and (2) we can see the relation between currents and positions. However, the motor input are voltages. Modeling each phase as an RL-circuit, the relation between the input voltages and the resulting currents can be described in the d-q reference frame as follows:

$$\dot{i_{d_0}} = -\frac{R_{d_0}}{L_{d_0}} \cdot i_{d_0} + \frac{1}{L_{d_0}} \cdot U_{d_0} \tag{3}$$

The equations for the d and q currents of the other motor segments are identical to equation (3) and can be written in compact form as:

$$\dot{\vec{i}} = \frac{R}{L} \cdot \vec{i} + \frac{1}{L} \cdot \vec{U} \tag{4}$$

with $\vec{i}$ (6×1) containing the six currents and $\vec{U}$ (6×1) the six voltages. Combining these equations with equations 1 and 2 results in the description of the system as a whole.

4. Control Strategy

In order to control the machine as a whole, the electrical and mechanical domain are considerd seperately. First, a quick current controller is designed, which is then combined in a cascaded control loop with a controller for the mass-spring system as well as the rotational speed controller. Schematically, the overall control scheme is depicted in figure 3. Note that the speed control and the q-currents are omitted here for simplicity's sake, as this part of the dynamics is considerably slower than the position dynamics. However, one can imagine that a similar control loop is used with the q rather than the d currents and the speed instead of the position.

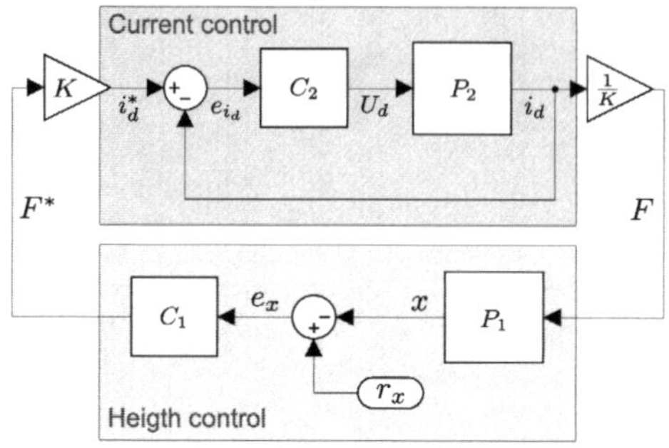

Figure 3: Schematic control overview

4.1 Current Control

Generally, controlling a current in an RL-circuit is relatively straight-forward and has been applied many times before. To this end, a classic PI-controller is used. The gain and integration constant are chosen based on the inductance and resistance of the coils. By design, the resistance and inductance on both the d and q axes are equal for each motor segment. Hence the same controller can be re-used. This PI-current control results in a settling time of roughly 0.5ms.

4.2 Height control

The most challanging part of the control scheme is the position (height) control. As the rotor is highly magnetic, it has a large negative stiffness, forcing it away from the operating point. This force has to be compensated very rapidly to remain in control. A PID-controller is designed with adequate gains to overcome this negative stiffness and keep the rotor in place. Figure 4 shows how the controller moves the unstable poles to the left-half of the complex plane, thus stabilizing the system. Note that the values on the imaginary axis are of the order 10^{-6}, meaning oscillations are almost completely ruled out.

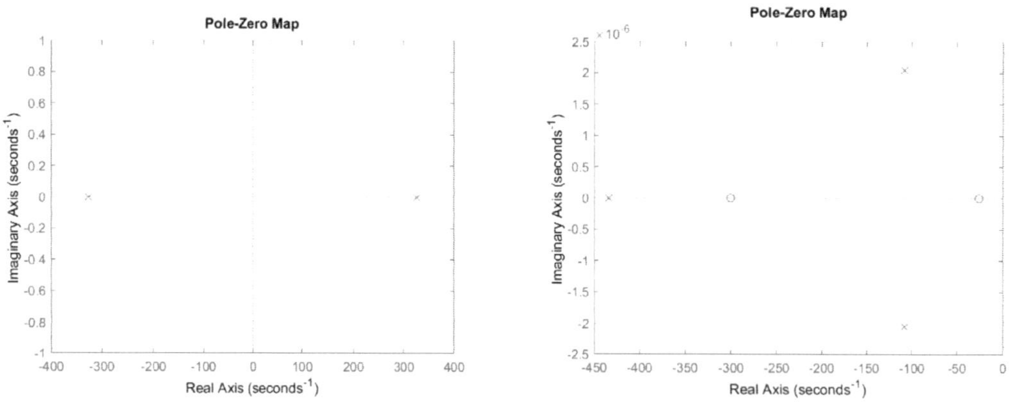

(a) Open loop poles *(b) Closed loop poles*

Figure 4: Movement of unstable poles by applying PID height-control

5. Implementation

As a prototype, a 4kW version of the rim-driven thruster is build with 12 permanent magnet pole pairs and 18 coils. For this proof of concept, the coils are connected in series and powered by a 60V power supply. Pulse Width Modulation is applied to supply the desired voltages. The power electronics are connected to a d-SPACE setup combined with MATLAB/Simulink to allow quick and accurate signal processing. The system operates at 16kHz and parameters can be adjusted real-time. Voltages, currents, x,y positions and angle are measured and used as control signals. Later the angular measurement is replaced with an observer, such that the encoder can be removed. The most important system parameters are summarized in table 1.

Stiffness	Mass	Current to force factor	Inductance	Resistance	Peak Flux	Nr of windings
c	m	K_f	L	R	$\hat{\Phi}$	n
-426000	4	5	$176e^{-6}$	$27.5e^{-3}$	$4.05e^{-4}$	28
$[N/m]$	$[kg]$	$[N/A]$	$[H]$	$[\Omega]$	$[Wb]$	[-]

Table 1: System parameters for a 4 kW prototype

Figure 5 shows the control setup connected to the latest prototype, which is a $20kW$ version of the rim-driven thruster. Testing of this larger machine is still ongoing and results will be available in due time.

Figure 5: 20 kW hubless rim-driven thruster and control setup

6. Results

Applying the proposed control strategy to the real machine allowed levitation of the rotor through its entire rpm range (0-1300 rpm). Figure 6 shows the orbit plots of the rotor at increasing rotational speeds. It can be observed that the movement around the equilibrium point increases slightly with an increase in rotational speed. Overall though, the distance from the origin does not grow beyond 0.1 mm, which is an acceptable deviation. The nominal gap from rotor to stator is 1mm, so there is no risk of physical contact. Figure 7 shows the corresponding d-currents of a single motor segment. As the deviations around the equilibrium point increase, the required currents to compensate for this movement also increase. Higher rotational speeds are not presented here, because the in-house test-rig does not contain a sufficient volume of water to use the machine's full power. However, the same machine has been used in the field as well. Here, attached to a vessel, the motor has been operated at full power without problems.

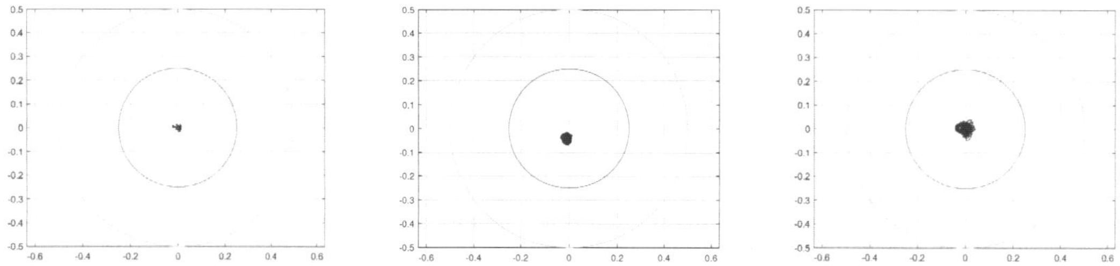

(a) 120 rpm (b) 360 rpm (c) 480 rpm

Figure 6: Orbit plots at various rotational speeds. (x and y axes represent displacement in [mm])

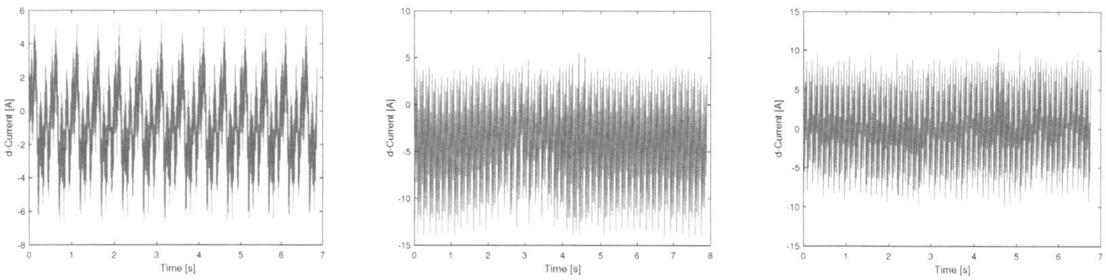

(a) 120 rpm (b) 360 rpm (c) 480 rpm

Figure 7: d-currents for one of the motor segments at various rotational speeds

7. Discussion

Using cascaded control loops in combination with a nine phase machine, an operational bearingless drive has been build that controls the rotors position in two dimensions as well as its rotational speed. Energy used for the bearing function is limited, but not insignificant. Further refinement of the control strategy should be investigated to try and reduce this effort. Also, the models that surved as the starting point for the controller design have been serious simplifications of reality. Extending the models to include coupling terms and non-linear effects of for example the rotor's position on electromagnetic properties could improve performance. Furthermore, the controller relies on highly precise, real-time measurements of the rotor's position. Sensors able to provide these signals in underwater conditions are costly. Hence, the possibility of replacing these sensors with an adequate observer is also a subject for further study.

8. Conclusion

All in all, a working prototype of a hub-less bearingless rim-driven thruster has been build and tested both in lab conditions and in the field. Results are promising and encourage further research and development steps. The 4kW machine has already been scaled up to 20kW, which can in the future be increased further. Both machines have been tested succesfully in a lab environment as well as in the field. Additional research will be conducted to further improve the bearing performance and reduce its energy consumption.

References

Ahad, M. A. & Ahmad, S. M. (2021), Investigation of a 2-dof active magnetic bearing actuator for unmanned underwater vehicle thruster application, *in* 'Actuators', Vol. 10, MDPI, p. 79.

Chen, J., Zhu, J. & Severson, E. L. (2019), 'Review of bearingless motor technology for significant power applications', *IEEE Transactions on Industry Applications* **56**(2), 1377–1388.

Sokolov, M. et al. (2021), 'Bearingless motors: Modeling and control'.

Wang, F., Zhu, Y., Wang, H. & Zhao, D. (2019), 'Design and analysis of a bearingless permanent-magnet motor for axial blood pump applications', *IEEE Access* **8**, 7622–7627.

Asymmetries in Planetary Touch-Down Bearings

Benedikt SCHÜßLER [a], Felix PÜSCH, Stephan RINDERKNECHT [b]
Technical University Darmstadt – Institute for mechatronic Systems, Otto-Berndt-Straße 2, 64287 Darmstadt, Germany,
a schuessler@ims.tu-darmstadt.de
b rinderknecht@ims.tu-darmstadt.de

Abstract

For certain applications, magnetic bearings are preferred over rolling element bearings, for example, due to the absence of mechanical friction leading to high efficiency and low maintenance needs. However, they also require touch-down bearings (TDB) to bear thpe rotor in case of a malfunction or overload. For vertical systems with high DN numbers, like outer rotor flywheels, the design of the TDB becomes a challenging task. For such systems, the planetary TDB can be applied. The suitability of this design has already been shown in the literature (Quurck et al., 2017; Quurck et al., 2018). However, there are multiple parameters influencing the performance of the planetary TDB. For example, if an asymmetric planetary TDB should be preferred over a symmetric one, and if this is the case which asymmetries increase the performance of the planetary TDB. Therefore, this paper investigates asymmetries in planetary TDB in a simulation study performed with the Matlab-based simulation environment ANEAS. The results of this study indicate that the performance of planetary TDB can be increased when the air gap of the individual bearing units differs. If the air gap from one bearing unit is reduced by 20 % the maximum force acting on the TDB was reduced, in the best case up to 52 %.

Keywords: *touch-down bearing; backup bearing; safety bearing; drop-down; simulation*

1. Introduction

Magnetic bearings have many advantages over conventional bearings like no mechanical friction and lower maintenance needs; hence they are preferred for some applications. However, they always need a mechanical bearing that bears the rotor at standstill or in case of malfunction of the magnetic bearing. Conventional touch-down bearings (TDB) for inner rotor systems consist of plain or rolling element bearings with a slightly greater diameter than the rotor diameter. Hence, in normal operation, there is a gap between the rotor and the TDB. To prevent a destructive, friction-induced backward whirling of the rotor in the TDB, rolling element bearings are preferred especially for vertical systems. If the product of the diameter and the rotational speed (DN number) reaches high values, conventional rolling element bearings can no longer be used, because the bearings cannot withstand the high centrifugal forces under rotation. Such high DN numbers are reached for example in outer rotor flywheels, where the rotor is a hollow cylinder mainly made of fiber-reinforced plastic. For such systems with high DN numbers (Penfield JR. and Rodwell, 2000) proposed using the planetary TDB. Similar designs are given in (Fonseca, et al., 2015) and (Lahriri and Santos, 2013). In the planetary TDB proposed by (Quurck et al., 2017), which is also the design under investigation in this paper, multiple small bearing units are distributed circumferentially around the stator. Figure 1 shows on the left a partial section view of a planetary TDB. Each TDB unit consists of a roller that gets in contact with the rotor in case of a drop-down and two rolling element bearings on the upper and lower end of the roller. The right of Figure 1 shows a schematic representation of the clearance of the planetary TDB and how it arises. The clearance of the planetary TDB is polygonal shaped and the number of corners depends on the number of TDB units in the planetary TDB. The boundary of the clearance is defined by the herpolhodes of the rotor rolling on the different TDB units.

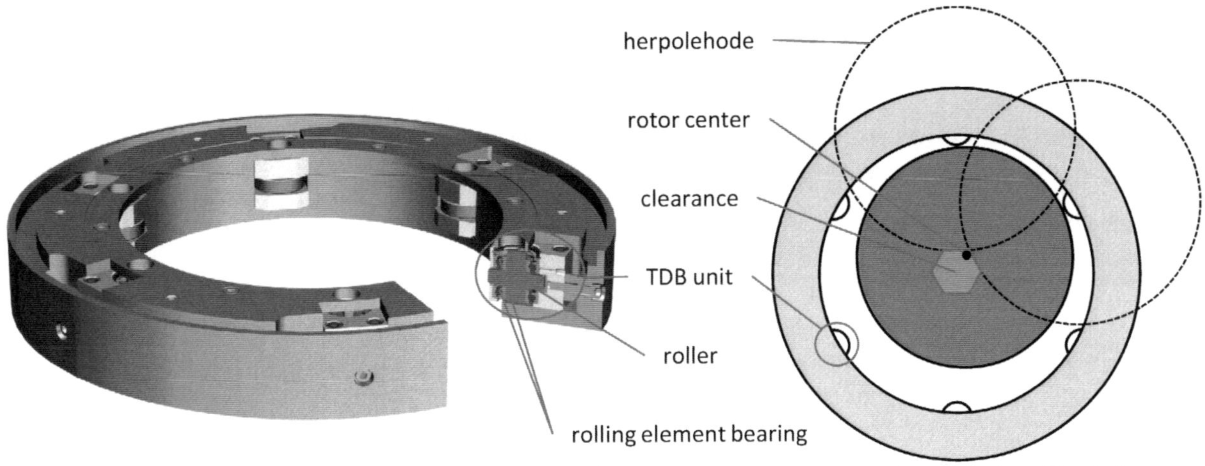

Figure 1 Left: planetary TDB of the TDB test rig with partial section of one TDB unit; right: schematic top view on planetary TDB with clearance of the rotor in the planetary TDB (not true to scale)

The advantage of the planetary TDB design is, that the rolling element bearing diameter is decoupled from the rotor diameter and hence, the planetary TDB is suited for high DN numbers. A further advantage is a whirl suppressing characteristic of the planetary TDB, which is shown with plain bearings in (Lahriri and Santos, 2013) and with rolling element bearings in (Schüßler et al., 2022). However, the rolling element bearings in the planetary TDB units are comparatively small to the overall system, which is why the forces may exceed the static load rating of the bearing. Hence, the planetary TDB must be carefully designed to withstand multiple drop-downs. In literature, the influence of different designs on the TDB loads is rarely investigated. Only a few guidelines are given on how the planetary TDB should be designed. For example, in (Simon, 2002) and (Schüßler et al., 2022) the number of elements in a planetary-like arrangement is investigated both in simulations and in experiments. In (Zülow and Liebich, 2009) a further planetary design is investigated theoretically for a horizontal rotor system. The investigated TDB configurations contain even a geometrically asymmetrical TDB design. The results show that the rotor run-out in the bottom of the clearance in the asymmetric configuration is more unsteady and therefore not preferred for the investigated horizontal system. In comparison to the investigations in the literature, this paper focuses on vertical systems. In addition, it investigates further asymmetries and analyses how they affect the TDB loads.

2. Method

The investigation of different asymmetries in planetary TDB is performed as a simulation study. The system chosen for the modeling and consequently for the investigation is the TDB test rig, with which multiple drop-down experiments for symmetric planetary TDB have already been performed (Quurck et al., 2018; Quurck, 2019). The principal behavior of these drop-down experiments at the TDB test rig can already be simulated with the MATLAB-based simulation tool ANEAS (Quurck, 2019). Due to this validation, this simulation tool is used for the present investigation, too. A short overview of the simulation software ANEAS is given and essential ideas of the modeling are shown. Afterward, two severity indicators are introduced, to compare the different simulation results of the conducted investigation. Figure 2 shows an overview of the asymmetries investigated in simulations. These asymmetries can either be geometric or they can be asymmetries in the stiffness and damping between the different TDB units.

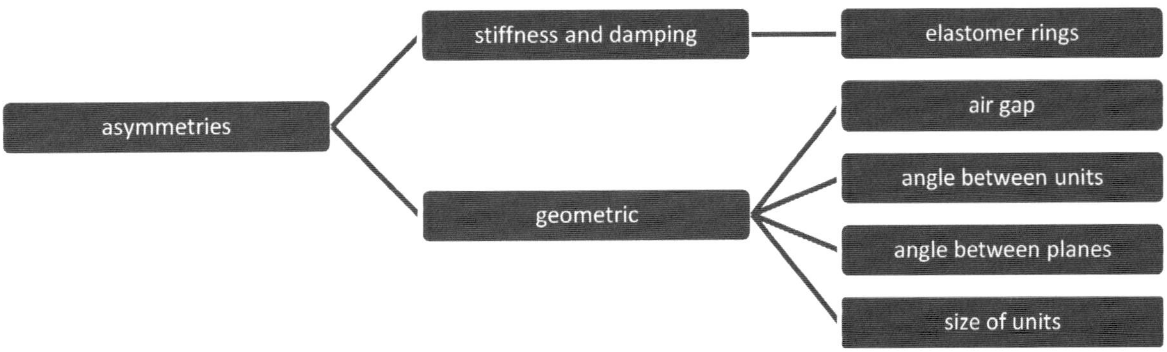

Figure 2 Investigated asymmetries.

In the real TDB, there are different possibilities to vary the stiffness and damping, one is the use of elastomer rings (ER). As shown by (Schüßler et al., 2021) the effectiveness of the elastomer rings is heavily influenced by their properties: on the one hand by the material on the other hand by the geometric properties. Hence, the investigation in this paper is based on the modeling and the results in (Schüßler et al., 2021). Based on this previous investigation the model parameters of fluorocarbon rubber (FKM) are used for the modeling of the elastomer rings. The geometric asymmetries are the size of the air gap, the distance or angle between the different TDB units around the stator, the angle between the two polygonal-shaped clearances of the two TDB planes, and the size of the TDB units. Figure 3 shows, how the different geometric asymmetries affect the TDB clearance.

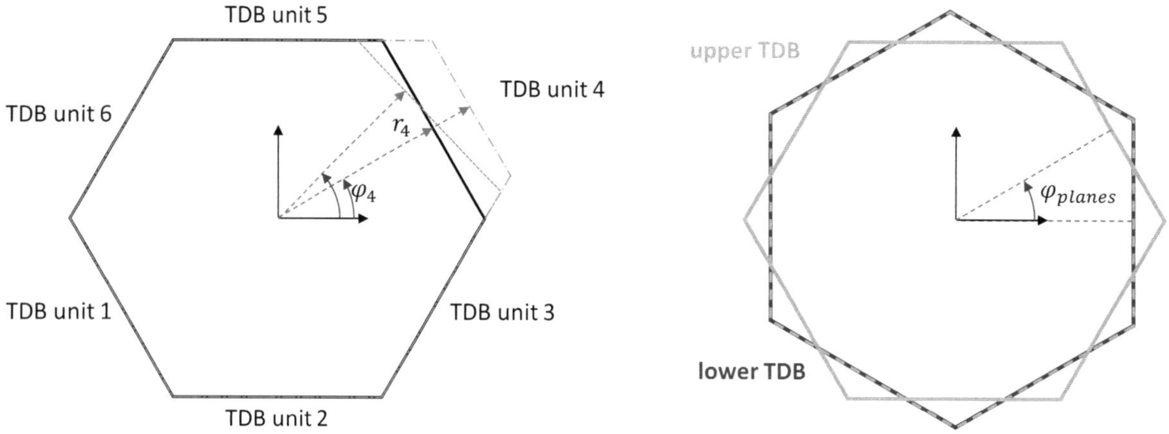

Figure 3 Change of the clearance of the planetary TDB resulting from the different asymmetries. Left: change due to air gap (cyan dashed-dotted line) and angle between TDB units (orange dashed line); Right: change due to angle between TDB planes (green: changed clearance of upper TDB plane compared to blue lower TDB plane)

The size of the TDB units affects the TDB clearance only slightly. The clearance boundaries are based on the herpolhodes of the rotor rolling on the TDB units as it is shown in Figure 1. If the size of the roller is changed, also the radius of the herpolehode changes, and hence, the clearance boundary changes as well. However, the radius is much bigger than the clearance. Therefore, the change of curvature is not seen in the clearance if the size of the roller is changed in a physically reasonable range.

In this investigation, simulations were conducted for TDB with 3, 4, 5, and 6 TDB units per TDB plane. A higher number of TDB units were investigated, neither for symmetric TDB nor for asymmetric TDB because it was shown in (Schüßler et al., 2022), that 6 TDB units are preferred over 8 TDB units. The investigation of asymmetries was conducted in two steps. First, the single simulations were conducted for a broad range of asymmetries to investigate the influence of the asymmetries in principle. Afterward, the promising asymmetries were deeper investigated in a parameter study, in which for the same asymmetry the simulation was conducted six times with different model parameters. In this way, a profound decision on the asymmetries is found.

3. Modeling

The simulation study is based on the simulation environment ANEAS. The software was originally developed by (Orth and Nordmann, 2002) for the investigation of magnetically levitated magnetic bearing systems, especially for the case of drop-down simulations in conventional TDB. Later it was extended by (Quurck et al., 2017) for simulating drop-downs in symmetric planetary TDB, too. In the present investigation, the model was adjusted for asymmetric planetary TDB. The model in ANEAS is based on the equations of motion for the rotor and the stator, which are coupled by the contact forces in case of a drop-down. The rotor and stator models are finite element models based on Timoshenko beam elements. The behavior in the axial direction is not modeled, in the investigated drop-downs only the radial contact and movements are of interest. The general equation of motion for the rotor and stator model is shown in (1). The systems are described by the rotational speed Ω, the mass matrix $\boldsymbol{M}$, the matrixes for the inner and outer damping $\boldsymbol{D}_{in}$, $\boldsymbol{D}_{in}$ and $\boldsymbol{D}_{out}$, the gyroscopic matrix $\Omega\,\boldsymbol{G}$ and the stiffness matrix $\boldsymbol{K}$. The displacements and tilting angles are described by $\boldsymbol{q}$. The external force vector $\boldsymbol{f}$ on the right side of the equation of motion contains for example the contact forces.

$$\boldsymbol{M}\,\ddot{\boldsymbol{q}} + (\boldsymbol{D}_{out} + \boldsymbol{D}_{in} + \Omega\,\boldsymbol{G})\,\dot{\boldsymbol{q}} + (\boldsymbol{K} + \Omega\,\boldsymbol{D}_{in}^{*})\,\boldsymbol{q} = \boldsymbol{f} \quad (1)$$

For the stator, this general equation of motion is simplified since the axial rotational degree of freedom is neglected, and hence, no gyroscopic effects occur. Because a drop-down in the TDB is a contact problem, the system is nonlinear.

Due to this nonlinearity, the right side of the model has to be evaluated in every time step of the time domain based simulation in ANEAS.

4. Severity Indicators

To compare the different drop-down simulations with each other, severity indicators are used. For planetary TDB the maximum normal force during the drop-down is of high relevance. Because of the small rolling element bearings in the TDB, the maximum force is likely in the range of their static load rating. Hence, one short-term contact can damage the bearing permanently. As a result, the first severity indicator chosen for the evaluation is the static safety S_o as shown in (2). The maximum normal force during the drop-down is expressed by $F_{N,max}$ while $P_{TDB,0}$ is the static load rating of the TDB unit.

$$S_o = \frac{F_{N,max}}{P_{TDB,0}} \tag{2}$$

The second severity indicator used for the evaluation is the bearing service life of the TDB. The TDB consists of multiple TDB units. The service life entire system is the lowest bearing service life of all TDB units $L_{10,h,min}$. The service life of one TDB unit $L_{10,h}$ is calculated according to (3) based on ISO 281 and (Wittel et al., 2009). C is the dynamic load rating of the TDB unit, P_m the average bearing load and n_m the average bearing speed. Neither the bearing load nor the bearing speed are constant during a drop-down and varies in a broad range. The effective values of these are calculated by weighted averaging of bearing speed n and normal force F_N for each time step i. The final time step of the simulation is denoted as ts and the proportion a time step has on the whole simulation time is denoted by a_i. The normal force acting on the TDB unit F_N is divided by the factor of two, because it is assumed that the force is distributed equally on two bearings (see Figure 1).

$$L_{10,h} = \frac{10^6}{60\, n_m}\left(\frac{C}{P_m}\right)^3 \quad \text{with} \quad P_m = \sqrt[3]{\frac{\sum_{i=1}^{ts} a_i n_i \left(\frac{F_{N,i}}{2}\right)^3}{n_m}} \quad \text{and} \quad n_m = \sum_{i=1}^{ts} a_i\, n_i \tag{3}$$

With this second severity indicator, the whole drop-down is taken into account, compared to the first severity indicator S_o where only a single value of a drop-down is considered.

5. Results

The basis for comparing the influence and the benefit of the asymmetries are simulations of symmetric configurations with 4, 5, or 6 bearing units in the TDB. These simulations are conducted without and with ER on all TDB units. These simulations are then compared to simulations where one asymmetry is introduced. When comparing severity indicators of symmetric and asymmetric concepts, the following results are obtained:

- If the angle between the TDB units is changed such that some units are closer to each other, and some are further away from each other, S_o is increased, especially for a high number of TDB units. However, $L_{10,h,min}$ is decreased because the bearing service life of individual bearings becomes lower.
- The second investigated asymmetry is a change in the air gap for individual bearings such that some bearings have greater and others have a smaller air gap to the rotor. In general, the results are similar to the results for the asymmetric angle between the bearing units. For air gap asymmetries $L_{10,h,min}$ varies only slightly, whereas the S_o is increased for all simulations with 5 or 6 bearing units.
- If the stiffness of the individual TDB units within the TDBs is changed, no improvement in S_o and $L_{10,h,min}$ occurs compared to the simulations where all bearing units have the same stiffness.
- Different stiffnesses in the lower and upper TDB plane lead neither to an improvement in $L_{10,h,min}$ nor in S_o.
- If the upper and lower bearing plane a rotated to each other by half of the angle between the TDB units $L_{10,h,min}$ decreases slightly, whereas S_o is increased for 4 and 5 TDB units without ER. Figure 4 shows the results for this asymmetry based on the two severity indicators.
- It was also investigated if it is reasonable to replace the highest loaded bearing unit in the TDB with a bigger one, which has a higher static and dynamic load rating. This led to the result, that another bearing unit in

the TDB became the highest loaded one. Finally, the highest S_o and $L_{10,h,min}$ are reached, if all bearing units are replaced by bigger ones.

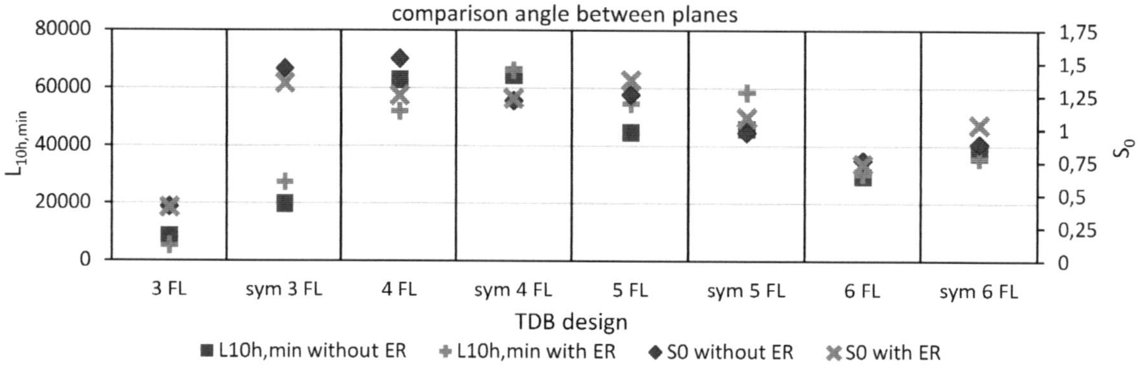

Figure 4 Results for the asymmetry "angle between planes"; one TDB plane is rotated by half of the angle between the individual TDB units

Based on these findings a parameter study with the most promising asymmetry has been performed. In the parameter study, three parameters have been varied between two values: the initial rotor speed (15,000 rpm; 20,000 rpm), the start direction towards the clearance boundary (+30 °; -30 °), and the coefficient of restitution influencing the contact damping (0.6; 0.8). A positive angle for the start direction leads to a first contact of the rotor with the TDB in the direction of a backward whirl, while a negative angle leads to a contact in the direction of a forward whirl. All combinations of these three parameters have been simulated in the parameter study. Hence, 8 simulations have been performed for all investigated TDB configurations. As a result of the first step of the simulative investigation, it was found that the highest values for the two severity indicators are reached, if all bearings are replaced by bigger ones. Therefore, in the parameter study bigger bearings of type HY S 6201 are used. This is the maximum bearing size, which can be mounted in the available assembly space in the TDB test rig.

The asymmetry selected for further investigation is the asymmetry, where in a 5-unit TDB the air gap at one TDB unit is reduced by 20 %. The results for this configuration are compared to the best symmetric case. In (Schüßler et al., 2021) it was shown that ER reduce the forces on the TDB and increase the bearing service life for a TDB with 6 TDB units. The first step of this investigation showed that the forces are even lower in a TDB with 4 TDB units with ER than in a system with 6 TDB units with ER. Therefore, the best symmetric case is the TDB with 4 TDB units and ER. Figure 5 shows the simulation results of the best symmetric configuration compared with the ones of the best asymmetric configuration. In every simulation S_o is higher for the asymmetric configuration, hence the maximum normal forces are lower in the asymmetric configuration. The calculated $L_{10h,min}$ is in the same range for the symmetric and the asymmetric configurations with the tendency to reach higher values for the asymmetric configuration.

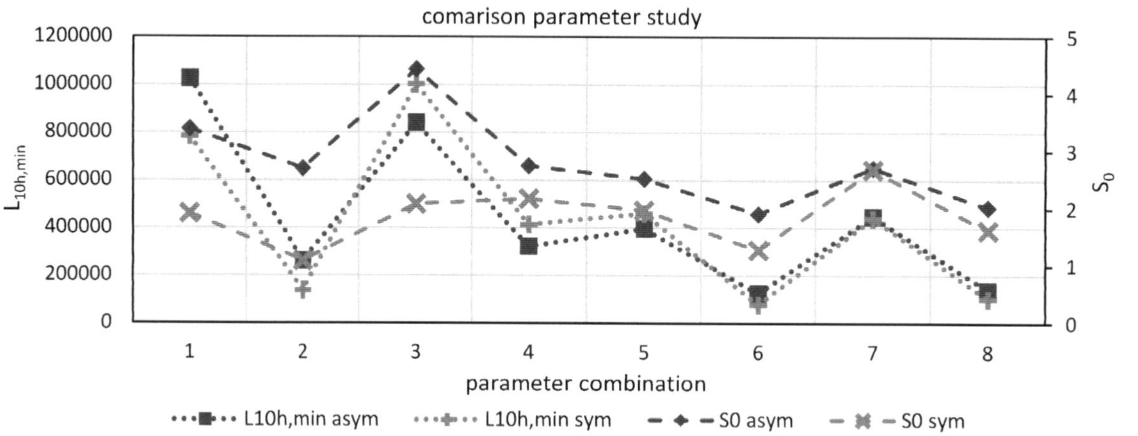

Figure 5 results of the parameter study for the severity indicators for the chosen symmetric configuration (red) in comparison to the investigated asymmetric configuration (blue)

In summary, it can be concluded that by using asymmetries the maximal forces on the TDB can be reduced significantly. For the investigated case in the parameter study, an average reduction in S_o of 52% was reached, whereas $L_{10,h,min}$ was only increased by 16%. Consequently, asymmetries can be considered in the design of planetary TDB to increase the performance of the TDB.

Acknowledgment

This research was funded by the German Federal Ministry for Economic Affairs and Energy, grant numbers 03ET6064A and 03EI3000A.

References

Fonseca, CALL, Aguiar RR and Weber HI (2015) Chaotic and stable orbits of a rotor in a non-conventional back-up bearing. A numerical and experimental approach. In: *SIRM 2015:* Magdeburg, Deutschland, 23.2.-25.2.

Lahriri S and Santos IF (2013) Experimental quantification of dynamic forces and shaft motion in two different types of backup bearings under several contact conditions. *Mechanical Systems and Signal Processing* 40(1): 301–321.

Orth M and Nordmann R (2002) ANEAS: A Modeling Tool for Nonlinear Analysis of Active Magnetic Bearing Systems. *IFAC Proceedings Volumes* 35(2): 811–816.

Penfield SR JR. and Rodwell E (2000) Auxiliary bearing design considerations for gas cooled reactors. In: *Proceedings of the International Atomic Energy Agency Technical Committee Meeting on Gas Turbine Power Conversion Systems for Modular HTGRs: (*ed International Atomic Energy Agency), Palo Alto, California, USA, 14–16 November.

Quurck L (2019) *Fanglagerung magnetgelagerter Schwungmassenspeicher*. Darmstadt: TUprints.

Quurck L, Franz D, Schüßler B, et al. (2017) Planetary Backup Bearings for High Speed Applications and Service Life Estimation Methodology. *Mechanical Engineering Journal* 4(5).

Quurck L, Viitala R, Franz D, et al. (2018) Planetary Backup Bearings for Flywheel Applications. In: *Proceedings of ISMB 16:* Beijing, China, August 13.-17.

Schüßler B, Hopf T and Rinderknecht S (2021) Simulative investigation of rubber damper elements for planetary touch-down bearings. *Bulletin of the Polish Academy of Sciences: Technical Sciences*. DOI: 10.24425/bpasts.2021.139615.

Schüßler B, Hopf T and Rinderknecht S (2022) Drop-Downs of an Outer Rotor Flywheel in Different Planetary Touch-Down Bearing Designs. *Actuators* 11(2).

Simon U (2002) Rotor–Stator–Kontakt in polygonförmigen Fanglagern. Dissertation.

Wittel H, Muhs D, Jannasch D, et al. (eds) (2009) *Roloff/Matek Maschinenelemente*. Wiesbaden: Vieweg+Teubner Verlag.

Zülow D and Liebich R (2009) Ein Außenrollenlager als Fanglagerkonzept für magnetgelagerte Rotoren. In: *SIRM 2009 - 8. Internationale Tagung Schwinungen in rotierenden Maschinen:* Wien, Öserterreich, 23.-25.2.

Bearingless Slice Motors with PM-free Rotor for Disposable Centrifugal Blood Pumps

Tadahiko SHINSHI, Zeqiang HE, Ren YANG, Naohiro SUGITA
Tokyo Institute of Technology, 4259 Nagatsuda-cho, Midori-ku, Yokohama, Japan, shinshi.t.ab@m.titech.ac.jp

Abstract

The principle, prototyping, and performance evaluation of two novel bearingless slice motors (BELSMs) designed for disposable centrifugal blood pumps are presented. Both BELSMs do not use permanent magnets in the rotor of the pump housing, thus reducing the cost of disposable components. One BELSM has twelve shared coils for rotation and magnetic levitation. The BELMS achieves passive stiffness in the axial and tilting directions using magnetic coupling generated in the axial direction. The stator structure is relatively simple but requires twelve amplifiers to drive the coils independently. The other BELSM has three-phase coils for rotation, two-phase coils for magnetic levitation, and magnetic couplings formed radially. The three-phase amplifier can drive the motor coils, and the two single-phase amplifiers can drive the suspension coils, simplifying the power circuit, but the stator structure is complex. A centrifugal pump was realized using the former BELSM. The maximum flow rate was 4.3 L/min, which was insufficient for the 5 L/min required for a blood pump. Due to the coil's large inductance and the amplifier performance limitation, it could not reach sufficient rotation. In the second BELSM, the pump's maximum torque and maximum speed have obtained the values required to achieve the target flow and pressure of the blood pump.

Keywords: *Bearingless motor, Centrifugal blood pump, Disposable rotor, Magnetic coupling*

1. Introduction

Centrifugal blood pumps are used in the treatment of acute heart failure and as a short- and mid-term bridge to heart transplantation or an implantable artificial heart. Recently, ECMO (Extracorporeal Membrane Oxygenation) systems consisting of an artificial lung and a centrifugal blood pump have been used to treat severe pneumonia caused by COVID-19. Centrifugal blood pumps with non-contact bearings have been developed to avoid hemolysis and thrombosis and improve durability for long-term use.

Centrimag™ is the world's first extracorporeal centrifugal blood pump equipped with a bearingless slice motor (BELSM) (Gruber, et al., 2016). It is used for circulation support (John, et al., 2007) and ECMO therapy (Azziz, et al., 2010). The impeller of the Centrimag™ is magnetically levitated and rotated with a slice rotor, featuring radial feedback control, passive support in the axial and tilt directions, and rotation by rotor-stator cores integrated with magnetic bearing and motor functions. The rotor, impeller, and pump housing are disposable for each patient in these centrifugal pumps for extracorporeal circulation. The rotor of Centrimag™ is composed of a neodymium ring magnet.

A magnet-free rotor is required to reduce the cost of the disposable part. In the research on bearing slice motors, reluctance motor-type ones without a permanent magnet in the rotor have been studied (Gruber, et al., 2017) (Noh, et al., 2019). We are developing BELSMs for extracorporeal centrifugal blood pumps that do not use magnets in the rotor. The initial prototype was designed using a coil that generates a magnetic bias flux to cause tilting and passive axial stiffness. The power consumption for generating the magnetic bias flux was not acceptable for clinical use (Rao, et al., 2015).

We are developing two other BELSMs with permanent magnets on the stator side to reduce power consumption and form magnetic couplings. This paper presents the structure, magnetic levitation, and rotation performance of those BELSMs with a permanent magnet-free rotor and its pump performance when applied to a centrifugal pump.

2. Methods

Two types of BELSM structures (Type A and Type B) based on reluctance motors are proposed. Both rotors have central teeth and upper and lower flanges. These magnetic flux bias paths of the magnetic couplings that form the restoring force and restoring torque in the axial and tilting directions are different. In Type A of Fig. 1, permanent ring magnets magnetized in the thickness direction are placed on the upper and lower surfaces of the stator teeth tip, respectively. The magnets generate the axial magnetic flux flow. The principle of restoring torque generation is shown in Fig. 1. The radial bias flux is varied using coils wound on each tooth of the stator to provide radial force for stabilization. The torque to rotate the rotor is generated by changing the circumferential magnetic flux distribution by the coils.

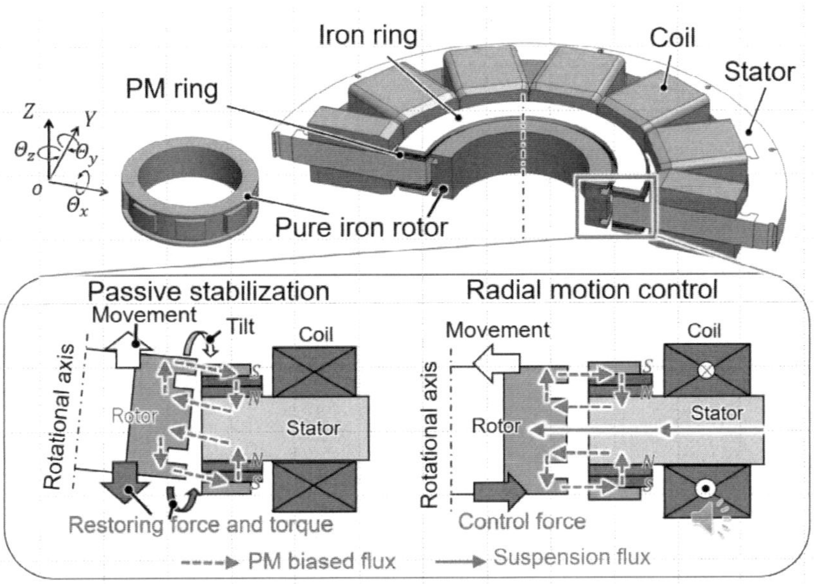

Figure 1 Type A BELSM having axial magnetic coupling and concentrated windings. The upper part is the configuration of the BELSM, lower part is passive and active suspensions (Shinshi, et al., 2018, Yang, et al. 2022)

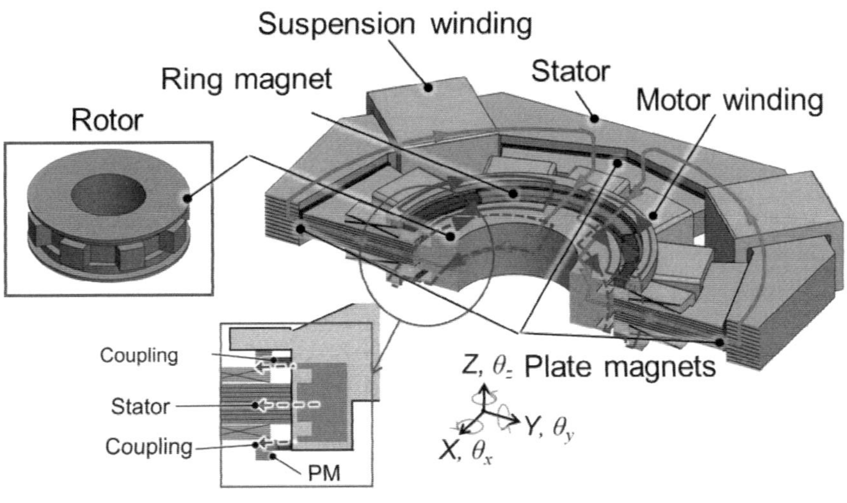

Figure 2 Type B BELSM having radial magnetic coupling, and motor and suspension windings. The upper part is the configuration of the BELSM, lower part is three-layered magnetic circuits (He, et al., 2022)

In Type B of Fig. 2, the magnetic coupling is formed in-plane by radially multipole magnetized ring magnets, generating restoring torque in the tilt direction and restoring force in the axial direction. To generate a magnetic bias flux in the magnetic motor circuit, plate magnets magnetized in the thickness direction are sandwiched between the inner and outer stator cores. Three layers of bias magnetic flux flow enhance the passive stiffness of the magnetic spring that supports the rotor.

Type B has advantages over Type A by separating the motor coils from the suspension coils. The inductance of one coil can be reduced, which lowers the drive output voltage of the amplifier. Another is that the number of amplifiers can be reduced. For example, if the stator has twelve teeth, Type A requires twelve independent single-phase amplifiers to apply independent currents to all coils. Type B also has twelve motor coils, but they are in sets of four and can be driven by a three-phase amplifier. The four coils for magnetic levitation can also be used in pairs, so only two single-phase amplifiers are required. The two types of BELSMs were designed using magnetic field analysis under the following requirements: rotor diameter of 50 mm, motor torque of 0.060 Nm or more, and bearing clearance of 1.5 mm. A centrifugal pump incorporating a Type A BELM is shown in Fig 3, and a Type B BELSM is shown in Fig. 4.

3. Results

Prototypes of Types A and B bearingless motors have been fabricated, and experiments on magnetic levitation and rotation have been successfully carried out, as shown in Figs. 3 and 4. The bearingless motor of Type A has already been integrated into a centrifugal pump, and measurements of flow rate and head pressure are in progress. The target rotation of 3000 rpm could not be reached, and the maximum flow rate was 4.3 L/min at 2300 rpm, which did not reach the minimum practical flow rate of 5 L/min. This is because the coil inductance is large, and the amplifier output voltage is insufficient to supply enough current to the coil at high-speed rotation.

Although the experiment was conducted outside the pump, a contactless rotation of 3000 rpm was successfully achieved for Type B. The experiment also confirmed that at 3000 rpm, a minimum of 0.06 Nm, which is the minimum required to drive the pump, can be generated. Type B BELSM will be integrated into centrifugal pumps and tested.

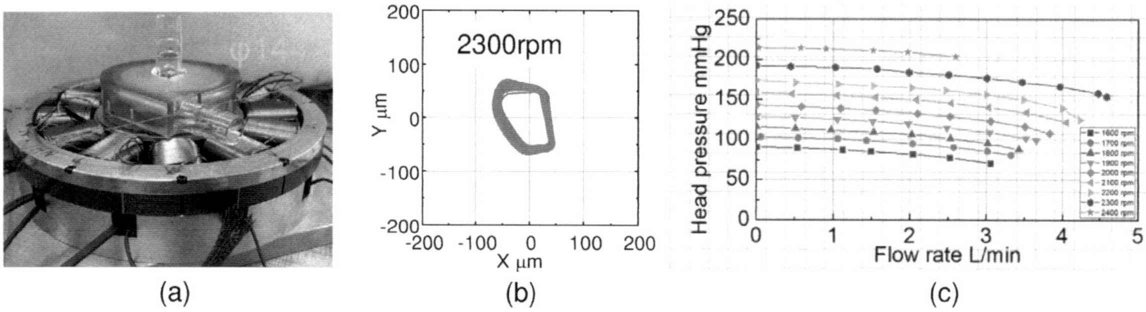

Figure 3 Fabricated Type A BELSM. (a) Prototype centrifugal pump incorporated with the BELSM, (b) Rotational accuracy at 2300rpm during a flow rate of 4.6 L/min, (c) HQ characteristics of the centrifugal blood pump.

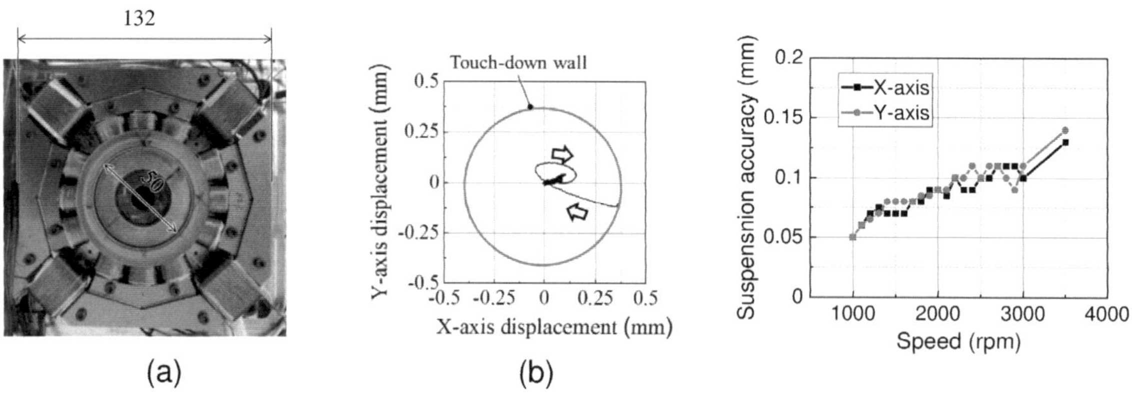

Figure 4 Fabricated Type B BELSM. (a) Prototype, (b)Start-up characteristics, and (c) Rotational accuracy levitating and rotating in the air.

4. Conclusion

The structures, prototypes, and performance of two BELSMs developed for disposable centrifugal blood pumps were presented. Both BELSMs used a rotor with only an iron core, without permanent magnets, to reduce the disposable part's cost. Both succeeded in magnetic levitation and rotation. However, the one with a simple stator structure, in which the suspension coil and motor coil were shared, had a large coil inductance, which caused problems in achieving high speed, and, when incorporated into a pump, did not achieve the pressure and flow rate requirements for a blood pump. The other type, in which the suspension and motor coils were separated, has a more complex stator structure, but the number of amplifiers for the drive can be reduced, and high-speed rotation can be achieved. In the future, the Type B BELSM is also aimed at being installed in centrifugal pumps and its evaluation.

This research is partly supported by Tsugawa Foundation, Japan, and the Research Center for Biomedical Engineering, jointly operated by Tokyo Medical and Dental University, Tokyo Institute of Technology, Hiroshima University and, Shizuoka University.

References

Azziz T A et al. (2010) Initial experience with CentriMag extracorporal membrane oxygenation for support of critically ill patients with refractory cardiogenic shock. J Heart Lung Transplant 29(1):66-71.

Gruber W and Silber S (2016) 20 years bearingless slice motor – its developments and applications. In: Proc. 15th Int. Symp. on Magnetic Bearings, Kitakyushu, Japan, August 2016, pp. 91–98.

Gruber W, Bauer W and Radman K (2017) Comparison of homopolar and heteropolar bearingless reluctance slice motor prototypes. Proceedings of the Institution of Mechanical Engineers, Part I: Journal of Systems and Control Engineering 231(5): 339–347.

He Z, Zhong J, Sugita N and Shinshi T (2022) Novel heteropolar bearingless slice motor with a PM-free rotor for a centrifugal blood pump application. In: JSME-IIP/ASME-ISPS Joint International Conference on Micromechatronics for Information and Precision Equipment Proceedings, Nagoya, Japan, August 2022, D3-1-06.

John R, Liao K et al. (2007) Experience with the Levitronix CentriMag circulatory support system as a bridge to decision in patients with refractory acute cardiogenic shock and multisystem organ failure. Cardiopulmonary Support and Physiology 134(2): 351-358.

Noh M and Trumper D L (2019) Homopolar bearingless slice motor with flux-biasing Halbach arrays. IEEE Transactions on Industrial Electronics 67(9): 7757–7766.

Rao J, Hijikata W and Shinshi T (2015) A bearingless motor utilizing a permanent magnet free structure for disposable centrifugal blood pumps. Journal of Advanced Mechanical Design, Systems, and Manufacturing, 9(3): JAMDSM0046.

Shinshi T, Yamamoto R, Nagira Y and Asama J (2018) A bearingless slice motor with a solid iron rotor for disposable centrifugal blood pump. In: Proc. Int. Power Electron. Conf., Niigata, Japan, May 2018, pp. 4016–4019.

Yang R, He Z, Sugita N and Shinshi T (2022) Low-Cost and Compact Disposable Extracorporeal Centrifugal Blood Pump Utilizing a Homopolar Bearingless Switched Reluctance Slice Motor, IEEE Access, 11:24353 - 24366.

Short-Circuit Fault-Tolerant Strategy for AMB Power Electronic Controller Based on Shared-Bridge Topology

Feng Hu, Dong Jiang, Yishuan Shuai, Jianfu Ding, Zicheng Liu
Huazhong University of Science and Technology, Luoyu Road 1037, Wuhan, China, fenghu@hust.edu.cn

Abstract

Short-circuit faults of power electronics devices are major reason for active magnetic bearing (AMB) system failures, posing a serious threat to system stability, especially during high-speed rotor rotation. Despite this, very few studies have explored switch short-circuit fault-tolerance in AMB system. To address this issue, this paper introduces a fault-tolerant control strategy for short-circuit faults in switching devices, based on shared-bridge topologies. Two set of shared-circuit topologies are used separately to control one coil of each axis, once the short circuit happens, the coils controlled by the faulty shared-bridge topology are controlled to constant current, while the coil controlled by the normal shared-bridge topology are controlled to bias current and control current. In addition, the current must to be reversed to ensure the controllability of the faulty shared-bridge topology. The method also includes short-circuit fault detection. Relevant fault-tolerant experiments have been done to verify the effectiveness of the proposed approach. This approach provides high reliability for AMB system even with the power electronics devices short circuit failure.

Keywords: *Active magnetic bearing, Power electronic controller, Short-circuit fault, Fault-tolerant control*

1. Introduction

The non-contact active magnetic bearings (AMB) are widely used in high-speed rotating machinery, as centrifugal compressors, flywheel energy storage and so on. In the above applications, reliability is undoubtedly a key performance indicator, and the fault-tolerant control is an effective way to improve that [1]. In order to solve the reliability of actuators, Maslen provides the mathematical basis for bias current linearization of magnetic bearing in [2]. This kind of approach is applied to various kinds of magnetic bearing [3]. Many scholars have also studied the method of magnetic bearing fault detection. A fault identification based on discrete time wavelet coefficients is presented in [4]. Cheng et al. [5] provides a fault-diagnosis method based on the equivalent slope of current. However, there are few studies on the fault tolerant control of power amplifiers.

In reference [6], the survey results show that power electronic devices in the converters are the most vulnerable component to fail, there are mainly fault types such as open circuit and short circuit, caused by overheating, overvoltage or overcurrent, etc. Reference[7] provided a reliable switching amplifier, which consists of two or more left and right hand half-bridges. The abundant redundancy of the half-bridge enables it to deal with open or short-circuits of switching devices. However, it doubles the number of switching devices, which significantly increases the cost and size of the amplifiers, and switching off the damaged half-bridge is also an additional burden. In order to solve this problem, reference [8] proposes an open-circuit fault-tolerant control strategy for AMB power electronic controller and fault-tolerant experiments had been carried out at different speed conditions, this fault tolerant method not only reduce the number of switching devices, but also makes the fault tolerant switching process smoother. However, this paper about the fault tolerance of AMB power electronic devices only involves the open-circuit fault so far, it's necessary to design a fault-tolerant control strategy for short-circuit fault.

Fig.1 shows the short-circuit fault-tolerant driving system in this paper, where the thrust AMB is driven by two H half-bridge topologies and two radial AMBs are driven by tow shared-bridge topologies. The shared-bridge topology consists of one bipolar bridge and four unipolar bridges under the normal working condition, each unipolar bridge is connected to one end of the coils and the other ends are connected to the bipolar bridge. It is well known that the electromagnetic force acting on a single degree-of-freedom(DOF) is the difference of the forces generated by two electromagnetic coils, referred to as A coil and C coil. So in order to achieve the short-circuit fault-tolerant control, one shared-bridge topology controls the A coils(x_{a1}, y_{a2}, x_{a3}, y_{a4}) and the other one controls the C coils(x_{c1}, y_{c2}, x_{c3}, y_{c4}). Then if the unipolar bridges are replaced by bipolar ones, the topology can both deal with open circuit and short circuit fault. Open-circuit fault tolerance can refer to reference [8], only short-circuit fault is considered in this paper. Fig.2 shows the core idea of short-circuit tolerance, when the short-circuit fault occurred to the top IGBT of bridge L_{a1}, driven topology will be rapidly switched from the normal mode to the fault-tolerant mode, where the bridge L_{a2} reverses, it doesn't matter that the winding current i_{a2} is reversed, because the electromagnetic force of AMB is independent of winding current direction.

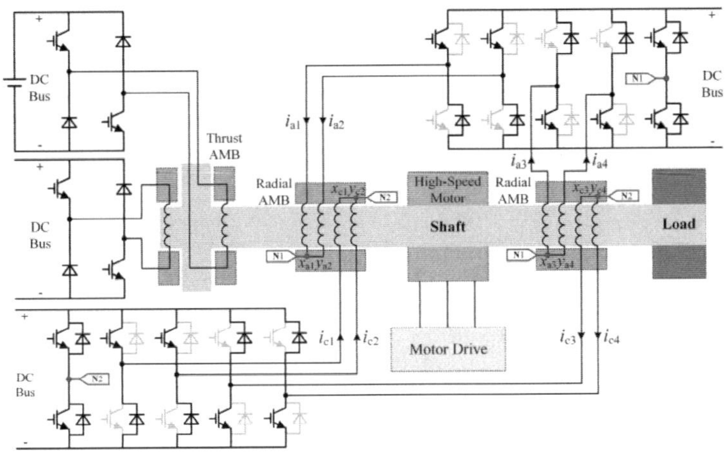

Figure 1 The short-circuit fault-tolerant drive system of AMB based on shared-bridge topology

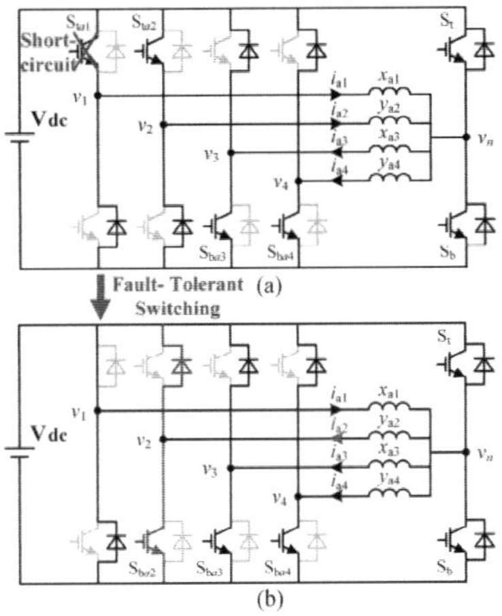

Figure 2 The fault-tolerant topology switching. (a): normal mode; (b): fault-tolerant mode

2. Fault-tolerant control strategy

2.1 Topology analysis in normal and fault mode

As shown in the Fig.3(a), the power electronic devices drawn with dotted lines are not working in normal mode. One end of coils are connected to four unipolar bridges separately and the other ends are connected the bipolar bridge together. Fig.4(a) shows the current control block diagram of shared-bridge topology in normal mode, the duty cycles of IGBTs on bridge L_{N1} are fixed at 0.5 and each control current(i_{a1}, i_{a2}, i_{a3}, i_{a4}) is decoupled. When the IGBT S_{ta1} occurs the short-circuit fault as shown in the Fig.3(b), the current i_{a1} will rise quickly and significantly deviate from the reference value since the midpoint voltage v_n of the bridge L_{N1} is fixed to $0.5v_{dc}$. If the control method is not changed at this time, the current control will fail, the rotor falls, and cannot work normally.

As shown in the Fig.4(b), in order to avoid the continuous rise of the current i_{a1}, the duty cycles of IGBTs on bridge L_{N1} are no longer fixed, but determined by the output of the PI controller (i_{a1}). The gate drive signal g_t needs to stay high level for a long time during a switching period to ensure that i_{a1} is controllable, which is exactly why the bridge L_{a2} needs to reverse, because it can greatly lower the impact of gate drive signal g_t on the wingding current i_{a2}. Although the above operations can ensure that each current is controllable, it will deteriorate the dynamic control performance of current. Therefore, the other core idea of the short-circuit fault-tolerant strategy is that the shared-bridge topology in which a short circuit occurred only generates the constant current, and the normal shared-bridge topology generates both bias current and control current.

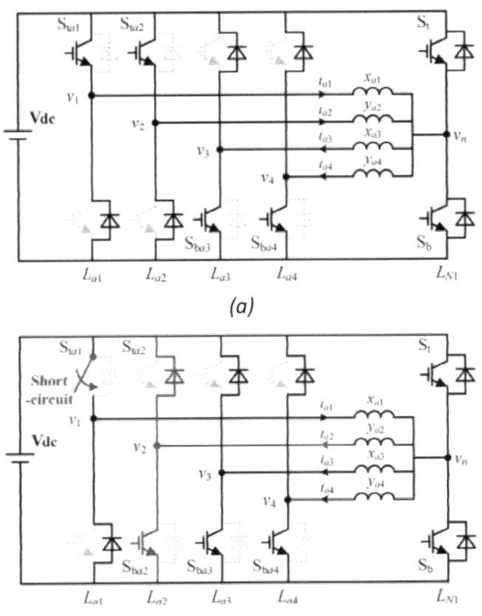

Figure 3 The power amplifier drive topology. (a): normal mode; (b): fault-tolerant mode

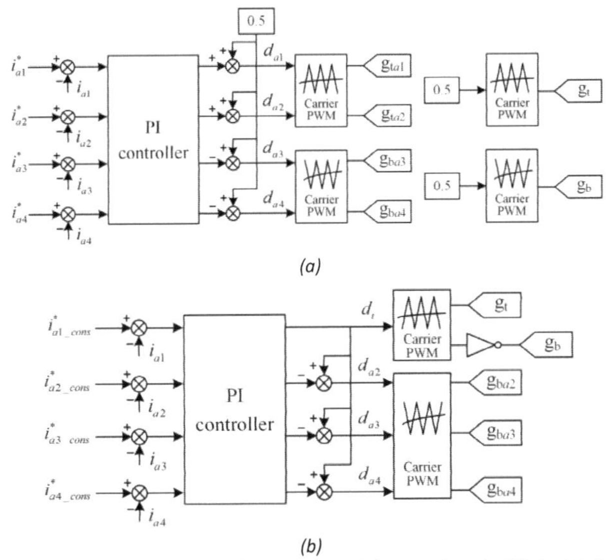

Figure 4 Control block diagram of current loop. (a): normal mode; (b): fault-tolerant mode

2.2 Fault analysis and detection

The duty cycles of IGBTs on bridge L_{N1} are fixed at 0.5 in the normal mode, so when the short-circuit fault occurred to the top IGBT of bridge L_{a1}, the winding current i_{a1} will rise rapidly, however, the winding current i_{c1} that controls the same axis doesn't change suddenly due to it is controlled by the other shared-bridge topology, mostly, the sum of i_a and i_c is constant in normal operation. Therefore, the sum of i_a and i_c can be used to identify whether a short circuit fault has occurred, and the location of the fault can be distinguished by the rise rate of the current.

2.3 Fault-tolerant control system

Fig.(5) shows the short-circuit fault-tolerant control system, the difference of this from a normal AMB control system is the inclusion of a fault detection module and a fault-tolerant current controller. When the fault detection module detects a short-circuit fault, the current controller switches from normal mode to fault-tolerant mode, thus achieving fault-tolerant control of the AMB.

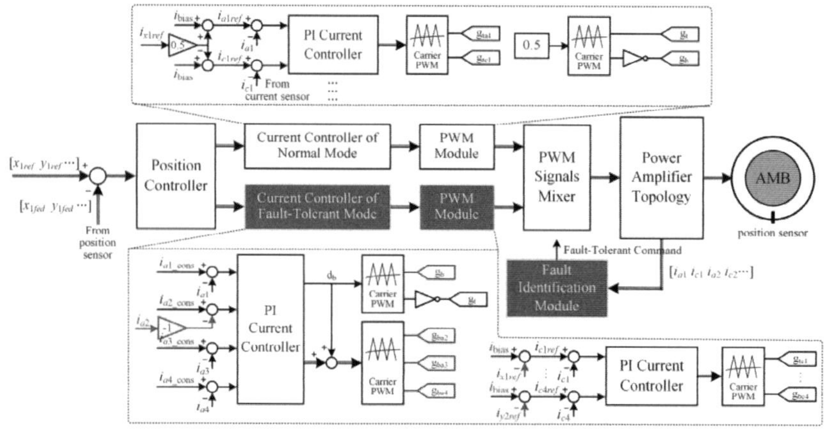

Figure 5 The short-circuit fault-tolerant control system diagram

Table 1 Main experimental parameters

Symbol	*Variable*	*Value*
Touch down bearing gap	g_0	250μm
AMB gap	s_0	500μm
DC voltge	V_{dc}	150V
Bias current	I_{bias}	5A
Switching frequency	f_s	20kHz
Rotor mass	m	5Kg
Coil inductance	L	14mh
Coil resistance	R	0.5Ω
The threshold value	I_{limit1}	11A
The threshold value	I_{limit2}	$V_{dc}/2L$

3. Experimental results

Experiments have been done to validate the effect of the short-circuit fault-tolerant control system, and the main parameters are listed in Table 1. Fig 6 shows the AMB fault-tolerant experimental results under static suspension after short-circuit failure of S_{ta1}, at 0.5s, the g_{ta1} is set to a constant high level to simulate the short-circuit fault of IGBT. It can be found the winding current i_{a1} starts to rise, when the sum of i_{a1} and i_{c1} is larger than i_{limit1} and the rise rate of i_{a1} is larger than i_{limit2}, the fault identification module identifies that fault, then sends a signal to switch the topology and change the mode of current controller. Fig.6 (a) and (b) indicates that after the short-circuit fault, the g_{ta2} is set to a constant low level, the bottom IGBT of L_{a2} starts to work, and the i_{a2} begins to flow in reverse. It can be also found that the winding current of the faulty shared-topology are controlled to a constant current, and the normal shared-bridge topology generates both bias current and control current like i_{c1}. Finally, fig.6 (c) shows that the rotor position returns to stable suspension after a transient fluctuation.

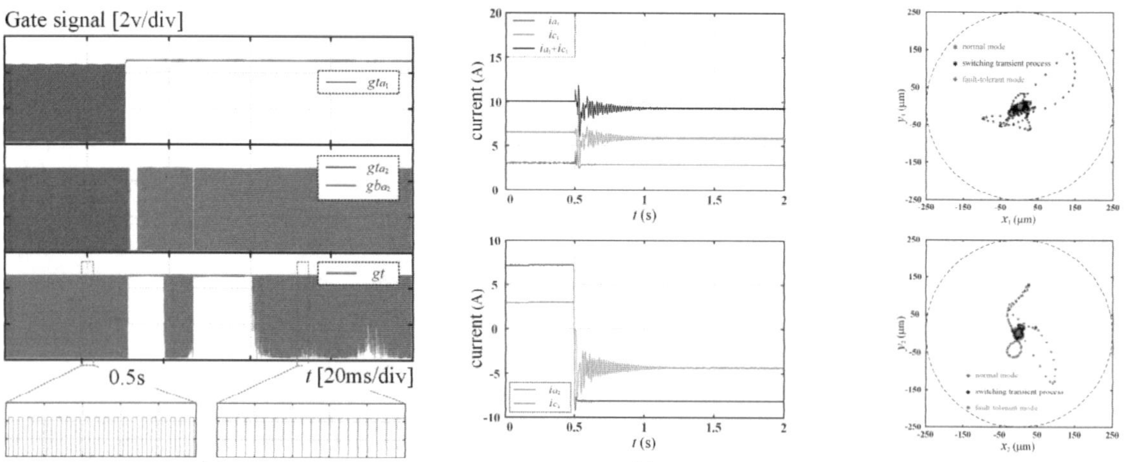

(a) *(b)* *(c)*

Figure 6 The short-circuit fault-tolerant experimental results under static suspension. (a): gate drive signals; (b): coil currents; (c): rotor displacements

Fig 7 shows the AMB fault-tolerant experimental results under rotor rotation after short-circuit failure of S_{ta1}, at 0.5s, the g_{ta1} is set to a constant high level to simulate the short-circuit fault of IGBT, and the relevant experimental results are basically consistent with the Fig.6.

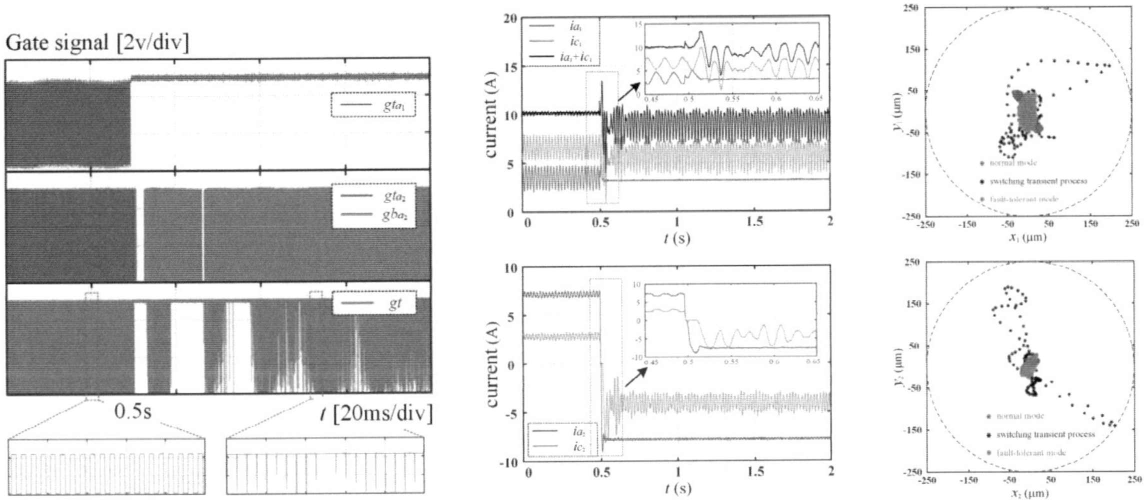

(a) (b) (c)

Figure 7 The short-circuit fault-tolerant experimental results under rotor rotation. (a): gate drive signals; (b): coil currents; (c): rotor displacements

4. Conclusion

This paper proposes a short-circuit fault-tolerant strategy for AMB power electronic controller based on shared-bridge topology, which is compatible with fault tolerance of open-circuit fault. By the way, the variable bias current control after short circuit is worth further study.

References

[1] G. Schweitzer, E.H.Maslen, H. Bleuler, and M. Cole (2010) Magnetic Bearings: Theory, Design, and Application to Rotating Machinery. Berlin, Germany: Springer.

[2] E. H. Maslen, et al. (1995) Fault tolerance of magnetic bearings by generalized bias current linearization. IEEE Transactions on Magnetics 31(3): 2304–2314.

[3] U. J. Na et al. (2000) Fault tolerance of magnetic bearings with material path reluctances and fringing factors. IEEE Transactions on Magnetics 36(6): 3939–3946.

[4] I. S. Cade et al. (2005) Fault identification in rotor/magnetic bearing systems using discrete time wavelet coefficients. IEEE/ASME Transactions on. Mechatronics 10(6): 648-657.

[5] X. Cheng et al. (2021) Design and implementation of a fault-tolerant magnetic bearing control system combined with a novel fault-diagnosis of actuators. IEEE Access 9: 2454-2465.

[6] Yang S, et al. (2011) An industry-based survey of reliability in power electronic converters. IEEE Transactions on industry applications 47(3): 1441-1451.

[7] A. Schulz, et al. (2005) A Reliable Switching Amplifier driving an Active Magnetic Bearing - Experimental Results. 2005 IEEE International Conference on Industrial Technology, Hong Kong, China: 389-394.

[8] Ding J, et al. (2022) Open-circuit fault-tolerant control and topology switching strategy of active magnetic bearing power electronics controller. IEEE Transactions on Power Electronics 37(9): 10928-10938.

Ten-Phase-Nine-Leg Topology for Five-Axis Active Magnetic Bearing Drive

Yixuan SHUAI [1], Dong JIANG [1], Zicheng LIU [1], Feng HU [1], Mingqu ZHOU [1]

1 Huazhong University of Science and Technology, No.1037, Luoyu Road, Hongshan District, Wuhan, China, yxshuai@hust.edu.cn

Abstract

Active magnetic bearing plays an essential role in high-speed rotational machine. Due to the need for multiple coils, optimal control performance necessitates power amplifier with numerous power electronics switches. However, a significant amount of power electronics switches greatly increases cost, volume and risk of switch failure. Therefore, this paper proposes a novel topology for optimizing the number of switches without affecting control performance. The basic properties and control strategies of the Ten-Phase-Nine-Leg topology are analyzed in Section 2. In simulation, results proved the feasibility of this topology, shown in Section 3. Finally, Section 4 concludes this article.

Keywords: *power amplifier, power electronics switches, Ten-Phase-Nine-Leg topology.*

1. Introduction

With the merit of no mechanical contact, no lubrication requirement, active magnetic bearing (AMB) has become the best solution to replace traditional mechanical bearings in high-speed rotating machinery [1]. As the core of AMB, power amplifier has more than ten switches to achieve AMB drive.

Half-H topology is the first type which uses independent H-bridge to control each winding. It requires a large number of power devices, leading to large controller, more loss and greater risk of switch failure [2]. Therefore, many experts have studied the topology based on switch multiplexing, like three-half-phases-legs, shared-bridge, reversed shared-bridge, series-end winding [3-5]. The above topologies at least need 11 legs to drive 5-axis AMB, which is still quite a lot. There is a basic requirement for topology design, that is, bias current of all magnetic bearing windings that control the radial are equal. Based on the above characteristic, four-phase-four-leg topology, shown in Figure 1(a), was introduced to drive four-axis AMB, which achieves the least switches usage [6]. However, this topology is limited by radial windings drive. For this reason, the most common type, five-axis AMB could not utilize such an imperfect scheme.

Therefore, also taking advantage of radial windings bias current equality, this article presents a ten-phase-nine-leg topology, shown in Figure 1(b). This topology achieves the minimum switches to drive five-axis AMB. Section 2 analyzes the principles of the proposed topology. The simulation results are displayed in Section 3. Finally, Section 4 concludes this article.

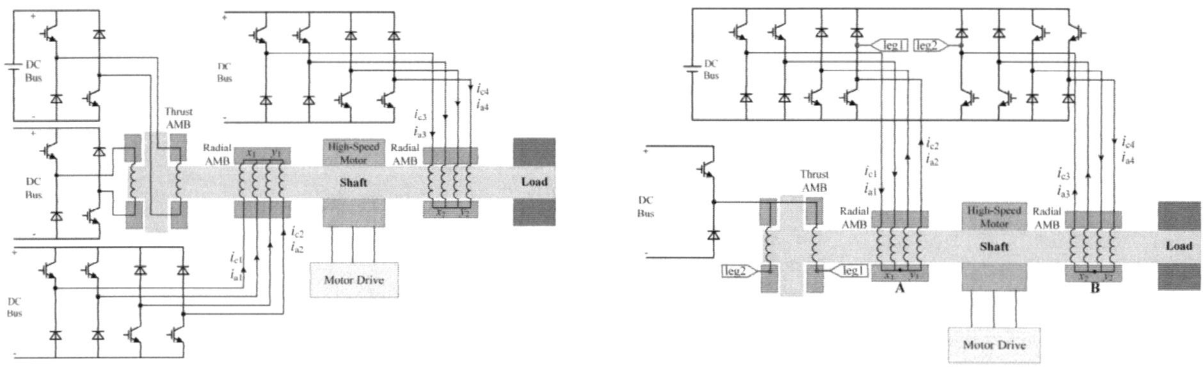

(a) (b)

Figure 1 Comparison of four-phase-four-leg topology(a) and ten-phase-nine-leg topology(b).

2. Principles of Ten-Phase Nine-Leg Topology

As shown is Figure 1, it is not difficult to see ten-phase-nine-leg topology also using feature of radial windings bias current. The point A and B have the KCL equations, expressed as:

$$\begin{cases} i_{a1} + i_{c1} = i_{a2} + i_{c2} = 2I_{bia} \\ i_{a3} + i_{c3} = i_{a4} + i_{c4} = 2I_{bia} \end{cases} \quad (1)$$

The overall diagram is shown in Figure 2.

It is worth mentioning that modulation of the topology is the key to drive. The procedure of multi-leg modulation is:

1) According to the output of the current loop and the shape of this topology, list the KVL equations, expressed as:

$$\begin{cases} u_1 - u_2 = u_{a1} - u_{c1} \quad u_2 - u_3 = u_{c1} + u_{a2} \quad u_3 - u_4 = -u_{a2} + u_{c2} \\ u_4 - u_5 = -u_{a5} \quad u_5 - u_6 = u_{c5} \\ u_6 - u_7 = -u_{a3} + u_{c3} \quad u_7 - u_8 = -u_{c3} - u_{a4} \quad u_8 - u_9 = u_{a4} - u_{c4} \end{cases} \quad (2)$$

2) With special modulation limit, like maximum DC voltage utilization or minimum switch loss, duty ratio of each leg can be obtained.

3) Through carrier-based PWM, PWM generated to control the switches.

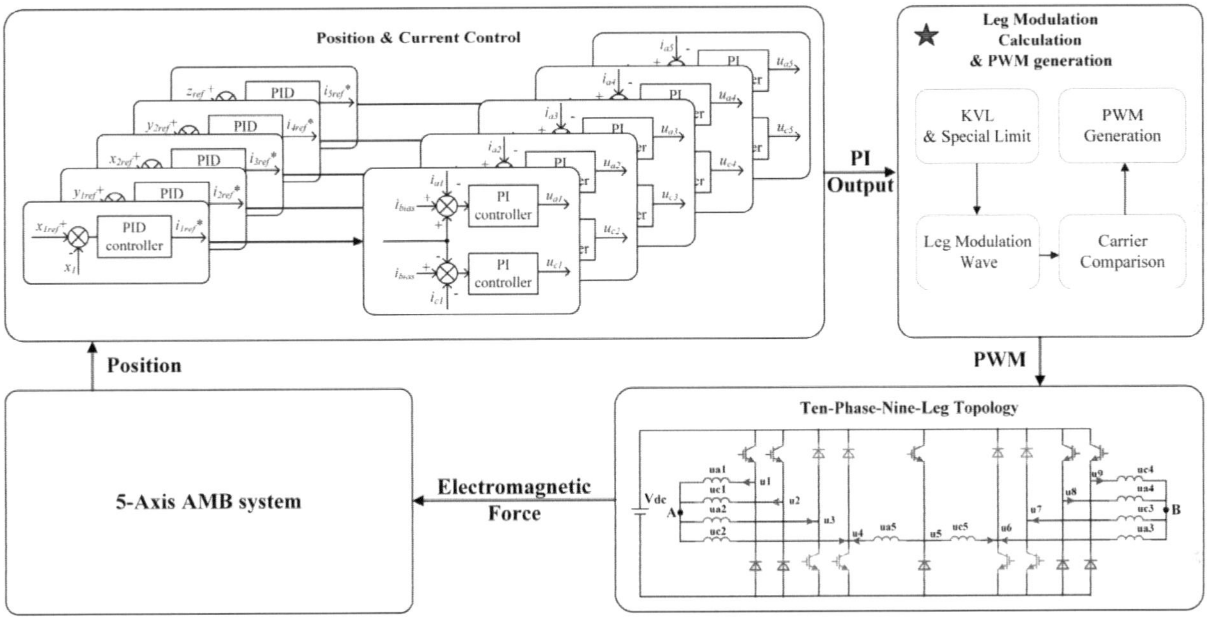

Figure 2 Overall diagram of ten-phase-nine-leg topology drive.

3. Simulation

To verify the proposed topology and performance of its drive, a simulation model was built, whose parameters are listed in Table 1. The following contains the waveforms of 5-axis currents and displacements under static suspension and high-speed rotation (12000 r/min) conditions.

Table 1 Simulation Parameters

	Parameter	Value
m	Rotor mass	10 kg
C_0	Airgap length	0.5 mm
I_0	Bias current	5 A
e	Eccentric distance	0.02 mm
f	Revolutions Per Minute	0~12000 r/min
L	Inductance	5 mH
R	Resistance	0.5 Ω
Vdc	DC-bus voltage	150 V

Figure 3(a) and Figure 3(b) respectively presented the currents under static and high-speed mode. In the figure, currents in all windings can be precisely controlled.

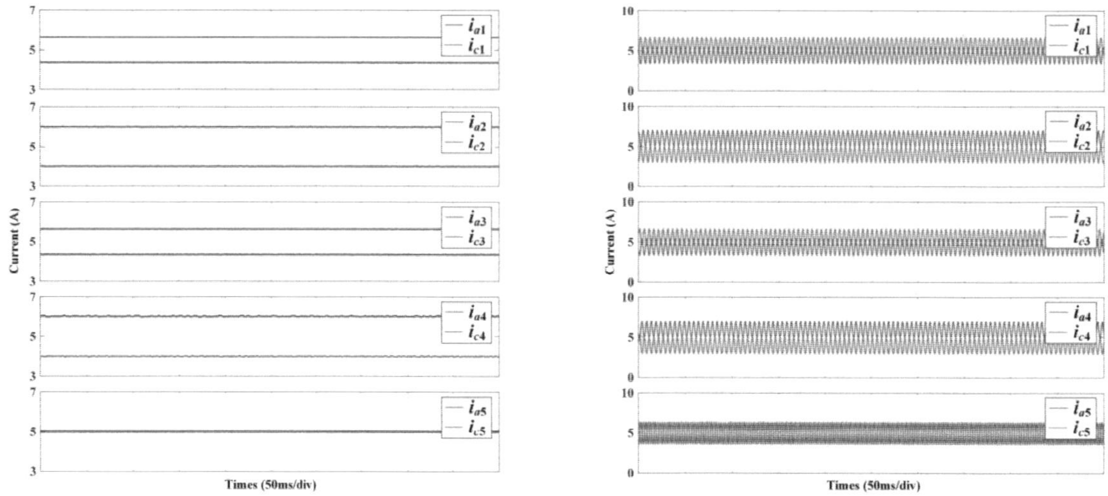

(a) (b)

Figure 3 Simulation results of currents in static and high-speed condition.

Figure 4(a) and Figure 4(b) respectively presented the positions under static and high-speed mode, which proved this topology has the ability to drive AMB in most applications.

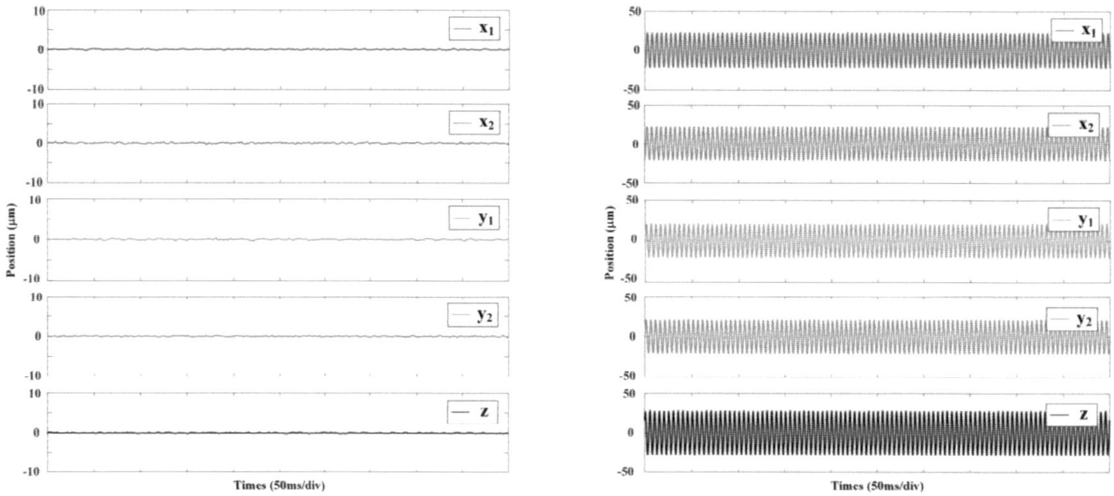

(a) (b)

Figure 4 Simulation results of displacements in static and high-speed condition.

4. Conclusions

In this article, a novel ten-phase-nine-leg topology is proposed to drive five-axis AMB. With only 9 legs, this topology significantly decreases the switches and increases controller power density to some extent. Simulations and results are shown that this scheme has excellent performance with precise displacement control.

References

[1] Denk J, et al. (2013) Industrialization of AMB Systems With Standard Drive Technology. IEEE Transactions on Industry Applications 49(2): 791-798.

[2] Wang J and Xu L (2009) Analysis and modeling of a switching power amplifier for magnetic bearing. In: 4th IEEE Conference on Industrial Electronics and Applications, May 25, 2009, pp. 2257-2261. IEEE.

[3] Jiang D and Kshirsagar P (2017) Analysis and control of a novel power electronics converter for active

magnetic bearing drive. IEEE Transactions on Industry Applications 53(3): 2222-2232.

[4] Jiang D, et al. (2020) Novel topologies of power electronics converter as active magnetic bearing drive. IEEE Transactions on Industrial Electronics 67(2): 950-959.

[5] Yang J, et al. (2021) Series-Winding Topology Converter for Active Magnetic Bearing Drive. IEEE Transactions on Industrial Electronics 68(12): 11772-11782.

[6] Hu Z, Jiang D and Sun H (2020) A Four-Phase Four-Leg Power Electronics Converter for Active Magnetic Bearing Drive. TRANSACTIONS OF CHINA ELECTROTECHNICAL SOCIETY 35(20): 4325-4335.

Less Losses in Active Magnetic Bearings

Bert-Uwe KÖHLER[a], Matthias LANG[b]
a HTW Berlin, Wilhelminenhofstr 75A, 12459 Berlin, bert-uwe.koehler@htw-berlin.de,
b DLR, Lieberoser Str. 13A, 03046 Cottbus, matthias1.lang@dlr.de

Abstract

An optimization method and a design for a reduction of eddy current losses in an active magnetic bearing (AMB) are presented. Since eddy current losses are a major source of energy dissipation in an AMB, their reduction may increase efficiency and operating range of the AMB, reduce cooling efforts and hence lower manufacturing and operational costs. With the proposed optimization, a reduction of eddy current losses of about 30 per cent is achievable.

Keywords: *Active Magnetic Bearings, Losses, Eddy Current, Optimization, Stator Design*

1. Introduction

In this paper, a method and a design example are presented that aim to reduce losses in an active magnetic bearing (AMB). Emphasis is put on demonstrating how the geometry of an AMB stator can be optimized in order to minimize the generation of eddy currents which are a major source of heat in the rotor part of the AMB. Reducing eddy currents and related losses may increase efficiency and operating range of the AMB and reduce cooling efforts as well as manufacturing and operational costs.

A typical heteropolar AMB design comprises a stator configured to provide a magnetic flux density and a rotor part configured to be magnetically attracted by the stator via the magnetic flux density. Stator and rotor are separated by an air gap. Based on Faraday's law of induction, the generation of eddy currents depends on the time derivative of the magnetic flux density acting upon the rotor, which usually consists of an electrically conductive material. In a heteropolar magnetic bearing, the magnetic flux density, as provided by the stator, generally includes a number of opposite polarization zones along an inner circumference of the stator. Therefore, during rotation, a point on the rotor usually passes a number of different polarization zones causing a number of more or less abrupt changes of the magnetic flux density in the point of the rotor and hence an induction of eddy currents. Details about AMB hardware and the generation of eddy current losses can be found in [1] and [2].

To reduce eddy current losses, it is commonly practiced to laminate the rotor portion of the AMB [2]. In addition to laminating the rotor (or at least a portion of the rotor), the method proposed in this paper aims at optimizing the AMB stator geometry with respect to eddy current losses in the rotor while maintaining, as an optimization constraint, a required minimum AMB load capacity. As it is shown in the sequel, a reduction of eddy current losses of about 30 per cent is achievable at a typical operating speed of a large machine.

2. Methods

2.1 Outline of the optimization procedure

The objective of the proposed optimization is to find a stator geometry that reduces the generation of eddy currents in a rotating rotor to a minimum.

The optimization procedure comprises the steps

a) defining the stator geometry as a function of stator geometry parameters and generation of a finite element model (FEM);
b) computing the value of a cost function for a given set of stator geometry parameters, wherein the cost function maps estimated eddy current losses to a scalar value;
c) varying the stator geometry parameters and repeating step (b) until a minimum of the cost function is reached.

The variation of the stator geometry parameters, can be performed with standard optimization methods, for example, sequential quadratic programming (SQP) and/or evolutional/genetic algorithms. Steps (a) and (b) are explained in the sequel. Optimization methods are available in numerical software packages, for example MATLAB [8].

2.2 Defining the stator geometry with geometry parameters and FEM modeling

For optimizing the AMB stator geometry it is, in a first step, required to establish a model of the AMB's magnetic circuit including a parametric model of the AMB stator. Model parameters determine at least a part of the geometric shape and material of a stator portion to be optimized and need to be accessible by an optimization routine.

In Figure 1a, an example of a part of an AMB's magnetic circuit is depicted, that is, a cross-sectional view of a quadrant of an AMB's rotor (1) and stator (2). The AMB stator (2) in Figure 1a comprises a nose (5) which is present in some AMB designs and which may be optimized to achieve a minimum of eddy current losses in the rotor.

The shape of the nose (5) may, for example, be modeled with a mathematical function such as a polynomial or a segment wise linear function or in some other parametrizable way. Parameters of the mathematical function serve as optimization parameters. Figure 1b shows an example for a polynomial and for a segment wise linear function.

The modeled AMB's magnetic circuit including stator and rotor geometry is subsequently used to establish a FEM model being used to calculate the magnetic flux density in the air gap (3). Based on the calculated magnetic flux density, the estimation of eddy currents (step b of the optimization procedure) is performed. The actual computation of the magnetic flux density can be performed with FEM software. In our investigation we used FEMM4.2 [7].

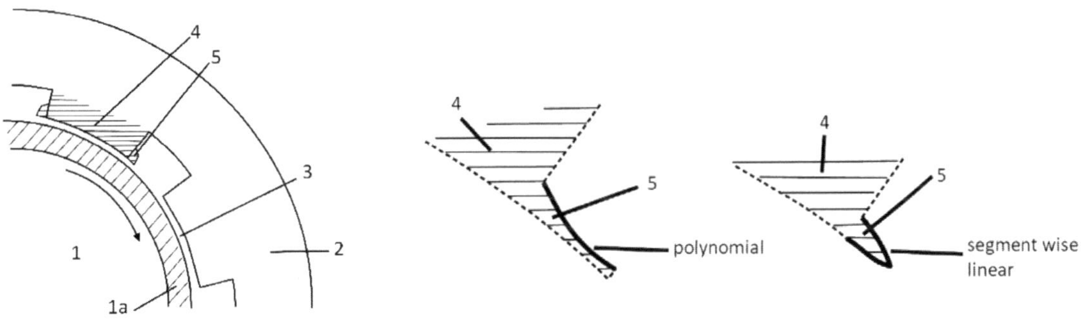

Figure 1a (left): Cross section of a non-optimized AMB: 1: rotor, 1a: rotor lamination; 2: stator, 3: air gap, 4: pole tooth; 5: nose; Figure 1b (right): parametrizations of the nose (5) - with a polynomial function or segment wise linear

2.3 Estimation of eddy current losses

The estimaton of eddy current losses has been investigated by several authors, for example [3,4,5,6]. Therein, it has been shown that eddy current losses depend on the time derivative of the magnetic flux density. Applied to the AMB setup as shown in Figure 1a, it is sufficient to consider only the normal component of the magnetic flux density. The term 'normal component of the magnetic flux density' refers to the magnetic flux density in the air gap in a direction orthogonal to the rotor surface which is often denoted as radial direction.

Great values of the time derivative of the normal component of the magnetic flux density may be experienced by a point on the rotor, when, during rotor rotation, the point on the rotor passes zones of different absolute values and/or polarizations of the magnetic flux density, wherein the local absolute value of the magnetic flux density depends on the local geometry of the magnetic circuit at the inner circumference of the AMB stator. In general, during rotor rotation, a change of the magnetic flux density as experienced by a point on the rotor may be rather abrupt.

When the circumferential course of the radial component of the magnetic flux density is transformed into Fourier domain, such abrupt changes, i.e. high values of the time derivative, correspond to large high frequency Fourier coefficients $F(jk)$. Hence, the generation of eddy current losses largely corresponds to these high frequency Fourier coefficients of the circumferential course of the radial component of the magnetic flux density.

In our approach we use an estimation formula for eddy current losses based on a method discussed in [4] where the Discrete Fourier Transform (DFT) of the circumferential course of the radial component of the magnetic flux density is applied as

$$Loss = \sum_{k=1}^{K} m_a \left| \frac{2 \cdot F(jk)}{1.5\ T} \right|^2 \cdot \left[\left(\frac{f_k}{50\ Hz} \right)^2 \cdot \sigma_{w,50} + \left(\frac{f_k}{50\ Hz} \right) \cdot \sigma_{h,50} \right], \tag{1}$$

wherein m_a is the mass of the relevant rotor portion, $f_k = k \cdot N/60$ (in Hz) with N as the rotational speed (in rpm), the frequency component $F(jk)$ is the kth Fourier coefficient (frequency component) of the circumferential course of the radial component of the magnetic flux density in the air gap, K is the total number of Fourier coefficients considered (with $\max\{K\} = \text{floor}(N_s/2)$ and N_s as the number of samples of the circumferential course of the magnetic flux density used for the Fourier transform), σ_w is the eddy current loss coefficient and σ_h is the hysteresis coefficient - both deducible from manufacturer's data sheets. The latter coefficients are material properties often determined for an excitation frequency of 50 Hz.

Equation (1) provides a quantitative expression of how the circumferential course of the radial component of the magnetic flux density is transformed into eddy current losses during rotation of the rotor. Since the circumferential course of the magnetic flux density depends on the AMB geometry, equation (1) can be used as a cost function in the optimization of the AMB's geometry.

As an optimization result, abrupt changes of the radial component of the magnetic flux density, and correspondingly high frequency Fourier coefficients $F(jk)$, are expected to be attenuated.

3. Verification Setup

The effectiveness of the proposed method is computationally verified for an AMB of a large machine with a nominal power of about 5 MW. The considered AMB has a load capacity of about 9 kN. Its stator has a cross-sectional diameter of about 400 mm. The finite element modeling is performed in two dimensions (2D) using [7].

The stator and rotor geometry are generated using the MATLAB interface of [7], wherein the nose of the stator is parametrized as a segment wise linear function. With [7], the circumferential course of the radial component of the magnetic flux density in the air gap, in total $N_S = 1500$ samples, is calculated in every iteration of the optimization and, based thereon, the eddy current losses are calculated using equation (1).

The actual optimization – step (c) of the optimization procedure – is performed using SQP-optimization of MATLAB's Optimization Toolbox [8] under the constraint of a minimum load capacity of 9 kN. The optimization is performed for a rotor speed of 6000 rpm.

4. Verification Results

Optimization result are shown in Figures 2a and 2b. Figure 2a shows the optimization result for the shape of the nose (5) modeled as a segment wise linear function. For the given setup, the shape of the nose (5) produces minimum eddy current losses in the rotor. With respect to the nature of the optimization algorithm, the identified minimum cannot be guaranteed to be a global minimum.

In Figure 2b, the curves of the radial component of the magnetic flux density along the rotor circumference are depicted for the optimized and the non-optimized case (in the non-optimized case the nose (5) is entirely omitted). The optimized curve (solid line) has a much smoother course compared to the course (dotted line) of the non-optimized nose indicating that, in the Fourier domain, the amplitudes of the high frequency Fourier coefficients of the optimized nose are attenuated. The load capacity of the optimized case is 96.4 per cent of the non-optimized one.

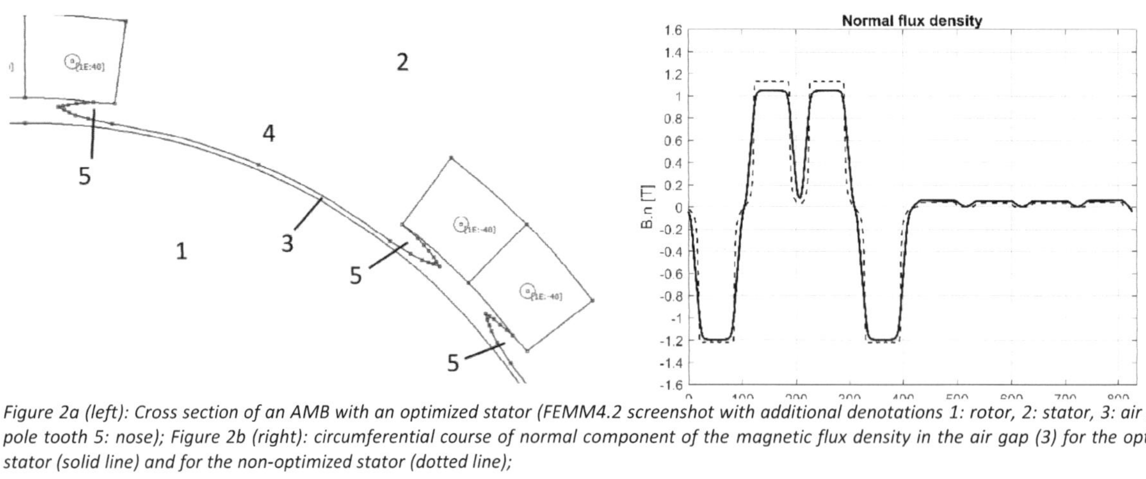

Figure 2a (left): Cross section of an AMB with an optimized stator (FEMM4.2 screenshot with additional denotations 1: rotor, 2: stator, 3: air gap, 4: pole tooth 5: nose); Figure 2b (right): circumferential course of normal component of the magnetic flux density in the air gap (3) for the optimized stator (solid line) and for the non-optimized stator (dotted line);

Figure 3a depicts the results of the verification in the Fourier domain. The dotted curve shows the absolute values of the Fourier coefficients up to frequency component $k = 300$ for the non-optimized case, the solid curve shows the corresponding values for the optimized case. The high frequency Fourier coefficients are clearly less for the optimized stator than for the non-optimized stator. The low frequency Fourier coefficients are in about the same range.

Figure 3b depicts the corresponding cumulated loss curves as a function of the frequency component and based on equation (1). Whereas the non-optimized Fourier coefficients contribute to the losses until approximately frequency component $k = 150$, in the optimized case there is no significant contribution above frequency component $k = 40$. For the assumed rotor speed of 6000 rpm, the eddy current losses are reduced by about 27 per cent. This value even increases for higher speeds: for example, the loss reduction for 12000 rpm (not depicted) is about 32 per cent.

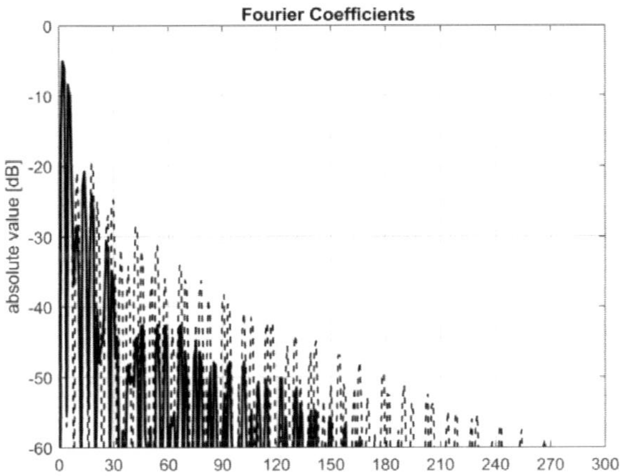

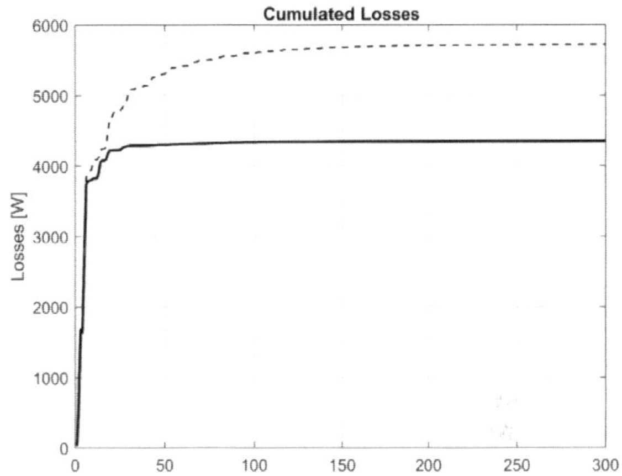

Figure 3a (left): Normal component of the static magnetic flux density along the rotor circumference optimized (solid line) and non-optimized (dotted line); Figure 3b (right): estimates of cumulated losses vs. frequency component optimized (solid line) and non-optimized (dotted line) – estimates for rotor speed 6000rpm using equation (1)

5. Discussion

The verification results clearly show that a reduction of eddy current losses is possible with an optimized stator design. For a typical rotational speed of a high-speed machine, the reduction of losses can reach values of about 30 per cent. The reduction of losses is due to a, compared to the non-optimized stator, smoother circumferential course of the radial magnetic flux density. The smooth circumferential course corresponds to a reduction of high frequency Fourier coefficients. Low frequency Fourier coefficients have values similar to the non-optimized case. This can be explained by the constraint with respect to a required minimum load capacity.

Looking at the proposed solution for the shape of the nose of the pole tooth there may be concerns regarding the mechanical stability of the optimized design. A possible solution for an improved mechanical stability may include a further, non-magnetic mechanical support in order to take up mechanical forces as depicted in Figure 4a (reference sign (6)).

Regarding the actual objective of the method, i.e. to make the circumferential course of the radial component of the magnetic flux density smoother, further optimization strategies targeting the same objective can be found. For example, a smoothing effect may also be achieved by implementing a permeability profile within the pole tooth. This might be implementable, for example, by means of 3D printing. An example for 3 permeability zones in a pole tooth is depicted in Figure 4b. In general, there could be more than 3 permeability zones and even a continuous transition or a continuous permeability gradient are conceivable.

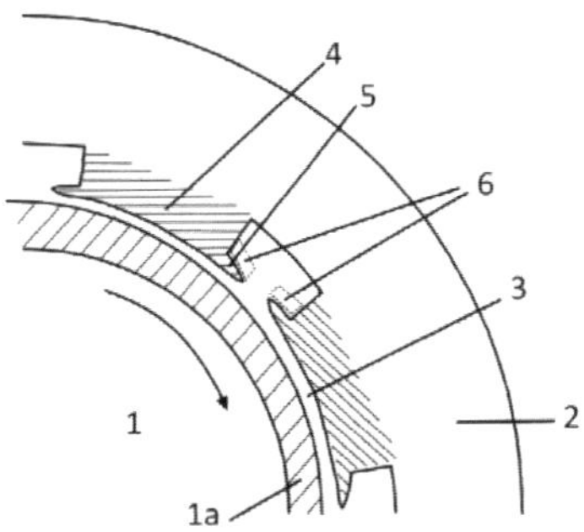

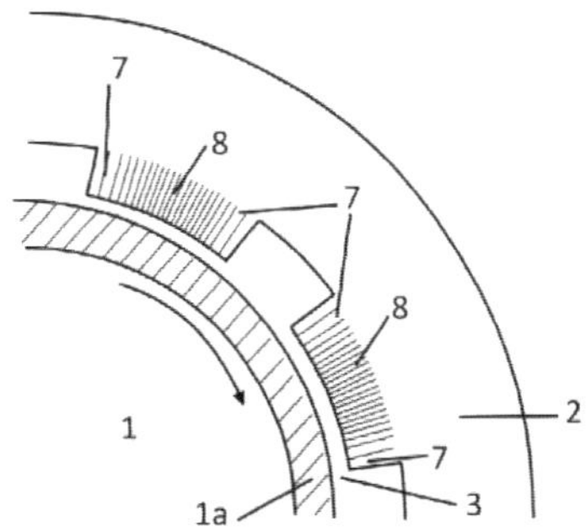

Figure 4a (left): Optimized AMB stator similar to Fig. 1b with additional mechanical support (6); Figure 4b (right): example for pole teeth with zones of different permeability – low permeability (7) and high permeability (8).

Another point to discuss is the estimation of the eddy current losses. Although there seems to be a general consent about impact factors, physics and methodology how eddy currents are to be estimated, there is currently, at least to the knowledge of the authors, no unique and commonly agreed estimation formula. So, even though the general trend should be independent from the actual formula being used, there may be differences with respect to the concrete results. In this sense, the proposed method for optimizing the magnetic bearing stator should be seen as a tool that needs to be adopted to the actual situation and design.

Furthermore, two examples of how a stator parametrization can be carried out are given in this paper. Other mathematical parametrizations may be available and may lead to different results.

6. Conclusion

The objective of the method proposed in this paper is to reduce eddy current losses in an AMB rotor. The applied strategy is to optimize the AMB stator geometry in order to make the circumferential course of the radial component of the magnetic flux density in the air gap smoother and to avoid, or at least to attenuate, abrupt changes of the radial component of the magnetic flux density for a point on the rotor moving in circumferential direction. The verification results have shown that the objective can be achieved by optimizing the shape of an AMB's pole tooth.

Depending on rotor speed, a significant reduction of eddy current related losses is possible. For a typical rotor speed of a large high-speed machine, a reduction by about 30 per cent is achievable.

The presented method may be generalized to the question about how an optimal AMB structure and/or what an optimal material (parameter) distribution within the AMB could look like. That is, which material with which property, for example permeability, must be placed where within the AMB. In this sense, the proposed method shows a path for further AMB optimization.

References

[1] A. Traxler, E.H. Maslen: Hardware Components. In: G. Schweitzer, E.H. Maslen (eds): Magnetic Bearings, Springer, Berlin, 2009

[2] A. Traxler: Losses in Magnetic Bearings. In: G. Schweitzer, E.H. Maslen (eds): Magnetic Bearings, Springer, Berlin, 2009

[3] P. Steinmetz: On the Law of Hysteresis, Transactions of the American Institute of Electrical Engineers, Vol. IX, No. 1, New York, 1892

[4] J. Reinert et al.: Calculation of Losses in Ferro- and Ferrimagnetic Materials Based on the Modified Steinmetz Equation, IEEE Transactions on Industry Applications, Vol. 37, No. 4, July/August 2001, pp. 1055-1061

[5] G. Bertotti: General Properties of Power Losses in Soft Ferromagnetic Materials, IEEEE Transactions on Magnetics, Vol. 24, No. 1, January 1988, pp. 621-630.

[6] G. Müller, K. Vogt, B. Ponick: Berechnung elektrischer Maschinen: Elektrische Maschinen 2, Wiley, 2007

[7] D. Meeker: Finite Element Method Magnetics, FEMM 4.2, available from https://www.femm.info/wiki/HomePage

[8] MATLAB, The MathWorks Inc., https://de.mathworks.com/products/matlab.html

A Series-Winding Topology Converter for the Magnetic Field Superimposed Magnetic Bearing

Jianfu DING, Dong JIANG, Zicheng LIU
Huazhong University of Science and Technology, No.1037, Luoyu Road, Hongshan District, Wuhan, China, djf@hust.edu.cn

Abstract

The power electronic drive is an essential component of a magnetic bearing system. It can rapidly change the current of the windings, thus controlling the dynamic changes of the electromagnetic force. This paper proposes a novel magnetic bearing power electronic drive based on a series-winding topology mainly used for magnetic field superimposed magnetic bearings. This topology can control the windings' bidirectional current and can be extended to an N-degree-of-freedom system with only 2N+2 switching devices. Compared to the conventional topology, this topology can reduce the cost of the drive, ensuring the current control capability. Moreover, this topology structure can evenly distribute the current of each winding between the bridge legs, reducing the current stress of the bridge legs compared to the common bridge-leg structure. The basic properties and control strategies of the series-winding topology are analyzed. Experiment and work verify the availability of the topology.

Keywords: *Magnetic Bearing, Series-Winding Topology, Power electronic drives*

1. Introduction

Magnetic bearing (MB) is a key technology of the high-speed rotational machine. The power electronic drive is an essential part of the MB system. Its performance directly determines the control performance of the suspension system and has a decisive impact on the control accuracy and maximum speed [1].

According to the output current direction, the topology of the MB switching power electronic drive can be divided into two categories: unipolar and bipolar. The output of the former is a single-direction flow, which is used to supply power to the current superimposed MB, while the latter can be a two-way flow, which is used to supply power to the magnetic field superimposed MB [2]. Many scholars have worked on unipolar switching power amplifiers [3]. For bipolar topology, the full-bridge circuit is the most widely used. However, each load current needs four switches for regulation, which is costly.

To solve this problem, scholars proposed a three-phase four-leg topology [4] and a five-phase six-leg topology [5]. The five-phase six-leg topology is shown in figure 1(a). The MB winding current is controlled by way of bridge leg multiplexing, which greatly reduces the cost. However, because the current stress carried by the common bridge leg results from the superposition of other bridge legs, the stress of the common bridge leg is too large, and the reliability and safety are reduced. Furthermore, [6] proposed a unipolar series-winding control topology to increase voltage utilization, ensuring better current control capability.

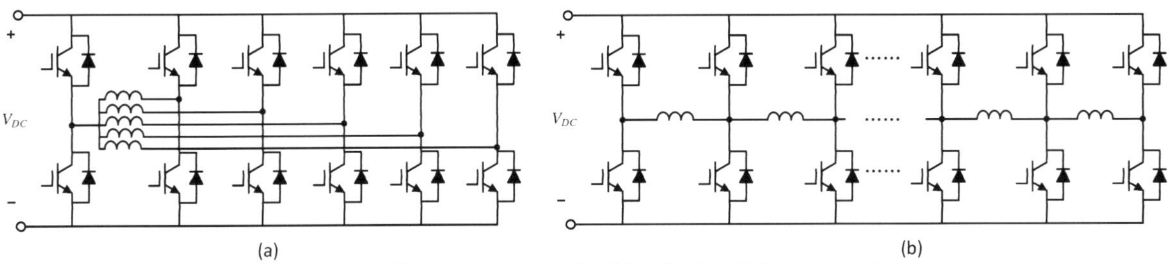

Figure 1 Five-phase six-leg topology(a) and series-winding topology (b).

The series-winding topology proposed in this paper is shown in figure 1(b), which can realize the control of bipolar current and evenly distribute the current stress of the bridge legs. The theory analysis and control method of the series-winding topology are presented in Section 2. To verify the availability of the series-winding topology, the experiments are carried out on the AMB test rig in Section 3. Conclusions are summarized in Section 4.

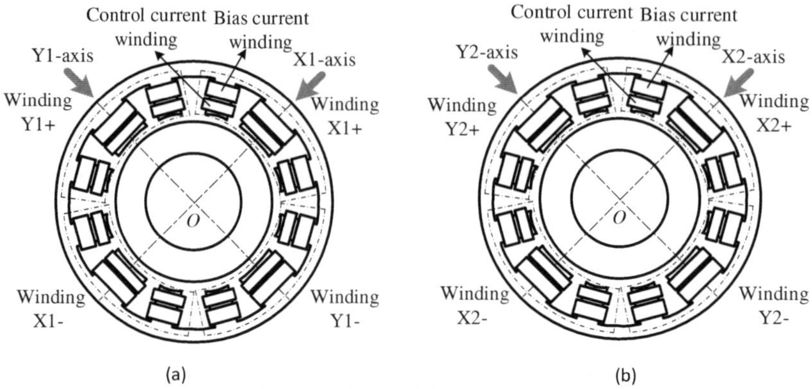

Figure 2 Front end magnetic bearing(a) and rear end magnetic bearing (b).

Figure 2 shows the magnetic bearing test rigs' winding structure and pole distribution. The whole system obtains two radial magnetic bearings. The magnetic bearings are 12-pole radial magnetic bearings. Each side of one axis consists of three poles that stagger 45 degrees to provide electromagnetic force together. Both bias and control windings are on each side of the magnetic poles. All the bias windings are connected in series and supplied with a constant bias current from an additional source. To improve the response speed of the control winding current, two control windings opposite to each other on the same axis are connected in reverse parallel.

2. Control method of Series-winding topology

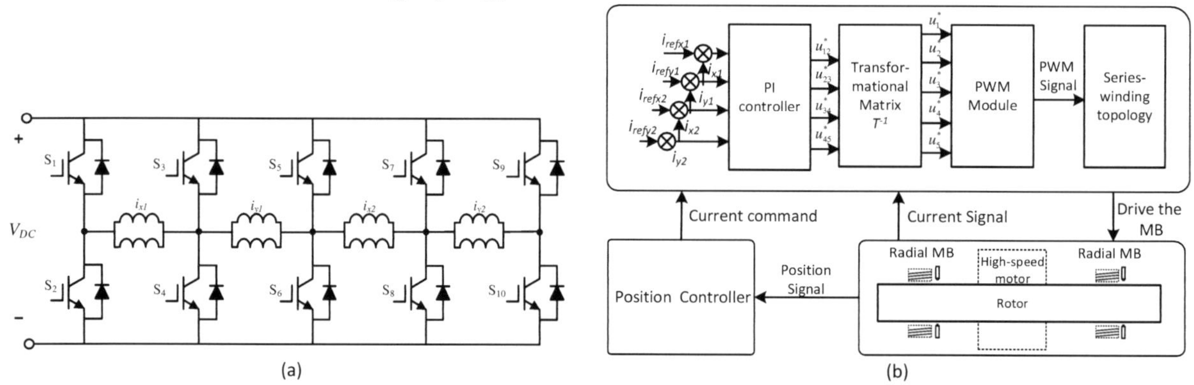

Figure 3 Series-Winding topology(a) and overall diagram of the MB system (b).

As shown in figure 3, only the differential mode current component of the MB needs to be controlled. The upper and lower switching devices of each bridge leg are conducted complementarily. The switching mode s_i = 1(i=1,2,3,4,5) is defined when the upper switching device is on and the lower switching device is off. Moreover, the upper switching device is off when the lower switching device is on, at which mode s_i = 0. So for each pair of control windings, the voltage modes at both ends include V_{DC}, 0, - V_{DC}. V_{DC} is the DC link voltage. It is easy to see that the current stress in the bridge legs at the two ends of the topology is the maximum value of the control current, and the current stress in the middle part of the bridge arm is up to twice the maximum value of the control current. Therefore, the current stress of the switching device is evenly distributed in each bridge leg compared to the common bridge leg structure.

Using pulse width modulated signals to drive switching devices. In the case of the average model, the voltage is controlled by the duty cycle of each switch device during a switching cycle. Variables d_1~d_5 represent the duty cycle of each bridge leg, and then the average voltages of each bridge leg midpoint u_1~u_5 are expressed as

$$\begin{bmatrix} u_1 & u_2 & u_3 & u_4 & u_5 \end{bmatrix}^T = \begin{bmatrix} d_1 & d_2 & d_3 & d_4 & d_5 \end{bmatrix}^T \times V_{DC} \quad (1)$$

In a control cycle, the relationship between the five winding currents i_{x1}, i_{y1}, i_{x2} and i_{y2}. The potentials u_1, u_2, u_3, u_4 and u_5 of each node are expressed in (2), where Z is the winding impedance.

$$i_{x1} = \frac{u_1 - u_2}{Z}, i_{y1} = \frac{u_2 - u_3}{Z}, i_{x2} = \frac{u_3 - u_4}{Z}, i_{y2} = \frac{u_4 - u_5}{Z} \quad (2)$$

Equation (2) can also be expressed in the form of a matrix

$$\begin{bmatrix} i_{x1} \\ i_{y1} \\ i_{x2} \\ i_{y2} \end{bmatrix} = \frac{1}{Z}\begin{bmatrix} 1 & -1 & 0 & 0 & 0 \\ 0 & 1 & -1 & 0 & 0 \\ 0 & 0 & 1 & -1 & 0 \\ 0 & 0 & 0 & 1 & -1 \end{bmatrix}\begin{bmatrix} u_1 \\ u_2 \\ u_3 \\ u_4 \\ u_5 \end{bmatrix} \tag{3}$$

To ensure that the potential at the midpoint of each phase leg is $0.5V_{DC}$ in the steady state, it is limited in (4)

$$0.5V_{DC} = \frac{u_1 + u_2 + u_3 + u_4 + u_5}{5} \tag{4}$$

The duty cycle of each switch device is controlled according to the instruction value, and then the winding currents are adjusted. Equivalent voltage values across the windings between two bridge arms, the corresponding current times of the impedance Z, are used as instruction voltages. It means that u_{12}^* , u_{23}^* , u_{34}^* and u_{45}^* are the outputs of the PI controller. Superscript "$*$" indicates that the parameter is the instruction value, and it is the per-unit value for the DC bus voltage V_{DC}. Let u_i^* indicates the midpoint voltage of the ith bridge leg. The potential voltage instructions of winding current instructions are obtained according to

$$\begin{bmatrix} u_{12}^* \\ u_{23}^* \\ u_{34}^* \\ u_{45}^* \\ 2.5 \end{bmatrix} = \underbrace{\begin{bmatrix} 1 & -1 & 0 & 0 & 0 \\ 0 & 1 & -1 & 0 & 0 \\ 0 & 0 & 1 & -1 & 0 \\ 0 & 0 & 0 & 1 & -1 \\ 1 & 1 & 1 & 1 & 1 \end{bmatrix}}_{\boldsymbol{T}}\begin{bmatrix} u_1^* \\ u_2^* \\ u_3^* \\ u_4^* \\ u_5^* \end{bmatrix} \tag{5}$$

The transformation matrix in Equation (5) is denoted $\boldsymbol{T}$. By solving for the inverse matrix $\boldsymbol{T}^{-1}$ of $\boldsymbol{T}$, the transformation relationship from the PI command value to the control voltage of each bridge leg can be obtained, as shown in Equation (6)

$$\begin{bmatrix} u_1^* & u_2^* & u_3^* & u_4^* & u_5^* \end{bmatrix}^T = \boldsymbol{T}^{-1}\begin{bmatrix} u_{12}^* & u_{23}^* & u_{34}^* & u_{45}^* & 2.5 \end{bmatrix}^T \tag{6}$$

The control block diagram for the current loop is shown in figure 3(b). This work illustrates the four degrees of freedom of current control. But this method can also be extended to achieve N-degree-of-freedom current control.

3. Experiment

To verify the feasibility of this series-winding topology, the experiment is carried out under Vdc=150V. The switching frequency is 20kHz. The specific parameters are shown in Table 1, and the experimental platform is shown in figure 5(a). The controller consists of two parts, the control board and the power board, which provide control commands and PWM signals and carry out current and position signal acquisition. The power board includes a power electronic topology and a drive circuit section to control the current after the windings are connected to the power board.

Table 1 Simulation Parameters

	Parameter	***Value***
g_0	Touch down bearing gap	200μm
s_0	AMB gap	400μm
I_0	Bias current	2 A
I_C	Control current	4A
f	Switching frequency	20kHz
Vdc	DC-bus voltage	150 V

A given sinusoidal current command controls the four winding currents. The frequency and amplitude are 200Hz is 3A, respectively. The experiment results of four symmetrical sine waves with given signals are shown in figure 4. The current command value is given at 0.026 seconds. All four control currents track the command signal well.

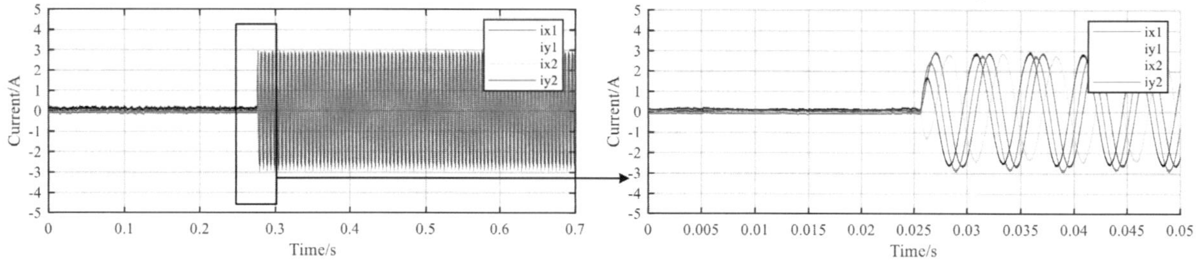

Figure 4 Current loop test results.

Suspension experimental tests were carried out on the experimental platform. Figure 5(b) shows the control winding current in four axes of magnetic bearing and the displacement of the rotor in the X1, Y1, X2 and Y2 axis. At first, the rotor leans against the touch-down bearing, then levitates to the center of the bearing. The suspension displacement fluctuations are controlled to within 20 μm.

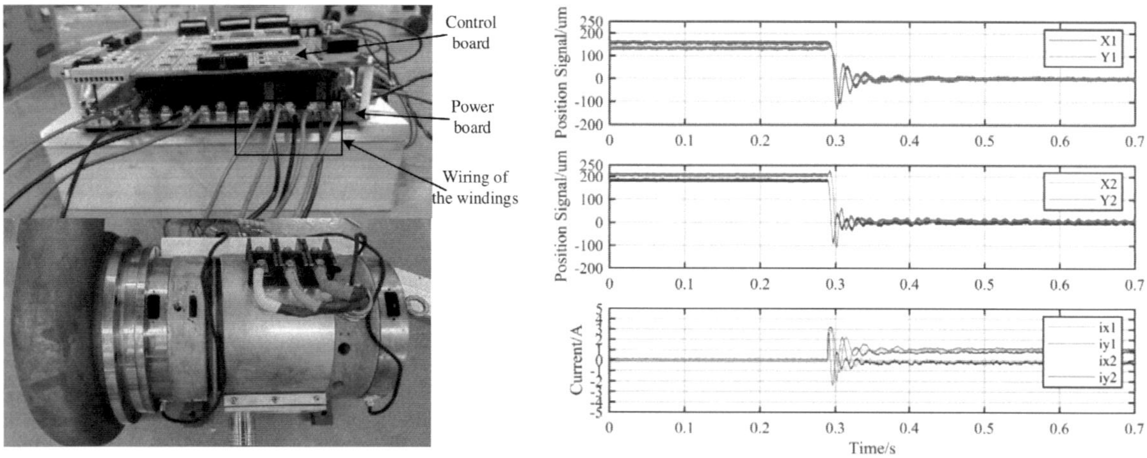

(a) (b)

Figure 5 Series-Winding topology(a) and overall diagram of the MB system (b).

4. Conclusions

This article proposes a novel series-winding topology to drive four-axis MB. With only ten switching devices, this topology significantly decreases the switches and the current stress is evenly distributed. The series-winding topology can extend to N-degree of freedom MB systems for multi-axis control. The experiment shows that this scheme can realize the current control of MB well.

References

[1] Chiba, et al. (2005) Magnetic Bearings and Bearingless Drives. Elsevier Press.

[2] Xie, et al. (1999) Coil-wrapping of Electromagnetic Bearings and Forms of the Power Amplifier. Mechanical Science and Technology (04):103-104.

[3] Jiang, et al. (2020) Novel topologies of power electronics converter as active magnetic bearing drive. IEEE Transactions on Industrial Electronics 67(2): 950-959.

[4] Li, et al. (2011) Analysis and Simplification of Three-Dimensional Space Vector PWM for Three-Phase Four-Leg Inverters. IEEE Transactions on Industrial Electronics 58(2): 450-464.

[5] Fei, et al. (2012) A Control Strategy of Novel Five-phase Six-leg Switching Power Amplifiers Applied in Magnetic Levitating Bearing Systems. Journal of the Chinese Society of Electrical Engineering 32(09):124-131+19.

[6] Yang J, et al. (2021) Series-Winding Topology Converter for Active Magnetic Bearing Drive. IEEE Transactions on Industrial Electronics 68(12): 11772-11782.

Influence of Eddy Current in Permanent Magnet Rotor on Suspension Forces and Experimental Verification in Bearingless Motors with Divided Winding

Masahide OOSHIMA [a], Masaki IHARA [a]

a Suwa University of Science,5000-1, Chino, Nagano, Japan, moshima@rs.sus.ac.jp

Abstract

This paper focuses on the influence of eddy current in the permanent magnet (PM) rotor on the suspension force in the bearingless motors with the divided stator winding. In the PM synchronous bearingless motors, the eddy current flows in the PM on the rotor at high speed drives, especially on its surface of PMs. As a result, the uncontrollable electromagnetic force is generated and interferes on the originally commanded suspension force. It makes the rotor levitation unstable and in the worst case, it is seriously concerned that the rotor stabilization unfortunately is not kept. The interference component is mainly Lorentz Forces generated in the PMs by the eddy current. The bearingless motor with 2-pole motor and 4-pole suspension windings (2t4s Belm) and the bearingless motor with 4-pole motor and 2-pole suspension windings (4t2s Belm) have been targeted and compared between the electromagnetic forces in both Belms by the authors. The eddy current in the PMs and the Lorentz Force have been computed by Finite Element Method (FEM) using a simulation software. The difference between 2t4s and 4t2s Belms in the electromagnetic forces, eddy current and Lorentz Force has been found using the FE models and discussed it in detailed. The reason why the rotor levitation is unstable in the 4t2s Belm model has been analytically derived by the simulation. In this paper, it is experimentally verified by a prototype machine of 4t2s Belm.

Keywords: *Magnetic bearing, Bearingless motor, Permanent magnet synchronous motor, High speed drive, Eddy current*

1. Introduction

In the Belms, the functions of electric motor and magnetic bearing are successfully integrated. Therefore, they have the remarkable advantageous features such as maintenance-free, long life-cycle and the drives in the special circumstance as in vacuum, high or low temperature can be performed. A lot of structures and rotor suspension control strategies have been proposed and the papers related to their technique have been published [1]. The Belms can be grouped into two by the winding configuration. One is the divided winding type, in which the suspension winding is independently wound in the stator core as well as the motor winding. Another is the combined winding type, in which only one kind of winding is wound in the stator core and both the torque and suspension force are generated by the excitation of its winding [2]. In this paper, the Belms with divided winding are focused. It is noted that the suspension winding with (n$\pm$2) pole is necessarily wound on the stator core for the motors with n pole [3]. As a result, the electromagnetic force can be properly generated in any radial direction and the rotor is stably levitated. In this kind of PM synchronous Belms, the difference between the angular speeds of revolving magnetic fields by PM on the rotor and MMF of the suspension winding is significantly caused. Because the pole number of suspension winding is different from that of motor winding. The eddy current in the rotor PMs is generated due to the difference between the angular speeds in accordance with the same principle as induction motors [4]-[6]. The current is induced in the PMs, especially on the surface of them by the eddy current. It results in the Lorentz Force generation and the interference on the originally commanded suspension force is caused. It is seriously concerned that the suspension force is not followed to its command and the rotor levitation is unfortunately unstable. Especially, the induced current is more as the rotational speed is up. Therefore, the rotor stabilization has to be paid attention in the high speed range.

The authors have focused on the 2t4s and 4t2s Belms and compared between the electromagnetic forces in both Belms by FE analyzed results using the above two models. In the 2t4s Belm, as a result, the influence of the eddy current on suspension force is not so much. In the 4t2s Belm, however, the electromagnetic force is much varied with the eddy current, especially in the high speed region. Therefore the rotor levitation is not stabilized. The reason why they are caused has been analytically made clear based on the FE analyzed results using a simulation software [5][6]. In next step, the experimental verification is necessary, in this paper, firstly the influence of eddy current on the electromagnetic force is experimentally verified by a prototype machine of 4t2s Belm. It is confirmed which it agrees with the simulation result or not.

2. Revolving Magnetic Field in Belm

The authors have theoretically derived the relationship between the suspension force and suspension winding current in the 2t4s and 4t2s Belms, respectively [1][6]. The theoretical equations of motor and suspension winding currents have been derived based on the relationship. As a result, the angular speeds of revolving magnetic fields produced by MMF of suspension winding for 2t4s and 4t2s Belms are shown in Table 1. In the 2t4s Belm, the angular speed of revolving magnetic field by MMF of the suspension winding is a half of that of the main revolving motor magnetic field and the minus means that it is slower than the angular speed of main motor revolving motor magnetic field. In the 4t2s Belm, on the other hand, the angular speed of revolving magnetic field by MMF of the suspension winding is twice of that of the motor. Therefore it is possible that the larger eddy current in the PM is generated in 4t2s Belm, especially in high speed region.

Table 1 Angular speed of revolving magnetic field produced by MMF of suspension winding. (2t4s and 4t2s Belms)

Model	*2t4s Belm*	*4t2s Belm*
Pole number of suspension winding	4	2
Mechanical angular speed of motor	ω	ω
Angular speed of suspension winding current	ω	$2\,\omega$
Angular speed of revolving magnetic field produced by MMF suspension winding	$-(1/2)\omega$	$2\,\omega$

3. Simulation result of suspension forces

The authors have verified by the simulation using the FEM software (JMAG, Ver.21.2, 2D, JSOL CORPORATION) that the current is induced in the rotor PMs and the suspension force is influenced by the induced current in the high speed range. Figure 1 shows the cross section and winding arrangement of 2t4sBelm and 4t2sBelm FE models and Table 2 shows the specification of FE models.

Firstly, the induced current density distribution in the rotor PMs (0-360 deg) in the axial direction for the 2t4s and 4t2s Belm FE models have been computed, respectively. Figure 2 shows the analyzed result of current density in PM in axial direction. The rotational speed is set at 12,000 r/min and the motor winding current is set to 0.5 A considering the inertia of rotor shaft at no load condition. The suspension force is commanded to x-positive direction. It is the waveforms of current density at the moment of 360 degree rotor angular position. It is found in the computed results that the frequency of current density of 4t2s Belm is a half of that in the 2t4s Belm. It is dependent on the number of poles of suspension winding. However, the peak value of current density in 4t2s Belm is a few times that of 2t4s Belm. Because the angular speed of revolving magnetic field produced by MMF of suspension winding is faster by 2 times than the rotor rotational speed as shown in Table 1. On the other hand, it is a half of the rotor rotational speed in the 2t4s Belm.

Next, Figure 3 shows the computed results of electromagnetic force in the 2t4s and 4t2s Belm FE models. The authors can compute the electromagnetic force in both conditions, in which the eddy current is considered or not on the employed simulation software. In each model, the electromagnetic force is computed under both of conditions without and with the eddy current. The suspension force is commanded in the x-positive direction in both models and the electromagnetic force is computed in x- and y-axis directions, respectively. Similar to Figure 2, the rotational speed is set at 12,000 r/min and the motor winding current is set to 0.5 A considering the inertia of rotor shaft at no load condition. In the 2t4s Belm of Figure 3 (a), the electromagnetic force is not influenced by the eddy current and approximately agrees with the suspension force command even if the eddy current is considered. In the 4t2s Belm of Figure 3 (b), on the other hand, when the eddy current is not considered, the electromagnetic force is generated approximately in accordance with the command. However, the electromagnetic force is generated also in y-direction under the consideration of eddy current, i.e., the mutual interference occurs between the x- and y-direction forces. In addition, also the x-direction force is varied with the eddy current. It is concluded in the result of Figure 3 that in the 4t2s Belm, the larger current is induced by the eddy current due to high frequency revolving suspension magnetic field and it much influences on the electromagnetic force generation.

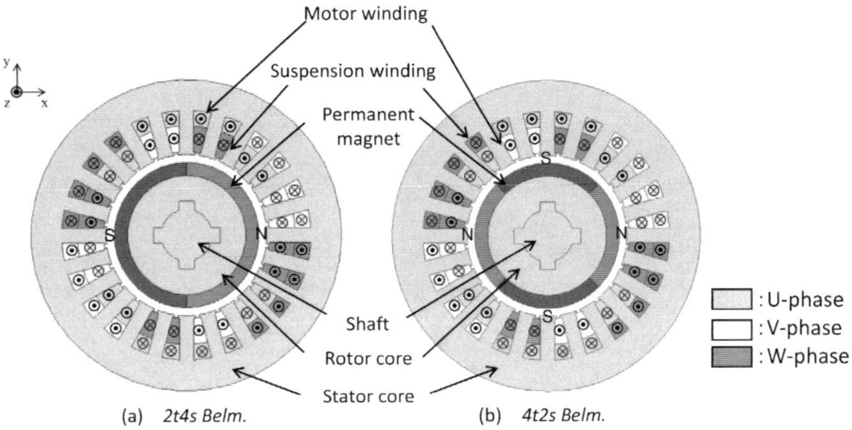

Figure 1 Cross section and winding arrangement of 2t4s and 4t2s Belm models.

Table 2 Specification of FE models. (2t4s and 4t2s Belms)

Stator core outer diameter		120 mm
Rotor core outer diameter		46 mm
PM thickness		5.0 mm
Airgap length		3.0 mm
Axial length		110 mm
Winding turns in series / slot	Motor	4
	Suspension	6
Rated current	Motor	13.6 A
	Suspension	9.42 A

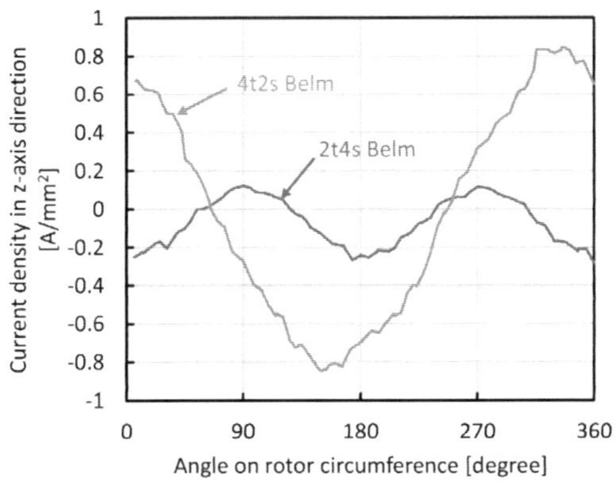

Figure 2 Analyzed result of current density in PM in axial direction at rotor angular position of 360 degree.

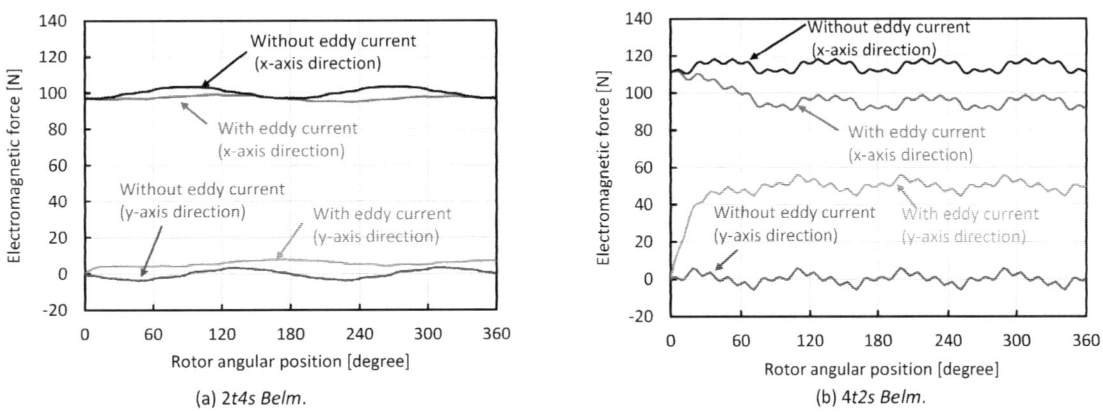

Figure 3 Analyzed results of electromagnetic force in x- and y- axis directions.

4. Experimental verification

4.1 Prototype machine

In order to verify the large variation of electromagnetic force caused by the eddy current and the mutual interference between the electromagnetic forces in x- and y-axes in the 4t2s Belm, a prototype surface-mounted PM synchronous type Belm is built. Figure 4 shows the structure and size of the prototype machine. It is a lateral type. The right side of the shaft is supported by a self-aligning ball bearing. In the Belm, the rotor radial position is actively controlled in two perpendicular radial axes. The touchdown bearing is equipped with the left side of shaft. The clearance of touchdown bearing is 100 μm. The rotor shaft is magnetically suspended without mechanical contact during the normal operation. It is touched down on the touchdown bearing only at standstill or emergency. The rotary encoder is equipped in the end of shaft right side.

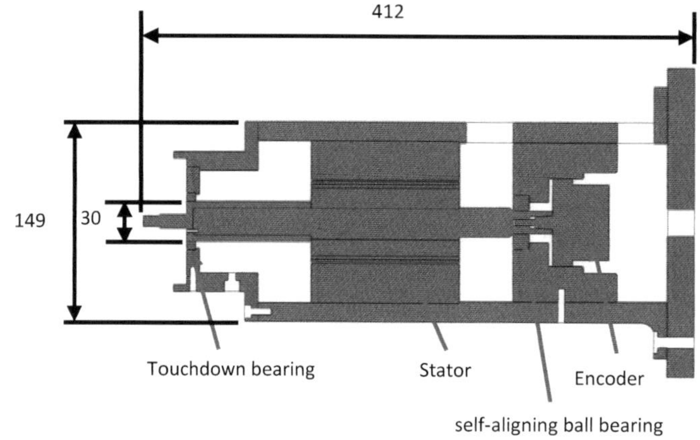

Figure 4 Prototype machine.

4.2 Interference component

When the experiment is carried out using the prototype machine shown in Figure 4, the prototype machine is arranged in lateral so that the vertical direction is aligned in x-axis and the below direction in vertical agrees in x-positive direction considering the experimental setup and the environment around it.

The electromagnetic force is computed once again because the direction of suspension force command is changed to the negative direction in x-axis. Figure 5 shows the computed x- and y-axis components of electromagnetic forces when the rotational speed is set at 12,000 r/min and the motor winding current is set to 0.5 A. The suspension force is commanded to x-negative direction. It can be seen in Figure 5 that similar to Figure 3, the electromagnetic forces are significantly varied by the influence of eddy current. The difference between the electromagnetic forces without and with the consideration of eddy current is caused. It is done in both x- and y-axis directions as shown in Figures 5 (a) and (b), respectively. The difference is defined as the interference component of electromagnetic force, in which the electromagnetic force without consideration of eddy current is subtracted from that with consideration of eddy current. It can be seen in Figures 5 (a) and (b) that the interference component is dc at steady state in both axes.

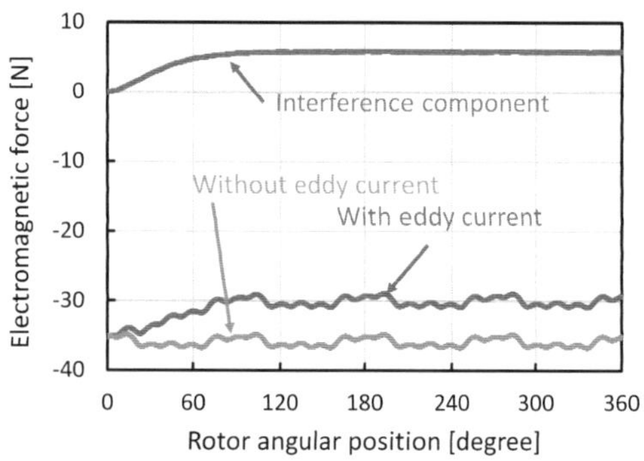

(a) *x-axis direction.*

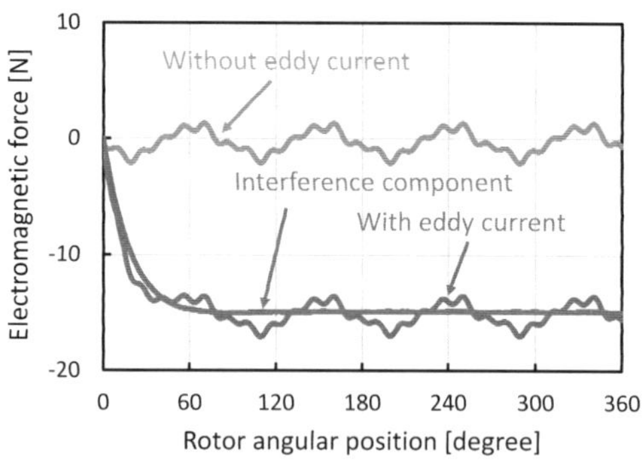

(b) *y-axis direction.*

Figure 5 Interference component of electromagnetic force in x- and y- axis directions.

4.3 Experimental test result

The electromagnetic force is generated so that it is followed to the suspension force command in normal operation of Belm. Therefore the generated electromagnetic force can be easily estimated by the command. To experimentally verify the influence of eddy current on the electromagnetic force and observe the interference component of electromagnetic force, the suspension force command in x- and y-axes in the magnetic suspension controller is measured in the magnetic suspension controller of the prototype machine, respectively. The position of prototype machine is set as mentioned in Sections 4.1 and 4.2. Therefore, as the gravity force is applied on the rotor shaft in x-positive direction, the suspension force is commanded to x-negative direction. Namely the experimental condition agrees with the simulation one shown in Figure 5 with respect to the force direction. Figure 6 shows the waveforms of suspension force command in x- and y-axes at the rotational speeds of 600 r/min and 12,000 r/min, respectively. It has been confirmed by the FE analysis that the eddy current does not influence on the electromagnetic force in the low speed region. Therefore the waveform of suspension force command at 600 r/min is also observed and compared with that at 12,000 r/min. In each waveform of Figures 6 (a) and (b), the ac component is caused by the rotor fluctuation due to the mechanical unbalance. As mentioned in Section 4.2, the interference component by eddy current is dc. Therefore the dc component of each waveform in Figures 6 (a) and (b) is derived and also described in these figures. It can be seen that there is a difference between the dc components of suspension force command at 600 r/min and 12,000 r/min. It is found also in the experiment that the electromagnetic force is influenced by the eddy current. The interference component is derived by subtracting the dc component of suspension force command at 600 r/min from that at 12,000 r/min. In the case of x-direction force in Figure 6 (a), the subtraction of the dc component of suspension force command at 600 r/min from that at 12,000 r/min is positive. It is estimated from this experimental test result that the interference component by eddy current is actually generated in x-negative direction. It is the same as that in y-axis direction as shown in Figure 6 (b).

Compared with the simulation results of electromagnetic force shown in Figures 5 (a) and (b), the estimated electromagnetic force from the experiment is obviously different in both the magnitude and direction as shown in Figures 6 (a) and (b). Therefore the electromagnetic force at the rotational speed of 12,000 r/min is simulated again so that the suspension force command agrees with one measured by the experiment in Figures 6 (a) and (b). The suspension winding current command is set in the FEA simulation software so that the same electromagnetic forces as Figures 6 (a) and (b) are generated. Figure 7 shows the simulation result of electromagnetic force when the suspension force is commanded to agree with one in the experiment shown in Figure 6. The rotational speed is set at 12,000 r/min. The interference component by the eddy current on the electromagnetic force is found in the y-axis direction of Figure 7 (b). In the x-axis direction of Figure 7 (a), however, the electromagnetic force with the consideration of eddy current almost agrees with one without consideration of eddy current. It seems that there is no interference component in the x-axis direction. It is accidentally caused depending on the magnitude and direction of suspension force command and the interference component by the eddy current. Actually the electromagnetic force is interfered by the eddy current.

The interference components estimated by the experiment and simulated by FEM are summarized in Table 3. The estimated interference component is calculated from the measured suspension force command by the experiment as shown in Figures 6 (a) and (b). It is defined as the subtraction of the dc component of suspension force command at 600 r/min from that at 12,000 r/min. It is noted that the direction of actual electromagnetic force is estimated in opposite of the force command. On the other hand, the simulated interference component is derived by the results shown in Figures 7 (a) and (b). The estimated interference component by the experiment does not agree with the simulated one. Because this discrepancy between the experiment and simulation results is caused by the error of eddy current in PMs. It is because the eddy current is dependent on the manufacture tolerance of the prototype machine and the employed PM characteristics.

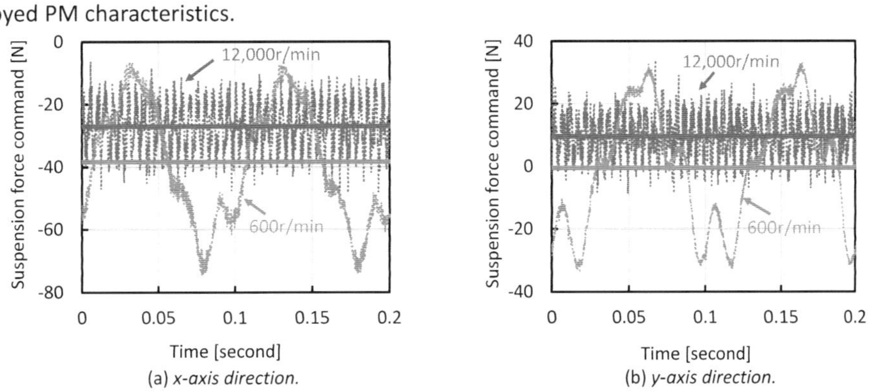

(a) *x-axis direction.* (b) *y-axis direction.*

Figure 6 Suspension force commands at 600 r/min and 12,000 r/min.

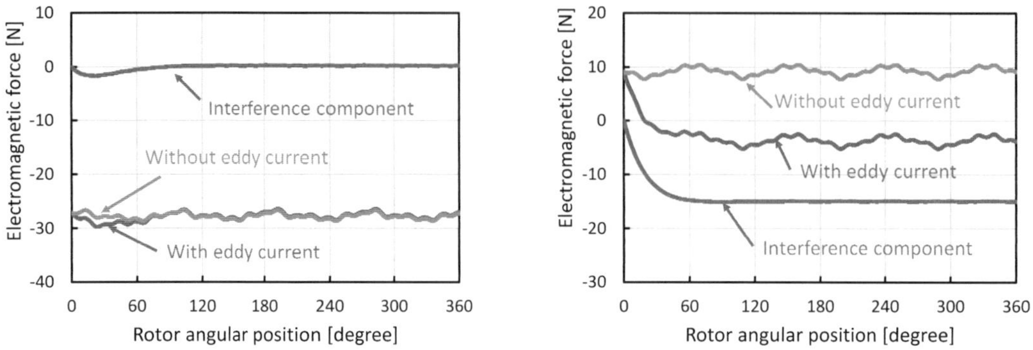

(a) *x-axis direction.* (b) *y-axis direction.*

Figure 7 Simulation result of electromagnetic force when the suspension force is commanded so as to agree with the experimental ones at 12,000 r/min.

Table 3 Interference components of electromagnetic force by eddy current

Axis	x	y
Force commands at 600 r/min [N]	-38.36	-0.3721
Force commands at 12,000 r/min [N]	-27.05	9.594
Estimated interference component [N]	-11.30	-9.966
Interference component by FEA [N]	0.2252	-14.94

5. Conclusion

In the Belm with 4-pole motor and 2-pole suspension windings, the magnitude and direction of electromagnetic force is varied with the eddy current in the PM, which is caused by the difference between the angular speed of revolving magnetic field by MMF of PM on the rotor and that by MMF of suspension winding. It is remarkable especially in the high speed region. In this paper, it is verified by the experiment test using the prototype machine as well as the FE analyzed result using the simulation software. The reason why the difference between the interference components of electromagnetic force derived by the experiment and simulation is caused should be solved in the next stage. In addition, a prototype Belm with 2-pole motor and 4-pole suspension windings will be built and it will be verified that the influence of eddy current on the electromagnetic force is less than that in the 4t2s Belm.

References

[1] Akira Chiba, Tadashi Fukao, Osamu Ichikawa, Masahide Ooshima, Masatsugu Takemoto and David G Dorrell (2005) Magnetic Bearings and Bearingless Drives: Newnes Publishers, ISBN 0-7506-5727-8.

[2] Syunsuke Kobayashi, Masahide Ooshima and M. Nasir Uddin (2013) A Radial Position Control Method of Bearingless Motor Based on d-q Axis Current Control, *IEEE Transactions on Industry Applications*, vol.49, No.4, pp.1827-1835, (2013).

[3] Y.Okada T. Ohishi and K.Dejima (1992) Levitation Control of Permanent Magnet (PM) Type Rotating Motor: Magnetic Bearings, Magnetic Drives and Dry Gas Seal Conf. & Exhibition, Alexandria, VA, USA, 29-30 July, 1992, pp.157-166. Proc. of Magnetic Bearings, Magnetic Drives and Dry Gas Seal Conf. & Exhibition.

[4] Masashi Sawada, Yuji Shindo, Tomoaki Tamiya, Yoshihiro Kawase, Tadashi Yamaguchi, Hirokatsu Katagiri, Yuki Ono and Takashi Kodera (2012) A Study on a Bearingless Drive of a Surface Permanent Magnet Synchronous Motor, *Conf. on Electrical Machines and Systems* (ICEMS 2012), Sapporo, Japan, Oct. 2012, LS2B-2.

[5] Toshihide Fukushima, Daiki Satou and Masahide Ooshima (2019) Influence of Pole Number on Magnetic Suspension Force in Surface-Mounted Permanent Magnet Bearingless Motors, Annu of IEEJ 2019, Sapporo (Japan), March 2019, p.5-013 (In Japanese).

[6] Toshihide Fukushima, Daiki Satou and Masahide Ooshima (2020) Combination of Pole Number of Motor and Suspension Windings and the Influence on Suspension Force in Bearingless Motors, *The 23rd International Conference on Electrical Machines and Systems* (ICEMS), Hamamatsu, Japan, 24-27 Nov. 2020, pp. 44-49.

How Magnetic Bearings Contribute to TotalEnergies Ambition Lessons Learnt and Expectations from Magnetic Bearings

Alain GELIN[a], Antoine LUCAS[a]

a TotalEnergies, Avenue Larribau, 64018 Pau Cedex, France, alain.gelin@totalenergies.com

Abstract

The purpose of this paper is to present how Active Magnetic Bearings contribute to TotalEnergies Net Zero Ambition in 2050. TotalEnergies is operating hundreds of Turbomachines worldwide to enhance Oil Recovery and to handle associated Gas from Liquefaction to Transportation businesses. By allowing physical segregation of processing streams from atmosphere, Active Magnetic Bearings help to avoid incidental hydrocarbon and thus CO_2 releases. The recent development of High-Speed Induction Electrical Motors in conjunction with the utilization of Active Magnetic Bearings offer many opportunities to reinvent the future in term of compression systems and to address carbon footprint reduction efforts. This paper will describe the technology of Integrated compressor solutions and will present some Green and Brown Field experiences. The paper will also focus on lessons learnt from a major Operator of Turbomachinery's and his expectations in term of Magnetic Bearings to meet future climate challenges.

Keywords: *Magnetic Bearings, Integrated Compressors, High Speed Electrical Motors, Lessons Learnt*

1. Introduction

TotalEnergies is a global multi-energy Company that produces, refines, and markets energies: oil and biofuels, natural and green gases, and electricity from renewables. Active in more than 130 countries, TotalEnergies put sustainable development in all its dimensions at the heart of its Projects and Operations. As a major player in the energy, Company's ambition is to reinvent the way energy is produced and consumed to get net zero by 2050, together with society, and to offer solutions to address the climate challenge.

TotalEnergies is active across the energy value chain from Production, Transportation and Processing into intermediate or finished products; to Storage and Distribution to meet the needs of individual and business customers. The Company is also developing carbon reduction projects for its own sites and for its customers by the deployment of solutions to enhance energy efficiency and to capture and store carbon (CCS and natural carbon sinks). To face climate challenges, TotalEnergy is moving forward, together, towards new energies and its energy mix will move from more than 90% Oil&Gas in 2020 to 25 % in 2050, see Fig.1 & 2.

Figure 1 - TTE Energy Mix

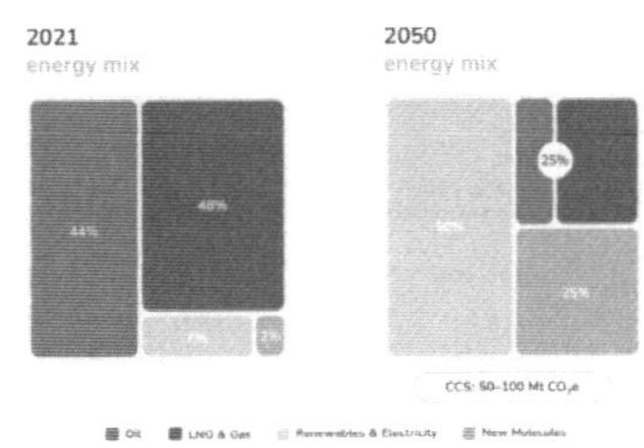

Figure 2 - Energy Mix from 2021 to 2050

2. From the origins... to the future with Integrated Compressor

As an example, the operation of a large FPSO (Floating Production Storage and Offloading) producing around 150 to 200 thousand barrels of oil per day with about 7 to 10 MMscfd of associated gas to be recompress for exportation or to enhance oil recovery, see Fig. 3 and 4, would require an installed power of about 100 MW. The corresponding CO_2 emission due to fuel gas consumption and combustion is estimated to roughly 600 kT CO_2 per year using open cycle gas turbines.

Another example of CO_2 emission is related to Liquified Natural Gas (LNG) applications. A liquefaction plant of 1 MTPA (Million Tonnes Per Annum) of liquefied natural gas using refrigerant compressors requires about 43 MW mechanical power, and when open cycle gas turbines are used to drive the refrigerant compressors, the corresponding CO_2 emissions are roughly 265 kT CO_2eq/y.

Figure 3 - FPSO in Operation in Angola

Figure 4 - FPSO in Operation in Nigeria

Both examples demonstrate the necessity to rethink oil and gas facilities design to reduce CO_2 and Greenhouse Gas (GHG) emissions, that Active Magnetic Bearings (AMBs) will contribute to meet.

Here below are some examples of actions already taken to reduce global emissions for either Greenfield or Brownfield Projects:

- Power generation based on Combined Cycle Gas Turbines (CCGTs) instead of Open Cycle to drastically reduced CO_2 emissions by roughly 30% thanks to lower fuel gas consumption
- Retrofit of Turbo Compressors by Electrical Compressors, and fed from renewable power or distant grid
- All electrical solutions including power generation based on Wind and/or Solar farms
- All new Projects are based on Zero routine flaring strategy that TotalEnergies is committed to achieve by 2030 on all its facilities, with Flare Gas recovery systems that are now mandatory
- Retrofit of installations with dedicated flare gas recovery systems. Many available technologies are proven such as static ejectors, dry screw compressors, liquid ring compressors, reciprocating or centrifugal compressors
- Rebundling of centrifugal compressors to avoid or to limit operation in recycle mode with excessive power consumption
- Dry Gas Seals (DGS) reliability and availability improvements. The DGSs with their associated conditioning and monitoring systems are very complex. Many failures have been reported that could have escalated to catastrophic events. As an example, the complete failure in 2009 of a DGS for a very high-pressure compressor, see Fig. 5 resulted of a large amount of released gas to the lube oil system. The process gas and lubricant oil mixture released by all ports and vents lines and reached the hot part of the exhaust of the 30 MW Gas Turbine leading to ignition. The fire propagated all around the package putting the integrity of the platform at risk.

Figure 5 - Dry Gas Seal Catastrophic Failure

Figure 6 - Fire ignition due to DGS failure

In 1987, TotalEnergies commissioned and put in operation a centrifugal compressor fitted with Active Magnetic Bearings (AMBs) and DGSs. This world first application, see Fig. 7 was installed in Lacq field in the southwest of France to handle sour gas containing about 20% of H2S toxic gas. To avoid any contamination of the AMBs, the DGSs of the compressor were fed with "clean and dry" fuel/seal gas. This compressor was decommissioned in 2005 after more 130 000 operating hours without any issue. Three (3) other centrifugal compressors with AMBs were installed on this field in the 90's for gas re-boosting. Many Turbo Expanders also equipped with AMBs are operated by TotalEnergies for natural gas treatment and dew pointing. These experiences on AMBs made the Company very confident about the technology.

In 1994, TotalEnergies commissioned and started a High-Speed Electrical Motor (HSEM) compressor package in Donges refinery in France (3600 kW @ 6300 rpm), see Fig 8. The technology of the induction motor is based on a fully laminated rotor stacked with tie rods allowing the reach very high peripheral speed around 270 m/s. The motor, still in operation today, is working at atmospheric pressure and is cooled using atmospheric air, a dedicated air/water heat exchanger is required. The rotor of this first High-Speed electrical solution was nevertheless fitted with hydrodynamic oil lubricated bearings.

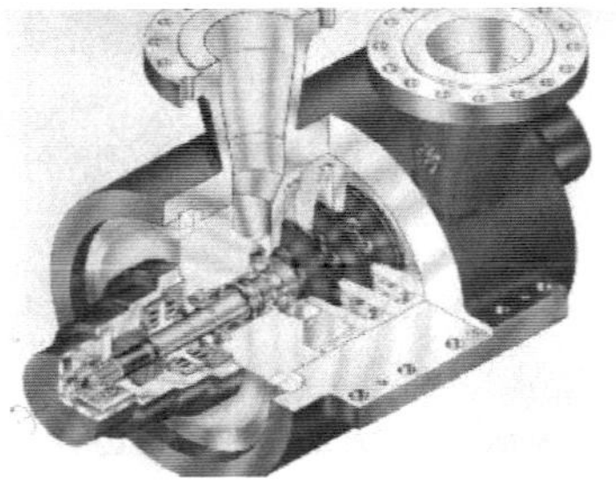

Figure 7 - KB-403 Lacq Compressor fitted with AMBs

Figure 8 – Donges High Speed Motor fitted with AMBs

More recently, and associating both AMB and HSEM technologies, TotalEnergies decided to move forward introducing the "so-called" Stand-Alone and Oil-Free solutions for high-speed motor compressor packages. The compressor is directly driven by the high-Speed motor without the requirement for a speed increaser gear box and the complete shaft line is supported by AMBs, see Fig.9. Without gear box and lubricated bearings, the complete lube oil system can be removed, and the package becomes less complex as some auxiliaries are removed. The HSEM and the AMBs are cooled using air at the atmospheric pressure. To be noted that the size of the atmospheric air cooler on top of the motor is larger than the motor core itself and it requires auxiliary motors for internal air circulation.

Unfortunately, the compressor is still equipped with DGSs and associated conditioning system to isolate the electro-magnetic components from the process gas. The DGS conditioning seal gas panel contains many auxiliary process equipment such as filters, scrubbers, booster, heater and more than 30 transmitters are used for the control, see Fig. 10.

In addition to that, an acoustic insulation hood or blankets are applied onto the motor to limit noise emission due the high speed and the thin frame walls of the atmospheric HSEM, see Fig. 11.

Figure 9 - HS Stand Alone Package

Figure 10 - DGS Monitoring Panel

Figure 11 - Package with noise insulation

3. Fully Hermetic and Integrated Package

The Dry Gas Seals with their associated seal gas conditioning skids represent the weak point of centrifugal compressors in term reliability, and roughly 80% of the lack of availability of centrifugal compressors for Upstream applications are linked to the malfunctioning of DGSs. In addition to that, the continuous leakage of process gas exiting the DGSs inboard seal is either released unburnt to the atmosphere or routed to the flare system contributing to Green House Gas (GHG) or CO_2 emissions. As a rule of thumb, the primary seal gas leakage for a compressor is roughly 5 Nl/min/bar (suction pressure to be sealed) and for High Pressure applications the corresponding emission is not negligible.

Thanks to the development of HSEMs see Fig. 12 with a fully laminated rotor inside a High-Speed balancing machine, and AMBs see Fig.13, the Stand-Alone solution for motor compressors can be made fully Integrated and Hermetic without any gas leakage, see Fig.14. To do so, the DGSs of the compressor are removed, and both AMBs and HSEM are directly cooled using the process gas at the suction pressure of the compressor.

Figure 12 - HSEM Fully laminated rotor with AMBs

Figure 13 - AMB static parts

Figure 14 – ICML during Factory Testing

Of course, the material used for HSEM and AMBs shall be able to withstand any corrosion attack from process conditions due to the potential presence of contaminant such as CO_2, H_2S, Mercury or liquid Water. A particular attention shall be paid during stand still conditions at the settle out pressure due to possible water condensation while the temperature of the pressurized gas inside the machine decreases below gas dew point.

Another concern is related to the High and Low voltage connections respectively for HSEM power supply and AMBs control/command system, and some deep qualification processes are required to fully validate the sealing and the tightness of these electrical passages and connectors.

In 2007, TotalEnergies was the first Oil & Gas Operator to qualify the technology from Baker Hughes Company (BH), so-called ICL™ for Integrated Compressor Line. The HSEM and the AMBs are respectively provided by GE/PC and SKF/S2M. The technology was first qualified for "Commercial, Dry and Clean" gas applications. Few years after the qualification was extended to "Water Saturated Gas" applications including some contaminants such as CO_2 and H_2S. More recently TotalEnergies extended furthermore the qualification to the "small scale" integrated compressors line from BH, so-called ICL-LV™, technology using Low Voltage permanent magnet HSEM running up to 30 000 rpm and targeting Flare Gas Recovery applications.

In 2022, TotalEnergies pronounced the qualification of the Hermetic Oil Free technology from MAN-ES company so-called Sealed -HOFIM™ to enhance its portfolio of hermetic solutions and favor competition.

The main advantages of the Integrated and fully Hermetic compressor packages are summarized as follow:

- No leakage of natural gas to the atmosphere or to the flare system
- No Lube Oil system with all auxiliaries, and no speed increaser Gear Box
- No Dry Gas Seal and associated conditioning system
- No requirement for Cooling Water nor Nitrogen
- Low noise package thanks to the high speed and the HSEM casing thickness
- Only few remaining instruments, instead of tenth, to monitor the package
- Drastic reduction of weight and footprint
- "Plug and Play" pre-commissioned packages
- Less spare parts
- Better reliability and availability compared to conventional packages
- Lower maintenance and associated downtime

Due to the process conditions, including but not limited to gas, pressure, temperature, the compressor architecture is always a Project specific and different cooling loops can be implemented. The impact of the cooling loop either "quasi-closed" or "fully close" shall be properly evaluated from a performance standpoint during the design phase of the Project as the windage effects are directly linked to the geometry, the velocity, and the gas density. The specific heat capacity of the gas is key parameter as it governs the cooling flow requirement. Full Load String Test (FLSTs) as strongly recommended for hermetic packages with AMBs, not only to check its mechanical behavior and integrity but also to confirm the performance in term of electrical power consumption. It is a standard requirement within TotalEnergies.

For all new TotalEnergies projects, either Brownfield or Greenfield, Onshore or Offshore, the integrated and seal less technology is selected as the base case. One limitation is coming from the suitability of the AMBs and HSEM components versus the process gas and associated contaminants, if any. Another limitation is coming from the high pressure to sustain the aerodynamics excitations coming from the gas density, as a rule of thumb, the suction pressure of the compressor is limited to roughly 100-120 bara. This limitation is also driven by the low stiffness of the AMBs compared to tilting pad bearings. As another rule of thumb, AMBs are roughly 10 times softer than hydrodynamic bearings. In addition, the higher the compressor pressure, the higher the windage motor losses are, and the integrated solution can be discarded due to excessive power consumption.

In 2015, TotalEnergies commissioned and started two (2) ICL Booster compressors of 4.5 MW in Bolivia, each handling 7.5 MSm3/d of process gas from 73 to 106 bara and running at 12000 rpm, see Fig. 15. The compressor is equipped with a single impeller directly clamped to the motor without any flexible coupling. The cooling gas loop is ensured thanks to a dedicated fan directly mounted on the Non-Drive End (NDE) side of the motor to circulate the process gas inside the motor/stator and the AMBs before to be reinjected hotter at the inlet of the compressor. After 9 years of operation, these ICLs will be retrofitted in 2024 from parallel operation (high flow / low pressure ratio) to "in series" mode (lower flow / higher pressure ratio) to better comply with the depletion of the gas field.

In 2016, another ICL compressor of 9 impellers in back-to-back arrangement was installed in the south of France for an ethylene transportation project, see Fig. 16. The compressor is a two-stage machine of 2.8 MW running at 11 000 rpm and handling 0.8 MSm3/d from 13 to 97 bara. The cooling principle for this multistage compressor is more conventional. The cooling gas is taken from a dedicated internal stage, without any fan, to circulate the ethylene through the HSEM and the AMBs before to be also rejected hotter at the inlet of the compressor.

Figure 15 – Both compressors in Operation at site In Bolivia

Figure 16 – Ethylene compressor in Operation

The construction of a new ICL is on-going for a Low-pressure phase Project in Brunei. The compressor of 10.8 MW and running at 8 500 rpm will handle 3.8 MSm3/d of gas from 13 to 59 bara. It will be a new world first as the AMBs and HSEM will be cooled and submerged in the Water Saturated Gas, impacting the technology and the material selection.

Many other Projects, all based on Integrated and Hermetic compressors are under discussion to meet TotalEnergies objectives in term of Carbon Footprint Reduction (CFR). Some of them are related to the electrification of existing facilities (either onshore of offshore) with the decommissioning of large Turbo Compressors and their replacement by Integrated and "Green electrical" packages. Other are related to the recovery of Flare Gas to directly decrease the CO_2 emissions.

4. Lessons Learn and Expectations from AMB systems

Thanks to the number of Rotating Equipment and Turbomachinery's operated by TotalEnergies and fitted with magnetic bearings, from turbo-expanders to electrical motors and including compressors. With more than 30 years' operating equipment with AMBs, TotalEnergies has developed a large experience on the technology.

The following lessons learnt and expectations from AMBs are based on Projects, qualification programs, testing activities, commissioning, and site operation:

4.1 Lack of stiffness and damping at low frequency

The machines equipped with AMBs may suffers of sub-synchronous vibrations with a very low frequency content. The frequency spread may vary from few hertz to tenth of hertz and corresponds to the frequency domain where the AMBs are very soft compared to conventional hydrodynamics bearings. The AMBs are usually ten time softer than sleeve or tilting pads lubricated bearings. These sub synchronous broadband vibrations are often observed for Turbo-Expanders is case of liquid formation due to gas expansion across some internal labyrinths of passages.

The common way to cure this sub-synchronous vibration is to work on the process conditions to avoid any liquid, but stiffer AMBs especially at low frequency would be beneficial to limit sub-synchronous activity during upset and transient conditions.

4.2 Sub-synchronous Instability – Forward but also Backward...

Some sub-synchronous rotor instabilities may also occur on the first natural frequency, either in forward or backward motion, for medium and high-pressure compressors. The API-617 for centrifugal compressors focusses on rotor stability as it is the main concern for HP compressors, reference is made to API-617 Part 1 Annex D. The stability analysis deals with the forward motion of the first natural frequency that could be self-excited by aerodynamic due to internal gas velocity in the direction of the rotation. To avoid the aerodynamic effect and to increase the damping (logarithmic decrement) of the first natural frequency, many stabilizing devices can be implemented to limit cross coupled stiffnesses affecting the forward mode:

- Impeller eye swirl breakers to limit and decrease the circumferential gas pre-swirl at seal entrance
- Axial balance drum equipped with either Honeycomb or Hole-Pattern seals with pressurized cells
- Shunt hole at the balance drum inlet to also cancel pre-swirl at the inlet.

All previous practices are well-known to ensure stability of HP compressors, but what is beneficial for the forward motion is detrimental for the backward one. TotalEnergies experienced a pure sub-synchronous instability of the first natural frequency in backward direction while the compressor equipped with AMBs was reaching the maximum pressure and maximum speed. To cure the issue, the internal parts of the compressor had to be modified removing some stabilizing devices for the forward mode to not destabilize the backward one. The transfer functions of the AMBs were also modified to bring additional damping in the frequency area of concern. The same HP compressor equipped with hydrodynamics tilting pad bearings would never have demonstrated any sub-synchronous instability in backward due to the higher impedance of the lubricated bearings (stiffness and damping) versus AMBs.

Recommendation is made for medium and high-pressure compressors fitted with AMBs to not only focus on forward stability analysis as per API requirement but also to take care of the backward motion first natural frequency. References is also made here to a very interesting paper (Bidaut, et al., 2006).

4.3 Load capacity and requirement for thrust magnetic bearing

The capacity of AMBs is much less compared to hydrodynamics bearings and some efforts are also recommended in that direction to accept higher loads, and this applies also in the axial direction for the active magnetic thrust bearings. A centrifugal compressor is always equipped with a thrust bearing to contain the aerodynamic residual load. Electrical motors are usually not equipped thrust bearing as the stator core magnetic centering is sufficient to axially locate the rotor in a stable position. For high pressure applications of integrated compressors with high density gas, the cooling flow rate may induce an axial load requiring also a thrust magnetic bearing. Having a thrust magnetic bearing on both compressor and motor linked together with a flexible coupling may induce mechanical and vibratory uncertainties during transient events as start-up, loading and coast down.

All efforts shall be deployed from a design standpoint to not installed a magnetic thrust bearing for high-speed motor.

4.4 Process gas compatibility and qualification

For Integrated and Hermetic solutions, and as the AMBs and HSEM are submerged inside the process gas, some efforts are still needed to make the components more robust and suitable whatever the contaminants. Qualification programs and acceptance criteria shall be mutually agreed between parties: AMBs & HSEM sub-supplier, Original Equipment Manufacturers (OEMs) and Operators.

4.5 Auxiliary bearings

Auxiliary bearings are used to support the rotor at rest and to protect the AMBs to avoid any contact with sensitive parts and laminations (actuators and sensors). The clearance of auxiliary bearings is typically 30% of the AMBs clearance and special roller bearings are usually used. The concept of “soft” and “hard” landings is often used by AMBs suppliers or OEMs, and it should be revisited. In addition, the justification for auxiliary bearing replacement shall be fully justified.

Furthermore, it is not recommended to perform any AMBs delimitation during Factory Acceptance Test, and this applies either during Mechanical Running Test of single equipment or during String Test of complete package.

4.6 Rotordynamics analysis

API-617 mainly refers to lateral unbalance response analysis. As different types of synchronous filters may be proposed to optimize the rotordynamics lateral behavior, the unbalance responses shall be run with and without synchronous filter including a narrative and a justification of their use. In addition to the stability analysis as specified in the API Standard, the Campbell’s diagram without any synchronous filter shall be part of the lateral analysis. All lateral natural frequencies, with associated mode shape, either forward or backward, and damping (logarithmic decrement) shall be clearly identified up to two (2) times the frequency corresponding to the trip speed and the analysis shall be run up to the trip speed. The lateral analysis shall also include the stiffness and damping properties of the AMBs versus the frequency and resulting from the Transfer Functions.

The axial rotordynamics analysis of the complete train shall be part of the torsional, and this applies independently of the number of thrust bearings present in the shaft line.

4.7 Tuning of AMBs

The tuning of AMBs, either during Factory Acceptance Test (FAT) or during site commissioning activities, represents a constraint in term of Project management and scheduling. The initial and pre-implemented Transfer Function within the AMBs control system shall be ONE used for the rotordynamics analysis as it governs the stiffness and damping characteristics. Any modification of the transfer function shall be justified and documented by the AMBs supplier and by the OEM responsible of the complete package. If some tuning activities are nevertheless required, the rotordynamics analysis needs to be rerun based on the true implemented transfer function. In addition, and in case of identical equipment and packages, the transfer function shall be the same whatever the package, and in case of bundle or rotor replacement at site, no AMB retuning is foreseen.

AMBs shall not be considered as "Black Box" and to make them fully democratized, the AMBs shall be as simple as hydrodynamics bearings in term of Operation.

4.8 Acceptance Criteria during Factory Acceptance Testing

The acceptance criteria for AMBs, as specified in the API-617 – Annex-E, mainly focus on the direct (global) vibrations and are only based on the auxiliary bearing clearances. These criteria should be revisited to also consider the following:

- one per rev (1X) component including its thermal vector change and repeatability check especially for electrical motors. If API-541 provide some rules it is recommended that the one per rev (1X) vector change does not exceed 25% of the direct overall criteria, and the reading shall be repeatable to confirm rotor thermal stability
- some limitation for low frequency broad band signature
- none sub-synchronous or super-synchronous instability corresponding to any undamped natural frequency shall be accepted
- some current limitations to anticipate any saturation issues. As a criterion, 75% of the maximum current that the amplifiers can deliver by axis is proposed

Criteria shall mutually agree between parties: AMBs & HSEM sub-supplier, OEMs and Operators prior FAT, and in any case, including package Full Load String Test, the criteria shall be lower that Alarm Limits at Site.

4.9 Spare Parts Strategy & Obsolescence

If it is recognized that no capital spare parts are recommended for the hardware of the magnetic bearing, either rotoric or statoric components, some other spare parts are highly recommended. Of course, one or two sets of auxiliary bearings including rotating sleeves are recommended. For what concern power electronic and control, cables, connectors, batteries if any... the capital spares shall be discussed on Project basis based on reliability numbers and considering obsolescence constraints.

4.10 Cost acceptability of the solution

As energy must remain affordable, the cost of the installations, thus the equipment is particularly challenged during project preliminary phases. Expensive technologies make projects non economical and unsuccessful, preventing the very same technology from being used, confronted with field use, and improved. To be successful, the AMB technology as any developing technology must offer technical benefits while remaining affordable to purchase and to maintain.

5. Conclusion

This publication gives an overview of TotalEnergies Strategy in term of evolution to comply with climate challenges with the objective to reduce CO_2 and GHG emission, the final goal being "Net Zero" in 2050.

Active Magnetic Bearings are clearly identified as part of the solution to address this exciting challenge.

Thanks to its experience for more than 35 years and the combination of Active Magnetic Bearings with High-Speed Electrical Motors, TotalEnergies is very confident that Integral and Hermetic compressors will be part of the future. The numerous advantages also make the solution very attractive for brown field offshore Projects including electrification and retrofit of turbo compressors.

This paper also provides some technical recommendations and expectations, and all the actors from AMBs suppliers to Operators passing thru OEMs must to keep in mind that to make the solution fully attractive, some efforts are still required in term of development, simplification and democratization of the "Black-Box".

References

API 617:2014 (2014) Axial and Centrifugal Compressors and Expander-Compressors.

API 541:2014 (2014) Form-wound Squirrel Cage Induction Motors – 375 kW (500 Horsepower) and Larger

Bidaut Y, Somaini R and De Lima Rugue MA (2019) Stability considerations of centrifugal compressors equipped with active magnetic bearings, Proc. 48th Turbomachinery Symposium, Houston, 2019

Magnetic Suspension System Using Persistent Current Flowing in Superconducting Coil

Mochimitsu KOMORI [a], Keigo MURAKAMi [a], Kaoru NEMOTO [a], Ken-ichi ASAMI [a]

a Department of Electrical and Electronic Engineering, Graduate School of Engineering, Kyushu Institute of Technology, 1-1 Sensui, Tobata, Kitakyushu, Fukuoka 804-8550, JAPAN, komori_mk@yahoo.co.jp

Abstract

Superconducting techniques are useful and applied to a superconducting magnetic suspension system. The system is composed of copper coils, a superconducting coil, a magnetically suspended object, a PID controller, a photo sensor, and power amplifiers. Control current flowing in copper coils and persistent current flowing in superconducting coil are used for suspending object and controlling object, respectively. Large gap trial for the suspension system is discussed, and static characteristics and dynamic characteristics of the suspension system are also discussed.

Keywords: Superconducting coil; Suspension system; Superconducting suspension

1. Introduction

Magnetic suspension techniques are very interesting because of mechanical contact-free. Generally, magnetic suspension techniques use electromagnets (EMs) in order to control the positions of suspended objects. These EMs usually need a lot of electric power and energy. In order to realize larger distance between EMs and suspended object, suspension systems need to generate much larger electromagnetic forces. Then, electromagnetic forces by permanent magnets (PMs) are tried to apply to reduce electric power (Kawamura and Takenaga, 2004), (Sawada and Kunimatsu, 2002). However, there is a technical limit for making much larger dimensions PMs in order to generate larger electromagnetic forces using PMs.

On the other hand, superconducting techniques are useful for various kind of levitation or suspension systems (Hull and Mulcahy, 1995), (Komori and Kobayashi, 1998). Especially, superconducting persistent current is suitable for suspension systems [5]. There is no electric power loss in superconducting coil because there is no electric resistance of superconducting coil.

2. Superconducting magnetic suspension system

An illustration of suspension system using superconducting persistent current in upper superconducting coil is shown in Figure 1. The suspension system is composed of upper copper coils, an upper superconducting coil, lower copper coils, a suspended object, an analog PID controller, a photo sensor and some power amplifiers. The critical current of the superconducting coil (100 turns) is 160 A (at 77K). The superconducting coil measures 10 mm in width, 103.4 mm in outer diameter and 60 mm in inner diameter. The copper coil (0.3 mm ×500 turns) measures 25 mm in

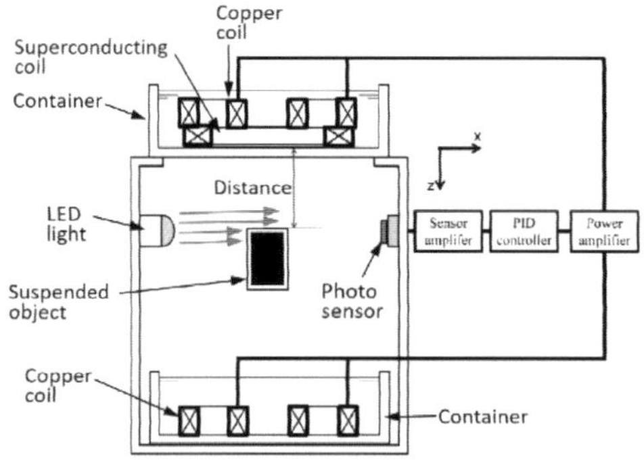

Figure 1 Schematic illustration of suspension system using persistent current in superconducting coil

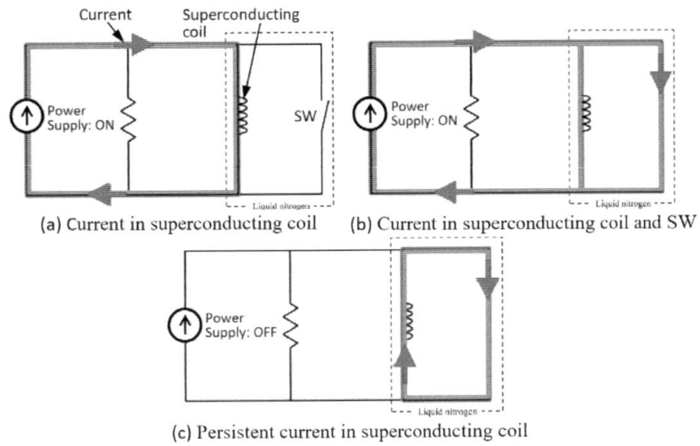

Figure 2 Electric circuit for persistent current mode.

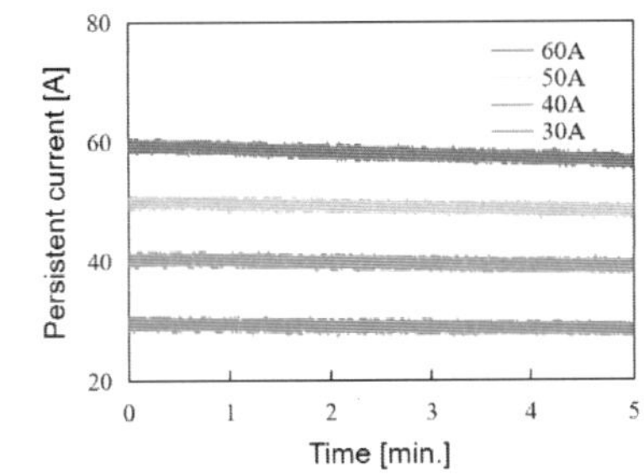

Figure 3 Relationship between persistent current and time for various initial currents of 30, 40, 50 and 60A.

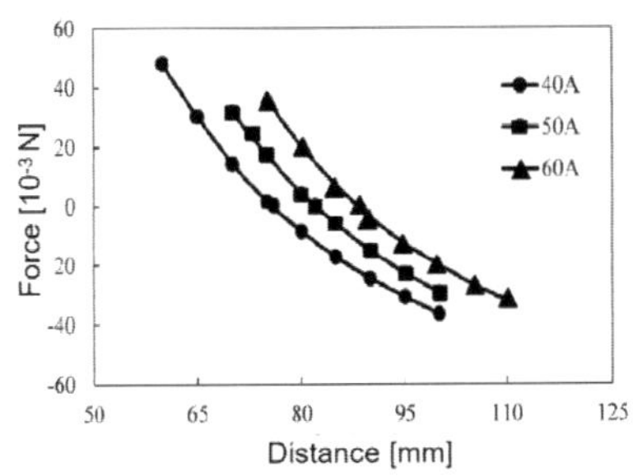

Figure 4 Relationship between suspension force and distance from a superconducting coil to the suspended object.

thickness, 96 mm in outer diameter and 60 mm in inner diameter. Two copper coils and the superconducting coil are installed in upper container. And two copper coils are also installed in lower container. The suspended object (96.3g, 30 ×40mm) has a PM (NdFeB, 70.5g, 20 ×30 mm) inside the plastic object. Detecting the displacement of suspended plastic object is carried out using a photo sensor.

Figure 2 shows the electric circuit for superconducting persistent current mode. After the persistent current (SW) becomes "OFF", the DC power applies a constant current to the superconducting coil in Figure 2(a). After the switch (SW) becomes "ON", constant currents flow through both the superconducting coil and the switch "SW" in Figure 2(b). Moreover, after the current from the DC power supply decreases, the constant current (persistent current) continues to keep the same current in the superconducting coil in Figure 2(c).

The relationships between superconducting persistent current and time for various initial currents in the time range from 0 to 5 min. are shown in Figure 3. Each persistent current from 30 to 60 A keeps the same current for more than 5 min. The persistent current at least for 5 min. is satisfactory to carry out some experiments.

The relationship between suspension force and distance for applied currents from 40 to 60 A is shown in Figure 4.

Each suspension force becomes smaller as the distance increases. The suspension force becomes larger with increasing applied current. Hereafter, persistent current 50 A is adopted in the experiments.

3. Experiments and discussions

Figure 5 shows the illustration of control forces given to suspended object. During experiments, these copper coils and superconducting coil are immersed in liquid nitrogen. Hereafter, the copper coils are used to control the displacement of suspended object and the superconducting coil is used to suspend the weight of object. In order to pass the same current through these coils, both upper two copper coils and lower two coils are connected in series. Thus in the case that polarities of upper coils and object are S and S respectively, the repulsive forces are produced. In the case that polarities of lower coils and object are S and N, the attractive forces are produced. These repulsive force and attractive force are forced to move the suspended object in the vertical direction.

The relationship between attractive force and applied current to copper coils is represented in Figure 6. In the experiment, the same current flows through these four copper coils in order to produce the same attractive force. Then, the total attractive force to the object is produced in the vertical direction. As shown in Figure 6, the attractive

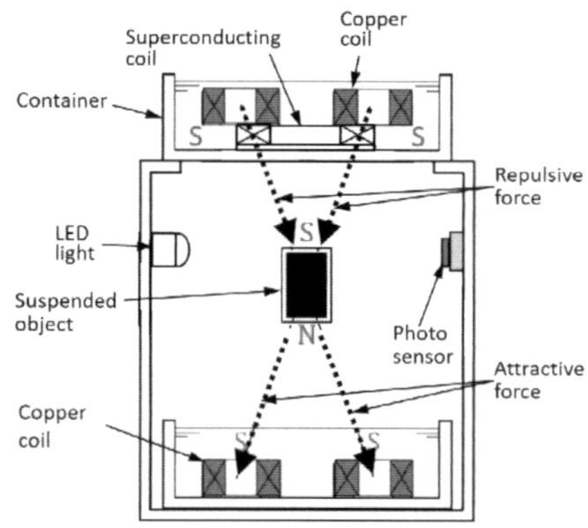

Figure 5 Schematic illustration of control force applied to suspended object.

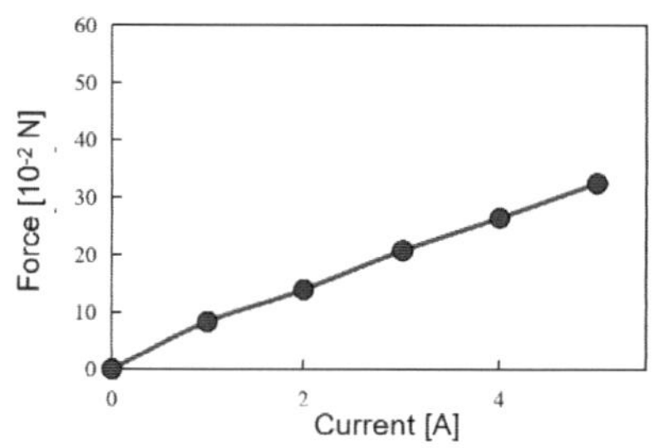

Figure 6 Relationship between attractive force and applied current to copper coils.

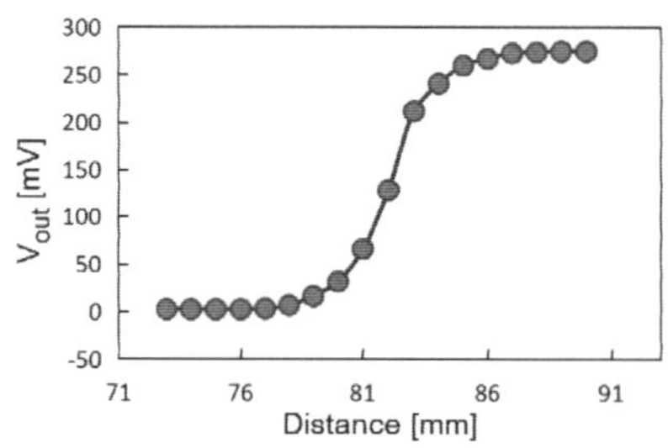

Figure 7 Relationship between output voltage and distance between superconducting coil and suspended object.

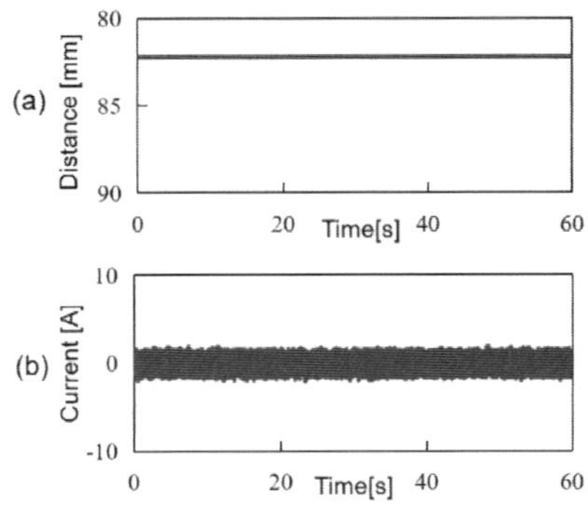

Figure 8 Relationships between (a) distance and time and between (b) current and time in the case of persistent current.

force is almost linear for applied current, which is applicable to controlling the suspended object.

Figure 7 represents the relationship between output voltage from displacement sensor and distance between suspended object and superconducting coil. The relationship near 82 mm is almost linear, which is corresponding to the sensor gain.

The results for the suspended object are represented in Figure 8. During the experiment, the superconducting persistent current of 50 A flows through the superconducting coil. The relationship between distance from superconducting coil bottom surface to suspended object and time is shown in Figure 8(a). The suspended object continues to keep the same distance for more than 60 s as shown in Figure 8(a). During the suspension, the displacement is less than 0.5 mm and small. Figure 8(b) represents the relationship between control current to copper coils and time. The amplitude of control current is less than 3 A. From the experimental results in Figure 8, it is found that the suspension system is working fine.

4. Summary

The suspension system using persistent current of 50 A is developed. The experimental results show that the object succeeds to be suspended at a distance of 82 mm and continues to keep the suspension for more than 60 s. During the suspension, the displacement is small and less than 0.5 mm and the amplitude of control current is less than 3 A.

As a result of experiments, it is found that the suspension system works well and seems to be very promising for some applications such a wind tunnel test equipment.

References

Kawamura Y, Takenaga T, Oh J-B, Takahashi T, Kwon C-K and Mizota T (2004), Development of a low electric power 40 cm class magnetic suspension and balance system, Journal of Wind Engineering, 1, 117-127.

Sawada H and Kunimatsu T (2002), Development of a 60cm Magnetic Suspension System, Journal of the Japan Society for Aeronautical and Space Sciences, 580, 188-195.

Hull J R, Mulcahy T M, Uherka K L, Abboud R G (1995), Low rotational drag in high-temperature superconducting bearings. IEEE Trans. Applied Superconductivity, 2, 626-629.

Komori M, Kobayashi H and Kumamoto M (1998), Hybrid high Tc superconducting magnetic bearing (SMB) system with no bias currents, The 6th International Symposium on Magnetic Bearings, 214-223.

Takase S, Komori M, Nemoto K, Asami K and Sakai N (2015), Basic study on magnetic levitation system using superconducting coil, JSME Mechanical Engineering Journal, 3, 1-8.

Data-driven modelling of touch-down bearing forces

Benedikt SCHÜßLER [a], Bastian TIGGES [b], Tuomas TIAINEN [c], Stephan RINDERKNECHT [d]

a Technical University of Darmstadt – Institute for mechatronic Systems, Otto-Berndt-Straße 2, 64287 Darmstadt, Germany, schuessler@ims.tu-darmstadt.de

b Technical University of Darmstadt – Institute for mechatronic Systems, Otto-Berndt-Straße 2, 64287 Darmstadt, Germany

c Aalto University School of Engineering - Department of Mechanical Engineering, Otakaari 24, 02150 Espoo, Finland, tuomas.tiainen@aalto.fi

d Technical University of Darmstadt – Institute for mechatronic Systems, Otto-Berndt-Straße 2, 64287 Darmstadt, Germany

Abstract

To reduce CO2 emissions, the share of renewable energies in the grid increases. At the same time, many sectors like the transport and the building sector are changing to be powered by electricity. Especially electric cars demand high peaks in current from the grid. Storage is needed to balance demand and supply of electric energy. Flywheels can be part of the solution as they can be charged and discharged with high power and do not suffer from losing significant capacity even after thousands of cycles. Minimum loss of energy is crucial for a flywheel therefor active magnetic bearings (AMB) are used. If a malfunction of the AMB occurs the rotor falls into a touch-down bearing (TDB). To decide whether further maintenance in case of a drop-down event is needed information about the forces stressing the TDB is important. To avoid costs for physical sensors soft sensors are a suitable solution. In this research, a data-driven soft sensor based on recurrent neural networks is created to calculate the forces during the drop-down event. As input data only the position of the rotor is used. A test rig with physical sensors applied to every TDB supplies the force data to train, validate, and test the soft sensor model. Three different network architectures are compared. The results show that the sensor can calculate whether the rotor hits a TDB and is also capable of predicting the peaks in the force signal.

Keywords: *soft sensor; AI; neural network; touch down bearing; backup bearing*

1. Introduction

To tackle climate change, the share of renewable energies in the grid increases. At the same time, the overall demand for electrical power rises, and consumers like charging stations for electric cars request high peaks in current from the grid. To balance production and demand storage is needed. Flywheels represent a suitable storage system for several occasions as they can be charged and discharged with high power and do not suffer from losing significant capacity even after thousands of cycles. To lower the loss of energy within the storage the rotor normally runs in an evacuated room and active magnetic bearings (AMB) are used to support the rotor. In the case of a malfunction of the AMB touch-down bearings (TDB) are applied to prevent a destructive rotor-stator contact elsewhere in the system. The design of TDB is a research topic due to the challenging conditions. TDB must endure high forces and rotational speeds caused by the rotor but also by working in evacuated areas which complicates lubrication and hinders heat dissipation. If a drop-down event occurs, one needs to decide whether maintenance of the TDB needs to be conducted. One way would be to disassemble the flywheel and do a visual check of the bearings. Another approach can be to check the forces over time that stressed the TDB during the drop-down event. The forces can be measured by using piezo electric sensors or strain gauges. In both ways, the sensors need to be applied to every TDB unit. When using planetary TDB this can mean that at least six sensors in the overall flywheel are required, which drives up the cost of the system.

Soft sensors or virtual sensors can overcome this problem. They use existing data and calculate a desired output variable. Soft sensors can be divided into model-driven and data-driven soft sensors (Jiang et al., 2021). The model-driven approach can be used if the knowledge about the physical relations between input and out variables is well known and understood. If this is not the case data-driven soft sensors come into play. They can be based on different methods like partial least squares (PLS) (Kano and Nakagawa, 2008), principal component analysis (PCA) (Zhang et al., 2020), or support vector machines (SVM) (Yan et al., 2004). In the last few years, soft sensors based on neural networks became more popular (Yuan et al., 2020). When it comes to the prediction of time-related data recurrent neural networks (RNN) show a good ability to understand and utilize the information which is embedded in the time series (Du and Swamy, 2019). A drawback of an RNN is the vanishing or exploding gradient problem especially when longer series of data is computed (Yuan et al., 2020). To overcome this problem, the long short-term memory (LSTM) architecture was developed by Hochreiter and Schmidhuber (1997).

The LSTM architecture was applied to several soft sensor applications. ElSaid et al. (2016) used an LSTM network to predict the vibration of an aircraft engine up to ten seconds into the future by using several input variables measured already by the engine. Bellone et al. (2020) showed that an LSTM is capable of estimating states like the NOx values of a combustion engine of a car as a real-time application. In the work of Miettinen et al. (2021) the displacement of rotors used in the paper industry was estimated by a bidirectional LSTM using the bearing reaction forces.

This paper presents an approach where a soft sensor based on different neural networks was trained to predict the forces on the TDB.

2 Method

This section describes the test rig to explain the overall setup and to show where the data used for the NN is measured. Afterward, the preprocessing of the data is explained. The different network architectures are further described and the methodology to analyze the results is explained.

2.1 Test Rig

To test the planetary TDB of the flywheel, a special TDB test rig for drop-down experiments in planetary TDB has been built at the Institute for Mechatronic Systems at the Technical University of Darmstadt. Even if the TDB test rig is built as a robust inner rotor system to withstand multiple drop-downs it has similar rotor dynamic properties as the flywheels (*Proceedings of ISMB 16*, 2018; Quurck et al., 2017; Quurck et al., 2018). The rotor has a mass of 18 kg and a maximal rotational speed of 20,000 rpm. This results in a surface velocity of the rotor at the planetary TDB of 230 m/s. Figure 1 shows a partial section view of the test rig without its base and a detailed view of the planetary TDB itself.

As the flywheels, the TDB test rig is levitated magnetically in all axes and driven by a permanent magnetic synchronous machine. For a drop-down experiment, the radial AMB are switched off, while the axial AMB remains active. This is because in the flywheel the axial levitation is performed by a passive permanent magnetic bearing which is not expected to fail. Since the purpose of the TDB test rig is to test the planetary TDB till failure the system has a secondary TDB which only gets in contact with the rotor if the planetary TDB has failed. The system is equipped with eddy current position sensors from eddylab, which are needed for the AMB and the drop-down evaluation. For a deeper analysis of the drop-down experiments, the system is equipped with multiple further sensors. For example, the force is measured at every TDB unit. In all, in this paper investigated drop-down experiments one TDB unit in each plane is equipped with a piezoelectric force sensor from pcb of type 211B. The charge amplifier is of type 5073A411 from Kistler. At the remaining TDB units, the forces in radial and tangential direction are measured with structurally integrated strain gauges. For the data acquisition, a system from National Instrument is used.

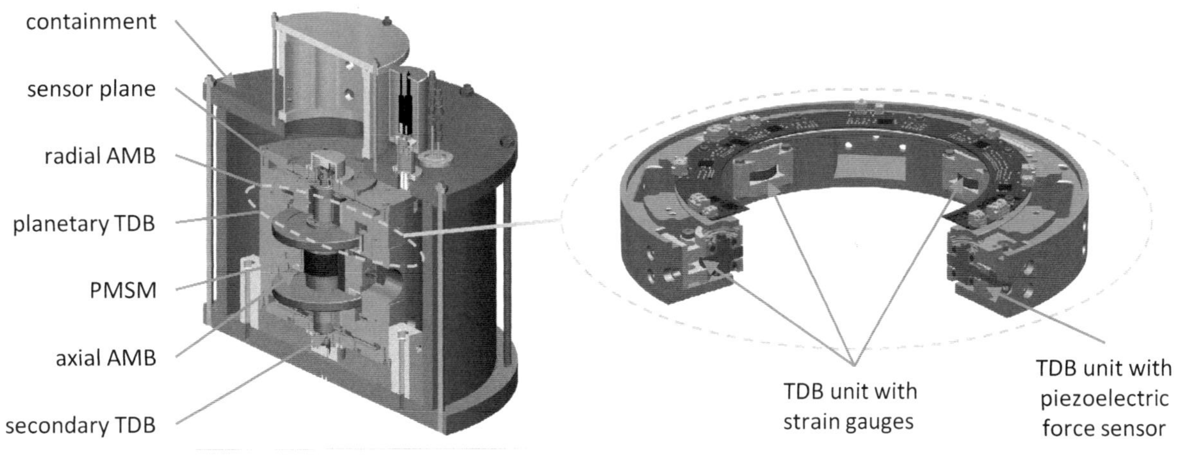

Figure 1 labeled partial section view of the TDB test rig without the base and a detailed view of the planetary TDB with 3 TDB units with strain gauges and one TDB unit with a piezoelectric force sensor.

2.2 Modeling

The test series that was used for the training, validation, and testing of the neural network contains 57 drop-down event experiments. The rotational speed when the drop-down was initiated varies between 2500 and 20,000 rpm. The force data given by the strain gauges were sampled at 20 kHz and the position data was sampled at 10 kHz. The length of one drop-down experiment varies with the initial rotational speed and goes up to 300 s, the higher peaks in force on the TDBs normally occur during the first 100 s. This leads to a maximum amount of 6,000,000 samples per sensor signal per test series. For the modelling the goal was to use only sensor signals for the soft sensor which are already available in flywheels. Therefore, even if the TDB test rig is equipped with multiple sensors, only the position measurement is used. For the training, validation, and testing 23 test series were used, 14 for training, five for validation, and four for testing.

The main usage of the soft sensor is to calculate the forces after a drop-down event to then decide whether maintenance is required. This also means that the sensor does not need to be real-time capable. The position data of the four sensors were fed in sequences into the soft sensor and the output of the soft sensor is a sequence of force data with the same length. It was investigated whether it is beneficial building one soft sensor that estimates the force on every TDB or build one soft sensor for each TDB. As the in- and output sequences of data are used the task can be described as a sequence-to-sequence regression task. The length of the input and output sequence was changed in length between 512 and 4096 samples to figure out what gives the best results. To increase the amount of training data the stride length between two sequences which are extracted from the test series was not equal to the length of the sequences. The stride length was changed in different tests from 32 samples to 1024 samples. Next to gaining more training data, the goal was to utilize high peaks in the force data in more than one sequence. As peaks in the force data can occur in more than one sequence it is important to not pick sequences randomly and distribute them to the training, validation, and test dataset. This would lead to a situation where the same data appears in training and testing. To avoid this case the different experiments were distributed into training, validation, and test data. Only 23 of the 57 experiments were chosen because the other ones were conducted with a lower rotational speed of the rotor at the drop-down event. To reduce computation effort sequences without any peaks in force were deleted. This was done by comparing the maximum force data of a sequence to a threshold. As there is noise on the force data this threshold was set at 500 N. As described before the sampling rate for the position data and the force data measured by the strain gauges are not the same. In this case, up-sampling for the position data was conducted. This was achieved by adding values between every other position data point by linear interpolation. In Figure 2 one can see an example sequence of the in- and output data. On the left, the orbit of the rotor during an example sequence is shown for the upper sensor plane. The green and red lines determine the part of the trajectory where the rotor hits the TDB and a force is measured. The measured force during the sequence can be seen on the right side of the figure. In this case, there is an offset on the sensor signal of about 100 N and some noise with an amplitude of around 50 N. The peak in Force reaches nearly 3200 N. The peak has a length of around 50 samples which represents 2.5 ms.

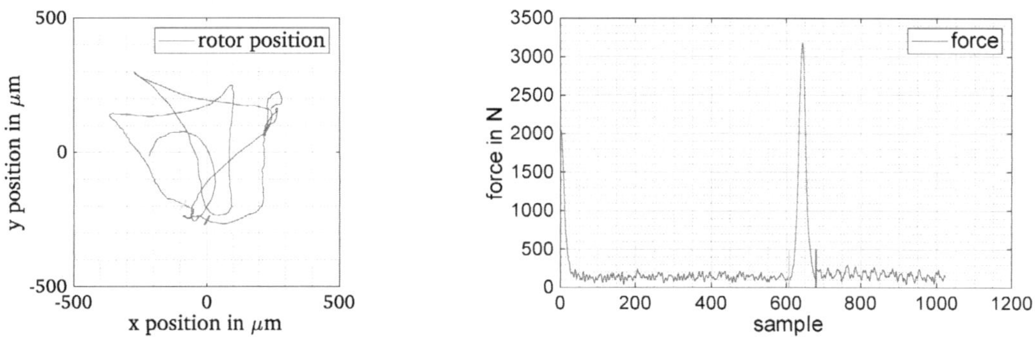

Figure 2 left: rotor trajectory in the upper sensor plane; right: measured normal force on TDB in the upper plane.

As the task is a sequence-to-sequence regression where the relation in time between different data points is important RNN were chosen. More specifically the LSTM and the gated recurrent unit (GRU) architecture were compared. In an RNN the output of one cell of the network is fed into the network in the next timestep again. This can lead to a vanishing or exploding gradient problem when it comes to training the network. Both the LSTM and the GRU architecture tackle this problem by adding a path where information can flow without being multiplied by any weights which could lead during the training to the vanishing or exploding gradient problem. The GRU architecture needs less computational effort which can be beneficial if a real-time application is on hand (Han et al., 2023). The bidirectional LSTM architecture is an addition to the normal LSTM architecture. Here, the input sequence is computed forward through and network and backward. In real-time applications, this can be crucial as a leg in time for the prediction appears. The number of layers and the number of hidden states were changed during the trials. After the RNN layers, fully connected layers were added to the network. In the end, the loss was determined. Here, the mean absolute error (MAE), the mean squared error (MSE), and the mean cubic error (MCE) were chosen. For the training of the networks, the Adam solver was used with a learning rate decreasing over time. The validation dataset was evaluated four times per epoch. The training stopped when the loss of the validation dataset did not decrease five times in a row.

To compare the results of the different architectures two methods were used. First, the overall root mean cubic error (RMCE) between the predicted values and the measured values from the test rig was computed according to (*1*). Here, n represents the total number of sequences and s the total number of samples per sequence. The RMCE was chosen because in the calculation of the lifetime of a grooved ball bearing the force affects the lifetime with the third power (Childs, 2019).

$$RMCE = \frac{1}{n\,s} \sqrt[3]{\sum_{1}^{n} \sum_{1}^{s} |(F_{measured_{n,s}} - F_{predicted_{n,s}})|^3} \quad (1)$$

Next to the RMCE further attention was paid to the height of the peaks in force. Therefore, the peaks were categorized into different groups according to their height. Every group has a range of 500 N, with the first group going from 500 to 1000 N. The number of groups depends on the height of the highest peak of the test data. Then, the peaks in the test data in every sequence were localized. The height of the peak and sample number of the location in the sequence were stored. Afterward, in the corresponding predicted sequence, the highest predicted value around the stored location was searched. For every group of peaks, the root mean squared error (RMSE) was computed by comparing the heights of the measured peaks from the test rig with the heights of the predicted peaks. Figure 3 shows an exemplary diagram of the accuracy of a network for different groups of peaks. In this case, the calculated RMSE for each group was divided by the force value of the group to gain a deviation in percentage. As the groups have a width of 500 N the median of the group boundaries was used to calculate the percental deviation. In the measurement shown in Figure 3, there was no peak with a value between 3500 and 4000 N and therefore there is no bar visible.

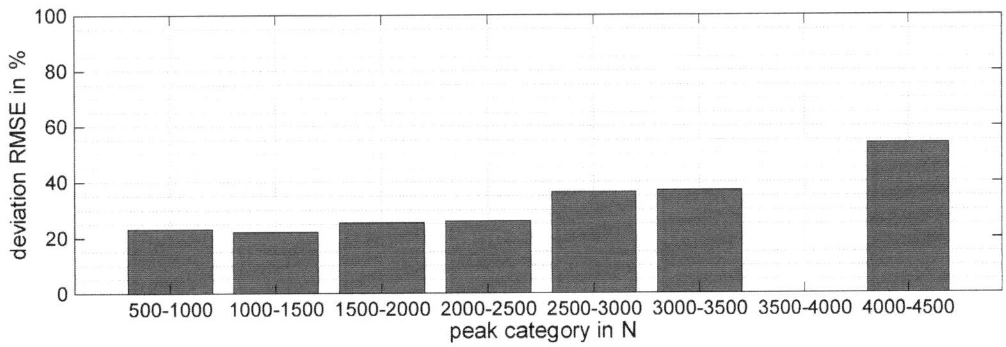

Figure 3 example of the accuracy in predicting peaks of different heights in percentage.

3 Results

Different architectures were compared on the data of the test rig. For the comparison the sequence length of the in- and output data was set to 1024 samples and the stride length between two sequences to 512 samples. The architectures can be seen in Table 1. "3 x 512 LSTM" means that three LSTM layers with 512 hidden units are used. "1 x 128 FC" represents one fully connected (FC) layer with 128 neurons. For every network, an MSE regression layer was added.

Table 1 network architectures

LSTM 1	LSTM 2	BiLSTM 1	BiLSTM 2	GRU 1	GRU 2
3 x 512 LSTM 1 x 128 FC 1 x 64 FC 1 x 32 FC 1 x 16 FC	3 x 256 LSTM 1 x 256 FC 1 x 128 FC 1 x 64 FC 1 x 32 FC 1 x 16 FC 1 x 8 FC	3 x 512 BiLSTM 1 x 128 FC 1 x 64 FC 1 x 32 FC 1 x 16 FC	3 x 256 BiLSTM 1 x 256 FC 1 x 128 FC 1 x 64 FC 1 x 32 FC 1 x 16 FC 1 x 8 FC	3 x 512 GRU 1 x 128 FC 1 x 64 FC 1 x 32 FC 1 x 16 FC	3 x 256 GRU 1 x 256 FC 1 x 128 FC 1 x 64 FC 1 x 32 FC 1 x 16 FC 1 x 8 FC
MSE regression layer					

In Figure 4, an example sequence for the force on one TDB is shown. The sequence is 1024 samples long which represents 51.2 ms. The force on the TDB measured in the test rig is visualized in blue and the force predicted by the network in orange. One can observe that especially the positions of the peaks in force are accurately predicted by the network. This means that the network can predict whether contact between the rotor and the TDB occurred or not. The heights of the peaks by trend seem to be predicted a bit lower compared to the measured ones. On the right side of Figure 4 the accuracy of the network for predicting peaks in different groups is visualized. The deviation is given in percentage as it was shown in Chapter 2.2. It is noticeable that the accuracy rises when the height of the peaks decreases. What needs to be mentioned is that the number of peaks that could be used to calculate the deviation in the prediction was rather low for the groups with high peaks. The value for the group from 4000 to 4500 N was only calculated on the one peak shown in the plot on the right. For the groups 2500 to 3000 N and 3000 to 3500 N, less than ten peaks were available.

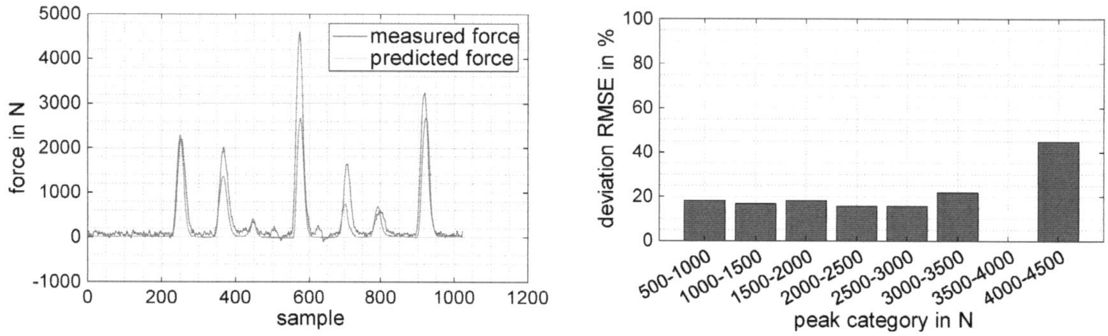

Figure 4 left: predicted (using 1. BiLSTM network) and measured force sequence of the test data; right: peak accuracy in percentage.

To compare the different architectures the overall RMCE was computed as described in Chapter 2.2. At the same time, the percental accuracies as shown in Figure 4 were computed for each network on the test data. Afterward, the percental accuracies for all six different architectures were visualized in one boxplot shown in Figure 5. The percental accuracy for the group of peaks from 4000 to 4500 N in Figure 4 of around 45 % can be spotted as an outlier value in Figure 5 in the box of the first BiLSTM network. On the right side of Figure 5, the overall RMCE for each network architecture is visualized. Here, one can see that the lowest RMCE values are achieved by the two BiLSTM networks.

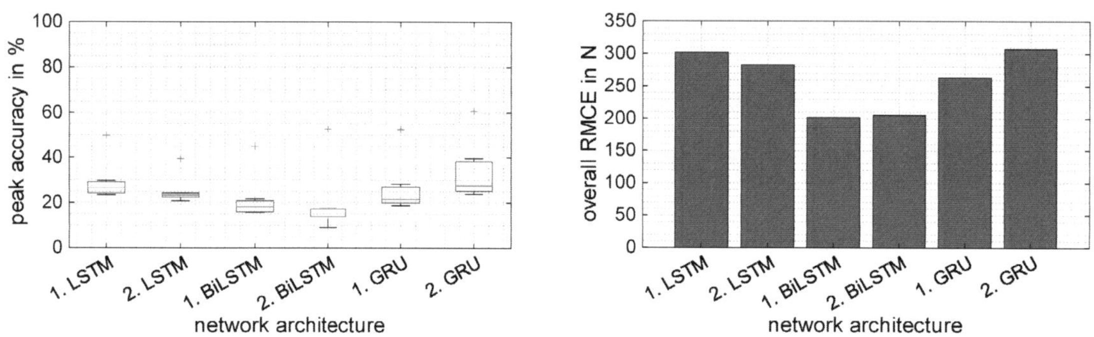

Figure 5 left: boxplot of the different peak accuracies of each network; right: overall RMCE of each network

4 Conclusion

This paper investigated utilizing soft sensors to measure the force on planetary TDB during drop down events. The data was collected on a test rig which was built for research on different planetary TDB designs. As input data for the soft sensor only the position data of the rotor in the two sensor planes was used. For the calculation of the force signal, different RNN were tested and compared by their overall RMCE and their accuracy in predicting peaks of the force signal on a test data set. Here, the accuracy of the network based on a BiLSTM structure performed best. The bidirectional path leads to leg in prediction which is not a problem in this case, because this is not a real-time application. An accuracy of predicting the heights of the peaks of around 20 % can be achieved for lower peaks. In these results, the accuracy for higher peaks goes down to 40 %. As the number of peaks with this height in test data was very low this value is less meaningful. The motivation of this work was to measure the force on the TDB to be able to take decisions on whether maintenance is necessary after a drop-down event. Therefore, a future research question could be, which force impact leads to damages in the system that need to be fixed before running the system again. In this work, the networks were trained on data measured on a test rig. In the future, it could be beneficial to use data that is gained from a simulation of the test rig for training and then testing the network on measured data from the actual test rig. This could be useful for developing flywheels where no sensors for the force on the TDB are applied.

Acknowledgment

This research was funded by the German Federal Ministry for Economic Affairs and Energy, grant numbers 03ET6064A and 03EI3000A.

References

Bellone M, Faghani E and Karayiannidis Y (2020) Comparison of CNN and LSTM for Modeling Virtual Sensors in an Engine. In: *SAE Technical Paper Series:* APR. 21, 2020. SAE International400 Commonwealth Drive, Warrendale, PA, United States.

Childs PRN (2019) *Mechanical design engineering handbook*. Oxford, Cambridge, MA: Butterworth-Heinemann.

Du K-L and Swamy MNS (2019) *Neural Networks and Statistical Learning*. London: Springer London.

ElSaid A, Wild B, Higgins J, et al. (2016) Using LSTM recurrent neural networks to predict excess vibration events in aircraft engines. In: *2016 IEEE 12th International Conference on e-Science:* Baltimore, MD, USA, 23.10. - 27.10., pp. 260–269. IEEE.

Han S, Yoon K, Park G, et al. (2023) Hybrid State Observer Design for Estimating the Hitch Angles of Tractor-Multi Unit Trailer. *IEEE Transactions on Intelligent Vehicles* 8(2): 1449–1458.

Hochreiter S and Schmidhuber J (1997) Long short-term memory. *Neural computation* 9(8): 1735–1780.

Jiang Y, Yin S, Dong J, et al. (2021) A Review on Soft Sensors for Monitoring, Control, and Optimization of Industrial Processes. *IEEE Sensors Journal* 21(11): 12868–12881.

Kano M and Nakagawa Y (2008) Data-based process monitoring, process control, and quality improvement: Recent developments and applications in steel industry. *Computers & Chemical Engineering* 32(1-2): 12–24.

Miettinen J, Tiainen T, Viitala R, et al. (2021) Bidirectional LSTM-Based Soft Sensor for Rotor Displacement Trajectory Estimation. *IEEE Access* 9: 167556–167569.

Proceedings of ISMB 16 (2018).

Quurck L, Franz D, Schussler B, et al. (2017) Planetary backup bearings for high speed applications and service life estimation methodology. *Mechanical Engineering Journal* 4(5): 17-00010-17-00010.

Quurck L, Viitala R, Franz D, et al. (2018) Planetary Backup Bearings for Flywheel Applications. In: *Proceedings of ISMB 16:* Beijing, China, August 13.-17.

Yan W, Shao H and Wang X (2004) Soft sensing modeling based on support vector machine and Bayesian model selection. *Computers & Chemical Engineering* 28(8): 1489–1498.

Yuan X, Li L and Wang Y (2020) Nonlinear Dynamic Soft Sensor Modeling With Supervised Long Short-Term Memory Network. *IEEE Transactions on Industrial Informatics* 16(5): 3168–3176.

Zhang B, Han Y, Yu B, et al. (2020) Novel Nonlinear Autoregression with External Input Integrating PCA-WD and Its Application to a Dynamic Soft Sensor. *Industrial & Engineering Chemistry Research* 59(35): 15697–15706.

Practical results of ADRC applied to Radial Position Control of a Bearingless Induction Machine

Rodrigo TEIXEIRA [a], Werbet SILVA [b], Jossana FERREIRA [a], José PAIVA [c], Andrés SALAZAR [a], Andre L MAITELLI [a]
a Universidade Federal do Rio Grande do Norte, Natal-RN, Brazil,
rodrigoandradeteixeira@gmail.com, jossana.ferreira@ufrn.br, andres@dca.ufrn.br, maitelli@dca.ufrn.br
b Instituto Federal de Brasília, Taguatinga-DF, Brazil, werbethluizz@hotmail.com
c Instituto Federal do Rio Grande do Norte, Natal-RN, Brazil, jalvaro.paiva@gmail.com

Abstract

A Bearingless Induction Machine (BIM) with split winding has an operation strongly dependent on well-designed control routines, since the magnetic shaft suspension is a naturally unstable system. Furthermore, this system is multivariable, non-linear, and it has coupling between variables. Because of this, linear position usually presents some limitations in dealing with the system complexity. In recent years the Active Disturbance Rejection Control (ADRC) has been used to control several types of systems and has shown good results in the control of systems with strong uncertainties and nonlinearities. The main idea of the ADRC is to group all the uncertainties of the plant, such as non-modeled dynamics, external disturbances, and variations in parameters, as an additional state variable to be estimated in real-time using the input and output variables. Thus, the additional state variable can be estimated and compensated in the control loop. In the work of Silva (2021) the control of a BIM with split winding with ADRC as position controller was proposed and results from a simulated system were shown. In this work, the results obtained by Silva (2021) were improved with experimental results, using a simplified version of the ADRC controllers and a BIM with a reduced driving scheme. The results showed it is possible to use the ADRC with a split winding BIM and will be the basis for future improvements on the control strategy.

Keywords: *ADRC, position controller, bearingless induction machine, split winding.*

1. Introduction

In the last two decades, magnetic bearing technology has attracted significant attention from professionals and researchers due to its potential to perform in high-speed applications in environments where maintenance routines are more complicated. Magnetic bearing technology has been applied in different types of electrical machines. The induction motor is often chosen due to its robustness, simplicity, and low cost.

In [2], an induction motor with magnetic bearings was proposed, where the group of coils that generates rotational torque also generates radial forces. This structure is known as split winding and simplifies operation in addition to improves the utilization of the motor's internal space. The same approach was used in [3] for a three-phase induction motor, resulting in a system capable of generating higher radial forces.

The magnetic suspension of a machine shaft is a naturally unstable dynamic with multiple variables, nonlinearities, and couplings [4]. In this sense, linear controllers may not achieve a sufficiently fast and stable response for proper suspension. The use of microprocessors and power electronic devices allows the application of advanced control theories such as predictive control, sliding mode control, neural network control, fuzzy control, and others [5]. In recent years, ADRC has proven to be a promising control alternative due to its simplicity and ability to handle model uncertainties, nonlinearities, and disturbances [6]. ADRC groups unmodeled dynamics, parameter variations, and external disturbances into a variable estimated in real-time by an extended state observer (ESO) and utilizes this estimation to compensate for their effects on the control loop.

In [1], the first phase of the project was carried out which aimed to apply the ADRC technique to the radial position control of an induction machine with split winding and magnetic suspension. On this occasion, the application of the linear and nonlinear ADRC structures in the shaft position control was simulated. This paper shows the practical results of applying an extended nonlinear state observer to a split-winding induction machine.

2. Bearingless Machine with Split Winding

In the stator of the prototype used, there are three groups of windings divided to generate radial force and torque. The coils are distributed in the stator as shown illustratively in Figure 1 (a). As an example, phase A is divided into A1 and A2. When the same current (I_a) flows through coils A1 and A2, a magnetic flux is created, as represented in Figure 1 (b) by solid blue lines. When the current increases by Δi_a in coil A1 and simultaneously decreases by Δi_a in coil A2, the flux density increases on one side of the rotor while it decreases on the opposite direction with the same magnitude. The unbalance in the original flux generated by Δia generates radial force in the positive direction of the axis X. Likewise, it occurs in the opposite direction when the additional current is added to coil A2 and subtracted from coil A1.

Extending this mode of operation to the three phases and decomposing the coordinate system of the position of the coils as shown in Figure 1 in the coordinate system xy, it is possible to control the radial position in the entire bounded plane for xy.

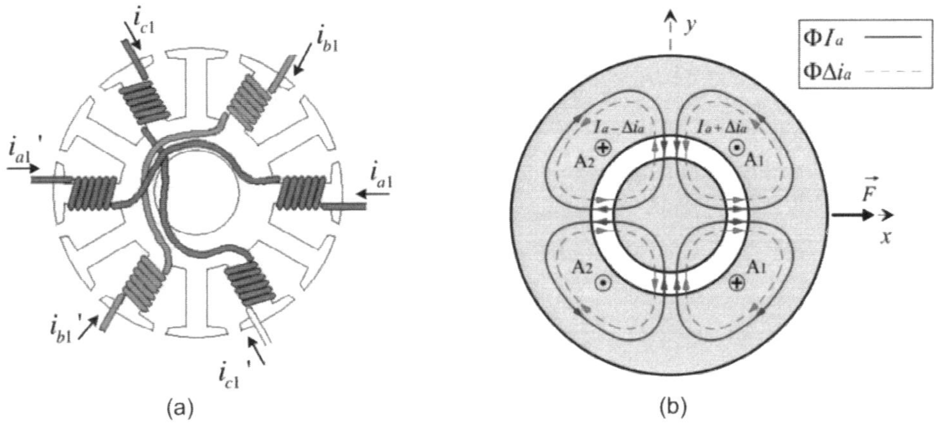

Figure 1 (a) windings distribution (b) radial forces generation.

The prototype used works vertically and is supported on the lower end of the shaft by a pivoted bearing that prevents radial and axial movements but allows angular movements. This type of system has two degrees of freedom for position control and the rotor displacement occurs in the directions of the x and y-axes.

From the application of the second law of motion to Figure 2, [7] developed a linear dynamic representation for the rotor radial displacement system. In this dynamic representation represented in Equation 1, $y(t)$ represents the radial displacement in each of the axes, $u(t)$ represents the input current signal conditioned by the control action, and $w'(t)$ is an output disturbance.

$$\ddot{y}(t) = 8374y(t) + 3680000u(t) + w'(t) \tag{1}$$

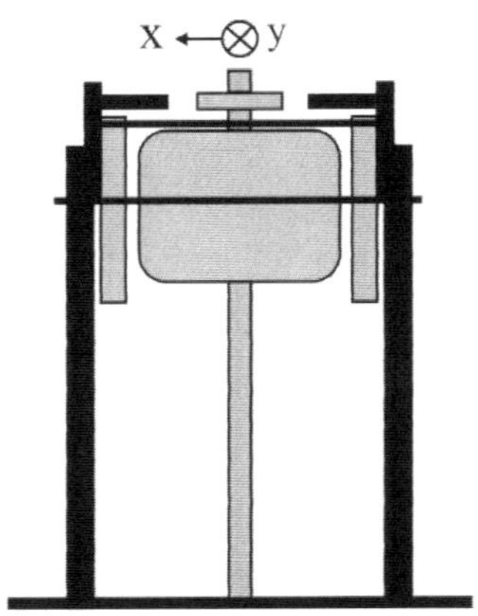

Figure 2 Machine diagram.

3. ADRC

The first version of the Active Disturbance Rejection Control (ADRC) was presented in full in Chinese by its inventor, Jingqing Han, in a work entitled "Auto disturbances rejection controller and its applications" in 1998 [8]. According to [9], "ADRC is a robust control strategy in which unknown dynamics and external disturbances are estimated and discarded in real-time, forcing the real system to behave as an approximate model. The idea is to combine internal and external disturbances as a variant parameter in the single time, the total disturbance, and use it as a fictitious state variable".

3.1 ADRC Concept

Although ADRC is applicable to systems of order n, linear and nonlinear systems, variant and time-invariant systems, and systems with multiple inputs and outputs, for the sake of simplicity its basic concepts have been shown using the second-order dynamics represented in Equation 2. In this equation, $y(t)$ represents the output position of a controlled object, b is a constant, $u(t)$ is the input force, $w(t)$ represents external disturbances, and $f(y(t),\dot{y}(t),w(t),t)$ represents the combined effects of internal dynamics and external disturbances on the object's acceleration.

$$\ddot{y}(t) = f(y(t),\dot{y}(t),w(t),t) + bu(t) \quad (2)$$

Adopting a disturbance rejection mechanism, the process represented in Equation 2 can be viewed as a double integrator plant, with gain b, and disturbed by $f(y(t),\dot{y}(t),w(t),t)$. The term $f(y(t),\dot{y}(t),w(t),t)$ is defined as generalized disturbance and in ADRC is the focus of the control design.

By specifying $\hat{f}$ as an estimate of the generalized disturbance at time instant t, the input represented in Equation 3 reduces expression 2 to the easily controllable double integrator of expression 4.

$$u(t) = (-\hat{f} + u_0)/b \quad (3)$$

$$\ddot{y}(t) = u(t) \quad (4)$$

The main difference between this approach and model-based approaches is that ADRC assumes no analytical expression for $f(y(t),\dot{y}(t),w(t),t)$.

3.2 Extended State Observer

The extended state observer proposed by J. Han [10] became practical through the tuning method proposed in [11] which simplified its implementation. The idea is based on using an extended state space model of Equation 2 that includes the generalized disturbance $f(y(t),\dot{y}(t),w(t),t)$ as an additional state. Considering the system output $y(t)$ as state $x_1(t)$, the output derivative $\dot{y}(t)$ as state $x_2(t)$, and the derivative of $\dot{y}(t)$ as extended state $x_3(t)$, this state becomes approximately equal to the generalized disturbance $f(y(t),\dot{y}(t),w(t),t)$.

Thus, a state observer with appropriately selected gains provides an estimation of the states $x_1(t)$, $x_2(t)$, and $x_3(t)$. Consequently, it provides an estimation of the generalized disturbance through the extended state $x_3(t)$. With the estimation of $f(y(t),\dot{y}(t),w(t),t)$ from the extended state $x_3(t)$ it is possible to use the input shown in Equation 3 and reduce the system to the double integrator of Equation 4.

The original structure of an ADRC controller is composed of a Tracking Differentiator (TD) block, an Extended State Observer (ESO), and a non-linear control structure block. In this work, the TD structure was not used.

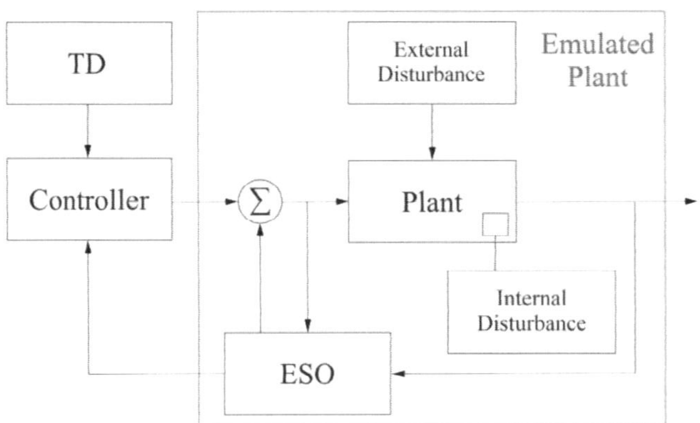

Figure 3 ADRC general diagram.

3.3 NLESO Structure

In the ESO structure, the generalized disturbance $f(y(t), \dot{y}(t), w(t), t)$ is considered an additional state of the system.

$$\begin{aligned} \dot{x}_1(t) &= x_2(t) \\ \dot{x}_2(t) &= x_3(t) + b_0 u(t) \\ \dot{x}_3(t) &= h \\ y(t) &= x_1(t) \end{aligned} \tag{5}$$

Where $x_3(t) = f(y(t), \dot{y}(t), w(t), t)$ is the augmented state and $h = \dot{f}$, its derivative. h is seen as an unknown disturbance. The above system can be rewritten in state space by:

$$\begin{aligned} \dot{x}(t) &= \boldsymbol{A}x(t) + \boldsymbol{B}u(t) + \boldsymbol{E}h(t) \\ y(t) &= \boldsymbol{C}z(t) \end{aligned} \tag{6}$$

Where:

$$\boldsymbol{A} = \begin{bmatrix} 0 & 1 & 0 \\ 0 & 0 & 1 \\ 0 & 0 & 0 \end{bmatrix}, \quad \boldsymbol{B} = \begin{bmatrix} 0 \\ b_0 \\ 0 \end{bmatrix}, \quad \boldsymbol{C} = [1 \quad 0 \quad 0], \quad \boldsymbol{E} = \begin{bmatrix} 0 \\ 0 \\ 1 \end{bmatrix}.$$

From the extended representation of the system, an observer can be constructed to estimate numerical values for $z_1(t)$, $z_2(t)$ and $z_3(t)$, which are respectively the estimated values of the states $x_1(t)$, $x_2(t)$ and $x_3(t)$. The work [12] proposed a nonlinear structure for the ESO. This structure is referred to as NLESO and can be written mathematically as:

$$\begin{aligned} \dot{z}_1(t) &= z_2(t) - \beta_1 fal(e, \alpha_1, d_1) \\ \dot{z}_2(t) &= z_3(t) - \beta_2 fal(e, \alpha_2, d_2) + b_0 u(t) \\ \dot{z}_3(t) &= -\beta_3 fal(e, \alpha_3, d_3) \end{aligned} \tag{7}$$

Where β_1, β_2 and β_3 are the observer gains and $fal(e, \alpha, d)$ is a nonlinear function, defined by:

$$fal(e, \alpha, d) = \begin{cases} |e|^{\alpha} sgn(e), & |e| > d \\ e/d^{1-\alpha}, & |e| > d \end{cases} \tag{8}$$

the value of e is the error between the system output and its estimate and α and d are turning parameters. The block diagram of the NLESO applied to a generic second-order plant is shown in Figure 4.

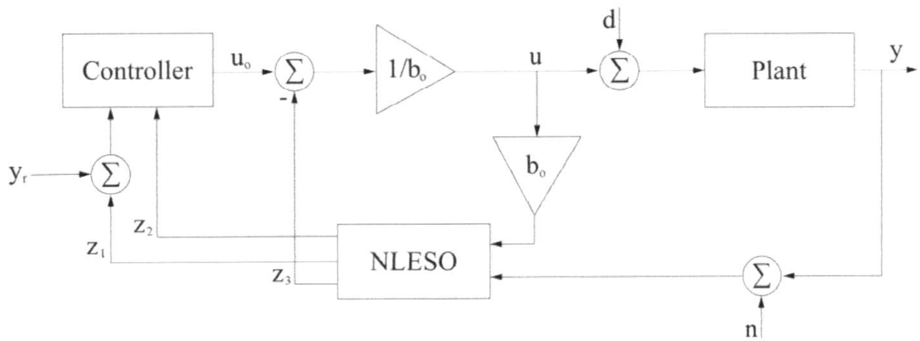

Figure 4 Block Diagram of an ADRC control loop.

4. Description of the System

Figure 5 provides an overview of the experimental bench used in the work. The equipment was grouped into the following blocks: monitoring and control, power supply, and bearingless machine. The bearingless machine block consists of the bearingless electric induction machine on which the tests were performed. The power supply circuit block comprises the entire infrastructure of electronic components and equipment responsible for providing the power supply at the appropriate levels to control the machine variables. The monitoring and control block consists of the electronic circuits, processors, algorithms, and monitoring devices that were necessary to implement the system's control and supervision routines. The variable i_{abxy} represents the four-phase electrical connections that allow the drive block to supply controlled electrical currents to the bearing motor. Figure 5 also indicates the reading of these i_{abxy} currents performed by the monitoring and control block. The variable Δxy represents the signals corresponding to the radial positions of the x and yy axes of the bearing motor using by the monitoring and control block. Finally, the arrow named *PWM* indicates the signals sent by the monitoring and control block to the drive block.

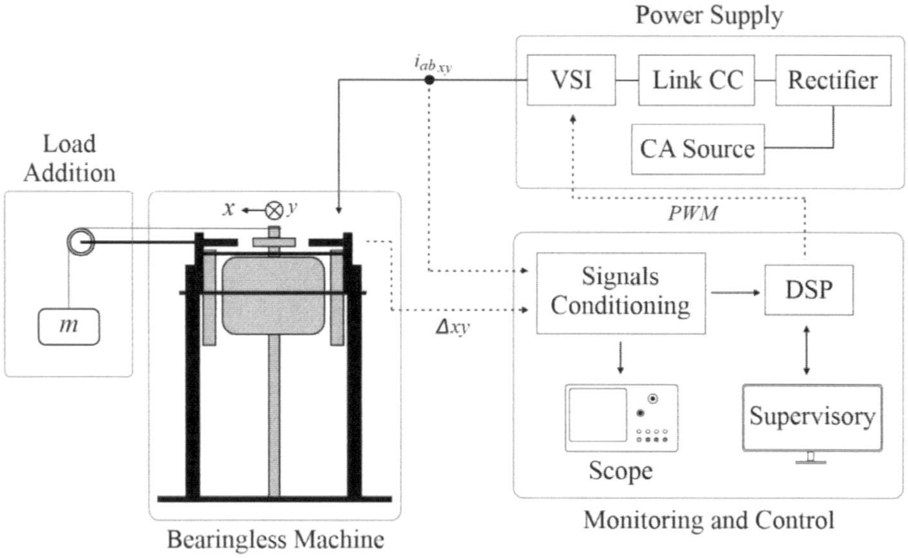

Figure 5 Components of the experimental bench.

Figure 6 shows a real image of the experimental bench. This structure is in the Laboratory in the Computer and Automation Engineering Department at UFRN.

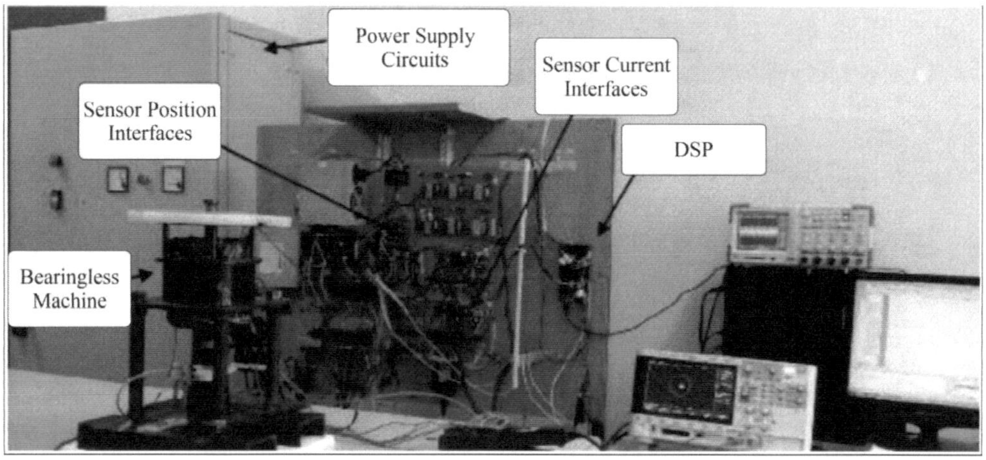

Figure 6 Experimental bench.

5. Results

From the simulated results by Silva (2021) it could be seen that ADRC with the linear and nonlinear structure are successful in providing a stable response to the system. However, the nonlinear structure of the ADRC exhibits a faster and more damped response in the presence of disturbances. For this reason, this structure was used in obtaining the practical results shown in this paper.

Figure 7 shows the response of the radial position control in the x and y directions with ADRC controllers in two different moments. At first, the y direction is set constant and the x position is variable (Figures 7(a) and 7(b)), then, the x direction is set constant and y position is variable (Figures 7(c) and 7(d)). The results obtained during this test show that step variations in one of the controlled axes do not have a significant influence on the dynamic behavior of the opposite axis.

To analyze the robustness of the controller to external disturbances an apparatus as shown in Figure 5 was used. The states z_1, z_2 and z_3 of the x and y axes were obtained at a time interval that allowed the instant of adding radial

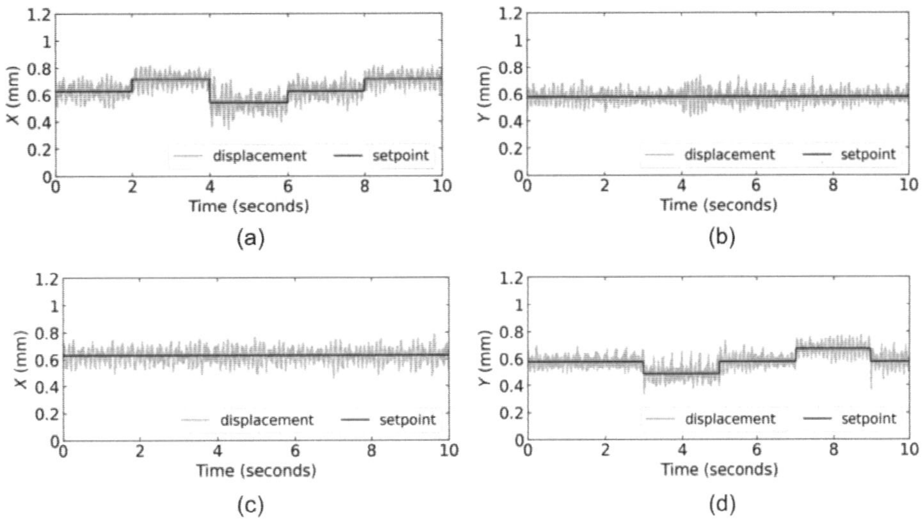

Figure 7 Rotor radial position with ADRC.
(a) X position with a constant reference (b) Y position with a constant reference.
(b) X position with a constant reference (d) Y position with a variable reference.

load to the machine shaft to be detected. In each graph in Figure 8, a dashed vertical line was inserted to indicate the instant of application of 150 grams aligned with the x-axis.

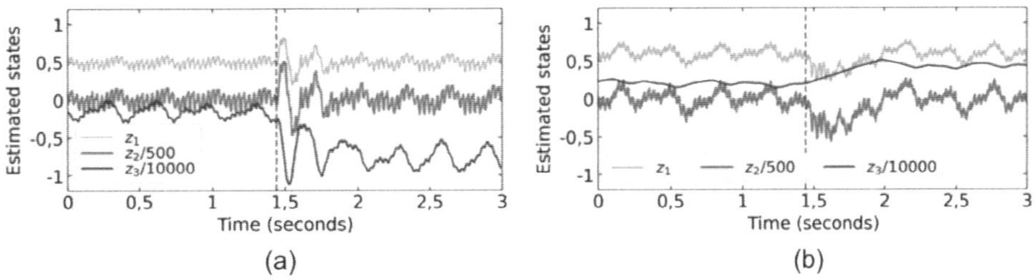

Figure 8 Estimated states behavior for NLESO under radial load application.
(a) X-axis states. (b) Y-axis states.

Considering the states z_1- x and y position, z_2 - derivative of z_1, and z_3 - derivative of z_2, it is possible to conclude by observing the z_1 states in Figures 8 (a) and (b) the controller was little influenced by the application of radial load, since the response was similar before and after the addition of the disturbance. The behavior of the states also indicated an aggressive reaction of the controller in the transient regime, which resulted in high z_2 values during this region in the graphs. Finally, the change in the mean value of z_3 indicates that the extended observer was able to identify the addition of the external disturbance inserted into the system.

6. Conclusions

In this paper, the application of the ADRC active disturbance rejection technique to a bearingless induction machine was analyzed. The results discussed were obtained using a nonlinear framework for the extended state estimator. The results for reference variation show good dynamic decoupling between the x and y axes. The test with load imposition showed that the implemented nonlinear observer was able to detect the generalized disturbance and maintained the average value at the output before and after load application. Finally, the results obtained show the success of the ADRC technique in the active estimation and rejection of disturbances in the radial positioning process of a split rotor bearingless machine.

7. Acknowledgments

This work was supported by Comissão de Aperfeiçoamento de Pessoal do Nível Superior (CAPES) and Programa de Pós-graduação em Engenharia Elétrica e de Computacão (PPgEEC/UFRN).

References

[1] Silva, Werbet, et al. "Application of Active Disturbance Rejection Control (ADRC) in the Radial Position Control of a Bearingless Induction Machine With Split Winding" Proc. 17th ISMB. 2021.

[2] A.O. Salazar and R.M. Stephan. A bearingless method for induction machines. IEEE Transactions on Magnetics, 29(6):2965–2967, 1993.

[3] Jossana Maria de Souza Ferreira. Modelagem de máquina de indução trifásica sem mancais com bobinado dividido. 2006.

[4] Zebin Yang, Chengling Lu, Xiaodong Sun, Jialei Ji, and Qifeng Ding. Study on active disturbance rejection control of a bearingless induction motor based on an improved particle swarm optimization–genetic algorithm. IEEE Transactions on Transportation Electrification, 7(2):694

[5] Zebin Yang, Jialei Ji, Xiaodong Sun, Huimin Zhu, and Qian Zhao. Active disturbance rejection control for bearingless induction motor based on hyperbolic tangent tracking differentiator. IEEE Journal of Emerging and Selected Topics in Power Electronics, 2019.

[6] Li Sun, Wenchao Xue, Donghai Li, Hongxia Zhu, and Zhi-gang Su. Quantitative tuning of active disturbance rejection controller for fopdt model with application to power plant control. IEEE Transactions on Industrial Electronics, pages 1–1, 2021.

[7] Francisco Elvis Carvalho Souza. Otimização da estrutura de acionamento para o controle de posição radial de um motor de indução trifásico sem mancais com enrolamento dividido. 2018.

[8] Huang, Deqing, Da Min, Yupei Jian & Yanan Li (2019), 'Current-cycle iterative learning control for high-precision position tracking of piezoelectric actuator system via active disturbance rejection control for hysteresis compensation', IEEE Trans. on Industrial Electronics.

[9] Ahmad, Saif, and Ahmad Ali. "Active disturbance rejection control of DC–DC boost converter: A review with modifications for improved performance." IET Power Electronics 12.8 (2019): 2095-2107.

[10] Han, Jingqing. "From PID to active disturbance rejection control." *IEEE transactions on Industrial Electronics* 56.3 (2009): 900-906.

[11] Gao, Zhiqiang. "Scaling and bandwidth-parameterization based controller tuning." *ACC*. 2003.

[12] Zhiqiang Gao. From linear to nonlinear control means: A practical progression. ISA transactions, 41(2):177–189, 2002.

Velocity Estimation on an Active Suspension for a Translational Electrodynamic Maglev System

Eugenio TRAMACERE[a], Marius PAKŠTYS[a], Luis Miguel CASTELLANOS[a], Renato GALLUZZI[b], Lorenzo BOSICA[a], Nicola AMATI[a], Andrea TONOLI[a],

a Department of Mechanical and Aerospace Engineering, Politecnico di Torino, 10129 Torino, Italy eugenio.tramacere@polito.it
b School of Engineering and Sciences, Tecnologico de Monterrey, Calle del Puente 222, Mexico City, 14380, Mexico

Abstract

The concept of Hyperloop was proposed over a decade ago. Since then, numerous efforts have focused on improving this technology towards commercial applications. Although promising, the electrodynamic levitation system is inherently unstable. However, instability can be mitigated by the introduction of a secondary suspension between the capsule and the bogie. To evaluate stabilization strategies, a dedicated test bench has been constructed. Nevertheless, these methods require a reliable estimation of the relative velocity between the capsule and the bogie for their successful practical implementation. This work focuses on the estimation of this velocity using a Kalman filter approach based on an augmented model of the electrical circuit of a voice coil actuator, which constitutes the secondary suspension of the presented electrodynamic levitation system. The proposed approach has been implemented and tested on a real-time ECU, demonstrating the capability of the method for applications involving active damping control.

Keywords: Hyperloop, electrodynamics, levitation, Kalman filter, velocity estimation

1. Introduction

The Hyperloop concept has emerged in the last decade as an ultra-high speed transportation system, with potentially low emissions. Urban connectivity as well as freight transportation may see overall improvements in terms of travel and lead time, motivating a research interest in this technology. The core operating principle exploited is passive electrodynamic levitation, where an array of permanent magnets interacts with the non-ferrous conductive track due to their relative translational velocity. Eddy currents generated in the track result in lift and drag forces experienced by the levitating pad. However, this levitation phenomenon is inherently unstable, and stability has been numerically achieved by introducing a secondary suspension that ensures a two degree of freedom (DOF) configuration (Galluzzi et al. (2020)).

A test bench has been constructed to study this electrodynamic levitation system, allowing for the validation of proposed control strategies. It is characterized by a copper ring mounted on an aluminium disk coupled to a brushless servomotor by means of a mechanical joint. The disk is oriented with its spinning axis vertical, so as to ensure a small footprint. The test bench structure is composed of transparent panels, and a cavity permitting equipment installation. Such equipment is constituted by two masses installed onto a micro-metric stage as shown in Figure 1a. The masses can be identified as per the equivalent quarter car model in Figure 1b: the sprung mass m_s associated with the capsule of a maglev train and the unsprung mass m_{us} associated with the bogie. Horizontal or lateral movement is prevented by flex hinges, which only permit vertical displacement. A first set of flex hinges connects the unsprung mass to the frame of the test bench (equivalent stiffness k_u). A second set of flex hinges is installed between the sprung and the unsprung masses (equivalent stiffness k_s). The combination of these flex hinges and a voice coil actuator (VCA) with an equivalent damping c_s constitutes the secondary suspension.

For a commercial application of this technology, passenger comfort must be guaranteed. The VCA provides

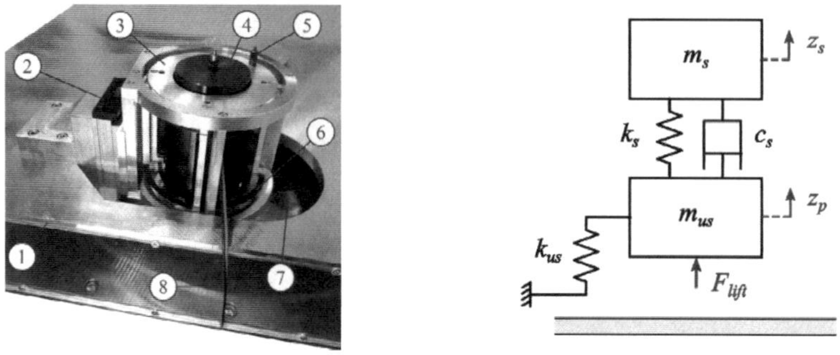

(a) (b)

Figure 1: Experimental test bench. (a) Rig setup constituted by (1) support structure; (2) micrometric stage; (3) unsprung mass; (4) sprung mass; (5) accelerometer [x2]; (6) flex hinges; (7) copper track; (8) voice coil winding terminals. (b) Equivalent quarter car model.

the possibility of implementing passive or active control, such that the oscillations of the sprung mass are attenuated. Passive control involves supplying a load-relieving voltage to the VCA, and exploiting the intrinsic damping due to eddy current generation in the structure of the actuator. In the studied application, the VCA functions as a passive damper. Active and semi-active control is of greater interest as several control strategies can be introduced to significantly improve comfort (Zhang (2019)). The Linear Quadratic Regulator (LQR) falls under an active control scheme, while the skyhook controller is a potential semi-active control strategy (Negash et al. (2021)). For their practical implementation in the context of the VCA, a reliable estimation of relative velocity between the sprung and the unsprung masses is required.

This work focuses on the estimation of the relative velocity using a Kalman filter approach, incorporating an augmented numerical model of the VCA. The use of a Kalman filter with a LQR control strategy is explored for active magnetic bearings by Schuhmann et al. (2006), demonstrating the potential for active control in a system with similar operating principles. A single measured variable is necessary for the Kalman filter to estimate relative velocity. Electrical dynamics of the VCA can be exploited to estimate the relative velocity in a sensorless approach. Noise in the estimation must be considered to verify the viability of the aforementioned approaches in active control strategies.

The remainder of the paper is organized as follows. Section 2 outlines the numerical method used to implement the Kalman filter. Section 3 shows the experimental results for the relative velocity estimation. Finally Section 4 concludes the work.

2. Method

The numerical model adopted to estimate the relative velocity between the two masses is represented by the electrical circuit of the VCA. The electrical domain of a voice coil actuator can be represented by a simple series circuit with impedance terms R_{vc} and L_{vc}, input voltage source V and a back electromotive force (back-EMF) $E_{vc} = K_v(\dot{z}_s - \dot{z}_p)$, where $\dot{z}_s$ represents the velocity of the sprung mass and $\dot{z}_p$ represents the velocity of the unsprung mass. Kirchhoff's voltage law is applied to the electric circuit of the VCA:

$$V = L_{vc}\frac{\mathrm{d}i_{vc}}{\mathrm{d}t} + R_{vc}i_{vc} + E_{vc} \tag{1}$$

During testing, the VCA is controlled with a voltage input applied by a TI™ F28379D LaunchPad™ card and a TI™ BOOSTXL DRV8323RS power stage. The current across the VCA is measured by a current sensor, as shown in Fig. 2. Subsequently, the measured current is processed in real-time by the dSPACE™ MicrolabBox that calculates the back-EMF value through Eq. 1. The disadvantage of this method is the use of the derivative of

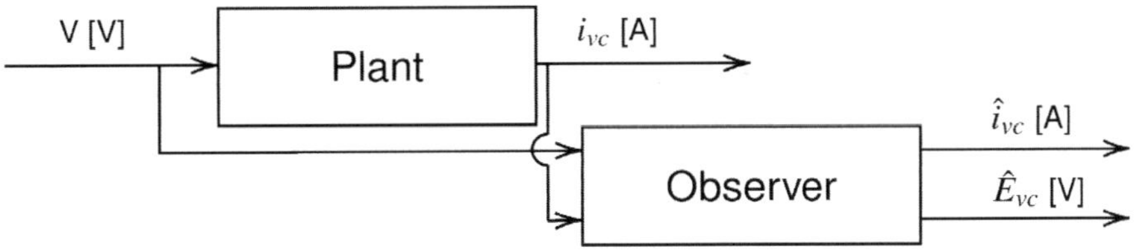

Figure 2: Scheme of the proposed observer.

the measured current which introduces noise in the calculated back-EMF signal. Using this signal as a feedback variable for active damping control may result in sub-optimal vibration attenuation.

Therefore, a disturbance observer based on Kalman filter is designed to suppress model parameter uncertainty and an external disturbance signal, as Figure 2 shows. Assuming the induced voltage E_{vc} as a disturbance with zero derivative (Chappuis et al. (2015)), the following expression can be written:

$$\frac{d}{dt}E_{vc} = 0 \tag{2}$$

The derivative of the back-EMF tends to zero because it is assumed that the relative velocity variation due to the mechanical dynamics of the electrodynamic levitation system is slower than the electrical dynamics of the VCA. Thus, the induced voltage, coupled with the relative velocity variable through the VCA velocity constant K_v, has a quasi-static behavior.

A continuous state space representation based on Eqs. 1 and 2 is assembled. The state vector ξ in Eq. 3 contains the voice coil current i_{vc} and the induced voltage E_{vc}. The input vector u in Eq. 4 contains voltage across the VCA terminals V. The output vector y in Eq. 5 contains VCA current. The dynamic matrix A_c, input gain matrix B_c, output gain matrix C, and direct link matrix D in Eqs. 6, 7, 8, and 9 describe the augmented system.

$$\xi = \{i_{vc}\ E_{vc}\}^{\mathsf{T}} \tag{3}$$

$$u = \{V\} \tag{4}$$

$$y = \{i_{vc}\} \tag{5}$$

$$A_c = \begin{bmatrix} -\frac{R_{vc}}{L_{vc}} & -\frac{1}{L_{vc}} \\ 0 & 0 \end{bmatrix} \tag{6}$$

$$B_c = \begin{bmatrix} \frac{1}{L_{vc}} & 0 \end{bmatrix}^{\mathsf{T}} \tag{7}$$

$$C = \begin{bmatrix} 1 & 0 \end{bmatrix} \tag{8}$$

$$D = \begin{bmatrix} 0 \end{bmatrix} \tag{9}$$

The augmented plant model can thus be introduced into the observer model. Equation 10 refers to the observer, from which the filter can be obtained. The dynamic matrix $\bar{A}$, input gain matrix $\bar{B}$, output gain matrix $\bar{C}$, and direct link matrix $\bar{D}$ in Eqs. 14, 15, 16, and 17 represent the state space of the Kalman filter. Note that the Kalman gain K_f is constructed with reference to the plant output. The plant output y identifies the measurement input of VCA current for the Kalman filter. Observer states $\hat{\xi}$ in Eq. 12 are identical to plant states, while the inputs $\bar{u}$ in Eq. 13 consist of the VCA voltage and the measured current $\hat{i}_{vc}$. The outputs y contain the estimated states as it is shown in Eq. 16. Finally, the relative velocity estimation v_{rel} can be obtained due to its dependency on the VCA velocity constant K_v as Eq. 18.

$$\dot{\hat{\xi}} = A\hat{\xi} + Bu + K_f(y - (C\hat{\xi} + Du)) \tag{10}$$

$$= \bar{A}\hat{\xi} + \bar{B}\bar{u} \tag{11}$$

$$\hat{\xi} = \{\hat{i}_{vc} \ \hat{E}_{vc}\}^{\mathsf{T}} \tag{12}$$

$$\bar{u} = \{V \ i_{vc}\}^{\mathsf{T}} \tag{13}$$

$$\bar{A} = A - K_f C \tag{14}$$

$$\bar{B} = \begin{bmatrix} B - K_f D & K_f \end{bmatrix} \tag{15}$$

$$\bar{C} = \begin{bmatrix} 1 & 0 \\ 0 & 1 \end{bmatrix} \tag{16}$$

$$\bar{D} = \begin{bmatrix} 0 & 0 \\ 0 & 0 \end{bmatrix} \tag{17}$$

$$v_{rel} = \dot{z}_s - \dot{z}_p = \frac{\hat{E}_{vc}}{K_v} \tag{18}$$

Noises associated to states and to measurements are considered in the covariance matrices Q and R as tuning factors:

$$Q = \begin{bmatrix} w_i^2 & 0 \\ 0 & w_E^2 \end{bmatrix} \tag{19}$$

$$R = v_i^2. \tag{20}$$

The weighting parameter w_i refers to noise on the estimated current. Noise on the measured current is considered by the weighting parameter v_i. The weighting parameter w_E refers to noise on the back-EMF. The Kalman estimator in Eq. 10 computes a state estimate $\hat{\xi}$ that minimizes the steady-state error covariance. The proposed filter has been discretized with a sample time of 2 ms and deployed onto the dSPACE™ MicrolabBox on which the relative speed measurement from the RL circuit and the estimate deriving from the Kalman filter can be compared in real-time. The chosen working frequency of 5 kHz ensures a reliable communication between the LaunchPad™ and the MicrolabBox, without incurring overruns. To reduce measurement noise, the analog current signal from the current sensor is acquired by the MicrolabBox synchronously when the H-bridge leg of the LaunchPad™ power stage is closed. Table 1 summarises all parameters necessary for the practical implementation of the discrete Kalman filter.

Table 1: Kalman filter parameters.

Parameter	Value	Unit
R_{vc}	1.8	Ω
L_{vc}	0.0152	H
K_v	25	Vs/m
w_i	0.001	A
w_E	0.0075	V
v_i	0.001	A

3. Results

Both methods based on back-EMF calculation and the Kalman filter estimation have been tested. The test is performed at a constant disk speed of 500 rpm. The two-mass system is positioned on the micrometric stage such that the flex hinges for the sprung and the unsprung masses do not exhibit any pre-deformation. Furthermore, the VCA operates by applying an offset voltage that causes the separation of the two masses. Figure 3 shows the measured current i_{vc} (black solid line) and the estimated current from the Kalman filter $\hat{i}_{vc}$ (red dotted line), highlighting the estimated current is completely based on the measured signal. The oscillation of the current with a period of approximately 8 Hz is caused by the periodicity of the irregularities in the copper track, which induces system oscillations with a frequency proportional to the rotor's speed.

Figure 4 shows a comparison between the relative speed measurement using the VCA electrical equation and the Kalman filter's estimated output. A relative velocity signal obtained from accelerometer sensors is also plotted, confirming the results obtained through both estimation methods. The relative velocity signal estimated by the Kalman filter is smooth and filtered, indicating the suitability of the proposed approach for active control. The weighting parameters w_i, w_E and v_i are treated as adjustable parameters. An increase in the ratio between w_i and w_E makes the filter slower in estimating the disturbance, while an increase in the ratio between v_i and w_E makes the estimator less sensitive to the output noise (Castellanos Molina (2020)). The obtained results show a good trade-off in balancing the three weighting parameters. Indeed, the RMS value of the error between the Kalman filter estimation and the RL estimation is 0.0024 m/s. Moreover, the maximum absolute error between the two estimations is 0.0077 m/s.

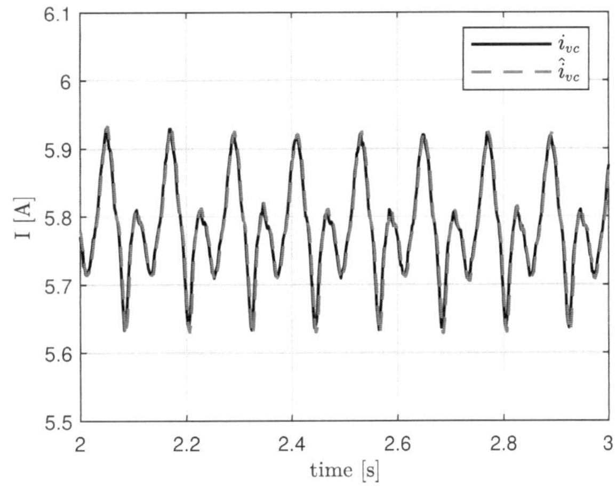

Figure 3: Current measured signal i_{vc} (black solid line) compared with the estimated current state from the Kalman filter $\hat{i}_{vc}$ (red dotted line).

4. Conclusions

The paper proposes a Kalman filter for relative velocity estimation of an electrodynamic levitation system with a VCA working as a secondary suspension system between the capsule and the bogie. The estimator is based on an augmented model of the electrical circuit of the VCA. Due to the augmented state, a relative velocity estimation with less noise compared to the calculation derived from the RL circuit can be achieved. Therefore, the relative speed signal can be fed back to close active damping control loops, i.e. reducing acceleration peaks on the sprung mass for comfort reasons. However, the proposed approach raises critical concerns. The relative speed estimation is accurate provided that the VCA's parameters do not change. Unfortunately, the VCA's resistance tends to increase during operation time. Moreover, the VCA's inductance changes too due to transient operations introduced by the irregularities of the copper track, thus meaning the proposed method does not compensate for plant model mismatches. Future research may employ an Extended Kalman filter strategy or real-time model identification approaches to counterbalance plant uncertainties and disturbances.

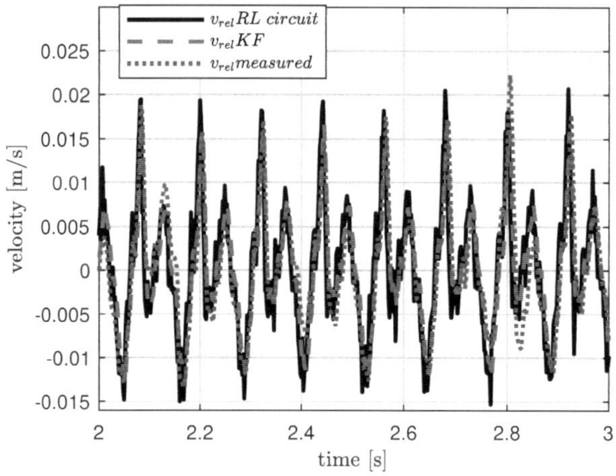

Figure 4: Relative velocity signal comparison between VCA RL circuit method (black solid line) and the Kalman filter approach (red dotted line).

References

Castellanos Molina, L. M. (2020), Offset-Free Model Predictive Control for Active Magnetic Bearings, PhD thesis, Politecnico di Torino.

Chappuis, B., Gavin, S., Rigazzi, L. & Carpita, M. (2015), 'Speed control of a multiphase active way linear motor based on back emf estimation', *IEEE Transactions on Industrial Electronics* **62**(12), 7299–7308.

Galluzzi, R., Circosta, S., Amati, N., Tonoli, A., Bonfitto, A., Lembke, T. A. & Kertész, M. (2020), 'A multi-domain approach to the stabilization of electrodynamic levitation systems', *Journal of Vibration and Acoustics* **142**(6), 061004.

Negash, B. A., You, W., Lee, J., Lee, C. & Lee, K. (2021), 'Semi-active control of a nonlinear quarter-car model of hyperloop capsule vehicle with skyhook and mixed skyhook-acceleration driven damper controller', *Advances in Mechanical Engineering* **13**(2), 1687814021999528.

Schuhmann, T., Hofmann, W. & Werner, R. (2006), 'Adaptive linear and extended kalman filter applied to amb with collocated position measuring'.

Zhang, Z. (2019), 'Applied adaptive controller design for vibration suppression in electromagnetic systems', *The Applied Computational Electromagnetics Society Journal (ACES)* pp. 567–576.

Rotor vibration control via active magnetic bearings and internal actuation

Gauthier FIEUX[a], Nicola BAILEY[a], Patrick KEOGH[a]

[a] University of Bath, Claverton Down, BA27AY Bath, United-Kingdom, p.s.keogh@bath.ac.uk

Abstract

Designing rotor systems for high rotating speeds requires careful consideration of vibration signatures. Active Magnetic Bearings (AMBs) improve the vibration response by removing physical contact between the rotor and the stator. Their performance is, however, limited by magnetic flux saturation and the amplifier switching frequency. This paper develops a rotor bend-control concept with an experimental test rig. It shows that it is possible to reduce the amplitude of oscillation of the demand control current of an AMB under PID control by up to 65% by applying a bend that counteracts the effect of unbalance. The optimal orientation of the bend is found using a phase sweep. It shows that the bend control has promising capabilities to control the vibration level and reduce the synchronous demand from the AMB.

Keywords: Active Magnetic Bearing, Vibration control, Active rotor

1. Introduction

Rotating machines are used for a wide variety of applications, from transport to energy and manufacturing. Rotor vibrations arise from manufacturing tolerances, operating conditions, and thermal effects (Keogh & Morton 1994). They degrade the performance of the machine, either by forcing larger clearances to accommodate rotor orbits or stressing the material which can lead to premature failure. The vibration condition of a passive rotor can be improved via balancing as described in the ISO standards (ISO 2004). More advanced balancing methods such as disks with heavy spots (Van de Vegte & Lake 1978) or automatic ball balancers (Thearle 1932) can mitigate variable balancing conditions during operation. However, balancing ultimately reaches performance boundaries at high rotation speeds. This is due to the onset of dynamic instabilities and increased difficulty to locate precisely the inertial spin axis. Passive rolling elements bearings are often used with squeeze film dampers (Bonello et al. 2004). They are, however, limited in speed and load by the oil whip phenomenon (Newkirk & Taylor 1925), which generates instabilities. Gas/foil bearings can reach higher speeds and become self-sustaining (Belforte et al. 2006) but are subject to wear at low speed or high loads (DellaCorte et al. 2004). Active Magnetic Bearings (AMBs) are capable of offering controllable stiffness and damping, and the application of finer additional control to filter out unwanted excitations (Bleuler et al. 2009). They can, however, only apply a force at discrete locations and are limited at higher frequencies by magnetic flux saturation (Kang & Palazzolo 2012) and amplifier switching frequency (Maslen et al. 1989). Application of forces in the rotating frame has been achieved using macro fibre composites (Vadiraja & Sahasrabudhe 2009), functionally graded material (Alexander et al. 2007), or external piezoelectric patches (Przybylowicz 2002). In these cases, a voltage is applied to the material or through the patches to generate a bending moment. However, the achievable force magnitude limits their efficiency away from resonant frequencies.

This paper presents a novel control method using an active rotor featuring wireless, low-frequency, internal bend control, supported by AMBs. This has the advantage that low-frequency actuation in the rotating frame, which effectively transforms to high-frequency synchronous control in the stationary frame of reference. The concept is presented, along with some preliminary results. A finite element model, modal response, expected deformation and a controller concept are presented in depth in Fieux et al. (2022).

2. Methodology

A hydraulic system with lead screw actuators is used to generate large internal rotor bending moments. The aim of this prototype illustrated in Figure 1 a) is to counter the bending generated by an unbalance on a rotor supported symmetrically by two AMBs. As the rotor is spun, a bowed shape is generated by the unbalance force. The bending mechanism is then used to straighten the rotor using symmetrical bending moments. Figure 1 b) shows a schematic of the actual test rig setup, with a main rotor supported by two AMBs and an auxiliary rotor containing all the electronics to control the bending actuator.

The main rotor is composed of a larger cross-section module in the centre and two smaller cross-section shafts on either side. Forces are applied on the flanges that connect the different cross-sections. These forces can be varied in magnitude, resulting in an equal and opposite internal bending moment applied to the rotor.

The rotor used for this test has a mass of 20 kg and a length of 1.2 m. The larger central section has an outer diameter of 166 mm and an inner diameter of 150 mm. The outer shafts are connected to the larger central module via flanges and couplings and they have an outer diameter of 30 mm and an inner diameter of 18 mm. The bending actuator uses lead screw actuators and a hydraulic circuit to distribute the forces to the flanges. The maximum bending moment that the actuator can apply is 190 N.m. The rotor is balanced via the influence coefficient method (Darlow 1987), and was limited to run below 2400 RPM.

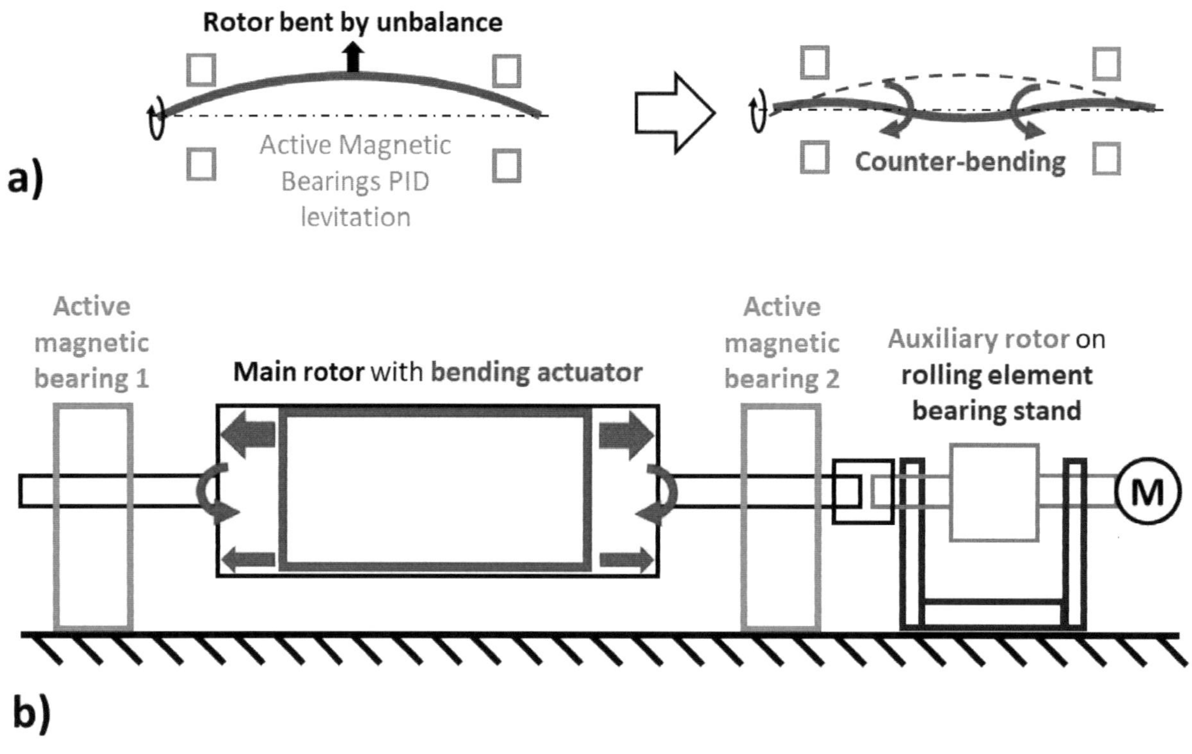

Figure 1: Rotor combined control concept.

3. Results

The rotor was spun at 800 RPM, supported by the AMBs under PID control. The current demand was monitored, isolating the contributions of the proportional and derivative controller action, since they are associated with the dynamic effects of rotor vibrations. The integral contribution of the control contributes only to the weight cancellation. To make the oscillatory current more readable, it is expressed in terms of a Root Mean Square

(RMS) value. Then, as the phase of the unbalance was unknown, the optimal bending phase was sought by applying rotor bend at 45° increments between 0° and 315°.

Figure 2 illustrates the current response of one axis of an AMB to this phase sweep at 800 RPM. The initial level of 700 mA between 0 s and 40 s is without any bend actuation, as is the final level between 220 and 250 s. Between 40 s and 160 s, the phase of the control leads to an increase in control current, whereas between 160 s and 220 s the phase decreases the current required. From a base level of 700 mA, the current required for the levitation can be reduced to 350 mA, which is a 50% reduction.

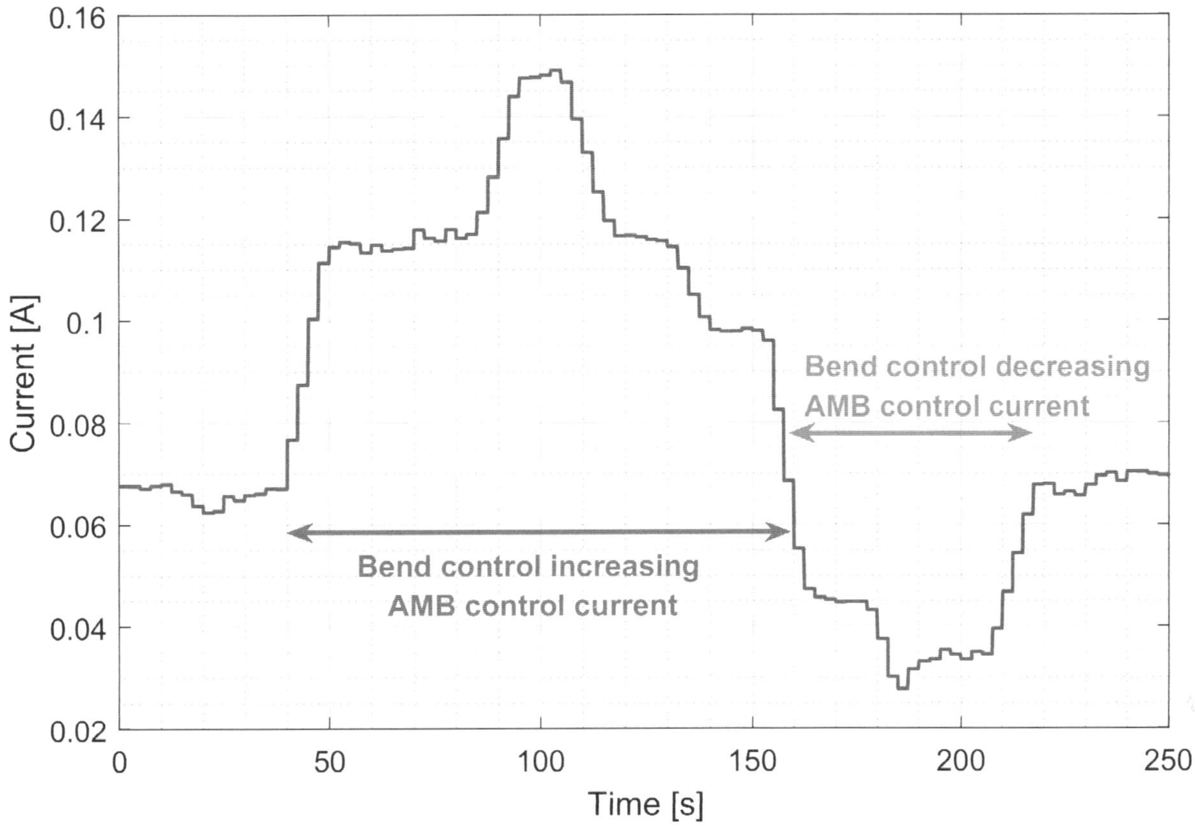

Figure 2: RMS of AMB current to applied bend phase sweep at 800 RPM with the AMBs under PID control. One axis of AMB1 displayed only.

Figure 3 displays the total control current required to maintain the levitation, at three different phases of bending. The recordings are over 100 ms at 5 s, 100 s and 190 s. Figure 3 a) shows the current associated with axis 1 from AMB1. The current for the uncontrolled, reference case oscillates between 340 mA and 515 mA (peak-to-peak amplitude of 175 mA). For the optimal bend control phase, it oscillates between 400 mA and 460 mA (peak-to-peak amplitude of 60 mA). This corresponds to a 66% reduction of the AMB1 control current amplitude of oscillation. In the axis 2 direction (Figure 3 b)), in the nominal case the current oscillates between 200 mA and 520 mA (peak-to-peak amplitude of 320 mA). For the optimal bend control phase, it oscillates between 260 mA and 435 mA (peak-to-peak amplitude of 175 mA). This corresponds to a 45% reduction of the control current amplitude of oscillation. Figure 3 c) and d), shows the corresponding results for AMB2. This time, the AMB current oscillation peak-to-peak amplitude is reduced from 310 mA to 150 mA in the axis 1 direction (-52%) and from 824 mA to 385 mA (-53%) in the axis 2 direction.

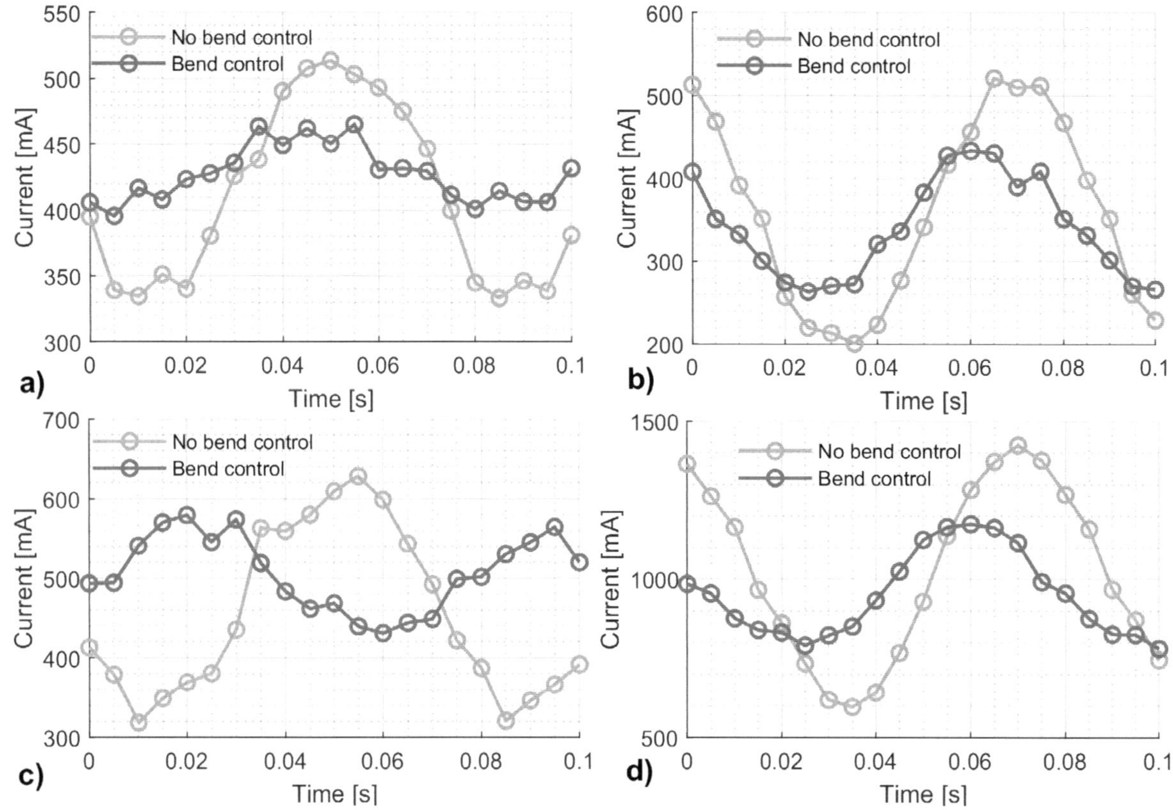

Figure 3: AMB control current response at 800 RPM comparing nominal, uncontrolled response to the controlled response at the optimal bend phase. a) shows AMB1 axis 1, b) AMB1 axis 2, c) AMB2 axis 1 and d) AMB2 axis 2.

4. Conclusions

This work has shown the potential to reduce the amplitude AMB control current by additional symmetrical internal rotor bend control. A prototype rotor has been manufactured and demonstrated that it was capable of reducing the dynamic part of the control current required for levitation by up 65% at 800 RPM. Reducing the amplitude of AMB current oscillation is of interest, considering that amplifier limitations at high frequencies are linked to the amplitude of the required current oscillation.

Further work is ongoing to characterise the rotor response in terms of rotor vibrations, as well as expanding these results for higher rotating speeds and different sources of vibration excitation.

References

Alexander, P. W., Brei, D. & Halloran, J. W. (2007), 'DEPP functionally graded piezoceramics via micro-fabrication by co-extrusion', *Journal of Materials Science* **42**(14), 5805 – 5814.
URL: *http://dx.doi.org/10.1007/s10853-007-1793-5*

Belforte, G., Raparelli, T., Viktorov, V., Trivella, A. & Colombo, F. (2006), 'An experimental study of high-speed rotor supported by air bearings: test rig and first experimental results', *Tribology International* **39**(8), 839 – 845.
URL: *http://dx.doi.org/10.1016/j.triboint.2005.07.013*

Bleuler, H., Cole, M., Keogh, P., Larsonneur, R., Maslen, E., Nordmann, R., Okada, Y., Schweitzer, G. & Traxler, A. (2009), *Magnetic bearings: Theory, Design, and Application to Rotating Machinery*, Springer.

Bonello, P., Brennan, M. J. & Holmes, R. (2004), 'A study of the nonlinear interaction between an eccentric squeeze film damper and an unbalanced flexible rotor', *Journal of Engineering for Gas Turbines and Power* **126**(4), 855 – 866.
URL: *http://dx.doi.org/10.1115/1.1787503*

Darlow, M. (1987), 'Balancing of high-speed machinery: Theory, methods and experimental results', *Mechanical Systems and Signal Processing* **1**(1), 105–134.
URL: *https://www.sciencedirect.com/science/article/pii/0888327087900872*

DellaCorte, C., Zaldana, A. R. & Radil, K. C. (2004), 'A systems approach to the solid lubrication of foil air bearings for oil-free turbomachinery', *Journal of Tribology* **126**(1), 200 – 207.
URL: *http://dx.doi.org/10.1115/1.1609485*

Fieux, G. A., Bailey, N. Y. & Keogh, P. S. (2022), 'Internal rotor actuation and magnetic bearings for the active control of rotating machines', *Actuators* **11**(2).
URL: *https://www.mdpi.com/2076-0825/11/2/57*

ISO (2004), Mechanical vibration – vibration of rotating machinery equipped with active magnetic bearings - Part 2: Evaluation of vibration, ISO 14839-2:2004, International Organization for Standardization, Geneva, Switzerland.

Kang, K. & Palazzolo, A. (2012), 'Homopolar magnetic bearing saturation effects on rotating machinery vibration', *IEEE Transactions on Magnetics* **48**(6), 1984 – 1994.
URL: *http://dx.doi.org/10.1109/TMAG.2012.2182776*

Keogh, P. & Morton, P. (1994), 'The dynamic nature of rotor thermal bending due to unsteady lubricant shearing within a bearing', *Proceedings of the Royal Society A: Mathematical, Physical and Engineering Sciences* **445**(1924), 273 – 290.
URL: *http://dx.doi.org/10.1098/rspa.1994.0061*

Maslen, E., Hermann, P., Scott, M. & Humphris, R. (1989), 'Practical limits to the performance of magnetic bearings: Peak force, slew rate, and displacement sensitivity', *Journal of Tribology* **111**(2), 331 – 336.

Newkirk, B. & Taylor, H. (1925), 'Shaft whipping due to oil action in journal bearings', *General Electric Review* **28**(8), 559–568.

Przybylowicz, P. M. (2002), 'Near-critical behaviour of a rotating shaft actively stabilised by piezoelectric elements', *Systems Analysis Modelling Simulation* **42**(4), 527 – 537.
URL: *http://dx.doi.org/10.1080/02329290290031305*

Thearle, E. L. (1932), 'Dynamic balancing machine'.

Vadiraja, D. & Sahasrabudhe, A. (2009), 'Vibration analysis and optimal control of rotating pre-twisted thin-walled beams using mfc actuators and sensors', *Thin-Walled Structures* **47**(5), 555 – 567.
URL: *http://dx.doi.org/10.1016/j.tws.2008.10.004*

Van de Vegte, J. & Lake, R. (1978), 'Balancing of rotating systems during operation.', *Journal of Sound and Vibration* **57**(2), 225 – 235.
URL: *http://dx.doi.org/10.1016/0022-460X(78)90581-3*

Design of a PCB integrated eddy current sensor with shield feature for radial rotor displacement measurement

Dominik WIMMER, Markus HUTTERER, Manfred SCHRÖDL

TU Wien, Institute of Energy Systems and Electrical Drives, Gußhausstraße 25-29, 1040 Vienna, Austria, dominik.wimmer@tuwien.ac.at

Abstract

Active magnetic bearings require position feedback to enable a stable control of a levitating rotor. The position information can be provided by explicit position sensors or the usage of self-sensing bearing topologies. In this study an investigation of a PCB integrated eddy current sensor for radial rotor displacement measurement is conducted. In many AMB applications a short rotor length is desired to handle flexible rotor characteristics at high rotational speeds. Therefore, the active magnetic bearings, the motor as well as the sensors shall obtain a small axial length. Conventional eddy current displacement sensors often obtain a cylindrical shape with a circular measurement spot on the rotor target material, which defines the minimum axial length of the sensor arrangement. The aim of this study is to design a low-cost PCB integrated eddy current sensor with an arc-shaped measurement spot and low axial dimension. In contrast to previous studies in this field, it is investigated if it is possible to achieve a feasible sensor design without the use of a coupled coil arrangement of excitation- and sense coils. This study covers the sensor design, measurements on a prototype as well as a performance comparison between a non-shielded and a shielded sensor design.

Keywords: AMB sensor, radial displacement measurement, eddy current sensor

1. Introduction

Active magnetic bearings (AMBs) are often used for contact-free rotor suspension where conventional rolling bearings are not able to meet the increasing demands of high-speed applications. The control loop of AMBs requires feedback of the rotor position to enable a stable control. The rotor position can be obtained by sensor-based or self-sensing approaches. Self-sensing approaches (Maslen 2006) are a very exclusive way to determine the rotor position but the methods are mostly very specific for a certain magnetic arrangement and cannot be easily transferred to other AMB topologies. In contrast, AMB sensors require additional construction space, which can increase the size and weight of an AMB system. Especially, a longer shaft can cause problems by higher thermal expansion in the axial direction and lower natural frequencies of flexible rotor characteristics. In contrast to inductive sensor designs, also eddy current sensors are well established for position measurement but do not have special requirements concerning the magnetic properties of the target material. Special applications like AMBs often require adapted sensor designs to fulfill demands concerning sensor integration, measurement range, bandwidth, linearity, noise and of course the cost factor. Previous eddy current sensor designs for radial rotor displacement measurements deal with widely concentrated windings (Štusák 2014, Wang et al. 2018) or with a coupled coil arrangement of excitation and measurement coils (Larsonneur & Bühler 2004, Grobler et al. 2017). This study covers the design of a PCB integrated sensor design with low axial dimension and an integral (non-concentrated) measurement over the rotor surface. Furthermore, a very close coupling between the sensor coils and the rotor is aspired to delimit the stray field area to a minimum to reduce the impact of a desired sensor shield cage.

The principle of eddy current sensors relies on flux expulsion in the target material, which can be mathematically described by the diffusion equation (Markovic & Perriard 2009). As a result, the penetration depth δ of alternating magnetic flux (exponential decay) is given by

$$\delta = \sqrt{\frac{2}{\mu \gamma \omega}}, \tag{1}$$

with the permeability μ, the conductivity γ and the angular frequency ω. The eddy currents in the target material influence the magnetic flux distribution of a coil with alternating excitation. Thus, the presence of a conductive material can be detected by the electrical coil quantities, which enables the use for sensor applications.

2. Sensor Design

Figure 1 shows an illustration of the aspired PCB integrated sensor design for two-axis displacement measurement. The sensor arrangement consists of a sandwich with a coil-PCB in the center which is surrounded by shield-PCBs. The coil-PCB contains two opposing sense coils for each degree of freedom (x, y). Therefore, the sense coil is realized as an air coil, which is wound between multiple layers of the coil-PCB. The coil-PCB is surrounding by shield-PCBs in order to achieve a suppression of unwanted EMI couplings of close-by bearing and motor coils. Furthermore, the sensor shield provides a well-defined boundary for the stray flux of the sense coils and suppresses an undesired influence of close proximity conductive or ferromagnetic elements.

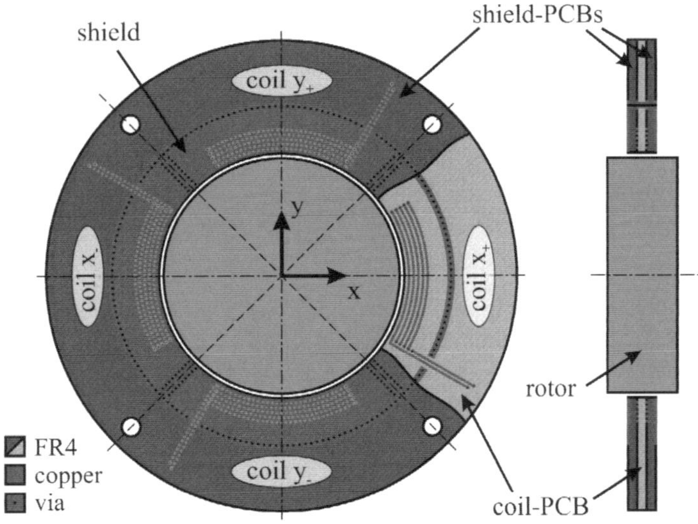

Figure 1: PCB integrated radial eddy current sensor design with optional shield-PCBs.

Table 1: Sensor design specification overview

Symbol	Parameter	Value
z_S	Axial sensor length (non-shielded design)	1.6 mm
$z_{S,shield}$	Axial sensor length (shielded design)	4.8 mm
$d_{S,o}$	Outer sensor diameter	100 mm
$d_{S,i}$	Inner sensor diameter	51.5 mm
d_R	Rotor diameter, rotor material: AW-6082	50 mm
$\Delta x_S, \Delta y_S$	Sensor range	±750 µm
$\Delta x_R, \Delta y_R$	Rotor displacement operational range	±400 µm
N	Turns per coil	16
$f_{carrier}$	Coil carrier frequency	1 MHz
f_{-3dB}	Sensor circuit bandwidth	22 kHz

3. Sensor Prototype

Figure 2 shows the sensor prototype, which meets the design specifications according to Table 1. The admissible sensor range is designed for radial rotor displacements of $\pm 400\,\mu m$. For safety reasons, $350\,\mu m$ additional air-gap was considered to avoid a rotor-sensor collision in touch down scenarios with a safety bearing. The sense coils were implemented in a 4 layer PCB with nested windings. The nested arrangement of the windings allows the integration of 16 windings within only $2\,mm$ radial space consumption. For this reason, it is ensured that even the outer windings are in close proximity to the target material. For achieving a shielding of the sense coils, the shield-PCBs can be stacked to the coil-PCB. Thus, the sense coils are surrounded by three-dimensional conductive chambers, which exhibit just one open side in direction of the measurement target.

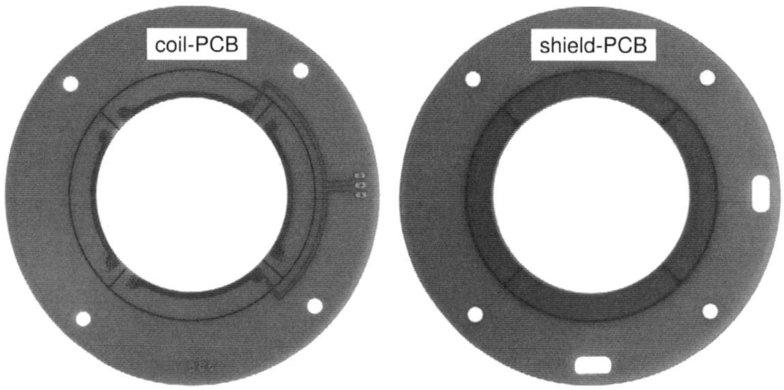

Figure 2: 4 layer coil-PCB prototype with optional 2 layer shield-PCB.

4. Impedance measurements

For obtaining a deeper insight into the frequency-dependent characteristic of the sense coil, the impedance was determined for selected coil-rotor distances (Fig. 3). It can be seen that the sense coil has an ohmic-inductive behavior in the depicted frequency range. An equivalent circuit fit at center position ($750\,\mu m$) with

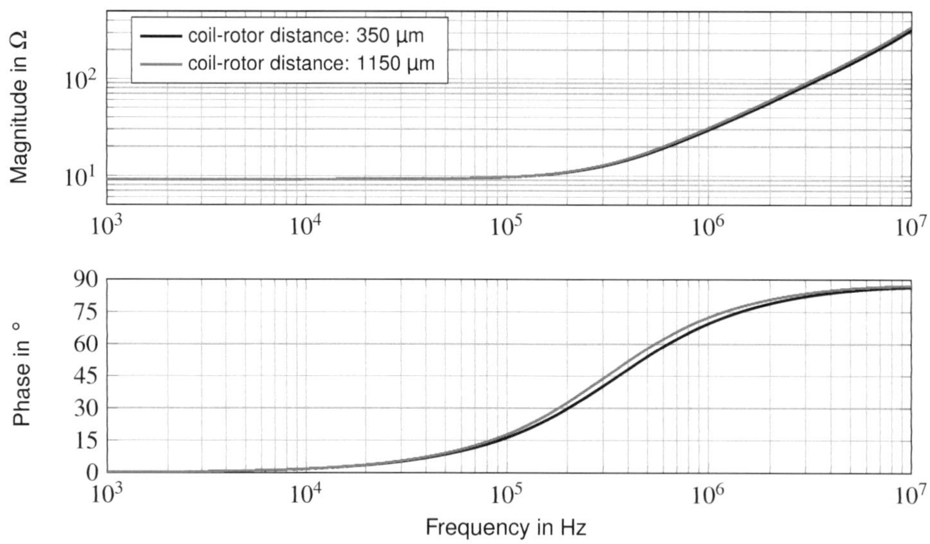

Figure 3: Sensor coil impedance for the boundaries of the nominal operation range (non-shielded design).

$R_{coil} = 9.17\,\Omega$ and $L_{coil} = 4.61\,\mu H$ results in a model magnitude error less than 1 % for frequencies smaller than 2 MHz. For the selection of the sensor operating frequency $f_{carrier}$ the frequency-dependent sensor sensitivity ζ was considered

$$\zeta(f) = \frac{\|Z_{1150\,\mu m}(f) - Z_{350\,\mu m}(f)\|_2}{0.5 \cdot \left(\|Z_{1150\,\mu m}(f)\|_2 + \|Z_{350\,\mu m}(f)\|_2\right)} \tag{2}$$

and is calculated by the impedance values at the boundaries of the nominal operational range. Figure 4 shows the sensor sensitivity as a function of the carrier frequency. It can be seen that the sensitivity is not significant below 40 kHz. The gradient of the sensitivity is very high at about 300 kHz, flattens at about 2 MHz and increases again for frequencies greater than 3 MHz. The carrier frequency was chosen as 1 MHz, which results in 5.2 % sensitivity. If a higher carrier frequency is desired, special care must be taken to parasitic winding capacitances

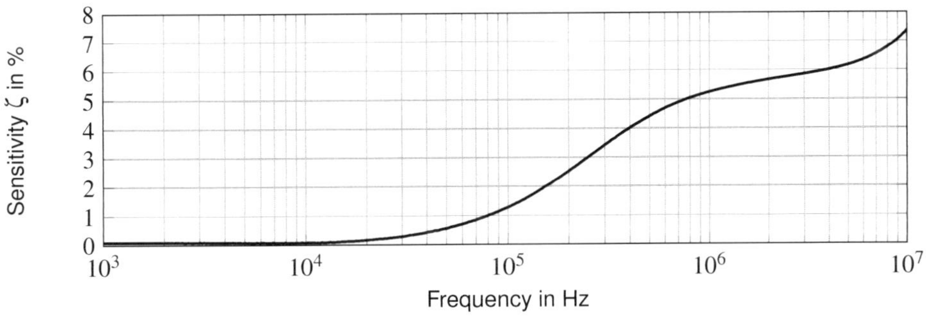

Figure 4: Sensor impedance sensitivity as a function of the carrier frequency (non-shielded design).

concerning PCB manufacturing tolerances and reproducibility. Figure 5 shows the coil characteristic at 1 MHz carrier frequency. It can be seen that the eddy currents in the target material cause a reduction of the impedance magnitude if the rotor is located closer to the sense coil. The overall low value of the impedance magnitude is a result of the small number of winding turns and must be considered in the sensor wiring and circuit design.

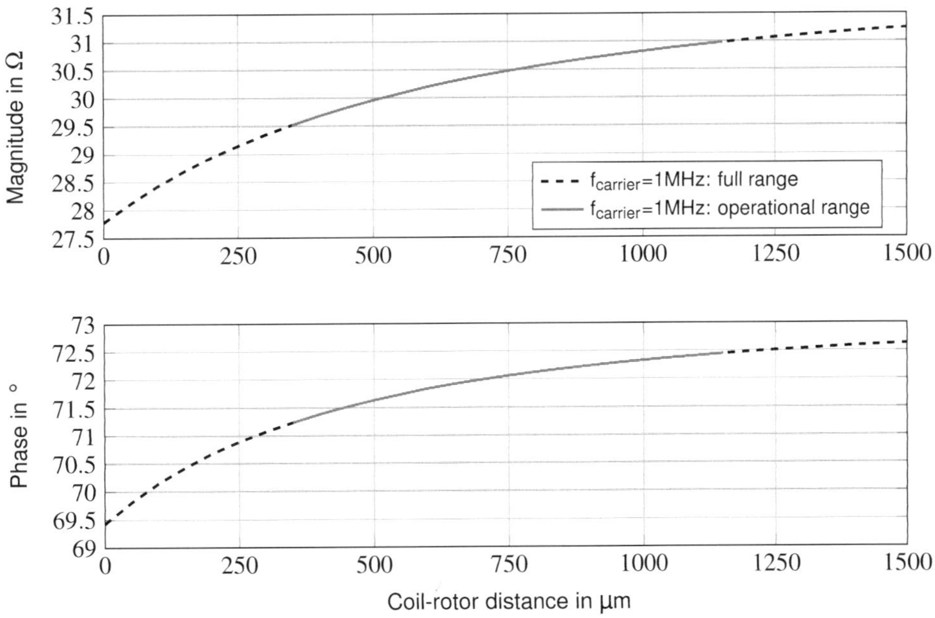

Figure 5: Sensor coil impedance at 1 MHz carrier frequency (non-shielded design).

5. Sensor Circuit and Performance Evaluation

Figure 6 shows an illustration of the electrical circuit for obtaining an output signal which is proportional to the rotor displacement. Two opposing sense coils are connected in a bridge configuration and are excited by a sinusoidal carrier waveform. The differential voltage of the bridge is amplified, band-pass filtered and demodulated to gain the displacement signal. Finally, the signal is low-pass filtered to suppress remaining

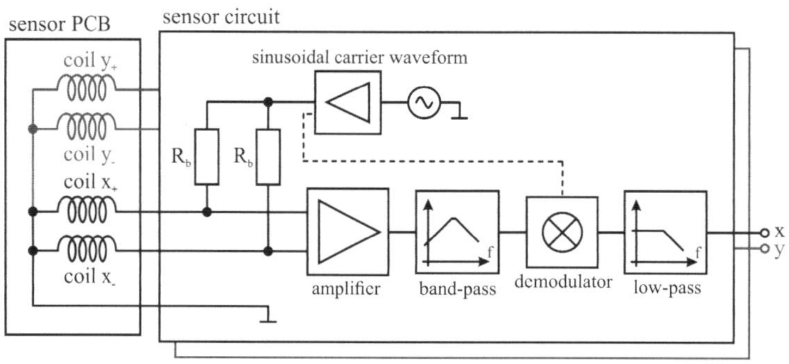

Figure 6: Sensor circuit with sinusoidal carrier waveform excitation and synchronous demodulation.

harmonics of the carrier frequency. Figure 7 and 8 show the output voltage of the sensor circuit for admissible rotor position setpoints as well as the linearity error and noise level, respectively. The non-shielded sensor design features a slightly lower linearity error. Although the shielded design has a lower sensitivity (Fig. 7), the noise

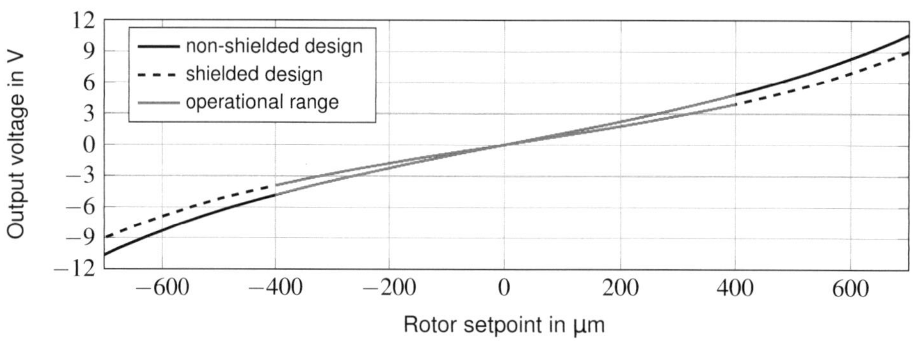

Figure 7: Sensor circuit output voltage as a function of the rotor position setpoint.

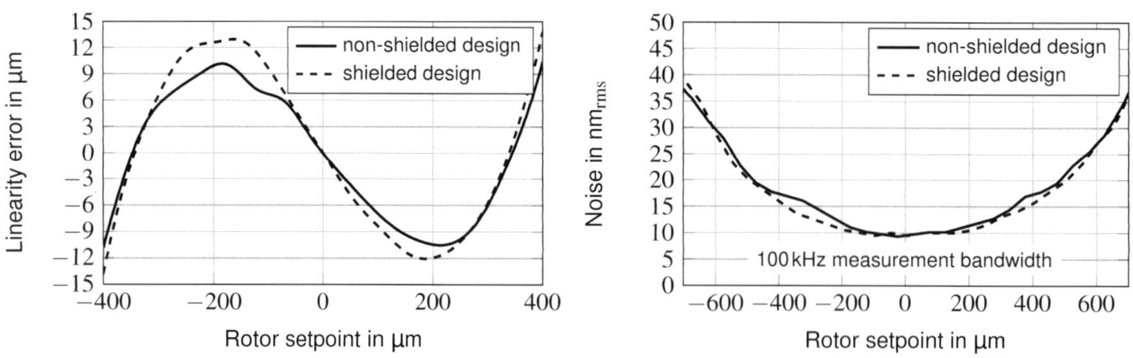

Figure 8: Linearity error and noise as a function of the rotor position setpoint (sensor + sensor circuit).

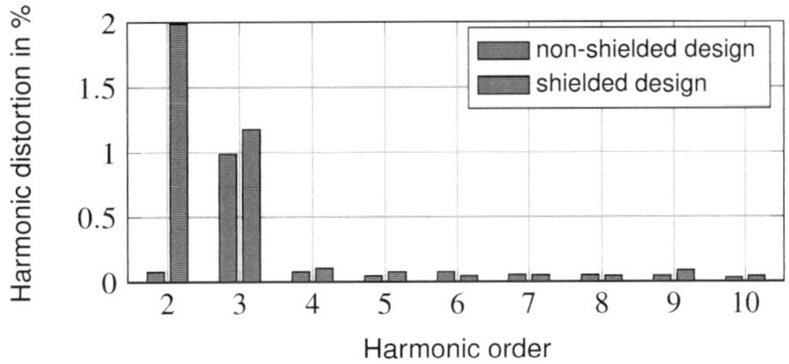

Figure 9: Harmonic distortions of a circular rotor trajectory with 662 μm diameter.

level of the designs is very similar to each other. For obtaining the sensor distortions for circular rotor trajectories, the rotor was 331 μm eccentrically mounted on a rotary table and was moved along a circular trajectory (Fig. 9).

6. Conclusion

This study covers the design and experimental verification of a PCB integrated eddy current sensor for radial position measurement. A major design objective was to achieve a sensor design with low axial dimension to enable the usage in AMB designs with limitations to the rotor length. The sensor design features an optional shield of the sense coils for the usage in high EMI environments or to suppress the influence of close-by conductive or ferromagnetic elements. Concerning applications with a high spatial temperature gradient, the sensor shield also enables a strong thermal coupling of the sense coils and avoids varying sense coil temperatures, which may lead to undesired signal drifts. Regarding sensor performance, the non-shielded as well as the shielded design obtains a linearity error below 15 μm and a noise level smaller than $20\,\mathrm{nm_{rms}}$ within the operational range of ± 400 μm rotor displacement. Although the shielded design has a smaller sensitivity, no degradations concerning sensor noise could be detected. Furthermore, a harmonic distortion analysis was performed for circular rotor trajectories which indicated low distortions. Summarized, both sensor designs showed a very good overall performance and seem to be suitable candidates for AMB applications. For future designs, it would be conceivable to integrate the sensor coils and the shield cage in a multilayer PCB design with a higher layer count.

References

Grobler, A. J., van Schoor, G. & Ranft, E. O. (2017), 'Design and optimisation of a pcb eddy current displacement sensor', *SAIEE Africa Research Journal* **108**(1).

Larsonneur, R. & Bühler, P. (2004), 'New radial sensor for active magnetic bearings', *ISMB9, 9th International Symposium on Magnetic Bearings, Lexington, Kentucky, August 3-6* .

Markovic, M. & Perriard, Y. (2009), 'Eddy current power losses in a toroidal laminated core with rectangular cross section', *International Conference on Electrical Machines and Systems, Tokyo, Japan, November 15-18* .

Maslen, E. H. (2006), 'Self–sensing for active magnetic bearings: overview and status', *ISMB10, 10th International Symposium on Magnetic Bearings, Martigny, Switzerland, August 21-23* .

Štusák, M. (2014), 'Eddy current sensors for magnetic bearings of the textile spinning machines', *ISMB14, 14th International Symposium on Magnetic Bearings, Linz, Austria, August 11-14* .

Wang, K., Zhang, L., Zheng, S., Zhou, J. & Liu, X. (2018), 'Analysis and experiment of self-differential eddy-current sensor for high-speed magnetic suspension electric machine', *IEEE Transactions on Industry Applications* **55**(3).

Modeling and Simulation of a Compressor Installed with Active Magnetic Bearings Exploiting Feedforward Control

Marius PAKŠTYS [a], Andrea TONOLI[a], Angelo BONFITTO[a]

[a] *Department of Mechanical and Aerospace Engineering – Politecnico di Torino, Duca degli Abruzzi 24, 10129 Turin, Italy (marius.pakstys@polito.it)*

Abstract

The present paper focuses on non-linear numerical modeling of active magnetic bearings (AMBs) in a compressor system. The rotor is modeled through a FEM approach and external disturbances are considered as constant and variable loads. Constant loads arise from nominal compressor operation and variable loads are shocks described by a known acceleration profile. Proportional-Integral-Derivative (PID) position controllers are coupled with feedforward (FF) control for the AMBs to mitigate variable load effects on the rotor. FF control is an adaptable static gain dependent on the versus of the imposed shock. Proportional-Integral (PI) current controllers reflect power electronics behaviour. Simulation results indicate that displacements for the most loaded radial AMB are 40% and 44% of the nominal air gap, for negative and positive shocks respectively. Displacements for the axial AMB are 30% and 54% of the nominal air gap for negative and positive shocks respectively. The implemented control scheme proves effective in mitigating the effects of known external disturbances.

Keywords: Active magnetic bearings, Non-linear modeling, PID, Feedforward control, Disturbance rejection

1. Introduction

Active magnetic bearings (AMBs) have become popular for use in rotor systems, due to little maintenance requirements over the span of their operational lifetime. Their implementation is well-suited where ease of access is not guaranteed. The absence of lubricants means that AMBs may be introduced in air conditioning systems. One such case is presented, where a compressor rotor mounted with a single-stage impeller is suspended by two cylindrical AMBs controlling radial displacement in four degrees of freedom (DOFs), and a thrust AMB sustaining axial disturbances in one DOF. Figure 1 indicates the features of interest. This work focuses on an analytical non-linear modeling approach for the electromechanical domain. Previous work has explored non-linear modeling, however quantities were obtained from lookup tables (Tomczuk & Wajnert (2018)). The compressor is mounted on a moving foundation subject to variable external disturbances, which are modeled based on a force estimation method. For radial disturbances, this method makes use of stator acceleration, rotor mass and relative position of AMBs with respect to the centre of mass (COM). In the axial direction, stator acceleration and rotor mass are used to estimate a force. Constant loads are considered as static contributions during modeling.

Stable and safe operation is ensured by the control strategy adopted. A Proportional-Integral-Derivative (PID) position controller with feedforward (FF) control is initialised for each AMB. The present study outlines a FF controller comprised of a static gain dependent on disturbance force direction, allowing for a straightforward application in all actuator axes. A previous study has implemented an adaptable acceleration feedforward, however in a single DOF application (Min Sig Kang & Woo Hyun Yoon (2006)). The behaviour of power electronics is considered by a Proportional-Integral (PI) controller.

The remainder of the work is organised as follows. Section 2.1 briefly discusses rotordynamic modeling. Section 2.2 presents non-linear modeling of the actuators and Section 2.3 indicates the modeling method for constant and variable loads. Section 3 highlights the control strategy adopted and the results of numerical simulations in Simulink®. Section 4 concludes the work.

2. Modeling

2.1 Rotordynamics

Flexural behaviour is modeled with a FEM approach employing DYNROT, a proprietary FEM-based code on MATLAB® developed by Politecnico di Torino. Sensors for radial displacements (x- and y-axes) are located in the immediate vicinity of each radial AMB. An equivalent impeller approximating inertial properties of the original impeller is used, allowing for a simplification in rotor discretisation. The generated state space is characterized by 246 inputs and 16 outputs, considering a nominal rotational speed of 45000 rpm. The first 6 inputs specify x- and y-axis forces at the actuator and impeller nodes. The outputs specify x- and y-axis displacements and velocities of actuator and sensor nodes. Well-known rotordynamic equations of motion serve as a reference (Chen (2015)). A model order reduction is performed in MATLAB® to reduce simulation time, with relevant dynamics conserved for a minimum order of 20. The state space is subsequently implemented into the overall Simulink® model. Dynamics in the z-axis are not present in the state space and are considered separately in the model.

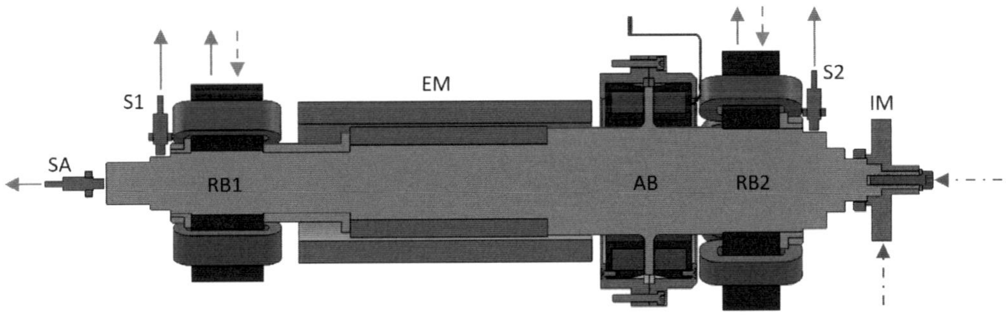

Figure 1: (a) Cross-section of compressor system displaying pertinent features. IM: equivalent impeller. RB1, RB2: radial bearings. AB: thrust bearing. SA: axial sensor. S1: RB1 radial sensor. S2: RB2 radial sensor. EM: electric motor. Outputs in state space model are denoted with solid blue arrows. Dashed red arrows are actuator inputs, and dashed-dot green arrows are impeller inputs.

2.2 Actuators

Actuators are modeled using a set of non-linear analytical expressions. A voltage-based method is used to model the AMBs exploiting Eqs. 1, 2, and 3 (Jeong et al. (2012)), with E denoting the voltage across actuator terminals.

$$E = L(x)\frac{\mathrm{d}i}{\mathrm{d}t} - \frac{L(x)}{(u_0 \pm x)}\dot{x}\cdot i + R\cdot i \tag{1}$$

$$L(x) = \frac{N^2\cdot\mu_0\cdot A}{2(u_0 \pm x)} \tag{2}$$

$$F(i,x) = \frac{N^2\cdot\mu_0\cdot A}{4}\cdot\frac{i^2}{(u_0 \pm x)^2} \tag{3}$$

Note that $L(x)$ is the inductance as a function of displacement x, and i is the coil current that includes the bias current. The terms N and μ_0 are the number of coil turns and vacuum permittivity respectively. The nominal air gap is u_0 and the active magnetic cross section is A. Coil resistance is represented with R, and λ is the magnetic flux. The chosen parameters and physical dimensions for this system yield continuous and peak force limits for the AMBs. Both radial AMBs have a continuous force limit of 400 N, and a peak force limit of 800 N. The axial AMB presents a continuous force limit of 600 N, and a peak force limit of 1100 N.

2.3 Disturbances

External disturbances are introduced as constant and variable loads. Variable loads are shocks following a known acceleration profile. An example is shown in Figure 5b, where the profile is superposed over the displacement. Peak acceleration is 96.1 m/s^2 with a duration of 11.49 ms. Three shocks with a negative versus and three with a positive versus are imposed on each axis of the COM. Stator acceleration in the axial direction is converted to a force using total mass of the rotor, $m = 9$ kg. Analogously, stator acceleration is used to estimate rotor disturbance load for the radial bearings, as presented in Figure 2a. Equation 4 indicates forces on the rotor, with $\ddot{q}_r$ and $\ddot{q}_s$ being general rotor and stator radial accelerations respectively. Net AMB force on the rotor is F_{AMB} and rotor weight is F_W. Relative position of AMBs with respect to rotor COM is used to split the inertial load stemming from stator acceleration, shown in Figure 2b. Forces on bearings RB1 and RB2 are given by Eqs. 6 and 7 respectively. The distances of bearings RB1 and RB2 relative to the COM are $a = 180$ mm and $b = 114$ mm respectively. This approach is possible if the external accelerations are known *a priori*.

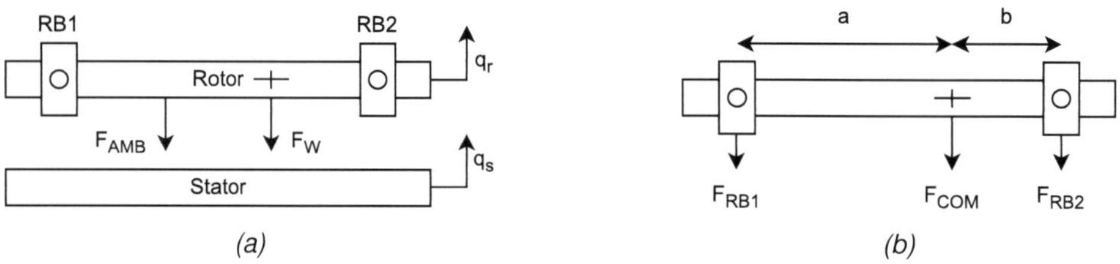

Figure 2: (a) Simplified free body diagram of rotor and stator. (b) Free body diagram of rotor indicating relative positions of radial AMBs with respect to COM.

$$m(\ddot{q}_r - \ddot{q}_s) = F_{AMB} + F_W \tag{4}$$

$$F_{COM} = m \cdot \ddot{q}_s \tag{5}$$

$$F_{RB1} = m \cdot \ddot{q}_s \frac{b}{a+b} \tag{6}$$

$$F_{RB2} = m \cdot \ddot{q}_s \frac{a}{a+b} \tag{7}$$

The static unbalance effect is introduced using a sinusoidal force term. It is assumed to act on the COM, hence the aforementioned procedure adopted for variable disturbances is used to assign load contributions to radial AMBs. The unbalance grade is $G = 2.5$ mm/s for compressors, as per ISO 1940-1 (ISO-1940-1:2003 (2003)). The corresponding maximum permissible eccentricity is $\varepsilon = 0.53$ μm.

Constant loads are implemented directly as known forces from compressor operation. Forces of 100 N in the x- and y-axes of the impeller node are considered as inputs in the state space model. A force of 300 N is imposed in the z-axis. Rotor weight contributes to the y-axis of each radial AMB and is split in the same manner as variable loads.

3. Control Strategy and Simulation Results

The strategy implemented for position control is a combination of feedback and feedforward logic. Feedback control is characterised by a PID controller (Anantachaisilp et al. (2012)), while FF control is an adaptable static gain that is exploited for shocks. The FF gain is sensitive to the versus of imposed acceleration, as constant net forces cause varying instantaneous loads in either positive or negative directions. The FF scheme is evidenced in Figure 3. Each AMB features its own PID and FF controller, describing a decentralised control architecture. A PI current controller is included for each AMB, with the output bounded by the DC bus voltage, $V_{DC} = \pm 50$ V. Figure 4 summarises the control layout implemented for the AMBs. Previous non-linear modeling is used as a reference (Tomczuk & Wajnert (2018)). Current and position controllers are modeled as filters, dependent on the Laplace variable s. The current controller C_c in Eq. 8 contains the proportional gain $K_{p,c}$, and the integral time constant $T_{i,c}$. The position controller in Eq. 9 contains the proportional gain $K_{p,p}$, the integral time constant $T_{i,p}$, and the derivative time constant T_d. A closing pole factor N_d is additionally included.

The parameters for position control are chosen on the basis of the rigid-body mode frequency in closed loop. The closed loop transfer function considers node displacement as the output, and external force on the actuator as the input. The parameters yield rigid body modes at 100 Hz with a gain of -125 dB for the radial AMBs. For the axial AMB, a frequency of 60 Hz and a gain of -120 dB is noted. The frequencies indicate a stiff behaviour of the system ensuring that the position controllers are operationally suitable for the provisioned external excitations. Current controller parameters are chosen such that they exhibit a zero at 5.3 Hz for the axial AMB, and 6.4 Hz for both radial AMBs.

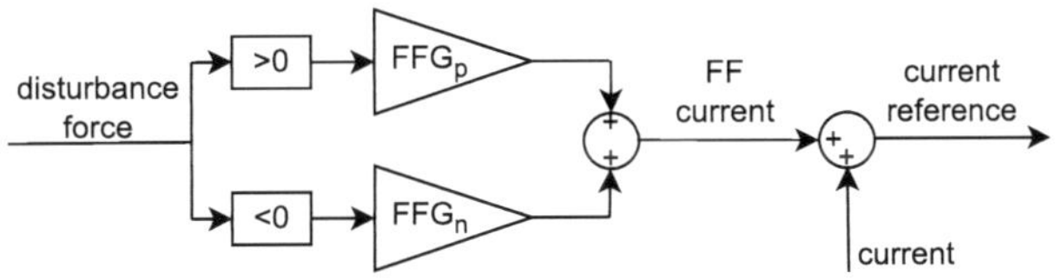

Figure 3: FF control logic for gain adaptation. FFG_p: gain for positive shocks. FFG_n: gain for negative shocks.

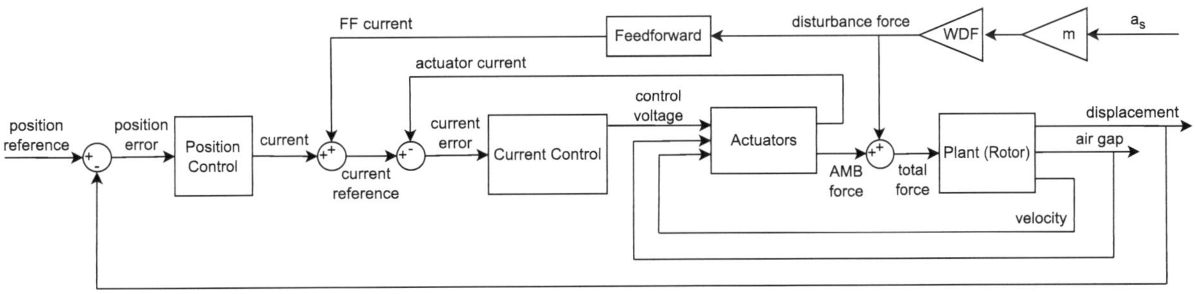

Figure 4: Non-linear control scheme for AMBs. a_s: stator acceleration. WDF: weight distribution factor calculated as per the relative position of radial AMB with respect to COM.

$$C_c = K_{p,c}\left(1 + \frac{1}{T_{i,c}s}\right) \tag{8}$$

$$C_p = K_{p,p}\left(1 + \frac{1}{T_{i,p}s} + \frac{T_d s}{1 + \frac{T_d}{N_d}s}\right) \tag{9}$$

The Simulink® model generates numerical responses due to combined constant and variable loads. Figure 5a highlights the net force experienced by RB1 in the y-axis. The first three peaks are positive net forces that counteract the negative shocks imposed. A constant force offset accounts for rotor weight and radial compressor load at the impeller node. Figure 5b displays the y-axis node displacement of RB1 due to the first negative shock, with a peak displacement of -0.15 mm, corresponding to 30% of the nominal air gap. For an imposed positive shock, node displacement covers 46% of the air gap.

Net force in the y-axis of RB2 is shown in Figure 6a. The continuous force limit is surpassed with a maximum duration of 12 ms. The force limit reflects the permissible current in the coils, such that actuator damage due to Joule losses is avoided. The duration is noted to be insufficient to induce overheating. The y-axis node response of RB2 is shown in Figure 6b with a peak displacement of -0.20 mm, referring to 40% of the nominal air gap. Displacement in the positive direction corresponds to 44% of the air gap.

Figure 7a indicates net forces exerted by the axial AMB. Shocks require a force request that surpasses the continuous force limit, however the duration of 11.6 ms is insufficient to induce coil overheating during operation.

The performance of AB is adequate, confirmed by node displacement due to the first negative shock in Figure 7b. A peak displacement of -0.15 mm is noted, corresponding to 30% of the nominal air gap. With a positive shock, node displacement covers 54% of the air gap. In all cases the PID and FF controller combination is observed to ensure a restoration of steady state displacement. All bearings suitably sustain constant loads, remaining below the continuous force limit.

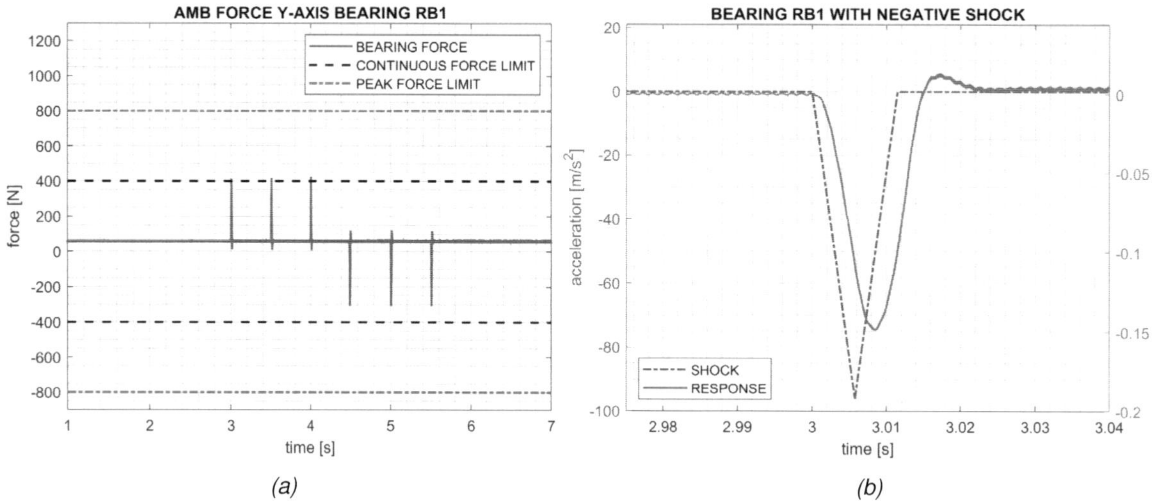

Figure 5: (a) Forces experienced by RB1 in y-axis due to shocks, constant radial compressor loads, and effect of static unbalance.

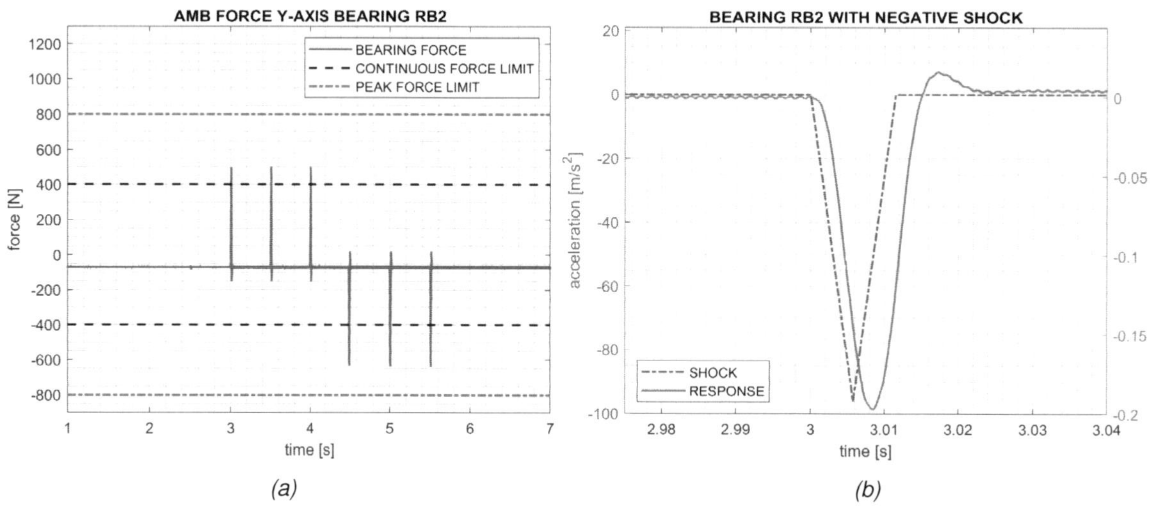

Figure 6: (a) Forces experienced by RB2 in y-axis due to shocks, constant radial compressor loads, and effect of static unbalance.

4. Final remarks

The modeling approach for a compressor rotor suspended by AMBs has been outlined, with non-linear analytical modeling for the electromechanical domain, and a FEM modeling approach for rotordynamic behaviour. Constant and variable loads are modeled, including rotor weight and static unbalance. An adaptive FF control

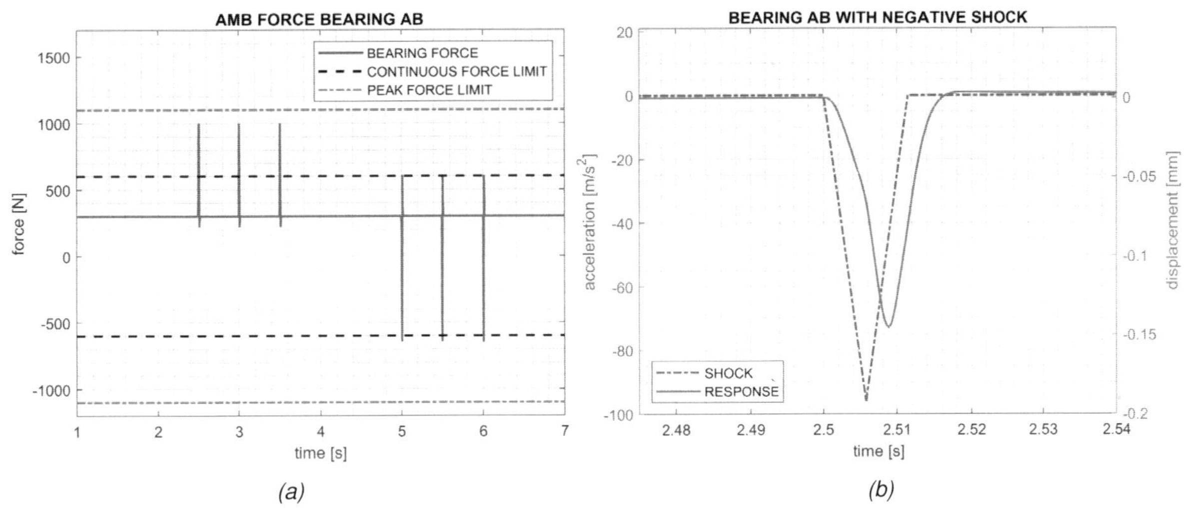

Figure 7: (a) Forces experienced by bearing AB in z-axis due to shocks and constant axial compressor load.

complements the PID position controller for each AMB. Numerical results indicate a stable response to imposed shocks, confirming the operational viability of the system. Knowledge of the shock profile is required, making this control approach suitable for well-studied applications. Deviations from the expected variable loading may not be adequately sustained by the controller, presenting a limitation to general applications. Further work involves an analysis of the robustness of such control. The choice of PID gains may be informed by a deeper exploration of system stability, observing the closed-loop poles and applying the Nyquist stability criterion. A more general force estimation method may also be included, to account for environments with unknown inputs. An experimental validation to assess the modeling method is also required.

References

Anantachaisilp, P., Lin, Z. & Allaire, P. (2012), Pid tuning methods for active magnetic bearing systems, *in* '13th International Symposium on Magnetic Bearings', Arlington, Virginia, USA, pp. 1–14.

Chen, W. J. (2015), *Practical Rotordynamics and Fluid Film Bearings*, 1 edn, Trafford Publishing.

ISO-1940-1:2003 (2003), Mechanical vibration – balance quality requirements for rotors in a constant (rigid) state, Standard ISO 1940-1:2003, International Organization for Standardization, Geneva, CH.

Jeong, H. H., Yun, S. N. & Yang, J. H. (2012), Control of magnetic bearing system, *in* R. Sehgal, ed., 'Performance Evaluation of Bearings', IntechOpen, Rijeka, chapter 8.
URL: *https://doi.org/10.5772/51185*

Min Sig Kang & Woo Hyun Yoon (2006), 'Acceleration feedforward control in active magnetic bearing system subject to base motion by filtered-X LMS algorithm', *IEEE Transactions on Control Systems Technology* **14**(1), 134–140.
URL: *http://ieeexplore.ieee.org/document/1564103/*

Tomczuk, B. & Wajnert, D. (2018), 'Field–circuit model of the radial active magnetic bearing system', *Electrical Engineering* **100**(4), 2319–2328.
URL: *http://link.springer.com/10.1007/s00202-018-0707-7*

Axial Position Control of a Lorentz-Force-Type Cylindrical Self-Bering Motor with Coreless Distributed Windings

Satoshi UENO[a], Koki MAEDA[a], Chengyan ZHAO[a]

a Ritsumeikan University, 1-1-1 Nojihigashi, Kusatsu, Shiga, Japan, sueno@se.ritsumei.ac.jp

Abstract

A Lorentz-force-type self-bearing motor rotates and supports a rotor using the Lorentz force generated by the magnetic flux from the rotor's permanent magnets and the current flowing in the stator windings. In the previous model, only rotational torque and radial forces were controlled by two coreless cylindrical six-phase distributed windings. This paper proposes a method to control the axial force using two diamond-shaped windings. The structure and control principle is introduced, then a prototype machine and the result of the levitation test are shown.

Keywords: Self-bearing motor, Lorentz force, Axial force, Levitation control

1. Introduction

A self-bearing motor (SBM) integrates a motor and magnetic bearings and is possible to reduce the overall size and cost of the device. As an application example, a blood pump using the SBM has been developed (Osa et al. 2021). Although various types of SBMs have been proposed, this study focuses on a Lorentz-force-type cylindrical SBM (L-SBM) with coreless six-phase distributed windings (Ueno et al. 2009). Conventional L-SMBs cannot control the axial force, then it is necessary to use the passive stability provided by a thin rotor (Steinert et al. 2014) or add an active magnetic bearing (Ren & Stephens 2005) to support the axial direction of the rotor. When passive stability is used, the shape of the rotor is limited, and when active magnetic bearings are used, the structure becomes complicated.

In this study, we propose a winding structure that enables control of the axial direction in addition to the motor torque and radial direction. By generating and controlling the axial force, an axial magnetic bearing can be removed, thus it is possible to further miniaturization and simplification. This paper describes the structure of the proposed motor, the principle and analysis of the bearing force

This paper describes the structure of the proposed motor, the principle of generating the bearing force, and the results of the bearing force analysis. The results of the levitation test on the prototype machine are also shown.

2. Structure and principle

2.1 Structure

Fig. 1 shows the structure of the proposed LSBM. Two coreless distributed windings are vertically arranged in the stator. The rotor consists of an iron core and permanent magnets, whose axial length overlaps the lower half of the upper coil and the upper half of the lower coil. In this paper, four PMs are attached, and the number of poles of the rotor is four.

Fig. 2 shows the development view of the stator winding. It consists of six-phase diamond-shaped windings and is rounded into a cylindrical shape to attach to the stator. Since the winding in this overlapping area is

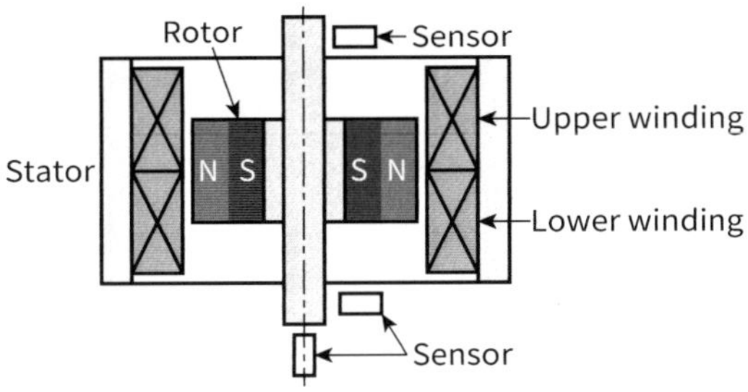

Figure 1: Structure of the proposed motor.

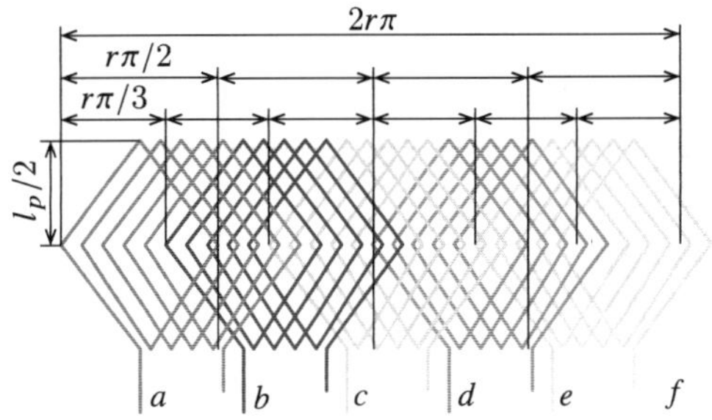

Figure 2: Development view of a six-phase winding.

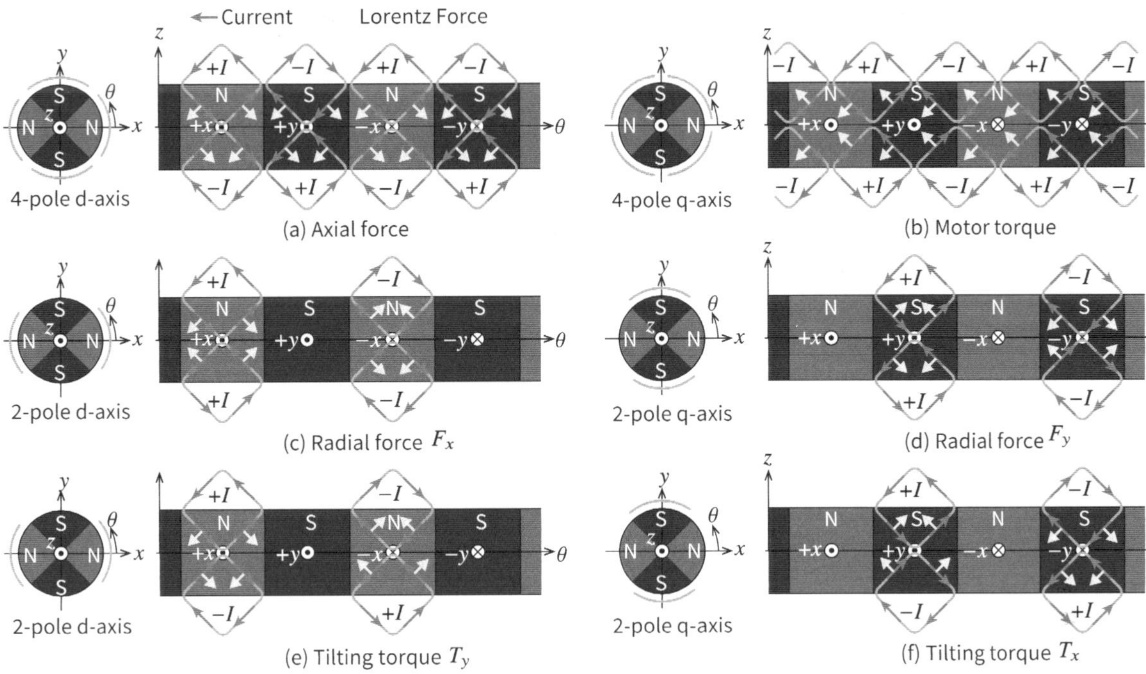

Figure 3: Generation of force and torque.

oblique, it can generate axial force. The width of a phase is $r\pi/2$, and the space between phases is $r\pi/3$, where r is the radius of the winding and l_p is the axial length of the PMs.

2.2 Generation of force and torque

The principle of control force generation is shown in Fig. 3. The right figure shows the top view, and the left figure shows the developed view. The red and blue represent the N- and S-pole, and the N-pole generating magnetic flux from the back of the paper to the front. For simplicity, only one winding in the d- and q-axis direction is shown. The green lines indicates the direction of the currents, and the yellow line indicates the direction of the Lorentz forces.

As shown in Fig. 3 (a), the axial force is generated by suppling the four-pole d-axis currents in opposite direction between the upper and lower windings. Since the tangential forces are cancelled, only the axial force is generated. The motor torque is generated by four-pole q-axis current as shown in Fig. 3 (b). Radial forces and tilting torques are generated by two-pole current as shown in Figs. 3 (c) - (f).

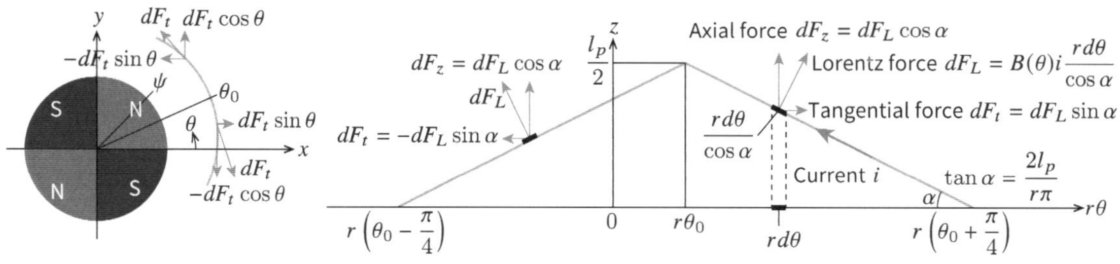

Figure 4: Coordinate of one-turn winding.

2.3 Analysis of force and torque

Next, the force and torque are analytically derived. Firstly, one-turn winding is considered and expanded into six-phase and N-turn winding.

The coordinate system is defined as Fig. 4. θ_0 is the angular position of the winding apex, and ψ is the rotational angle of the rotor. Assuming the magnetic flux density distribution to be

$$B(\theta) = B_r \cos 2(\theta - \psi) \tag{1}$$

where B_r is the amplitude of the magnetic flux density. Each phase current is expressed as

$$i_k = I_z \cos 2\left(\psi - \frac{\pi}{3}k\right) + I_m \sin 2\left(\psi - \frac{\pi}{3}k\right) + I_x \cos 2\left(\psi + \frac{\pi}{6}k\right) + I_y \sin 2\left(\psi + \frac{\pi}{6}k\right) \tag{2}$$

where $k = 0, 1, 2, 3, 4, 5$ which corresponds to phase a, b, c, d, e, and f.

The Lorentz force at θ is calculated as Fig. 4, then the axial force by the one-turn winding is calculated as

$$F_{z0} = \int_{\theta_0 - \frac{\pi}{4}}^{\theta_0 + \frac{\pi}{4}} dF_z = rB_r \cos 2(\psi - \theta_0)\, i \tag{3}$$

Each phase winding is placed with the angular space of $\pi/3$, then the axial force of the one-turn six-phase winding becomes

$$F_{z1} = rB_r \sum_{k=0}^{5} \cos 2\left(\psi - \theta_0 - \frac{\pi}{3}k\right) i_k = 3rB_r \left(I_z \cos 2\theta_0 + I_m \sin 2\theta_0\right) \tag{4}$$

When the turn number is N, the angular space of each turn is $\pi/(3N)$. Then, the angular position of n-th winding is

$$\theta_n = -\frac{\pi(N-1)}{6N} + \frac{\pi}{3N}(n-1) \tag{5}$$

where the center of a-phase is set to 0. Since the force acted to the rotor is the reaction of the Lorentz force, the total axial force is calculated as

$$F_z = -3rB_r \sum_{n=1}^{N} \left(I_z \cos 2\theta_n + I_m \sin 2\theta_n\right) = -\frac{3\sqrt{3}rB_r}{2\sin\frac{\pi}{3N}} I_z \tag{6}$$

I_z generates a downward force in the lower winding while an upward force in the upper winding. Therefore, the opposite current should be supplied to control the axial force.

The tangential force of the one-turn winding is calculated as

$$f_{t0} = -\int_{\theta_0 - \frac{\pi}{4}}^{0} df_t + \int_{0}^{\theta_0 + \frac{\pi}{4}} df_t = \frac{2B_r l_p}{\pi} \sin 2(\psi - \theta_0)\, i \tag{7}$$

Calculating the rotational torque of the six-phase winding in the same way as the axial force, we have

$$T_{z1} = -r\frac{2B_r l_p}{\pi}\sum_{k=0}^{5}\sin 2\left(\psi-\theta_0-\frac{\pi}{3}k\right)i_k = \frac{6rB_r l_p}{\pi}\left(-I_z\sin 2\theta_0 + I_m\cos 2\theta_0\right) \tag{8}$$

and the rotational torque of N-turn winding is calculated as

$$T_z = \frac{6rB_r l_p}{\pi}\sum_{n=1}^{N}\left(-I_z\sin 2\theta_n + I_m\cos 2\theta_n\right) = \frac{3\sqrt{3}rB_r l_p}{\sin\frac{\pi}{3N}}I_m \tag{9}$$

The rotational torque can be controlled by I_m and is not affected by other currents.

The radial forces of the one-turn winding are calculated as

$$f_{x0} = \int_{\theta_0-\frac{\pi}{4}}^{0} df_t\sin\theta - \int_{0}^{\theta_0+\frac{\pi}{4}} df_t\sin\theta = -\frac{B_r l_p}{3\pi}\left\{(2+\sqrt{2})\cos(2\psi-3\theta_0) - 3(2-\sqrt{2})\cos(2\psi-\theta_0)\right\}i \tag{10}$$

$$f_{y0} = -\int_{\theta_0-\frac{\pi}{4}}^{0} df_t\cos\theta + \int_{0}^{\theta_0+\frac{\pi}{4}} df_t\cos\theta = \frac{B_r l_p}{3\pi}\left\{(2+\sqrt{2})\sin(2\psi-3\theta_0) + 3(2-\sqrt{2})\sin(2\psi-\theta_0)\right\}i \tag{11}$$

Then the forces of the six-phase winding are

$$\begin{aligned} f_{x1} &= -\frac{B_r l_p}{3\pi}\sum_{k=0}^{5}\left\{(2+\sqrt{2})\cos(2\psi-3\theta_0-\pi k) - 3(2-\sqrt{2})\cos\left(2\psi-\theta_0-\frac{\pi}{3}k\right)\right\}i_k \\ &= -\frac{3(2-\sqrt{2})B_r l_p}{\pi}\left(I_x\cos\theta_0 + I_y\sin\theta_0\right) \end{aligned} \tag{12}$$

$$\begin{aligned} f_{y1} &= \frac{B_r l_p}{3\pi}\sum_{k=0}^{5}\left\{(2+\sqrt{2})\sin(2\psi-3\theta_0-\pi k) + 3(2-\sqrt{2})\sin\left(2\psi-\theta_0-\frac{\pi}{3}k\right)\right\}i_k \\ &= \frac{3(2-\sqrt{2})B_r l_p}{\pi}\left(I_y\cos\theta_0 - I_x\sin\theta_0\right) \end{aligned} \tag{13}$$

and the forces with N-turn winding are

$$f_x = \frac{3(2-\sqrt{2})B_r l_p}{\pi}\sum_{n=1}^{N}\left(I_x\cos\theta_n + I_y\sin\theta_n\right) = \frac{3(2-\sqrt{2})B_r l_p}{\pi\sin\frac{\pi}{3N}}I_x \tag{14}$$

$$f_y = -\frac{3(2-\sqrt{2})B_r l_p}{\pi}\sum_{n=1}^{N}\left(I_y\cos\theta_n - I_x\sin\theta_n\right) = -\frac{3(2-\sqrt{2})B_r l_p}{\pi\sin\frac{\pi}{3N}}I_y \tag{15}$$

The radial forces can be controlled by I_x and I_y, respectively.

3. Experimental verification

To verify that the proposed method can control the bearing forces, a levitation test was conducted on a test machine.

The test machine is shown in Fig. 5. For simple fabrication, rectangular magnets are used. Windings are made by hand winding, and attached to the casing made from engineering plastic. Fig. 5 (c) is the appearance without the upper displacement sensors. The amplitude of the fundamental component of the magnetic flux density is approximately 0.32 T from the FEM results, then the axial force is calculated as 0.82 N/A for each winding from Eq. (6). The requirement for levitation, 1.62 N, can be generated.

Fig. 6 shows the control system. Five displacement sensors are installed. Linear amplifiers are used to supply the current. Fig. 7 shows the controllers. PD controllers with approximate differentiators are used. The

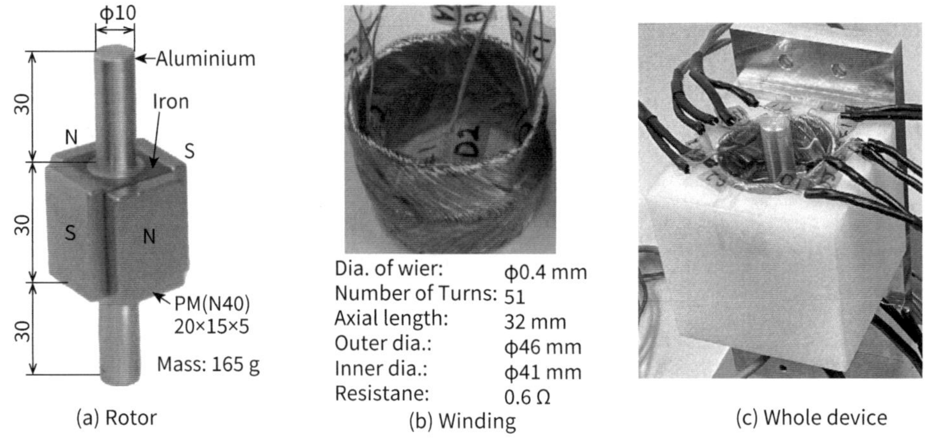

Figure 5: Test machine.

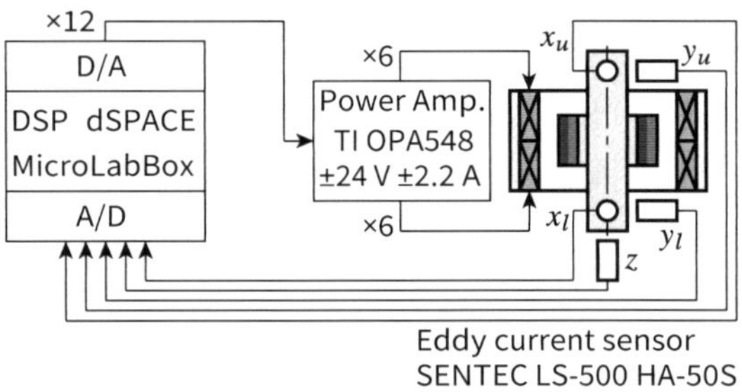

Figure 6: Control system.

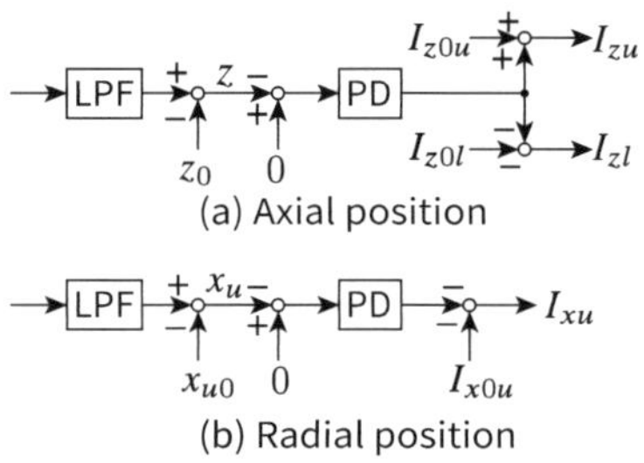

Figure 7: Controller.

Table 1: Parameters of controller

Proportional gain	2 A/mm	Break frequency of PD	500 Hz	Sampling time	0.1 ms
Differential gain	0.01 As/mm	Cutoff frequency of LPF	250 Hz		

controllers are converted to the discrete transfer functions by the Tustin method and implemented in the DSP. The upper and lower windings use different bias currents in the axial position control. Since the upper winding generates the attractive force while the lower winding generates the repulsive force, the rotational direction will be unstable if the same current is applied. Therefore, we reduce the bias current in the lower winding to stabilize the rotational position. The radial positions x_u, y_u, x_l, and y_l are controlled separately as shown in Fig. 7 (b). The gains, reference positions, and bias currents are tuned by trial and error in the experiment. The parameters of the controller are shown in Table 1. The same values are used for all PD controllers.

The startup responses are shown in Fig. 8. The displacement of the rotor is well controlled, and the rotor weight is supported by the stator currents. I_{z0u} and I_{z0l} were set to 1.4 A and 0.9 A, respectively. Since the average is 1.15 A, the axial force gain is 1.4 N/A, which is the almost same value as the theoretical value. These results confirm that the rotor can be actively supported in five axes by the proposed method.

4. Conclusions

A Lorentz-force-type self-bearing motor that controls both radial and axial forces is proposed, and its feasibility is demonstrated by theoretical analysis and experiments.

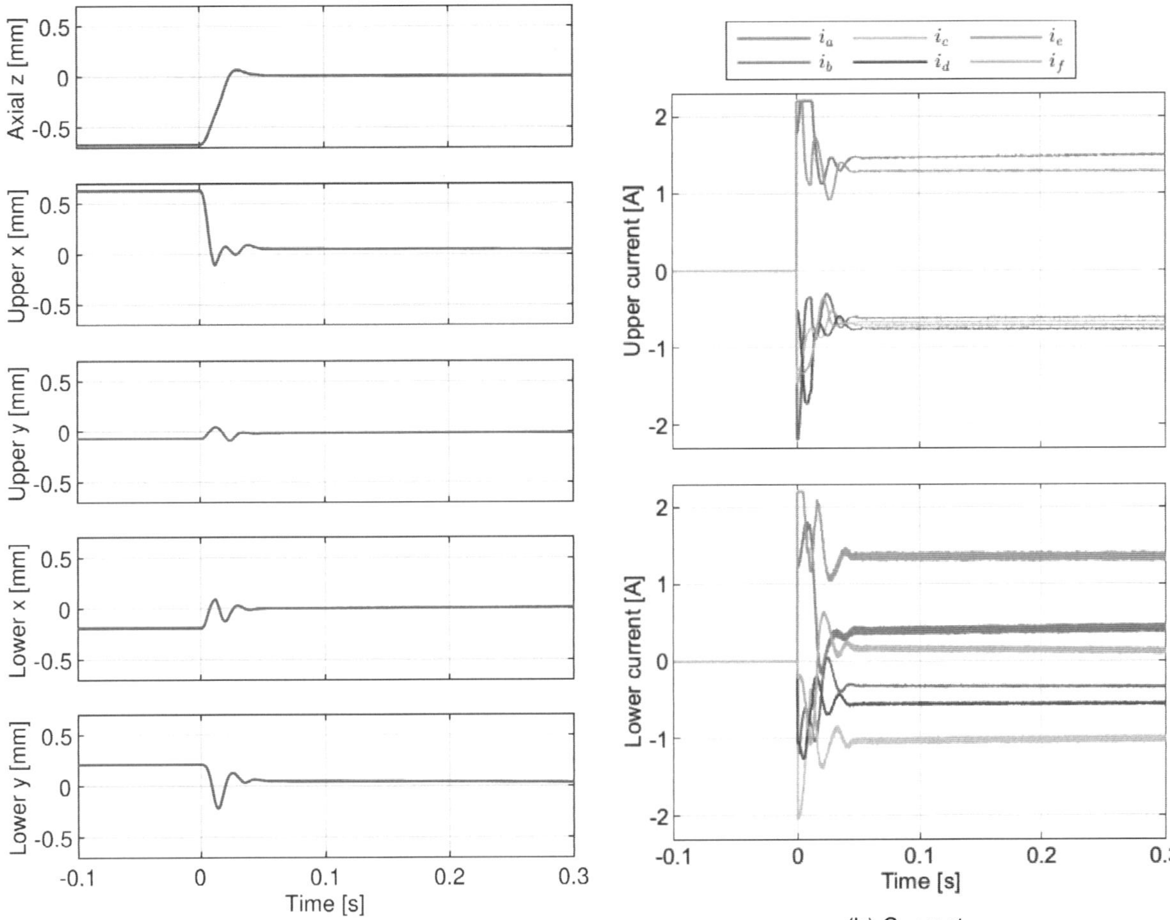

(a) Displacement (b) Current

Figure 8: Startup responses of the rotor displacement and currents.

Acknowledgment

This work was partially supported by Tsugawa Foundation.

References

Osa, M., Masuzawa, T., Yamaguchi, K. & Tatsumi, E. (2021), 'Double stator axial gap type ultra-compact 5-dof controlled self-bearing motor for rotary pediatric ventricular assist device', *IEEE Transactions on Industry Applications* **57**(6), 6744–6753.

Ren, Z. & Stephens, L. (2005), 'Closed-loop performance of a six degree-of-freedom precision magnetic actuator', *IEEE/ASME Transactions on Mechatronics* **10**(6), 666–674.

Steinert, D., Nussbaumer, T. & Kolar, J. W. (2014), 'Slotless bearingless disk drive for high-speed and high-purity applications', *IEEE Transactions on Industrial Electronics* **61**(11), 5974–5986.

Ueno, S., Uematsu, S. & Kato, T. (2009), 'Development of a Lorentz-Force-Type Slotless Self-Bearing Motor', *Journal of System Design and Dynamics* **3**(4), 462–470.

Model Order Reduction of an axial magnetic bearing

Jérôme TOMEZYK[a], Sebastian RODRIGUEZ[b], Yves DUPUIS[a], Joaquim DA SILVA[a]

[a] *SKF Magnetic Mechatronics, 2 rue des champs, 27950 Saint-Marcel, France, jerome.tomezk@skf.com*
[b] *Laboratoire PIMM, Arts et Métiers institute of technology, CNRS, CNAM, HESAM Université, 151 Boulevard de l'Hôpital, 75013 Paris, France, sebastian.rodriguez_iturra@ensam.eu*

Abstract

We introduce a model order reduction (MOR) method applied to the simulation of an axial active magnetic bearing (AMB) which is based on Finite Element discretisation of non-linear magneto-dynamic Maxwell's equations coupled with circuit equations and a moving part (rotor that can only move in one axis). In order to speed up the computation time, we studied a MOR based on the Proper Orthogonal Decomposition (POD) to reduce the order of the problem coupled with the Discrete Empirical Interpolation Method (DEIM) to deal with the nonlinearity. We introduce a way to consider the rotor's displacement, which do not require remeshing, and is compatible with such a method. The technique is evaluated on a 2D axisymetrical axial magnetic bearing model. This technique allows a speed up about a minimun of 2 times faster than the full size model.

Keywords: Finite Element Method, Non linear materials, Eddy currents, Proper Orthogonal Decomposition, Discrete Empirical Interpolation Method

1. Introduction

Due to the mechanical constraints, axial active magnetic bearing (AMB) are usually composed of solid stator and rotor part. Under voltage excitation varying at high frequency, Eddy currents are generated within axial AMB which highly limits its bandwidth. These Eddy currents coupled with the use of non linear magnetic materials make it necessary to run Finite Element (FE) simulation. However the computational time required to solve such a physical process can become prohibitive due to the complexity of the problem.

Model Order Reduction (MOR) techniques can be used to accelerate the computation time. Among the different techniques one can mention the use of the Proper Orthogonal Decomposition (POD), where the reduced basis is constructed "offline" after a Sigular Value decomposition (SVD) of FE solutions, in order to be used later in an fast "online" resolution step. In addition, MOR techniques have also been used for low-rank approximation of nonlinear terms. In particular, there is the Discrete Empirical Interpolation Method (DEIM), which evaluates the nonlinearity only on a few points in order to interpolate the remaining points, which, when combined with the POD method applied to the solution, allows to drastically reduce the computational costs of nonlinear solvers.

Different works have proposed the use of these methods to accelerate the simulation of a three-phase transformer (Henneron & Clenet 2015) or a rotating electrical machine (Montier et al. 2016). For this reason, in the present work we propose to use a POD-DEIM approach to construct a reduced model of an axial AMB with axial rotor displacement, magnetic saturation and eddy currents.

The paper is structured as follows. Section 2. introduce the governing equations of the considered problem. Section 3. gives details on how to construct the reduced model. Following Section 4. shows numerical results and finally, Section 5. provides conclusions and perspectives.

2. Problem formulation

The mathematical formulation used in the modeling of the axial AMB corresponds to the Magneto-Dynamic Maxwell's equation. First, the equation is described in the continuous form.

2.1 Continuous formulation

In order to solve Maxwell's magnetodynamic equations, the vector potential formulation is considered. The vector potential $\mathbf{A}$ is defined as $\mathbf{B} = \nabla \times \mathbf{A}$, such that the strong formulation is given as follows

$$\nabla \times (\nu(B)\nabla \times A) + \sigma\frac{\partial A}{\partial t} - i_1 N_1 - i_2 N_2 = 0 \quad (1)$$

$$\frac{\partial \phi_j}{\partial t} + R_j i_j = u_j \text{ for } 1 \leq j \leq 2 \quad (2)$$

where N_j corresponds to the unit current density, i_j the current, u_j the voltage and R_j the resistor of the jth coil, additionally, ν corresponds to the magnetic reluctivity and σ the electrical conductivity. The equations (2) are the circuit equations, to impose a voltage at the terminal of the stranded coils, then ϕ_j correspond to the magnetic flux. The force is computed with the Maxwell's tensor. Figure 1 shows one half of the magnetic bearing, with two coils that are connected in series. This strong formulation is then discretize using the FEM, which is detailed in the following section.

2.2 Semi-discrete formulation

We note $X \in \mathbb{R}^{N_{FE}}$ the finite element discretization of A, then the equations (1)-(2) becomes

$$M^{\nu}(X)X + M^{\sigma}\frac{dX}{dt} - \sum_{j=1}^{N_{coil}} i_j N_j = 0 \quad (3)$$

$$N_j^T\frac{dX}{dt} + R_j i_i = u_j \text{ for } 1 \leq j \leq N_{coil} \quad (4)$$

The exponent T corresponds to the transposition. In the following, the influence of rotor displacement is detailed.

2.3 Movement of the rotor

To take into account the rotor displacement, the mesh is deformed in the same spirit as in (Hasan et al. 2018). Thus, for a displacement in the y direction, all the rotor and air nodes are translated, except for the airgap, where an affine transformation is performed to keep a conform mesh (see figure 1). As a consequence, M^{σ} is independent of the displacement since σ is null in the air and the rotor is just translated, not deformed. Therefore, the stiffness matrix can be divided in two parts

$$M^{\nu}(X) = M^{\nu_{NL}}(X) + M_z^{\nu_0}$$

with $M^{\nu_{NL}}(X)$ a matrix independent from the position y and $M_y^{\nu_0}$ the stiffness matrix in the air. In contrast to the techniques that require a remesh for each time step, this way to consider the movement allows to have the number of degree of freedom N_{FE} constant and does not require any projection map during the resolution of the temporal problem. The fully discretized model can now be described.

2.31 Fully discretized model

A backward Euler is used for the time discretization with constant time step δt and Newton-Raphson procedure for the non linearity is considered. In this sense, the residual is defined as follows:

$$R\begin{pmatrix} X^n \\ i_1^n \\ i_2^n \end{pmatrix} = \begin{pmatrix} M^{\nu_{NL}}(X^n)X^n + M_y^{\nu_0}X^n + M^{\sigma}\dfrac{X^n - X^{n-1}}{\delta t} - i_1^n N_1 - i_2^n N_2 \\ N_1^T\dfrac{X^n - X^{n-1}}{\delta t} + R_1 i_1^n - u_1^n \\ N_2^T\dfrac{X^n - X^{n-1}}{\delta t} + R_2 i_2^n - u_2^n \end{pmatrix} \quad (5)$$

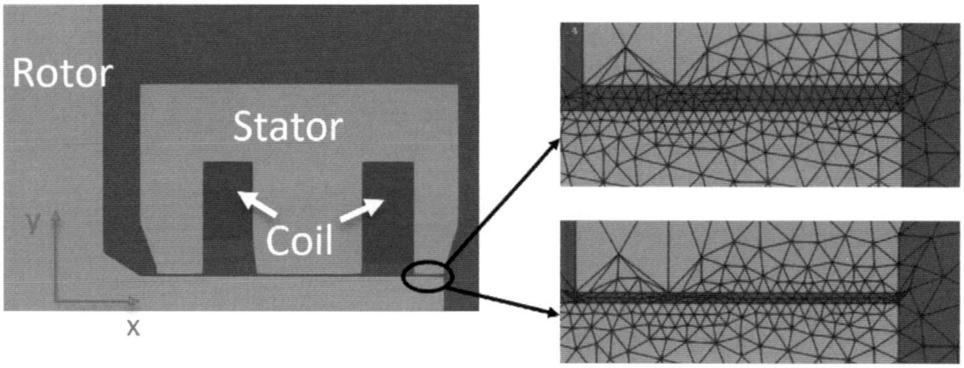

Figure 1: Half of axial magnetic bearing (in axisymmetric settings) with zoom on airgap in case of displacement.

Then, for each time step, one must find $\begin{pmatrix} X^n \\ i_1^n \\ i_2^n \end{pmatrix}$ such that $R\begin{pmatrix} X^n \\ i_1^n \\ i_2^n \end{pmatrix} = 0$. Then the Newton-Raphson procedure corresponds to iterate until convergence of the term

$$\tilde{X}_k = \tilde{X}_{k-1} - (J(\tilde{X}_{k-1}))^{-1} R(\tilde{X}_{k-1}) \tag{6}$$

with $\tilde{X}_{k-1} = \begin{pmatrix} X^{n-1} \\ i_1^{n-1} \\ i_2^{n-1} \end{pmatrix}$ and $J(\tilde{X}_{k-1}) = \dfrac{\partial R}{\partial \tilde{X}_{k-1}}$ the jacobian of R. We can write

$$J\begin{pmatrix} X^{n-1} \\ i_1^{n-1} \\ i_2^{n-1} \end{pmatrix} = \begin{pmatrix} M_y^{\nu_0} + (1/\delta t)M^{\sigma} & -N_1 & -N_2 \\ (1/\delta t)N_1^T & R_1 & 0 \\ (1/\delta t)N_2^T & 0 & R_2 \end{pmatrix} + \begin{pmatrix} J_{NL}(X^{n-1}) & 0 & 0 \\ 0 & 0 & 0 \\ 0 & 0 & 0 \end{pmatrix} \tag{7}$$

with $J_{NL}(X) = \dfrac{\partial M^{\nu_{NL}}(X)X}{\partial X}$.

3. Reduced Model: POD and DEIM/G-POD

3.1 POD

The first step in reducing the model is to obtain a reduced basis that is complete enough to represent the solution of the problem properly. Thus, following the POD strategy, a matrix of snapshots is compute as follows:

$$M_{snap} = [X_1 \cdots X_N] \in \mathbb{R}^{N_{FE} \times N}, \tag{8}$$

which contains on each column the solution of the full order model at N time steps.
Then, the Singular Value Decomposition (SVD) is applied to this matrix to obtain the orthogonal matrices $U \in \mathbb{R}^{N_{FE} \times N_{FE}}$ and $V \in \mathbb{R}^{N \times N}$ and the rectangular diagonal matrix $\Sigma \in \mathbb{R}^{N_{FE} \times N}$ such that

$$M_{snap} = U \Sigma V^T \tag{9}$$

We can define the reduce basis $\Psi = [U_1, \cdots, U_{n_{POD}}]$. As the matrix Σ contains the singular values σ_i sorted in decreasing order, then we can choose n_{POD} such that $\dfrac{\sigma_{n_{POD}}}{\sigma_1}$ is smaller than a tolerance for example.

We can project the FE system onto this reduced basis, and then replace M^{ν}, M^{σ} and N_i by

$$M_R^{\nu_{NL}}(\Psi X_R) = \Psi^T M^{\nu_{NL}} \Psi,\ M_R^{\sigma} = \Psi^T M^{\sigma} \Psi,\ M_{y,R}^{\nu_0} = \Psi^t M_y^{\nu_0} \Psi \text{ and } N_{j,R} = \Psi^T N_j, \tag{10}$$

and then the new unknown of the reduced problem is $X_R \in \mathbb{R}^{n_{POD}}$. To compute $M_R^{\nu_{NL}}(\Psi X_R)$, we have to come back to the full order space, and build a full size matrix to then project it in reduced space, that is computational expensive.

3.2 DEIM/G-POD

The second step is then to reduce the computation time of the nonlinear term. To this end, we use DEIM, a method which have been proposed first in (Chaturantabut & Sorensen 2010). The basic idea is, in place of compute the full matrix $M^{\nu_{NL}}$ and then project it on the POD basis, we only compute a few number of rows of this matrix and then deduce the matrix via an appropriate linear operator. We start by taking a matrix of snapshot for the nonlinear term and take its SVD,

$$G_{snap} = [M^{\nu_{NL}}(X_1)X_1 \cdots M^{\nu_{NL}}(X_{N_{Newton}})X_{N_{Newton}}] = U_G \Sigma_G V_G^T, \tag{11}$$

where we save the nonlinear term for each time step and each step of the Newton-Raphson algorithm. Using the algorithm given in (Chaturantabut & Sorensen 2010), we get a reduce basis $\Phi \in \mathbb{R}^{N_{FE} \times n_{DEIM}}$ and a list $p \in \mathbb{R}^{n_{DEIM}}$ of the rows on which we will evaluate the nonlinear term. We can then build the row selection matrix $P \in \mathbb{R}^{N_{DEIM} \times N_{FE}}$

$$P = \begin{pmatrix} e_{p_1}^T \\ \cdots \\ e_{p_{n_{DEIM}}}^T \end{pmatrix} \in \mathbb{R}^{N_{FE} \times N_{DEIM}} \text{ with } e_i \in \mathbb{R}^{N_{FE}} \text{ such that } e_i(j) = \delta_{ij}.$$

Hence the approximation of the reduced non linear matrix is given by

$$M_R^{\nu_{NL}}(\Psi X_R) \simeq \Psi^T \Phi (P\Phi)^{\dagger} P M^{\nu_{NL}}(\Psi X_R) \Psi = M_{DEIM} M^{\nu_{NL}}(\Psi X_R)|_p \Psi \tag{12}$$

$$J_{NL,R} \simeq M_{DEIM} J_{NL}(\Psi X_R)|_p \Psi \tag{13}$$

where $\dagger$ corresponds to the pseudo-inverse, $M_{DEIM} \in \mathbb{R}^{n_{POD} \times n_{DEIM}}$ is a matrix which we have to compute once and $M^{\nu_{NL}}(\Psi X_R)|_p \in \mathbb{R}^{n_{DEIM} \times N_{FE}}$ (resp. $J_{NL}(\Psi X_R)|_p$) corresponds to the non linear matrix (resp. jacobian), but only for the row contains in p.

Remark: A good approach to increase the stability of the method is to use the Gappy-POD (G-POD) which only consists in evaluating the nonlinear term on more points than the basis contains vectors, ie $\Phi \in \mathbb{R}^{N_{FE} \times n_1}$ and $p \in \mathbb{R}^{n_2}$ with $n_2 > n_1$. It is for this generalization that there is a pseudo-inverse in the definition of M_{DEIM}.

3.3 Reduced model

Now, we have all the key ingredient to write the reduced problem. Like the full order model, we solve the POD-DEIM model using the Newton-Raphson algorithm to handle the nonlinear term, then for each time step, we have to solve

$$R_R \begin{pmatrix} X_R^n \\ i_1^n \\ i_2^n \end{pmatrix} = \begin{pmatrix} M_R^{\nu_{NL}}(\Psi X_R^n) X_R^n + M_y^{\nu_0} X_R^n + M^{\sigma} \dfrac{X_R^n - X_R^{n-1}}{\delta t} - i_1^n N_{1,R} - i_2^n N_{2,R} \\ N_{1,R}^T \dfrac{X_R^n - X_R^{n-1}}{\delta t} + R_1 i_1^n - u_1^n \\ N_{2,R}^T \dfrac{X_R^n - X_R^{n-1}}{\delta t} + R_2 i_2^n - u_2^n \end{pmatrix} = 0 \tag{14}$$

The algorithm 1 gives a description of the resolution scheme.

Algorithm 1 POD-DEIM

Input: Ψ, p, M_{DEIM}, t, z, U, X_0
Output: X_k, i_1^k and i_2^k for $1 \leq k \leq N$
  for $k = 1 : N$ **do**
    $\tilde{X} = \begin{pmatrix} X_{k-1} \\ i_1^{k-1} \\ i_2^{k-1} \end{pmatrix}$
    Compute $M_{y_k}^{\nu_0}$ and compute $M_{y_k,R}^{\nu_0} = \Psi^t M_{y_k}^{\nu_0} \Psi$. ▷ Take in account the displacement
    while Newton does not converge **do**
      Compute $M_R^{\nu_{NL}} = M_{DEIM} M^{\nu_{NL}}(\Psi \tilde{X}_{1:N_{FE}})|_p \Psi$ and $J_{NL,R} = M_{DEIM} J_{NL}(\Psi \tilde{X}_{1:N_{FE}})|_p \Psi$.
      $\delta X = -J_R^{-1} R_R(\tilde{X})$
      $\tilde{X} = \tilde{X} + \delta X$
    end while
    $X_k = \tilde{X}_{1:N_{FE}}$, $i_1^k = \tilde{X}_{N_{FE}+1}$, $i_2^k = \tilde{X}_{N_{FE}+2}$
  end for

4. Application

4.1 Fixed rotor

As a first test, rotor's movement is not considered. To get a sinusoidal flux, we use a simple P-controller with saturation to control the voltage ($Vdc = 70V$ and $f_s = 1kHz$) and the simulation starts from a steady state (i.e., $i_1 = i_2 = 15A$ for the initial timestep). We can see on the figure 2 the flux, current and force for the FE and the POD-DEIM solutions. Computation time is about 38m18s for the full order model while it is of 35m for the POD model and 7m15s for the POD-DEIM model.

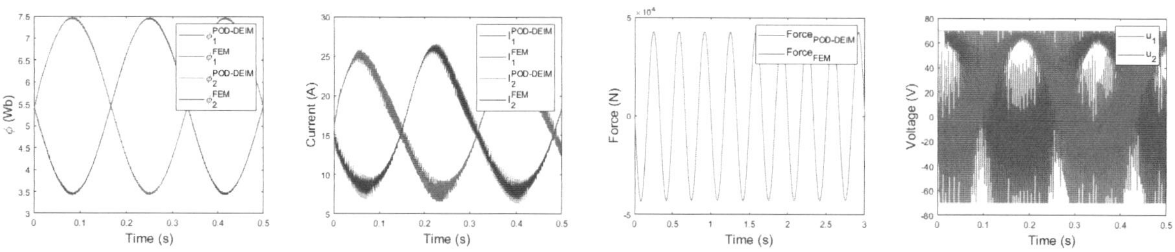

Figure 2: Flux, current and force ($n_{POD} = 15$ and $n_{DEIM} = 90$)

4.2 Constant voltage

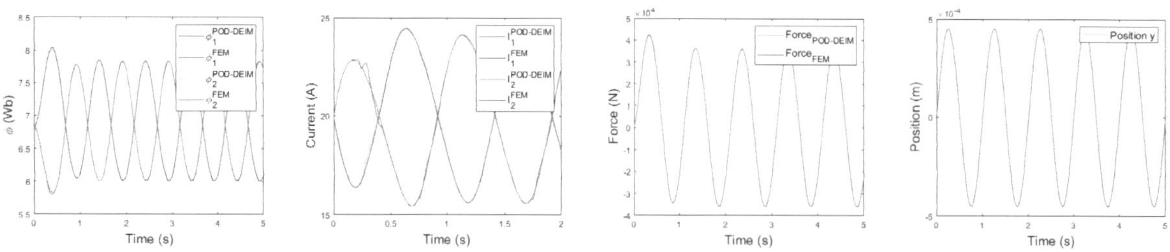

Figure 3: Flux, current and force ($n_{POD} = 10$ and $n_{DEIM} = 90$) for fixed voltage and sinusoidal displacement

In this case, the voltages stay at the same level during the whole simulation, such that $u_j = 20R_j$ $(j = 1,2)$ and the rotor displacement follows a sinusoidal curve. Figure 3 shows the results for this case. The computation time is here about 53m14s for the full order model, 45m47s for the reduced model with POD and 11m14s for that with POD-DEIM.

4.3 Concatenation of POD's basis

In this example, we combine both of the POD basis in order to obtain a new one that can handle a new scenario. For this test, we control the voltage to get a sinusoidal force ($Vdc = 70V$ and $f_s = 1kHz$). As the set point is the force, we have to add a mass-spring equation to address a physical movement (without taking into account the stiffness contact between stator and rotor parts). The DEIM parameters are computed using the data from the current test, which requires that n_{DEIM} is large to be able to reach the end of the simulation. The computation time is 54m15 for the full model, 47m56s for the POD and 27m28 for the POD-DEIM (figure 4).

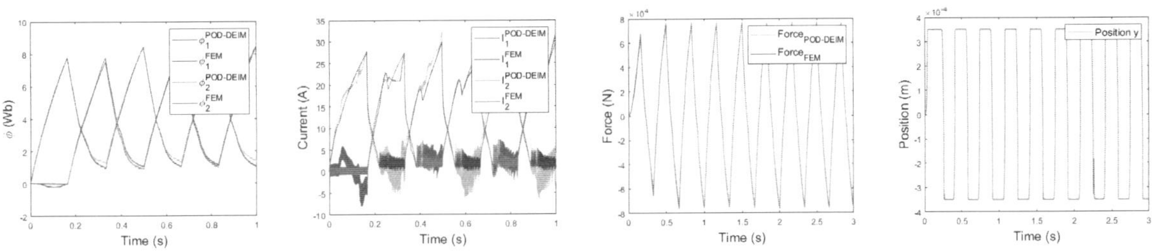

Figure 4: Flux, current, force and position ($n_{POD} = 25$, $n_{DEIM} = 1000$)

5. Conclusion and perspective

In this paper, we introduced a MOR technique based on the POD-DEIM applied to the simulation of an axial AMB, which takes into account Eddy current, magnetic saturation and rotor's displacement. We showed that this method allows to accelerate considerably the resolution time. However, some instabilities can appear when using the DEIM-GPOD, which makes the Newton-Raphson's algorithm fail to converge. These instabilities become more and more important when considering the rotor's displacement. In other hand, the POD approach alone is quite robust, but allow a speed up about only 10%.
In future work, we will try to circumvent this issue by updating the DEIM basis/linear operator during the simulation.

References

Chaturantabut, S. & Sorensen, D. C. (2010), 'Nonlinear model reduction via discrete empirical interpolation', *SIAM Journal on Scientific Computing* **32**(5), 2737–2764.

Hasan, M. R., Montier, L., Henneron, T. & Sabariego, R. V. (2018), 'Matrix interpolation-based reduced-order modeling of a levitation device with eddy current effects', *IEEE Transactions on Magnetics* **54**(6), 1–7.

Henneron, T. & Clenet, S. (2015), 'Model-order reduction of multiple-input non-linear systems based on pod and dei methods', *IEEE Transactions on Magnetics* **51**(3), 1–4.

Montier, L., Henneron, T., Clénet, S. & Goursaud, B. (2016), 'Transient simulation of an electrical rotating machine achieved through model order reduction', *Advanced Modeling and Simulation in Engineering Sciences* **3**(1), 1–17.

Comparison of Rotor Magnetic Circuit Topologies for Passively Levitated Self-bearing Machines

Adrien ROBERT, Joachim VAN VERDEGHEM, Bruno DEHEZ

Mechatronic, Electrical Energy, and Dynamic Systems (MEED),
Université catholique de Louvain (UCLouvain), Louvain-la-Neuve, Belgium.
Emails: {adrien.robert, joachim.vanverdeghem, bruno.dehez}@uclouvain.be

Abstract

Self-bearing motors can reach high rotation speeds thanks to the rotor magnetic levitation. In order to improve the compactness of such a motor, a fully passive levitation achieved with an electrodynamic thrust bearing has recently been proposed and studied. However, the proposed topologies do not integrate magnetic circuit in the rotor, or only at a significant distance from the winding, although it can improve the performance of the motor and the suspension through the impact it can have on the inductance coefficients of the winding. In this context, this paper aims to compare different rotor magnetic circuits topologies on the performance of axial flux self-bearing motors based on electrodynamic suspensions.

Keywords: Self-bearing motor, Electrodynamic suspension, Magnetic circuit.

1. Introduction

Self-bearing machines combine the drive and magnetic suspension functions in a single structure, making them ideal for applications requiring high rotation speed and power density (Chen et al. 2020, Pei et al. 2022). These motors use active methods to ensure magnetic levitation along each degree of freedom (DOF), which requires sensors, controllers and power electronics to operate. Reducing the number of actively controlled DOFs is therefore one of the current challenges for self-bearing motors to improve their compactness, reliability and cost. From this perspective, recent research has proposed theoretical and experimental studies of fully passively levitated self-bearing motors, with axial (Van Verdeghem & Dehez 2021) and radial (Rubio et al. 2023) flux topologies, based on an Electrodynamic Thrust Self-Bearing Machine (ETSBM). Both are based on a Surface-Mounted Permanent Magnet (SMPM) rotor whose ferromagnetic parts are located at such a distance from the winding that they almost do not affect the inductance coefficient. However, the presence of a magnetic circuit could have a positive impact on the ETSBM performance by increasing the winding inductance, which directly determines the electrodynamic suspension force at low speed, and making it dependent on the rotor position.

On these premises, a model that takes into account the impact of the ferromagnetic circuit on the dynamic behaviour of the motor, through the evolution of the inductance coefficients with the axial and angular positions of the rotor, has recently been proposed (Robert et al. 2023, Van Verdeghem et al. 2022). In addition to the new reluctant torque created by the saliencies, it reveals the appearance of new reluctant and cross terms in the expression of the axial stiffness, which could lead to a suspension force modulation through the injection of d-axis current. A simplified variant of this model has been derived to investigate the impact of ferromagnetic yokes called Back Iron (BI)(Van Verdeghem et al. 2022), attached to the axial flux rotor behind the armature winding. It shows an effective improvement of the driving torque, but the investigated topology is without magnetic saliencies. In this context this paper aims to evaluate the impact of such saliencies on the motor and suspension performance of the ETSBM.

2. General principle

The general structure of the ETSBM and its working principle are depicted in Figure 1. When the rotor is axially centred (Figure 1(a)), the PM flux linkages of the upper and lower windings are equal, which means that the motor current I_M, injected by the power supply, is evenly distributed between both windings. Due to the machine symmetry, this current generates only a driving torque $T_{\theta,M}$, regardless of the axial position of the rotor.

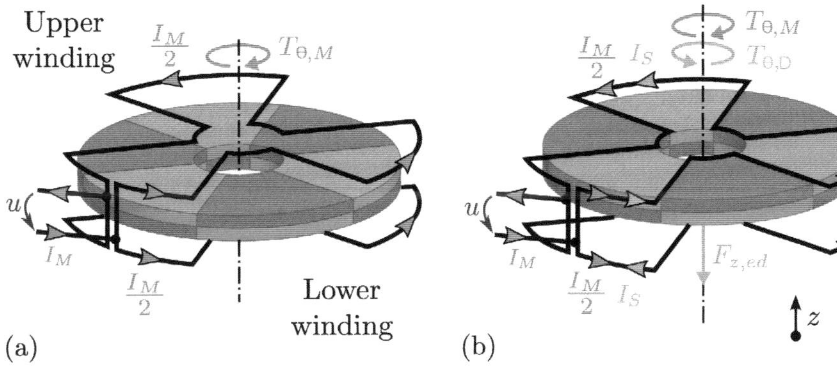

Figure 1: ETSBM operation, ironless topology and only one phase represented. (a) Centered position (z = 0). (b) Decentered position ($z \neq 0$). Inspired from (Van Verdeghem & Dehez 2021).

In the case of an axial displacement of the rotor (Figure 1(b)), a suspension current I_S arises due to the imbalance in both the back electromotive force and the impedance between the upper and lower windings. The different circulation direction of this current induces an axial restoring force F_z as well as a parasitic torque $T_{\theta,D}$ that is superimposed on the driving one. This parasitic component is also called "drag torque" because it is generally opposed to the driving torque.

3. Performance criteria

The restoring force depends linearly on the axial displacement, which allows to define an axial stiffness k_z. Its general expression, given below in the Park-Concordia frame, has been introduced in (Robert et al. 2023):

$$k_z = \frac{\omega^2}{\omega_e^2 + \omega^2} \cdot \underbrace{\left(\frac{3K_{\theta,z}^2}{L_{S,d}} + \sqrt{6} \cdot K_{\theta,z} \cdot \frac{L_{d,z}}{L_{S,d}} \cdot I_{M,d} + \frac{L_{d,z}^2}{2L_{S,d}} \cdot I_{M,d}^2 + \frac{L_{q,z}^2}{2L_{S,q}} \cdot I_{M,q}^2 \right)}_{k_z^\infty} \quad (1)$$

The stiffness depends on the direct and quadrature motor currents $I_{M,d}$ and $I_{M,q}$, the direct and quadrature suspension inductance $L_{S,d}$ and $L_{S,q}$, the axial gradient of the latter $L_{d,z}$ and $L_{q,z}$, the axial gradient of the flux amplitude $K_{\theta,z}$, the rotation speed ω, and the electrical pole defined as $\omega_e = \frac{R}{p\sqrt{L_{S,d} \cdot L_{S,q}}}$ with the phase resistance R and the pole pair number p.

Equation 1 shows that the axial force increases with the rotation speed up to reach its asymptotic value k_z^∞. The first component of this asymptotic stiffness is the electrodynamic one and is the only force component acting in ironless ETSBM. As soon as the rotor contains a ferromagnetic circuit, the inductance coefficients vary with the axial position, creating additional stiffness components which depend on the motor currents. Therefore, an ETSBM with a ferromagnetic circuit adds a possibility of stiffness modulation through the variation of the load angle α_{load}, unlike to the ironless motor where the suspension is completely independent from the power supply.

The suspension performance of an ETSBM can be evaluated using two terms. First, the asymptotic stiffness k_z^∞ defined in Equation 1, which is the maximum reachable stiffness. Second, the electrical pole ω_e, representative of the stiffness evolution: the lower the electrical pole, the faster the stiffness reaches its asymptotic stiffness and then the higher the suspension at low speed. Therefore, significant suspension performance results from a combination of an important asymptotic stiffness and a small electrical pole.

The general expression of the total torque developed by the motor is the following:

$$T_\theta = \underbrace{\sqrt{\frac{3}{2}}p \cdot I_{M,q} K_\theta \quad + \quad p \cdot \frac{I_{M,q} I_{M,d}}{2} \cdot (L_{M,d} - L_{M,q})}_{T_{\theta,M}} \quad + \quad T_{\theta,D}(z^2, \omega) \tag{2}$$

where are introduced the flux amplitude in centered position K_θ and the direct and quadrature motor inductances $L_{M,d}$ and $L_{M,q}$. As stated in Section 2.the total torque is composed of the driving torque $T_{\theta,M}$ and a parasitic torque $T_{\theta,D}$. This latter evolves with the rotation speed up to reach an asymptotic value and is proportional to the square of the rotor axial displacement. For clarity reason, its expression is not developed here but can be found in (Robert et al. 2023). In centered position, the only existing torque is therefore the driving one, composed of the classical electrodynamic and reluctant torque components.

4. Investigated topologies

Two rotor topologies are investigated, the classical SMPM and the Spoke Type Permanent Magnet (STPM). Both topologies, represented in Figure 2 with their characteristic dimensions, are studied with and without a back-iron (BI), the latter being part of the rotor to avoid an axial detent force resulting in a negative stiffness.

5. Methodology

The comparison of the different topologies presented previously is based on the two performance criteria of Section 3., the torque in centered position and the asymptotic stiffness. To that end, a NSGA-II algorithm is used to optimise each topology according to both criteria. These bi-objective optimisations are performed constraining the Joule losses P_J, the PM volume V_{PM}, and the load angle α_{load}. Equations 1 and 2 require the flux and inductance coefficients which are obtained through 3D FEM simulations. The parameters used in the optimisation are summarised in Table 1.

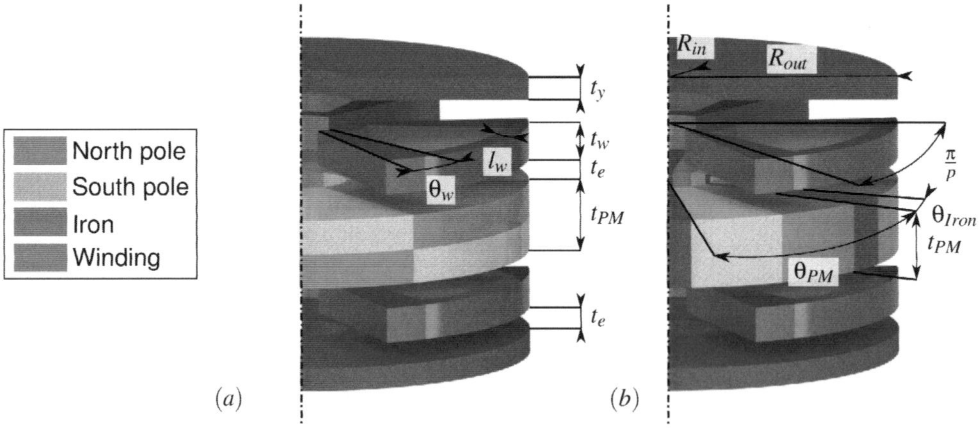

Figure 2: Dimension parameters for both topologies with back-iron, only one phase represented (a) Surface mounted permanent magnet rotor (b) Spoke type rotor.

Parameter	Symbol	Unit	Value for SMPM	Value for STPM	Parameter	Symbol	Unit	Value for SMPM	Value for STPM
Optimized parameters									
Number of pole pairs	p	-	[15,40]		Winding thickness	t_y	mm	[3,8]	
Outer radius	R_{out}	mm	[20,150]		Inner radius	R_{in}	mm	[10,90]	
PM angular opening	θ_{PM}	rad	/	$\left[\frac{1}{10}\frac{\pi}{p}, \frac{4}{5}\frac{\pi}{p}\right]$					
Fixed parameters									
Back iron thickness	t_y	mm	7.5		Air gap thickness	t_e	mm	0.75	
Angular opening coefficient of windings	c_o	-	0.75		Number of phase	N	-	3	
Joule losses	P_j	W	12.5		Load angle	α_{load}	rad	$\frac{\pi}{2}$	
Volume of PM	V_{PM}	cm^3	5						
Depending parameters									
Iron angular opening	θ_{Iron}	rad	/	$\frac{\pi}{p} - \theta_{PM}$	Iron thickness	t_{PM}	mm	$\frac{V_{PM}}{\pi \cdot (R_{out}^2 - R_{in}^2)}$	$\frac{V_{PM}}{p \cdot \theta_{PM} \cdot (R_{out}^2 - R_{in}^2)}$
Winding angular opening	θ_w	rad	$c_o \cdot \frac{\pi}{N \cdot p}$		Winding width	l_w	mm	$R_{in} \cdot \frac{\theta_w}{1-\theta_w}$	

Table 1: Optimizations parameters of both topologies

6. Results

The comparison is first made on the asymptotic stiffness directly provided by the optimization algorithm, and therefore investigated on the full speed range.

6.1 Comparison on asymptotic stiffness

Preliminary analysis can be made on the Pareto fronts obtained from the optimisation process and shown in Figure 3. They reveal that the SMPM and STPM topologies reach quite the same suspension performance when they are optimised for torque in the range of 0.25 Nm, but that BI has a global negative impact.

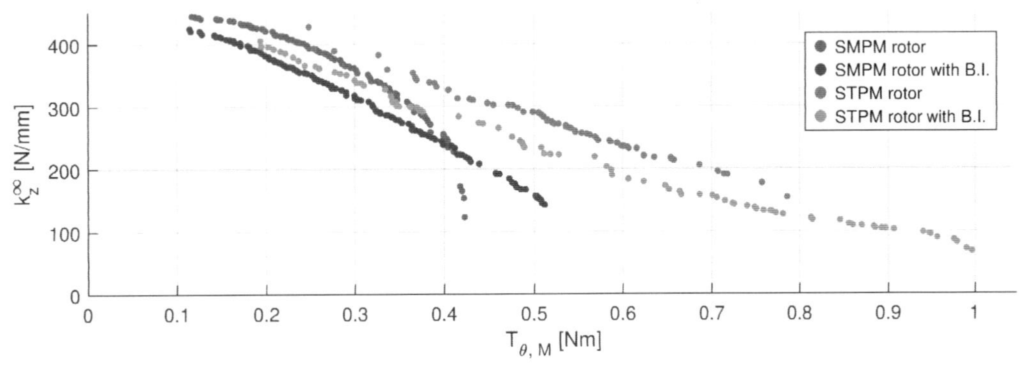

Figure 3: Pareto fronts reached for each topology at fixed permanent magnet volume and Joule losses.

When optimised for higher torques, both the SMPM and the STPM topologies without BI have a faster decrease in asymptotic stiffness than those with BI, to the point where they become less efficient at high torques. However, compared to SMPM, the STPM topopolgies without BI perform better as keep their advantage over topologies with BI for larger torque ranges.

Conversely, when optimised for lower torques, both the SMPM and the STPM topologies without BI see their suspension performance criteria increase less rapidly than those with BI, so that they tend to become less performant at low torque. When comparing STPM and SMPM and without taking into account the presence of BI, the first topology still provides better asymptotic suspension than SMPM one for a given torque. However, SMPM topologies reach wider low torque range, where the best suspension performances are obtained.

These results also show that topologies with BI outperform their counterparts without BI in terms of the maximum torque they can generated, as well as STPM topologies compared to SMPM. the opposite is true for the maximum reachable suspension: the topologies without BI perform better, especially the SMPM one.

6.2 Comparison on full speed range

Previous observations have been made considering the asymptotic stiffness, whereas the stiffness at the nominal operating point is intrinsically dependent on the speed and the electrical pole, as established by Equation 1. Then the question may arise as to whether these observations remain valid regardless of the speed regime. Figures 4(a) and 4(b) show, respectively for the topologies without and with BI, the evolution of the stiffness of the different machines of the Pareto front as a function of the operating speed. Moreover Figures 4(c) and 4(d) show the evolution of the stiffness difference Δk_z between the STPM and SMPM topologies for common torque range of both topologies and in parallel with the evolution of their electrical poles ω_e.

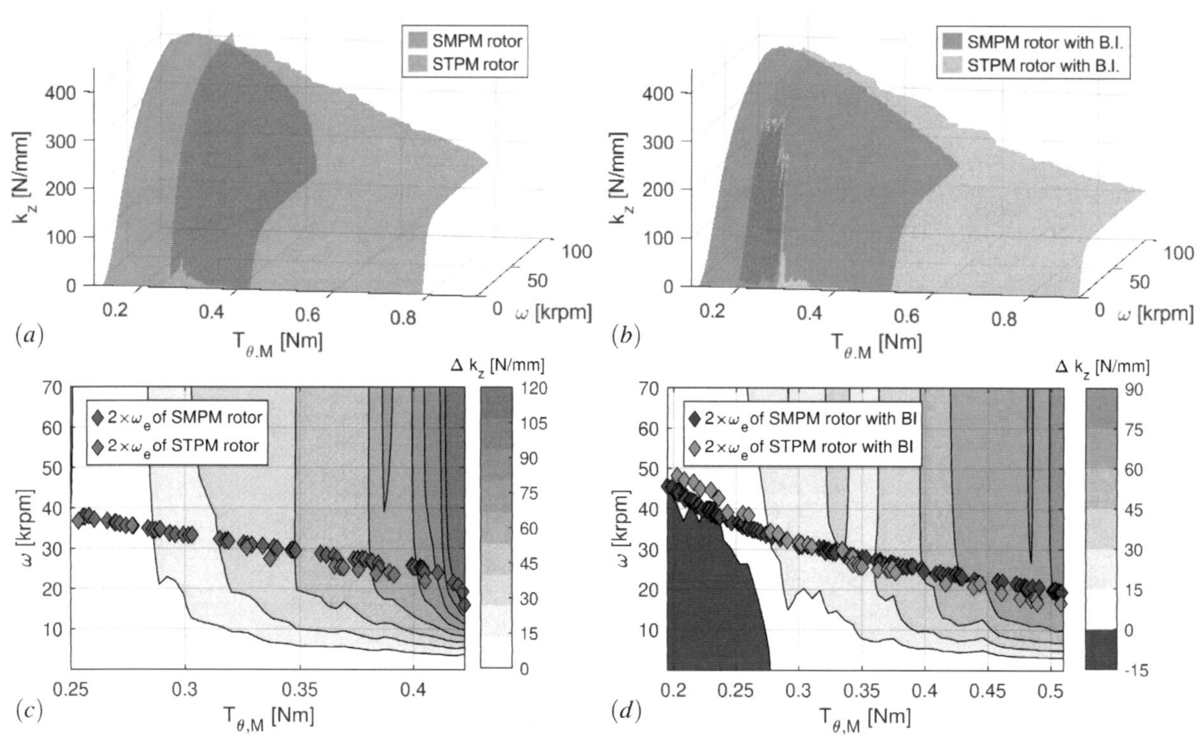

Figure 4: (a) and (b) Evolution of the stiffness with the rotation speed for Pareto front motors of SMPM and STPM topologies, respectively without and with BI, for rotation speed up to 100 [krpm]. (c) and (d) Difference in stiffness between the STPM and SMPM topologies, respectively without and with BI, for rotation speed up to 70 [krpm] and torque range achieved by both topologies. Both also show double the value of the electrical poles ω_e *of each Pareto front.*

Figures 4(a) and (c) reveal that the observations made on topologies without BI are still valid: for the torque range achieved by both topologies, the STPM always provides the best suspension capability. They also validate the principle of optimising and comparing both topologies based on asymptotic stiffness.

However, for topologies with BI optimised for the low torque range, the SMPM achieves a higher stiffness than STPM when the rotation speed does not exceed 50 krpm. This is explained by the lower value of the electrical pole of the SMPM topology for these low torque, as shown in Figure 4(d). For torques higher than 0.28 Nm, STPM topology combines higher asymptotic stiffness and lower electrical pole, which ensures better suspension performance than SMPM over the whole speed range.

7. Conclusion

An optimisation process, aiming to maximize the torque and asymptotic stiffness of an ETSBM, was applied to different rotor topologies with fixed permanent magnet volume and Joule losses. This showed that STPM topologies performed better for applications requiring high torque, while SMPM topologies performed better for applications requiring high stiffness. In addition. It also revealed that topologies without BI achieved a better compromise between torque and stiffness than those with BI. Furthermore, a study of the impact of speed on the induced stiffness achieved by both topologies in the common torque range showed that the STPM topology always provides the best suspension capability except in one case, for topology with BI at low rotation speed and torque, where the SMPM topology provides greater stiffness thanks to their lower electrical pole. Future work will investigate the influence of the different geometric parameters as well as the impact of the load angle, which could exploit the ferromagnetic saliences to increase the torque capability of the STPM rotor.

Acknowledgement

This publication is supported by "La Région wallonne" as part of the funding of a FRIA grant.

References

Chen, J., Zhu, J. & Severson, E. L. (2020), 'Review of Bearingless Motor Technology for Significant Power Applications', *IEEE Transactions on Industry Applications* **56**(2), 1377–1388.

Pei, T., Li, D., Liu, J., Li, J. & Kong, W. (2022), 'Review of Bearingless Synchronous Motors: Principle and Topology', *IEEE Transactions on Transportation Electrification* **8**(3), 3489–3502.

Robert, A., Herrman, S., Van Verdeghem, J. & Dehez, B. (2023), 'Dynamical Modeling of Passively Levitated Electrodynamic Thrust Self-Bearing Machines considering impact of ferromagnetic circuit', *To be submited to IEEE Transaction on Magnetics* .

Rubio, G. C., Hothur Komal, V. C., Fujii, Y. & Chiba, A. (2023), 'Experimental Verification of Passive Axial Electrodynamic Suspension in a Bearingless Motor', *IEEE Open Journal of Industry Applications* pp. 1–12.

Van Verdeghem, J. & Dehez, B. (2021), 'Fully Passively Levitated Self-Bearing Machine Implemented Within a Reaction Wheel', *IEEE Transactions on Industry Applications* **57**(6), 5782–5795.

Van Verdeghem, J., Robert, A., Herrman, S. & Dehez, B. (2022), Impact of Ferromagnetic Yokes on Axial Flux Passively Levitated Self-Bearing Machines, *in* '2022 25th International Conference on Electrical Machines and Systems (ICEMS)', pp. 1–6.

Design and manufacturing of permanent magnet bearing rings for high speed applications

Tan Tan [a], Daniel F. Förster [a], Ulrich Pabst [a], Stephan Polachowski [a], Michael Butzek [a], Ghaleb Natour [a, b]
a Central Institute of Engineering, Electronics and Analytics –Engineering and Technology (ZEA-1), Forschungszentrum Jülich GmbH 52425 Jülich, Germany, t.tan@fz-juelich.de
b Faculty of Mechanical Engineering, RWTH Aachen University, 52056 Aachen, Germany

Abstract

We provide an overview of the challenges for the design of magnet rings for high-speed permanent magnet bearings focusing on manufacturing aspects. We analyze an assembly of a NdFeB magnet ring with a hub and a bandage, both made of Ti-6Al-4V (titanium grade 5). Thereby, we consider bandage thicknesses, tolerances and assembling processes. This is supported by simulations using Finite Element Method (FEM) analysis in ANSYS workbench 2020. We analyze the stresses during a press-fit joining process between the magnet and the bandage and subsequent consequences for the stresses on the system for rotation speeds up to 120,000 RPM. Finally, we consider safety aspects by implementing a cracking of the magnet at high rotation speed.

The results of the simulations show that the stresses on the magnet are the limiting factor of the rotation speed in all cases analyzed here. For an idealized system, mainly the bandage thickness and the required rotation speed define suitable tolerances for the fitting of the bandages and magnets. For a real system, residual stresses of the joining process and ovalization effects of the bandage decrease either the allowable speed or the allowable tolerances, thereby considerably influencing the required manufacturing precision. A safety analysis shows that the hub-magnet-bandage assemblies analyzed here are safe with respect to magnet cracking even at high speed.

Keywords: *permanent magnet bearing rings, finite element method, manufacturing and tolerance, fail safe behavior*

1. Introduction

In several cases, special instruments for scientific applications require the use of magnetic bearings. The most common reasons are cases where maintenance is limited or not possible, or where a precision of motion control is required which is only achievable by frictionless bearings. Scientific applications often also demand for highest reliability and compatibility with ultra-high vacuum environments. Radial shear force permanent magnetic bearings (PMBs) with active axial stabilization (Fremerey, 1989) have a relatively low complexity and fulfil these requirements. Thus, they are often used for such purposes, e.g. for choppers and selectors (Cammarata, et al., 2009), (Förster, et al., 2015). The ever-increasing demands of the scientific applications require new optimized PMB designs, especially to achieve very high speeds far beyond 60,000 RPM. As sintered NdFeB magnets have a low tensile strength of only 80 to 90 MPa (IEC 60404-8-1:2015), the most crucial point to enable such high speeds using PMBs is the thorough design of the rotating magnet rings with suitable bandages.

There exist several publications which describe and analyze different design of PMBs with respect to the magnetic forces and bearing stiffnesses. While for high speed permanent magnet motors mechanical analysis was published by several authors (Shen, et al., 2022), we were not able to find publications which deal with manufacturing aspects of PMBs with respect to high speed operation.

2. Design of the high-speed permanent magnet bearing rings

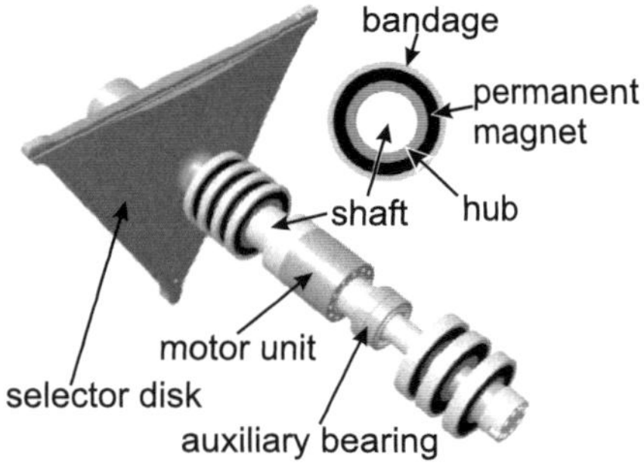

Figure 1 *Rotor of an X-ray pulse selector. The selector disk has a diameter of 194 mm and runs up to 60,000 RPM.*

The considered radial shear force PMBs consist of attractively stacked axially magnetized rings. In the simplest case, such a radial bearing is built by one rotating ring on the shaft and two rings fixed in the housing. The stiffness of such bearings depends on the size and magnetic properties of the magnet rings. It can be considerably increased by using multi-magnet-ring assemblies (Lang and Lembke, 2006). Furthermore, the stiffness can be increased by increasing the amount of the axially stacked magnet rings, i.e. by increasing the axial length of the bearings.

Fig. 1 shows an example rotor of approx. 1 kg mass for an x-ray pulse selector where the main radial bearing consists of stacks of four rotor and four stator magnet rings with a total radial stiffness of about 180 N/mm. The inset in Fig. 1 shows a front view of the magnet ring. The permanent magnets are reinforced by a bandage to withstand the centrifugal load and are mounted on a hub for the fixation to the shaft. These hub-magnet-bandage assemblies are the only rotating parts of these PMBs. Thus, their design is defining the maximum permissible speed.

Therefore, we investigate the design of the hub-magnet-bandage assemblies with the focus on their mechanical stability. We analyze by the aid of FEM analysis (I) the load at high speeds (up to 120,000 RPM), (II) the manufacturing process to find the necessary interference fit considering manufacturing tolerances, and (III) the fail-safe behavior in case of a magnet fracture.

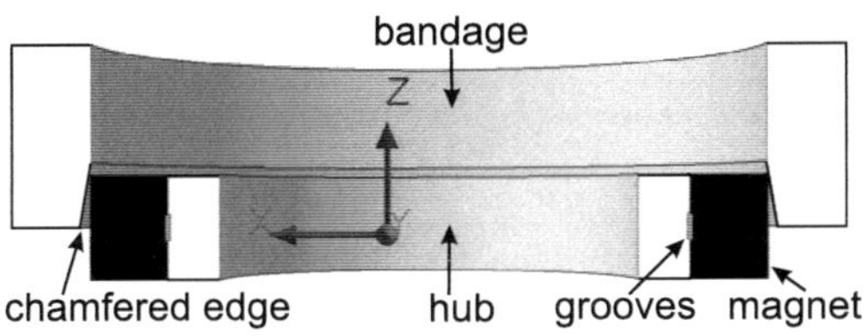

Figure 2 Cross-sectional view of the hub-magnet-bandage assembly during the axial press-fit process. The orientation of the coordinate system of the simulations is indicated for comparison, y-axis pointing towards the reader.

The assembling starts at the hub-magnet interface. Between hub and magnet, a low clearance fit H5/h5 is used and the gap is filled with epoxy resin. In the magnet and hub surface, 1 mm wide and 0.2 mm deep grooves are cut facing each other (compare Fig. 2). Filled with epoxy resin, these grooves considerably increase the mechanical interlocking against axial displacement.

For the contact between the bandage and the magnet ring, an interference fit is necessary. The required amount of interference for certain rotor speeds will be analyzed in section No. 4. To assemble interference fit parts, typically shrink-fitting or press-fitting is used.

Shrink fitting is problematic in this case due to the properties of NdFeB magnets. First, the thermal expansion of magnetically oriented NdFeB magnets is anisotropic. It is only substantial in the easy magnetization axis. For directions perpendicular to the easy magnetization axis, the thermal expansion coefficient is very low or even negative (Kumar, et al., 2021). Thereby, the details of the thermal expansion behavior depend on the exact composition of the magnet and the details of the manufacturing process (Rabinovich, et al., 1996). In consequence, cooling of the magnet cannot be considered as a suitable method to shrink the diameter of an axially magnetized NdFeB magnet ring.

Second, the magnet material is heat sensitive and without shrinkage of the magnet it would be necessary to heat up the bandage by several hundred K to establish substantial interference. Even if this process would have no influence on the material properties of the bandage, the temperature at the interface to the magnet would be considerably higher than the operating temperature of high remanence NdFeB magnets. This imposes the risk of partial demagnetization of the magnet in the vicinity of the contact surface.

As a consequence, press-fitting is chosen and further analyzed in detail. To enable the assembling, the bandage is manufactured with a larger axial length and equipped with a chamfered edge (compare Fig. 2). After the bandage is completely pressed onto the magnet, the excess part of the bandage with the chamfered edge is removed by further machining.

The magnet rings for the rotor shown in Fig. 1 have an inner diameter of 26 mm, an outer diameter of 34 mm and an axial length of 4 mm. The related hub-magnet-bandage assemblies are qualified for operation up to 60,000 RPM. By using an improved multi-magnet-ring stator design, it is possible to achieve a similar bearing stiffness for a reduced size of the rotor magnets with an inner diameter of 20 mm, an outer diameter of 26 mm, thereby considerably reducing the centrifugal load on the magnet material. Thus, in the following we use this magnet size for the hub-magnet-bandage assemblies and their optimization for highest possible speeds.

3. Details of the simulation model

The simulation is carried out in Ansys workbench 2020 using purely elastic material models. We use 2D models for the press-fitting process between bandage and magnet considering different interferences. To investigate the bandage ovalization and cracked magnets, 3D models are used. For all simulations, we define frictional contacts between the outer surface of the magnet ring and the inner surface of the bandage. As friction coefficients we chose 0.1 as we assume a lubricated joining process. All other contacts are defined as bonded.

Due to the expected high mechanical load we use titanium grade 5 as bandage material for our simulations. Thus, we can assume that the round bars for manufacturing the bandages have at least an 0.2% offset yield strength of $R_{p0.2}$ = 828 MPa (ASTM B348/B348M-19). For simplicity, we use the same material for the hub.

4. 2D Simulation for magnet ring with interference fit

As starting point, we investigate only the influence of prestress for the hub-magnet-bandage assemblies using 2D simulation. Therefore, we built an ideal model without considering any influence of joining, manufacturing tolerance, and magnet ring defect. We simulate the assembly with different outer diameters from 29 mm (1.5 mm bandage thickness, interference 70 µm), 30 mm (2 mm; 60 µm), 31 mm (2.5 mm; 55 µm) and 32 mm (3 mm; 50 µm) while maintaining a similar joint pressure. Due to the low tensile strength for sintered NdFeB magnets, the stresses in the tension-sensitive

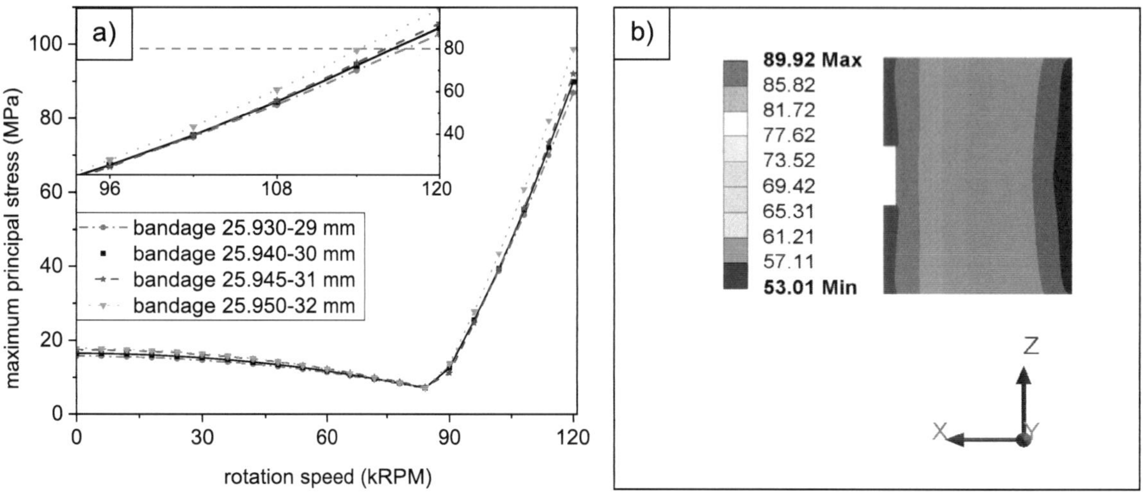

Figure 3 *a) Rotation speed dependent maximum principal stresses of the magnet ring with different bandage outer diameters (2D simulation). The inset is a zoom into the high-speed region. b) Distribution of the maximum principal stress for the magnet ring at 120,000 RPM for the bandage with diameter 25.94-30 mm, color scale given in MPa (2D simulation). The coordinate system is indicated, y-axis pointing towards the reader (compare to Fig. 2).*

magnet rings are the limiting factors for the permissible rotation speed of the hub-magnet-bandage assemblies, as will be shown later on. Hence, we focus here only on the differences in the magnet stresses.

The simulation results in Fig. 3a show that the maximum achievable speed decreases with increasing bandage thickness. Furthermore, for thinner bandages the maintained joint pressure requires larger interferences which increases the acceptable tolerance window. These effects make thinner bandages preferable, whereby the lower thickness limit would be given by the stress inside the bandage as the strain increases because of the larger interferences.

However, there is another effect which has to be considered and which increases with decreasing bandage thickness. Form manufacturing experience we know that relief of internal stresses leads to an ovalization when the bandage is cut off after lathing. This effect is more pronounced for thinner bandages and analyzed further in Sec. 6 using 3D simulations. As consequence, we have chosen 2 mm as bandage thickness for all further simulations, although a thinner bandage would be preferable with respect to the maximum achievable speed and the acceptable tolerance window for manufacturing.

The course of the maximum principal stress in Fig. 3a can be understood considering the different directions of the stresses inside the magnet. Under joint pressure the magnet is radially compressed, leading to axial strain due to the Poisson effect. The subsequent tensile stresses in axial direction reach their maximum at the 1 mm wide grooves. This contribution is highest at zero speed and decreases with increasing speed due to the reduction of the joint pressure. To the contrary, the tangential stresses (in circumferential direction) are initially compressive but increase fast with increasing speed, becoming tensile and the dominant contribution above 84,000 RPM. Fig. 3b shows the distribution of the maximum principal stress for the magnet ring for the final simulation step at 120,000 RPM. The maximum is located in the area of the inner edge.

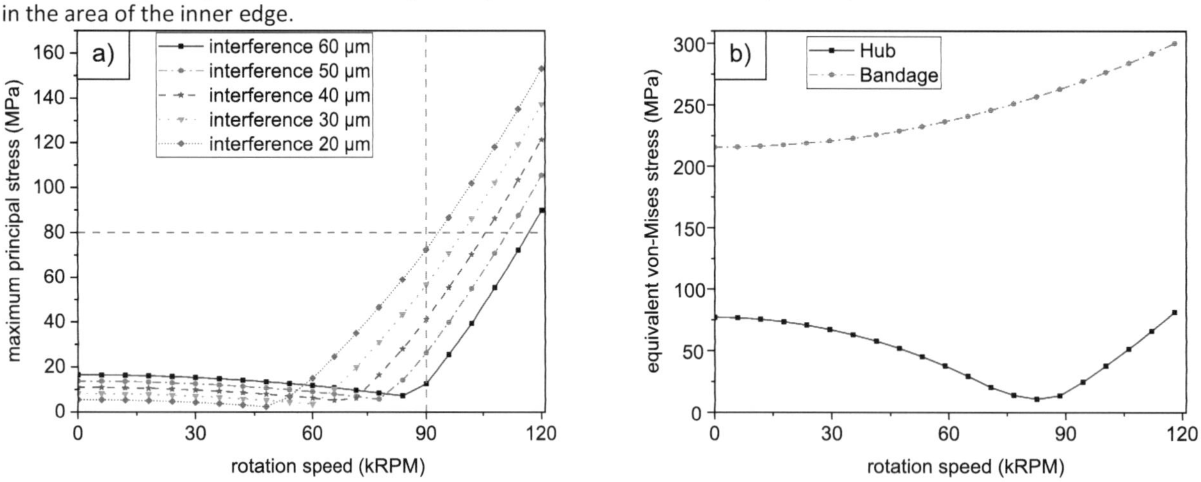

Figure 4 *a) Rotation speed dependent maximum principal stresses of the magnet ring for different interferences between 20 μm and 60 μm (2D simulation). b) Equivalent von-Mises stresses of hub and bandage for 2 mm bandage thickness and 60 μm interference (2D simulation).*

To analyze suitable tolerance windows for manufacturing, we calculate the maximum principle stresses also for lower interferences of 20, 30, 40, and 50 µm as shown in Fig. 4a. Interestingly, if 80 MPa would be used as tensile stress limit for the magnet, the permissible speed for 60 µm interference would be 114,000 RPM and every reduction of the interference by 10 µm would result in a reduction of the permissible speed of about 6,000 RPM.

As mentioned above, the load on bandage and hub are uncritical for all analyzed cases. For example, Fig. 4b shows the equivalent von-Mises stress on hub and bandage for 2 mm bandage thickness and 60 µm interference. Considering $R_{p0.2}$ = 828 MPa and a safety factor of 2 for ductile materials (Rennert, et al., 2012), the stresses remain well below the safety limit.

5. Simulation of joining process

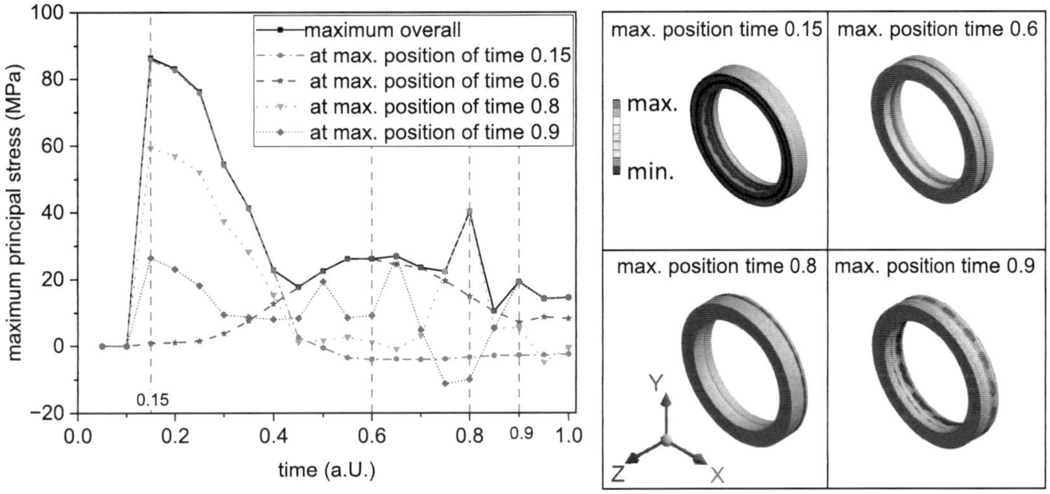

Figure 5 Maximum principal stress on the magnet ring for 2 mm bandage thickness and 60 µm interference (3D Simulation) and local max. stresses in the magnet ring during the joining process. The orientation of the coordinate system of the simulations is indicated.

As explained in Sec. 2, we use press-fitting to assemble the bandage on the magnet ring. To avoid damages to the magnet during assembling we also analyze this joining process by simulations. Fig. 5 shows the resulting maximum principal stress on the magnet ring. The black line represents the course of the absolute maximum during the joining process which requires press-forces up to 1260 N. For better understanding of these stress changes during joining, we investigate the four different local max. stresses.

The color-coded magnet rings in Fig. 5 visualize the different max. positions of the stresses. For simplified viewing, we use the results of 3D simulations which we have implemented for the following sections and which lead to similar results as the 2D simulations. For better orientation, the coordinate system is indicated.

Initially, when the magnet is compressed by the bandage at the frontside, the max. stresses occur at the backside of the magnet at the inner edge (time 0.15). At this position, the stress reaches the highest value during the joining process. But later it relaxes here and remains at a low level (red dash-dotted line). At simulation time 0.4 the position of the maximum changes to the inside of the groove (purple dashed line) where it reaches the max. at time 0.6. When the stress relaxes at this position, the maximum changes again to the outer edge of the magnet's backside (time 0.8). At time 0.9, the max. stress is distributed at the outer diameter of the magnet's backside half (close to the excess part of the bandage), but remains at relatively low levels.

After the press-fitting is completed, the simulation continues with the removal of the excess part of the bandage with the chamfered edge. After this step, the stress at the outer diameter relaxes and the maximum is again located at the groove region (not shown).

When the simulation is continued with rotation after the joining process, there are considerable differences to the idealized system discussed in Sec. 4. As can be seen in Fig. 6a, the maximum principal stress is initially slightly lower than for the ideal case, but the fast increase starts about 12,000 RPM earlier. The reason for this behavior is a strong asymmetry of the tangential stresses. Fig. 6b shows the minimum principal stress in the magnet ring after joining, which is solely compressive and oriented in the tangential direction (y-axis of the simulation coordinate system). As remainder

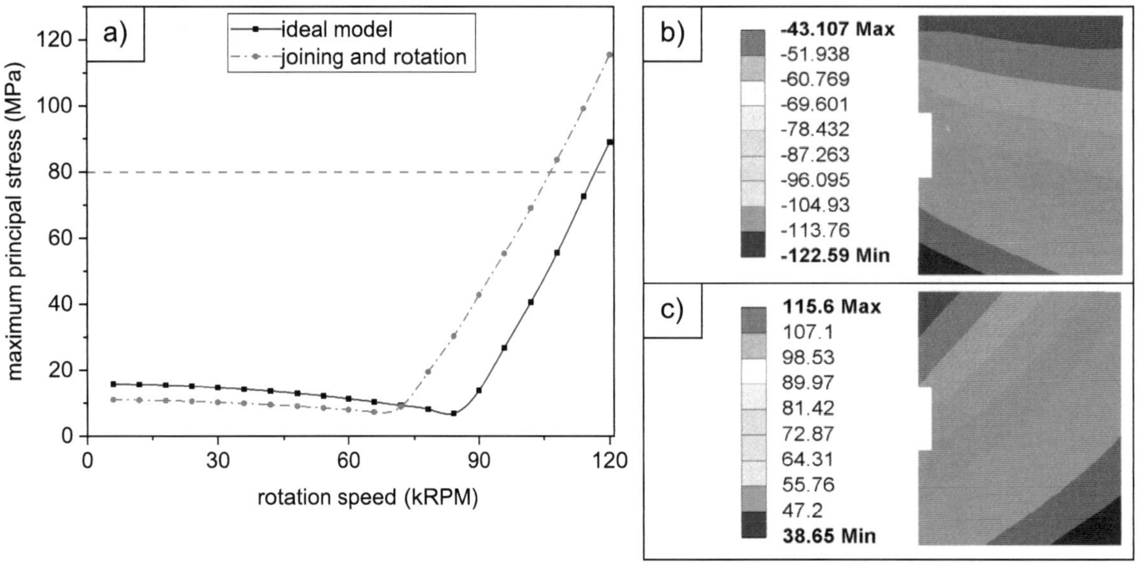

Figure 6 a) Rotation speed dependent maximum principal stresses of the magnet ring after joining compared with the ideal model. b) Distribution of the minimum principal stress on the magnet ring after joining. c) Distribution of the maximum principal stress on the magnet ring at 120,000 RPM. All results for 2 mm bandage thickness and 60 µm interference. Color scales of 2D simulations given in MPa (compare to Fig. 3b).

of the joining process, the tangential stress is considerably higher (less compressive) at the frontside (at the top of the image) of the magnet, where the press-fitting process started. This asymmetry is partially conserved with increasing speed and a difference of about ± 25 MPa between frontside and backside remains at 120,000 RPM as can be seen in Fig 6c (compare to Fig. 3b). Please note, that the maximum principal stress in Fig. 6c is tensile and solely oriented in tangential direction.

Again, the stresses in hub and bandage are well below the safety limit for all conditions. The max. equivalent von-Mises stress for the hub is 190 MPa and occurs during the joining process while it is 312 MPa for the bandage at 120,000 RPM.

6. Manufacturing tolerances: bandage ovalization

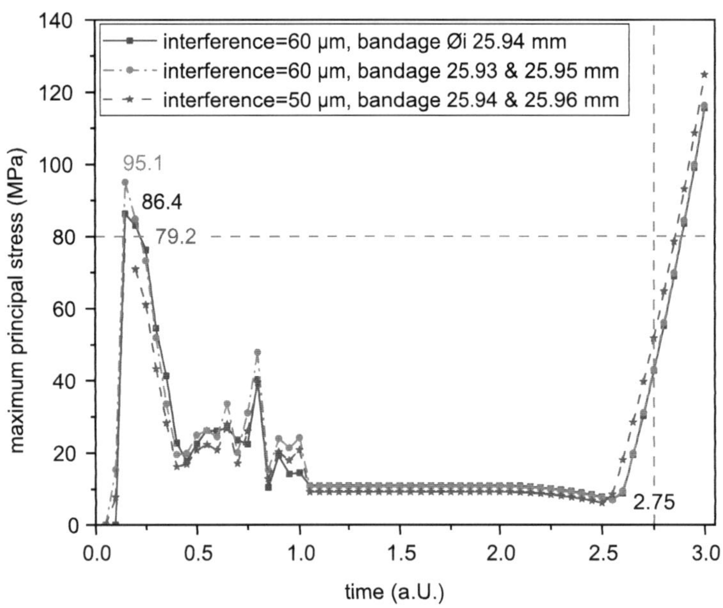

Figure 7 Maximum principal stress on the magnet ring during the joining process and subsequent rotation for 2 mm bandage thickness and 60 µm interference (black line), 60 µm interference with 10 µm ovalization (dash-dotted red line) and 50 µm interference with 10 µm ovalization (dashed purple line).

As already mentioned in Sec. 4, the bandage ovalizes when it is cut off after lathing due to internal stress relieve. This causes a circumferential asymmetry of the pressures during the joining process. To estimate possible effects, we simulate the joining process also for a slightly elliptically shaped bandage. Therefore, we model an elliptically shaped interference with a major axis of 70 µm and a minor axis of 50 µm (considered as 60 µm interference with 10 µm ovalization) for comparison with the perfectly circular 60 µm interference. All other parameters remain unchanged.

Fig. 7 shows the maximum principal stress on the magnet ring during the joining process and subsequent rotation for 2 mm bandage thickness and 60 µm interference for a perfectly circular bandage (black line) and for 10 µm ovalization (dash-dotted red line). The stress maximum increases from 86 MPa to 95 MPa, which is above the max. magnet's tensile strength of 90 MPa. To compensate for the increased pressure maximum, the interference can be reduced. For example, the dashed purple line shows the resulting stresses for a reduced interference of 50 µm, again with 10 µm ovalization. The stress maximum decreases from 95 MPa to 79 MPa. This comes to the cost of increased stress during rotation.

While the ovalization does not significantly influence the stresses during rotation, the reduced interference shifts the fast stress increase during rotation to lower rotation speeds. Thus, due to the necessity of countermeasures to compensate higher joining stresses, ovalization leads to a reduction of the permissible speed for the magnet. For hub and bandage, the max. equivalent von-Mises stresses only increase slightly due to ovalization by approximately 3 % at joining and at 120,000 RPM (for interference 60 µm with 10 µm ovalization). Thus, they remain still well below the safety limit for all conditions.

7. Fail-safe behavior

As part of the safety analysis of the high speed magnetic bearing a failure analysis of the hub-magnet-bandage assemblies is performed. While hub and bandage are made of a ductile metal alloy and are stressed well below their margin of safeties, the tensile stress limit of the magnet is already exceeded at some points of the simulations. Due to the brittle nature of the material, the possibility of undetected magnet defects due to manufacturing flaws or local damage during the joining process requires an analysis for the outcome of magnet cracking at high speed, even for a proper design with a sufficient margin of safety for the tensile magnet stresses.

To simulate the crack in the magnet, we separate the magnet ring in two parts and set bonded connections between them. We define the point of time when the magnet ring cracks and then delete one bonded connection in an additional simulation step. For example, we simulate a crack for a hub-magnet-bandage assembly with 60 µm magnet-bandage interference at 90,000 RPM. With a tensile stress of 43 MPa, the stress on the magnet can be considered to be within the margin of safety for the magnet and thus the corresponding hub-magnet-bandage assembly as suitable for continuous operation at this speed. Due to the crack, the maximum principal stress on the undamaged part of the magnet ring increases only marginally by less than 1 MPa. For the hub, the equivalent von-Mises stress increases after cracking from 41 MPa to 72 MPa. And for the bandage, it increases from 277 MPa to 287 MPa. Thus, all stress increase after cracking is uncritical. As there is also still a contact pressure between bandage and magnet of 27 MPa, the assembly will be held together and can be considered as safe with respect to magnet cracking at the rated operation speed.

8. Conclusion and Outlook

We elucidate several manufacturing-relevant aspects of hub-magnet-bandage assemblies for high-speed permanent magnet bearings. The results of the FEM simulations show that the stresses on the magnet are the limiting factor of the rotation speed in all cases. Remaining stresses of the press-fit joining process and ovalization effects of the bandage decrease allowable tolerances, thereby considerably influencing the required manufacturing precision. A safety analysis shows that the hub-magnet-bandage assemblies remain intact if the magnet cracks at high speed.

For a better understanding of the joining process, we will complement the simulation results by measurements. We will determine the coefficient of friction for unlubricated joining, lubricated joining and joining with adhesive by measuring the pressing force during joining. We will also make microsections of magnets and bandages before and after joining to determine changes of the roughnesses at the contact surfaces and analyze consequences of localized tensile stress above the magnet's strength limit during the press-fitting. Finally, we will perform simulations and experiments for bandages made of carbon fiber-reinforced polymers.

References

ASTM B348/B348M-19 (2019) Standard Specification for Titanium and Titanium Alloy Bars and Billets.

Cammarata M, Eybert L, Ewald F, Reichenbach W, Wulff M, Anfinrud P, Schotte F, Plech A, Kong Q, Lorenc M, Lindenau B, Raebiger J, and Polachowski S (2009) Chopper system for time resolved experiments with synchrotron radiation. Review of Scientific Instruments 80: 015101.

Förster DF, Lindenau B, Leyendecker M, Janssen F, Winkler C, Schumann FO, Kirschner J, Holldack K, and Föhlisch A (2015) Phase-locked MHz pulse selector for x-ray sources. OPTICS LETTERS 40(10): 2265–2268.

Fremerey JK (1989) Radial shear force permanent magnet bearing system with zero power axial control and passive radial damping. In: Schweitzer G (ed) Magnetic Bearings, Springer-Verlag, pp. 25-31.

IEC 60404-8-1:2015 (2015) Magnetic materials - Part 8-1: Specifications for individual materials - Magnetically hard materials.

Kumar R, La Rocca A, Vakil G, Gerada D, Gerada C, and Fernandes BG (2021) Significance of Anisotropic Thermal Expansion in High Speed Electric Machines Employing NdFeB Permanent Magnets. Energies 14: 7558.

Lang M and Lembke TA (2006) Design of Permanent Magnet Bearings with high stiffness. In: ISMB10, 10th International Symposium on Magnetic Bearings, Martigny, Switzerland, 1–4 August 2006, pp. 1–4.

Rabinovich YM, Sergeev VV, Maystrenko AD, Kulakovsky V, Szymura S, and Bala H (1996) Physical and mechanical properties of sintered NdFeB type permanent magnets. Intermetallics 4: 641-645.

Rennert R, Kullig E, Vormwald M, Esderts A, and Siegele D (2012) Rechnerischer Festigkeitsnachweis für Maschinenbauteile aus Stahl, Eisenguss und Aluminiumwerkstoffen. In: FKM-Richtlinie, 6. überarbeitete Auflage, Frankfurt: VDMA-Verlag.

Shen Q, Zhou Z, Liao X, Wang T, He X, and Zhang J (2022) Design and Analysis of the High-Speed Permanent Magnet Motors: A Review on the State of the Art. Machines 10: 549.

Generalized Modal Decoupling Control for Active Magnetic Bearings

Pascal Zeugin
Mecos AG
Zurich, Switzerland
pascal.zeugin@mecos.com

Beat Aeschlimann
ZHAW
Winterthur, Switzerland
aesc@zhaw.ch

Michael Hubatka
OST
Rapperswil, Switzerland
michael.hubatka@ost.ch

***Abstract*—Control design for unstable, uncertain, and highly coupled MIMO (Multiple Input and Multiple Output) systems such as AMB (Active Magnetic Bearing) systems is extremely challenging and requires sophisticated synthesis strategies. While postmodern control methods such as μ-synthesis may be able to solve such problems, they require an accurate mathematical model of the system and specialized software tools for the controller synthesis. The resulting control algorithms are typically of excessively high order and therefore computationally demanding and challenging to implement.**

In contrast, the control culture of the magnetic bearing engineering community also offers classical, essentially single-input, single-output (SISO) methods to produce reliable and robust solutions to a wide variety of control problems. One well-known approach is translational-tilting modal-transformation control. However, this approach is limited in cases where the first bending mode is close to the desired closed-loop rigid body modes.

This paper describes an approach for hand-synthesized, generalized modal decoupling control design for magnetically levitated rotors. The basic idea of this method is to use the singular value decomposition (SVD) to transform the decentralized inputs and outputs into generalized modal coordinates. This greatly simplifies the design of the controller, which in turn allows for much more sophisticated rotor dynamic designs by independently controlling the decoupled mode. This can be viewed as a generalization of the known decoupling transformations.

The transformation works with both model- and measurement-based frequency response functions, in contrast to the known approaches in the literature, and is shown to excel where common classical control methods fail. Results and comparisons are demonstrated by an experimental evaluation of the frequency domain control performance of a supercritical gas turbine.

***Index Terms*—MIMO-control, uncertain systems, robust control, AMB-system, rotor dynamics, Singular Value Decomposition**

I. Introduction

Magnetic bearing systems for rotating machinery are a prime example of robust MIMO control because they inherently involve many conflicting performance objectives. The system to be controlled exhibits significant variations in plant dynamics over the operating speed range, which is commonly referred to as a plant with high uncertainty. Essentially, this is the archetypal challenge for robust control techniques such as μ-synthesis, which, in contrast to classical and modern control methods, explicitly deals with uncertainty [1]. However, there are several requirements that must be met, such as

- Availability of a detailed plant model with uncertainty description
- Proprietary, sophisticated software tools
- Well-trained control engineers with a strong mathematical background
- Computing power for controller synthesis
- Numerically robust and efficient controller implementation.

Despite the fact that model-based postmodern robust control techniques are exceptionally powerful, if one or more of these requirements are not met, these methods cannot be applied. On the other hand, there are the proven classical control methods that have reliably produced robust solutions to a wide variety of control problems for decades.

The simplest and most intuitive way to tackle the control of an active magnetic bearing system is the so-called decentralized control, where each sensor feeds back to the actuator in the same degree of freedom [2]. However, in some systems this introduces significant off-diagonal coupling terms in the resulting transfer matrix, which in turn makes the task of control design very difficult. Zhang et al. introduced a method of modal decoupling control for systems with strong gyroscopic effects [3], where the big advantage is that the translational and tilting modes can be influenced almost independently of each other in different control channels. In other words, the speed-dependent gyroscopic effects of nutation and precession occur only in the tilt-mode control channel, where they are relatively easy to handle.

A similar technique is shown by Hutterer, Hofer, and Schrödl in [4]. However, these approaches are limited because they fail when the first bending mode is close to the closed-loop rigid body modes. In summary, no successful approaches to decoupling modes other than rigid body modes have been reported.

The following sections present the main ideas of the new method in the order of a typical engineer's workflow. The workflow starts by analyzing and highlighting the limitations of existing decoupling methods. This is followed by a brief summary of the mathematical foundations of the new method and its physical implications with respect to rotor dynamics. Finally, a case study demonstrates the application of the technique and evaluates its control performance using a real-world example.

II. Limitations of Translational-Tilting Control

Well-known decoupling methods such as translational-tilting control implement MIMO control while preserving the ability to interpret feedback parameters as physical quantities in a SISO-like manner, similar to local feedback. This control structure exploits the physical effect that the parallel and conical modes of the rigid body system can be decoupled using static transformation matrices, as shown in the Figure 1 for a plane through x and z coordinates (xz-plane). By transforming the controller input signals $\{x_{seA}, x_{seB}\}$ with T_{in} so that the parallel and conical modes can be detected separately, these modes can also be controlled separately. The controller output signals, which physically correspond to the torque τ and the concentrated force f_x with respect to the rotor center of gravity S, then only need to be transformed into suitable forces $\{f_{xA}, f_{xB}\}$ in the bearing planes A and B with T_{out} [5].

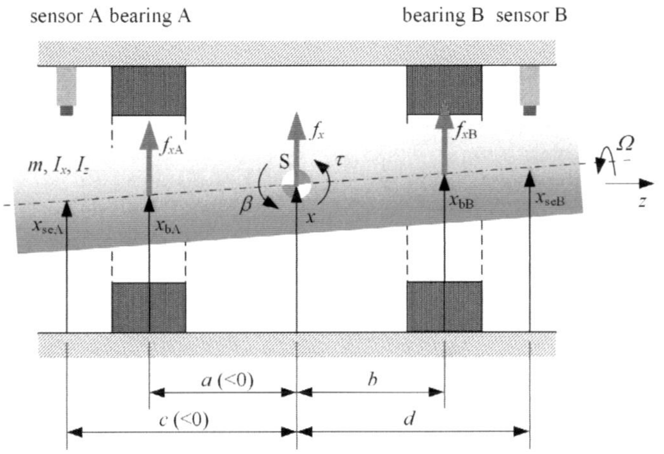

Fig. 1: Rigid rotor with bearing magnets and sensors as well as displacements and forces [5].

In this way, modal control of the rigid body modes is achieved. Figure 2 shows the corresponding control architecture.

To stabilize the rigid body modes in an AMB system, the controllers $G_{p/c,x/y}$ need sufficient positive phase (PD control, lead element) around the crossover frequency of the open-loop system. This results in high controller gains at higher frequencies, which is undesirable due to noise amplification and a destabilizing effect on the flexible modes. To overcome these problems, the controller features a low-pass, i.e. roll-off filter.

In a typical rotor design, the first flexible mode is well separated from the rigid body modes. If the desired closed loop bandwidth is around 50 Hz and the flexible mode is above 500 Hz, it is possible to use a steep roll-off to avoid excitation of the bending mode. If the bending mode is below 500 Hz, using a roll-off will compromise the positive phase to stabilize the rigid body mode. Moreover, this may not be admissible for all rotor designs as this procedure may violate the requirements for amplification factor and separation margin of the mode,

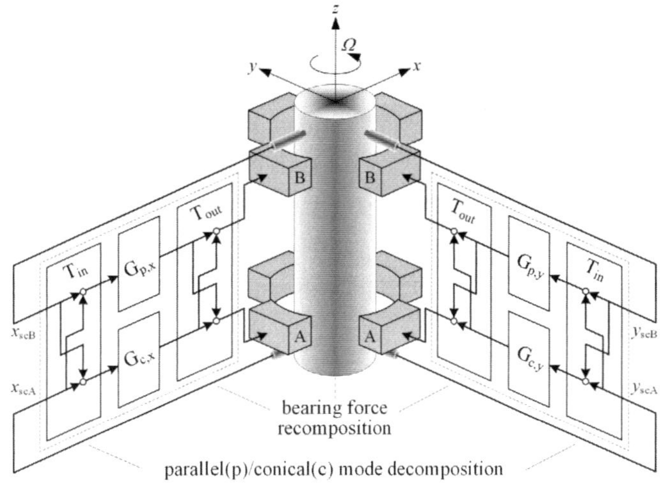

Fig. 2: Feedback structure for decoupled control of parallel and conical modes [5].

which are essentially designed to ensure that the rotor does not operate close to a ligthly-damped critical speed [6]. A possible, but theoretical solution is to use a positive phase in the controller to actively dampen the bending mode. This however results in even more gain in the controller, more noise amplification, potential actuator saturation, and excitation of other bending modes.

This means that the translational-tilting control architecture fails in applications where the first bending mode is too close to the rigid body modes.

A more sophisticated approach would be to separate the bending mode into a dedicated control channel. In the other control channel, the bending mode should be neither controllable nor observable.

III. Plant Model

The general plant model is represented by a small gas turbine with a maximum continous speed (MCS) of 35000 rpm. The rotor model G_p comprises 95 beam elements for the shaft and 4 lumped masses at the corresponding nodes to model the impellers. Before building up the complete system model, the rotor model is exported as mass, gyroscopic and stiffness matrices (MGK matrix form) and transformed into modal coordinates. This is important to model the rotor's speed-dependency resulting from its gyroscopic properties. Apart from the rotor, the full system model includes the actuator properties, amplifier dynamics, computational delay, negative stiffness of the motor and sensor dynamics. There are four force input stations $f = [f_{Ax}, f_{Ay}, f_{Bx}, f_{By}]^T$ at the locations of the magnetic bearings and four displacement output stations $r = [r_{Ax}, r_{Ay}, r_{Bx}, r_{By}]^T$ at the locations of the radial sensors. This is visualized in Figure 3, where the rotor and its modal analysis is shown.

The first flexible mode is at 226 Hz at standstill and the corresponding forward mode is crossed at around 270 rps. This needs a special unbalance force counteracting control (UFCC) [9] algorithm, which is not part of this paper.

The dynamics of the rotor is modeled as a transfer function matrix $G_p(j\omega)$

$$r(\omega) = G_p(j\omega) f(\omega). \tag{1}$$

The bode magnitude plots are shown in figure 9.

IV. Generalized Modal Decoupling Control

To achieve generalized modal decoupling control, Singular Value Decomposition (SVD) is particularly useful because of its close relationship to signal amplification in systems. It is extremely important to have a thorough understanding of the SVD, so the following section is intended to be a concise summary of its mathematical foundations and physical implications with respect to rotor dynamics. Despite the fact that all of this theory is readily available in the literature, an effort has been made here to gather the best information and present it in a way that makes it accessible to the reader.

Let's start with induced matrix norms. Consider the equation

$$r = Gf. \tag{2}$$

If $f \in \mathbb{C}^m$ is the input vector and $r \in \mathbb{C}^p$ the output vector, then $\|r\|_2/\|f\|_2$ is said to be the "amplification" or "gain" of the constant complex matrix $G \in \mathbb{C}^{p \times m}$. The maximum gain for all possible input directions is of special interest and is given by the *induced norm* defined as

$$\|G\|_2 := \max_{f \neq 0} \frac{\|r\|_2}{\|f\|_2} = \max_{f \neq 0} \frac{\|Gf\|_2}{\|f\|_2}. \tag{3}$$

That is, the induced 2-norm gives the largest possible amplification of the constant matrix G. Therefore we need to find the direction of the vector f that maximizes the ratio $\|r\|_2/\|f\|_2$, which is a constrained optimization problem, namely finding

$$\max_{\|f\|_2^2 = 1} \|r\|_2^2. \tag{4}$$

Applying the method of Lagrange multipliers, to seek the critical points for the functional

$$\mathcal{L}(f, \lambda) = r^T r - \lambda \left(f^T f - 1\right) = 0 \tag{5}$$

$$= f^T G^T G f - \lambda \left(f^T f - 1\right) = 0. \tag{6}$$

Differentiating $\mathcal{L}$ with respect to f and λ yields

$$\frac{\partial \mathcal{L}}{\partial f} = 2\left(G^T G f - \lambda f\right) = 0 \tag{7}$$

$$\frac{\partial \mathcal{L}}{\partial \lambda} = f^T f - 1 = 0. \tag{8}$$

The stationary point satisfies the eigenvalue problem

$$G^T G f = \lambda f. \tag{9}$$

The eigenvalues λ of $G^T G$, subsequently referred to as σ_i^2, are strictly non-negative because $G^T G$ is a symmetric matrix. The positive roots σ_i are the so-called *singular values* of G. The largest of these singular values is the maximum amplification possible for the matrix G and is therefore $\|G\|_2$.

The normalized eigenvectors of $G^T G$ are v_i. Because $G^T G$ is symmetric, these eigenvectors are all perpendicular to each other and form a unitary matrix $V = [v_1 \; v_2 \; \ldots \; v_m]$ with the property $V^{-1} = V^H$.

Defining the normalized vectors

$$u_j = \frac{Gv_j}{\|Gv_j\|_2} = \frac{1}{\sigma_j} Gv_j, \tag{10}$$

it can be shown that $U = [u_1 \; u_2 \; \ldots \; u_p]$ is also a unitary matrix with $U^{-1} = U^H$. Together with the matrix Σ, which is a diagonal matrix containing the non-zero, positive singular values σ_i. These are arranged in descending order along the diagonal, i.e

$$\sigma_1 \geq \sigma_2 \geq \ldots \sigma_m. \tag{11}$$

The SVD can be written as

$$GV = U\Sigma \tag{12}$$

or in the matrix factorization form

$$G = U \Sigma V^H. \tag{13}$$

In the two-dimensional space, the singular values have a nice graphical interpretation. Consider the matrix

$$G = \begin{pmatrix} 1.5 & 0 \\ -1.1 & 1.1 \end{pmatrix}.$$

The objective is to find the direction of the vector f, given $\|f\|_2 = 1$, which maximizes $\|Gf\|_2$. The norm constraint on f implies that f lies on the unit circle, like shown in figure 4. As the vector f is moved on the unit circle, $r = Gf$ describes an ellipse. To ensure SVD nomenclature consistency, v_1 corresponds to the direction of f with the maximum gain σ_1, whereas v_2 is referred to as the direction of f associated with the minimum gain σ_2.

The length of major and minor axes of this ellipse are the singular values σ_1 and σ_2 of the matrix G, respectively. To put it simply, out of all possible input directions the input vector v_1 creates the largest possible response in the direction of u_1.

This holds also if $G(j\omega)$ is a frequency depending matrix. For every fixed frequency ω there exist a SVD

$$G(j\omega) = U(j\omega)\Sigma(j\omega)V^H(j\omega). \tag{14}$$

That is, if G is frequency dependent, the singular values and singular vectors of the system will also have frequency dependence.

There is a nice physical interpretation of the SVD for a rotor model. Taking the model from section III evaluated at MCS and augmenting it by other force input stations at the beginning of the shaft, at the center of gravity and at the end of the shaft yields the following force vector

$$f_{\text{aug}} = \begin{bmatrix} f_{\text{beg},x} \\ f_{\text{beg},y} \\ f_{\text{cog},x} \\ f_{\text{cog},y} \\ f_{\text{end},x} \\ f_{\text{end},y} \end{bmatrix}.$$

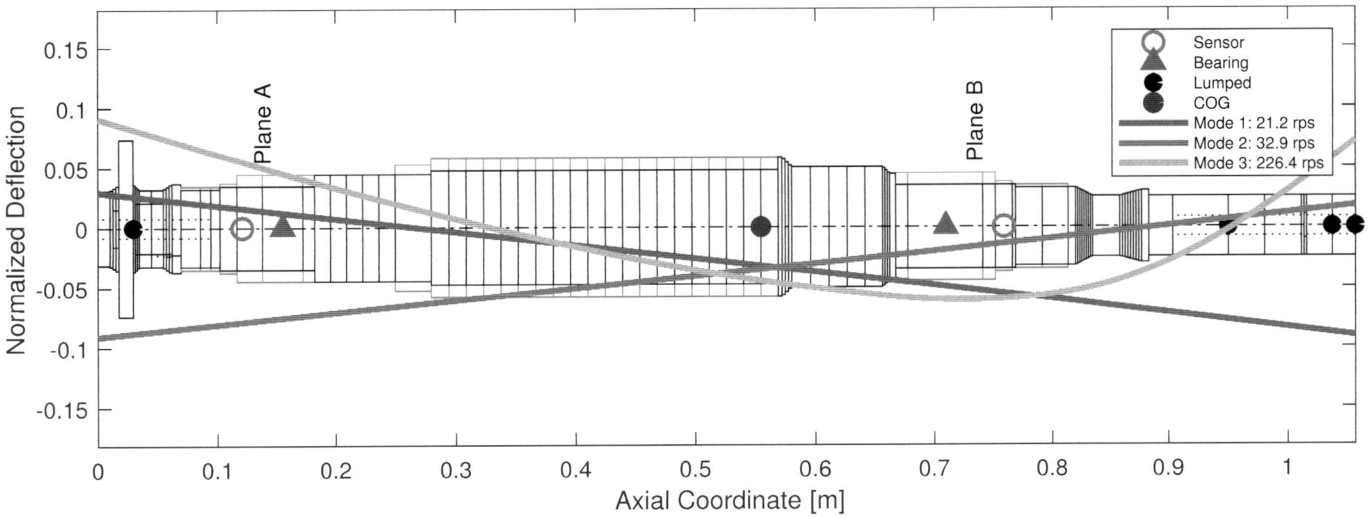

Fig. 3: Bending mode shapes of the corresponding natural frequency of the rotor. Both natural frequency and bending mode shape were calculated with a low bearing stiffness of $1 \cdot 10^6$ N/m to simulate free-free conditions.

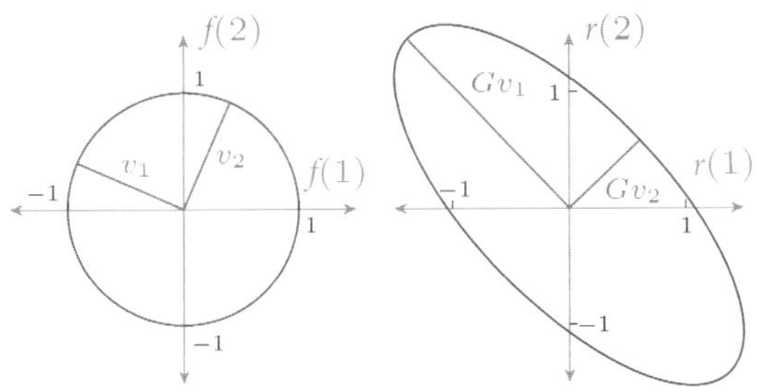

Fig. 4: Graphical interpretation of the singular value decomposition: The input f ($\|f\|_2 = 1$) is plotted on the left side, whereas the output $r = Gf$ is plotted on the right side

Such input forces are typical to excite the first bending mode [6]. Similarly, radial sensors are assumed to exist at every of the 95 nodes of the model.

$$r_{\text{aug}} = \begin{bmatrix} r_{\text{node1},x} \\ r_{\text{node1},y} \\ \vdots \\ r_{\text{node95},x} \\ r_{\text{node95},y} \end{bmatrix}.$$

This is expressed by the augmented plant transfer matrix

$$r_{\text{aug}}(\omega) = G_{\text{aug}}(j\omega) f_{\text{aug}}(\omega).$$

Now if the plant transfer matrix is evaluated at its natural frequency, i.e. the first bending eigenfrequency $G_{\text{aug}}(j\omega_n)$, a complex valued matrix G is obtained. Applying the SVD yields a complex left singular vector $u_1 = [u_{\text{node1},x}, u_{\text{node1},y}, \dots]^T$ which can be visualized by plotting the real and imaginary parts of its x- and y-elements, respectively, against the axial coordinates of the nodes (see Figure 5). Comparing Figure 5 with the mode shapes in 3 reveals that u_1 corresponds to the eigenshape of the first bending mode.

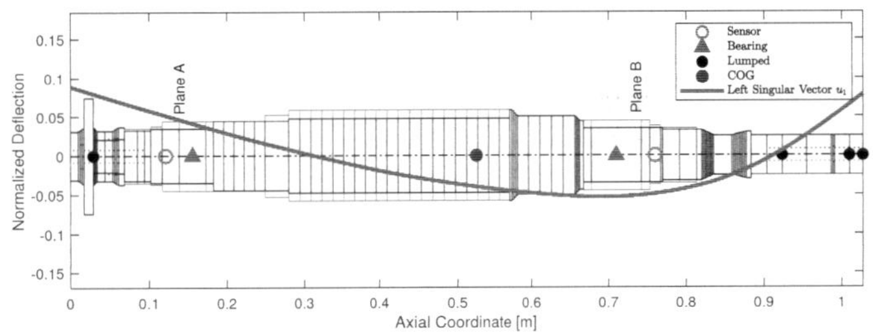

Fig. 5: Physical interpretation left singular vector u_1 of $G = G_{\text{aug}}(j\omega_n)$

Similarly, the right singular vector $v_1 \in \mathbb{C}^{2\cdot 3}$ can be thought of as some force distribution like unbalance along the rotor [7]. Out of all possible input directions, the force distribution which leads to maximum excitement of the rotor is where the force acting on the rotor's center of gravity is 180° out of phase with respect to forces at the rotor ends. Naturally, this will result in the largest possible response, which is exactly the bending mode as can be seen in Figure 6, where again the real and imaginary parts of the x- and y-elements of the singular vector is plotted against the respective axial node coordinates.

After resetting the displacement output stations and force input stations of the general plant model to the four locations of the radial sensors and magnetic bearings, respectively, the generalized modal decoupling control transformations can be found by examination of the left and right singular vectors u_1 and v_1, of the SVD of $G = G_p(j\omega_n)$, where ω_n denotes the frequency of the mode to be decoupled.

If $f \in \mathbb{R}^m$ and $r \in \mathbb{R}^p$, the transformation matrices to be found are represented by $T_{\text{U}} \in \mathbb{R}^{p\times p}, T_{\text{V}} \in \mathbb{R}^{m\times m}$. However, there is no loss of generality in assuming $m = 4$ and $p = 4$, as in the general plant model G_p. Now to create the new output vector which separates the natural frequency ω_n from its complement or null space, a transformation matrix T_{U}

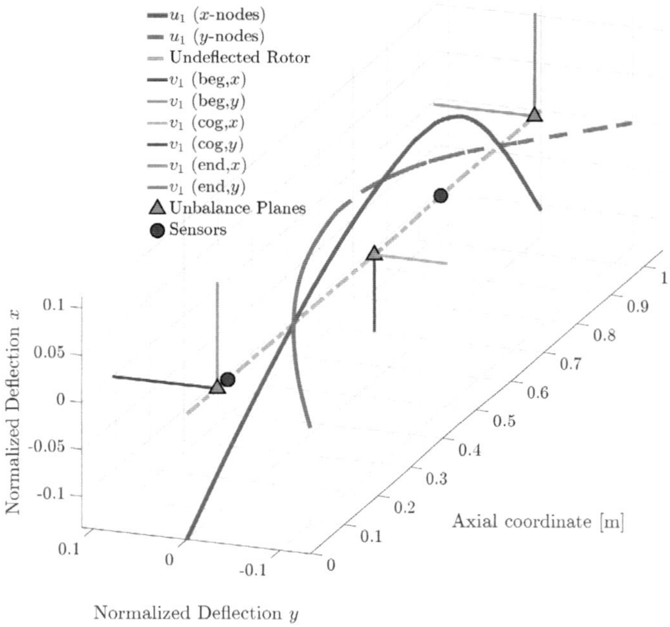

Fig. 6: Physical interpretation right singular vector v_1 of $G = G_{\text{aug}}(j\omega_n)$ and the resulting left singular vector u_1. Note that at resonance, the unbalance force and displacement are 90° out of phase.

which satisfies

$$\begin{bmatrix} r_{\text{mode},x} \\ r_{\text{mode},y} \\ \overline{r}_{\text{mode},x} \\ \overline{r}_{\text{mode},y} \end{bmatrix} = T_{\text{U}} \begin{bmatrix} r_{\text{A}x} \\ r_{\text{A}y} \\ r_{\text{B}x} \\ r_{\text{B}y} \end{bmatrix}. \tag{15}$$

needs to be found. $r_{\text{mode},x/y}$ are the transformed control input channels that contain the desired mode shape, and $\overline{r}_{\text{mode},x/y}$ are the transformed control channels that do not observe it.

Given that $u_1 = [u_{\text{A}x}, u_{\text{A}y}, u_{\text{B}x}, u_{\text{B}y}]^T$, it holds that the matrix from (16) satisfies (15)

$$\begin{aligned} u_{\text{ref},j} &= \frac{u_{\text{A}j}}{\|u_{\text{A}j}\|},\ j \in \{x, y\} \\ \hat{T} &= \begin{bmatrix} u_{\text{A}x} \cdot u_{\text{ref},x} & 0 & u_{\text{B}x} \cdot u_{\text{ref},x} & 0 \\ 0 & u_{\text{A}y} \cdot u_{\text{ref},y} & 0 & u_{\text{B}y} \cdot u_{\text{ref},y} \end{bmatrix} \\ T_{\text{U}} &= \begin{bmatrix} \text{Re}\left(\hat{T}\right) \\ \ker\left(\text{Re}\left(\hat{T}\right)\right) \end{bmatrix} \end{aligned} \tag{16}$$

where ker(.) is the kernel or null space of a matrix. The main idea behind converting complex values in u_1 to real values in T_{U} is to generate real frames that are aligned as closely as possible to the given complex frames. This can be done with the ALIGN algorithm of Kouvaritakis [8]. However here, for ease of understanding, the idea and the corresponding algorithm is outlined using the dot product. Since the transformation matrix must be real-valued, the complex-valued entries of the singular vectors must be interpreted accordingly. By treating the first entry of the singular vector as the normalized reference vector in the complex plane, all entries can be aligned to this vector via the dot product.

Figure 7 graphically illustrates the entries of the left singular vector u_1 of $G = G_p(j\omega_n)$, which represent the rotor deflection of the bending mode shape at the sensor locations.

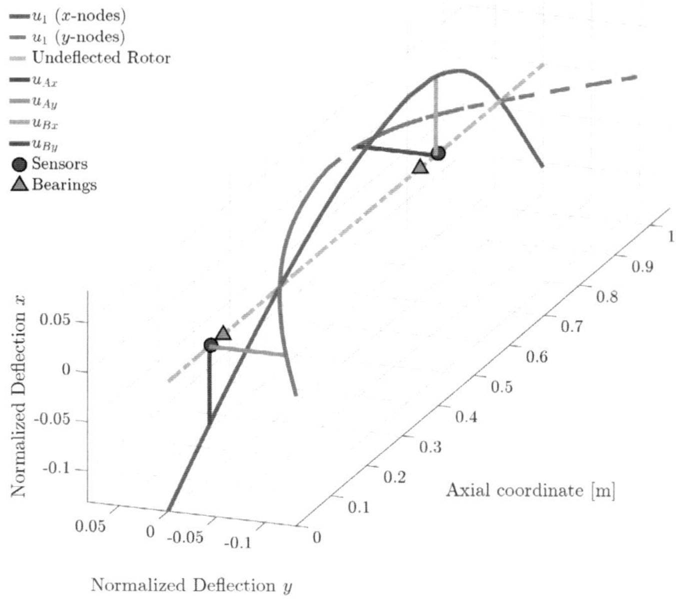

Fig. 7: Physical interpretation left singular vector u_1 of $G = G_p(j\omega_n)$

Similarly, given $v_1 = [v_{\text{A}x}, v_{\text{A}y}, v_{\text{B}x}, v_{\text{B}y}]^T$ a new input vector which directly influences the targeted mode shape and the corresponding complement or null space is found by (17) and (18), respectively.

$$\begin{bmatrix} f_{\text{mode},x} \\ f_{\text{mode},y} \\ \overline{f}_{\text{mode},x} \\ \overline{f}_{\text{mode},y} \end{bmatrix} = T_{\text{V}} \begin{bmatrix} f_{\text{A}x} \\ f_{\text{A}y} \\ f_{\text{B}x} \\ f_{\text{B}y} \end{bmatrix} \tag{17}$$

$$\begin{aligned} v_{\text{ref},j} &= \frac{v_{\text{A}j}}{\|v_{\text{A}j}\|},\ j \in \{x, y\} \\ \tilde{T} &= \begin{bmatrix} v_{\text{A}x} \cdot v_{\text{ref},x} & 0 & v_{\text{B}x} \cdot v_{\text{ref},x} & 0 \\ 0 & v_{\text{A}y} \cdot v_{\text{ref},y} & 0 & v_{\text{B}y} \cdot v_{\text{ref},y} \end{bmatrix} \\ T_{\text{V}} &= \begin{bmatrix} \text{Re}\left(\tilde{T}\right) \\ \ker\left(\text{Re}\left(\tilde{T}\right)\right) \end{bmatrix}^T \end{aligned} \tag{18}$$

Figure 8 graphically illustrates the entries of the right singular vector v_1 of $G = G_p(j\omega_n)$ representing the input forces of the magnetic bearings which lead to the maximum excitement of the rotor. This algorithm may be generalised for a setup with many control planes, which allows to decouple multiple modes (see Equation 20). The letters A to Z represent the control planes and n denotes the number of mode frequencies

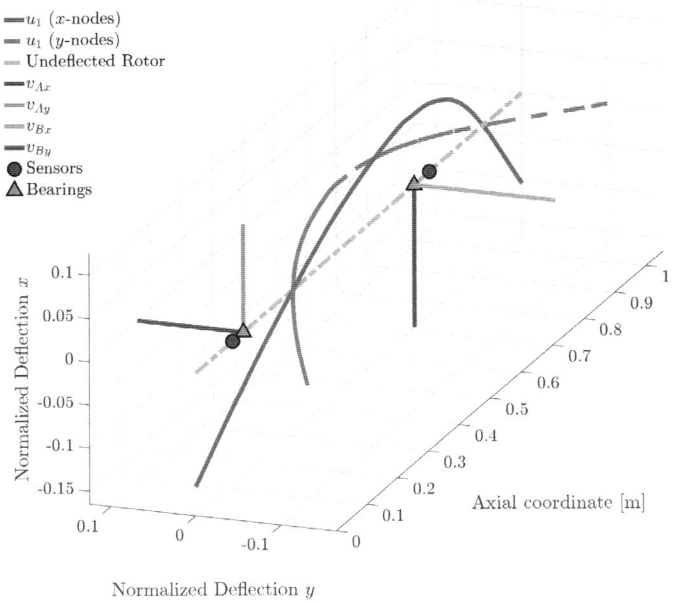

Fig. 8: Physical interpretation left singular vector v_1 of $G = G_p(j\omega_n)$. Note that at resonance, the unbalance force and displacement are $90°$ out of phase.

to be decoupled. In other words, the transformation matrix essentially consists of the blocks $T_{i,k} \in \mathbb{R}^{2\times 2}$, nicely stacked upon each other, where $i \in \{A, \ldots, Z\}$ and $i \in \{1, \ldots, n\}$. The input transformation T_V is obtained analogously when the left singular vector u is swapped with the right singular vector v.

V. Experimental Results

In this section the results achieved by the introduced decoupling control technique are investigated. The plant that is subject to this case study is a small gas turbine. The maximum continous speed (MCS) of the machine is 35000 rpm (583 rps) thus representing a supercritical application (MCS is above the the first flexible bending mode frequency). The model of the rotor borne by magnetic bearings is given in section IV. Its frequency response at standstill and maximum continous speed, respectively, can be found in the Bode plot in Figure 9 shown in blue.

In this section this model is used to design a hand-synthesized position controller, which meets the generally known and obvious objectives such as a sufficient stability margin and an adequate handling of external loads. It is vitally important to emphasize that the craftsmanship and optimization of hand-synthesized controllers is generally referred to as loop shaping[1] and is largely considered an art for systems with this level of complexity. As a result, the assessment of optimality of such controllers is non-trivial. However, the most straightforward and intuitive approach for designing a controller is to attempt to make the AMB act like conventional mechanical bearings at the actuator locations, that is, feedback control is being performed locally for each bearing unit and separately for each bearing axis. However, this approach, often referred to as decentralized control, over-simplifies the problem and must be adapted and extended for realistic rotor systems. Applying the Tilting-Translational coordinate transformation results in the brown frequency domain characteristics shown in Figure 9. It is not hard to see that the transformed plant features little to no benefits over the decentralized plant as the cross-coupling terms are still substantial and there is no useful mode separation. The golden transfer function is obtained by applying the generalized modal decoupling control transformation proposed in this paper, where the plant transfer matrix is evaluated at the first bending eigenfrequency (see Figure 3). The critical frequency is detected separately and can therefore be controlled separately, which greatly facilitates the control design. The design of a controller that meets the robustness and performance criteria of generalized modal decoupling control plant $G_{p,gmd}$ is fairly easy whereas the design that targets the tilting-translational control plant $G_{p,tt}$ is quite tedious and unable to meet the necessary performance criteria. This generally results in a conservative controller being unable to keep the rotor in position, i.e. centered under rotation. The position controller of the magnetic bearings must be able to maintain stability and performance under the influence of all plant uncertainties such as the splitting of eigenfrequencies due to gyroscopy, changes in suction pressure and gas composition to name a few. The ISO standard 14839 addresses this robustness issue and imposes limits on the peak value of the output sensitivity function $S_e = (I + L)^{-1}$. The standard recommends that the diagonal elements of S_e have a peak value below 3 for newly commissioned machines. However, this does not guarantee high performance of the closed-loop, i.e. small rotor orbits, good disturbance rejection, therefore other criteria must be introduced. Here the dynamic compliance G_f is used which is the relation between rotor displacement and external force[2]. Obviously, G_f must be as low as possible for all frequencies. Unlike the specification of S_e, there is no absolute upper limit for the compliance since it is dependent on the range of bearing force. As a result, the design goals can be summarized as follows:

- $\|S_e(i, i, j\omega)\|_\infty \leq 3,\ i \in \{1, \ldots, 4\}$
- $\|G_f(j\omega)\|_\infty \leq \epsilon,\ \epsilon$ as small as possible

Now the controller design can be divided into the following steps.

- Due to the decoupling properties of the proposed transformation, the off-diagonal elements can be neglected.
- Due to rotor symmetry, only two SISO controllers must to be designed for the four diagonal elements of the transformed plant $G_{p,gmd}$.
- The controller is designed in order to robustly stabilize the plant, that is, fulfilling the generalized Nyquist

[1]This classical approach aims to shape the magnitude of the open-loop function $L(j\omega) = G_pG_c$, where G_c is the feedback controller to be designed and G_p is the plant.

[2]Dynamic compliance refers here to $G_f(j\omega) = G_p(j\omega)S_u(j\omega)$, where $S_u(j\omega)$ is the sensitivity function at the plant input.

$$u_{\text{ref},j}(\omega) = \frac{u_{\text{A}j}(\omega)}{\|u_{\text{A}j}(\omega)\|},\ j \in \{x, y\} \tag{19}$$

$$T_{\text{U}} = \begin{array}{l} T_{i,k}^{2\times 2}\left\{\begin{bmatrix} u_{\text{A}x}(\omega_1)\cdot v_{\text{ref},x}(\omega_1) & 0 & \cdots & u_{\text{Z}x}(\omega_1)\cdot v_{\text{ref},x}(\omega_1) & 0 \\ 0 & u_{\text{A}y}(\omega_1)\cdot v_{\text{ref},y}(\omega_1) & \cdots & 0 & u_{\text{Z}y}(\omega_1)\cdot v_{\text{ref},y}(\omega_1) \\ \vdots & \vdots & \ddots & \vdots & \vdots \\ u_{\text{A}x}(\omega_n)\cdot v_{\text{ref},x}(\omega_n) & 0 & \cdots & u_{\text{Z}x}(\omega_n)\cdot v_{\text{ref},x}(\omega_n) & 0 \\ 0 & u_{\text{A}y}(\omega_n)\cdot v_{\text{ref},y}(\omega_n) & \cdots & 0 & u_{\text{Z}y}(\omega_n)\cdot v_{\text{ref},y}(\omega_n) \\ \multicolumn{5}{c}{\ker\left(\text{Re}\left(\tilde{T}\right)\right)} \end{bmatrix}\right\}\tilde{T} \end{array} \tag{20}$$

- criterion for any plant in the set of transfer functions represented by uncertainty.
- Ensure that the design goals are met for any plant in the set of transfer functions represented by uncertainty.

Figure 10a and 10b show the controller which meets the design goals, in the centralized and decentralized coordinate system, respectively. During the machine commissioning, it was gradually accelerated to reach its maximum continuous speed. At various rotational speeds, relevant control signals were measured, and frequency domain measurements of the control plant and sensitivity S_e were conducted. This exercise ensured the stability of the control system at any rotational speed. In order to deal with umodeled dynamics, adjustments were made to further enhance the controller by augmenting it with numerous tricks such as notch and phase bump filters. Figure 11 shows the measured sensitivity functions S_e at standstill and at MCS. Evidently, the peak values of all diagonal elements at both rotation speeds are below 3, which implies that the closed-loop exhibits excellent robustness against plant uncertainties and falls into zone A, according to ISO 14839 [9].

VI. Conclusion

In this paper a generalized modal decoupling control method has been proposed, which essentially is able to split up the system into modal parts and its null space via coordinate transformation. The functionality of this transformation was proved by experiments on a supercritically operated small gas turbine. The transformation coordinate system rendered the task of control design substantially easier because it achieves true mode separation. This allows to stabilize the first bending mode over a great frequency range without compromising the other flexible modes on the modal channel, whereas the requirement to actively stabilize any flexible modes is completely eliminated on the complementary channel. Essentially, the proposed method is a generalisation and improvement of existing decoupling methods such as tilting-translational control, therefore it is widely applicable to the vast majority of magnetic bearing control problems.

References

[1] *Mu-Synthesis for Magnetic Bearings: Why Use Such a Complicated Tool?*, ser. ASME International Mechanical Engineering Congress and Exposition, vol. Volume 9: Mechanical Systems and Control, Parts A, B, and C, 11 2007. [Online]. Available: https://doi.org/10.1115/IMECE2007-43910

[2] H. Bleuler, "Decentralized control of magnetic rotor bearing systems," Ph.D. dissertation, ETH Zurich, Zürich, 1984, diss. Techn.Wiss. ETH Zürich, Nr. 7573, 0000. Ref.: Schweitzer, G. ; Korref.: Mansour, M..

[3] Q. Zhang and C.-S. Zhu, "Modal decoupling control for active magnetic bearing-supported flywheel rotor system," *Zhendong Gongcheng Xuebao/Journal of Vibration Engineering*, vol. 25, pp. 302–310, 06 2012.

[4] M. Hutterer, M. Hofer, and M. Schrödl, "Decoupled control of an active magnetic bearing system for a high gyroscopic rotor," in *2015 IEEE International Conference on Mechatronics (ICM)*, March 2015, pp. 210–215.

[5] G. Schweitzer and E. H. Maslen, *Magnetic bearings: theory, design, and application to rotating machinery*. Springer, 2010.

[6] "API Standard 617 Axial and Centrifugal Compressors and Expander-compressors," 2014.

[7] C. H. Cloud, G. Li, E. H. Maslen, L. E. Barrett, and W. C. Foiles, "Practical applications of singular value decomposition in rotordynamics," *Australian Journal of Mechanical Engineering*, vol. 2, no. 1, pp. 21–32, 2005. [Online]. Available: https://doi.org/10.1080/14484846.2005.11464477

[8] J. EDMUNDS and B. KOUVARITAKIS, "Extensions of the frame alignment technique and their use in the characteristic locus design method," *International Journal of Control*, vol. 29, no. 5, pp. 787–796, 1979. [Online]. Available: https://doi.org/10.1080/00207177908922732

[9] ISO Central Secretary, "Mechanical vibration – vibration of rotating machinery equipped with active magnetic bearings. – Part 3: Evaluation of stability margin," International Organization for Standardization, Geneva, CH, Standard, 2006. [Online]. Available: https://www.iso.org/standard/39057.html

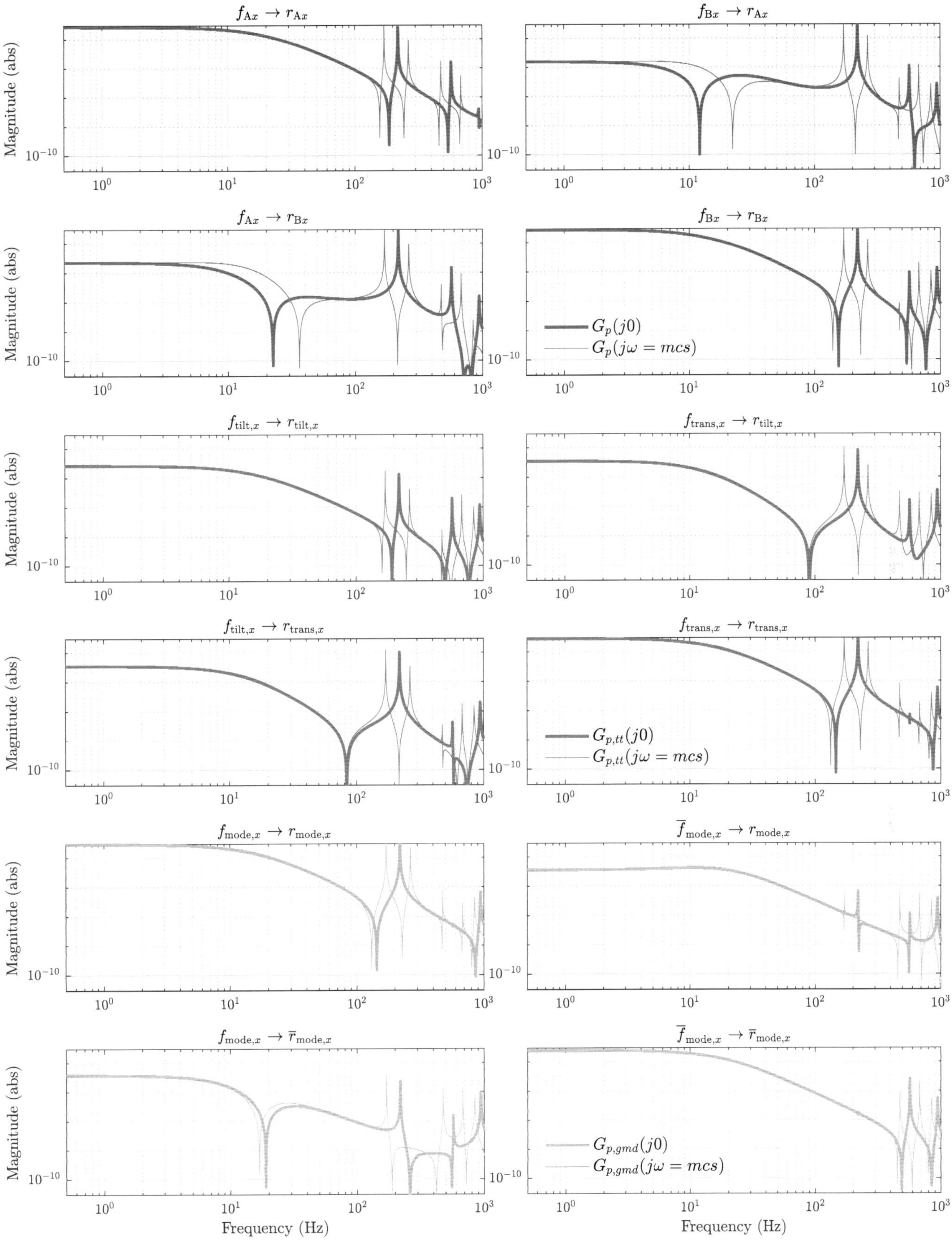

Fig. 9: Bode plot of the plant models (x-plane only). Top (blue): Initial 2x2 plant without decoupling. Middle (brown): Plant decoupled by T_{in} and T_{out} (Tilting-Translational). Bottom (gold): Plant decoupled by T_V and T_U (SVD). It is easy to see that the flexible mode is present in the main and off-diagonal channels in G_p and $G_{p,tt}$. For $G_{p,gmd}$ the mode is only present in the first channel and has a very low coupling to the other channel. The transformation has been tuned to be best at MCS for the forward mode.

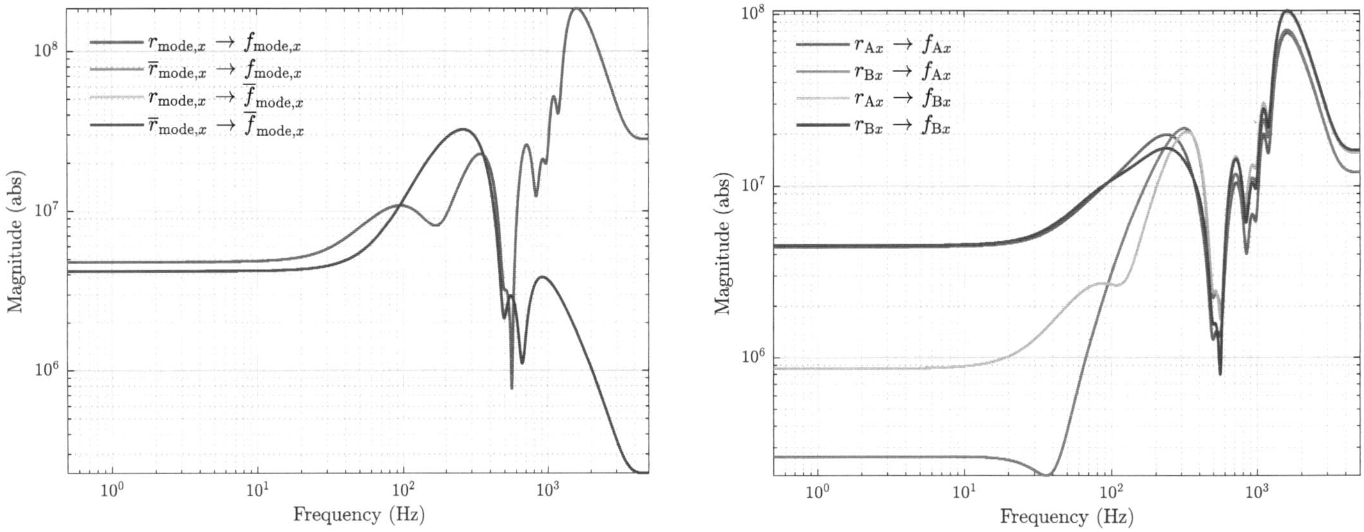

(a) Centralized controller which has no coupling between the modal decoupled channels. The controller is tuned by hand to fulfill the performance and robustness requirements for whole the speed range of the machine.

(b) Decentralized controller, generated by applying the modal decoupling T_{V} and T_{U} to the decentralized controller, showing significant and frequency dependent coupling between the sensor and actuator planes A and B.

Fig. 10: Controller in centralized (SVD) and decentralized coordinates, without integrator action. Plots are made for the x-plane. The y-plane uses the same controllers. It is worth to highlight that the controller needed to be augmented it with numerous tricks such as notch and phase bump filters to cope with spurious effects suchs as temperature depending housing resonances. Evidently, this reinforces the argument for the generalized decoupling control method as generally known methods would not provide room for corrective action, when the control design for the rotor itself is next to impossible.

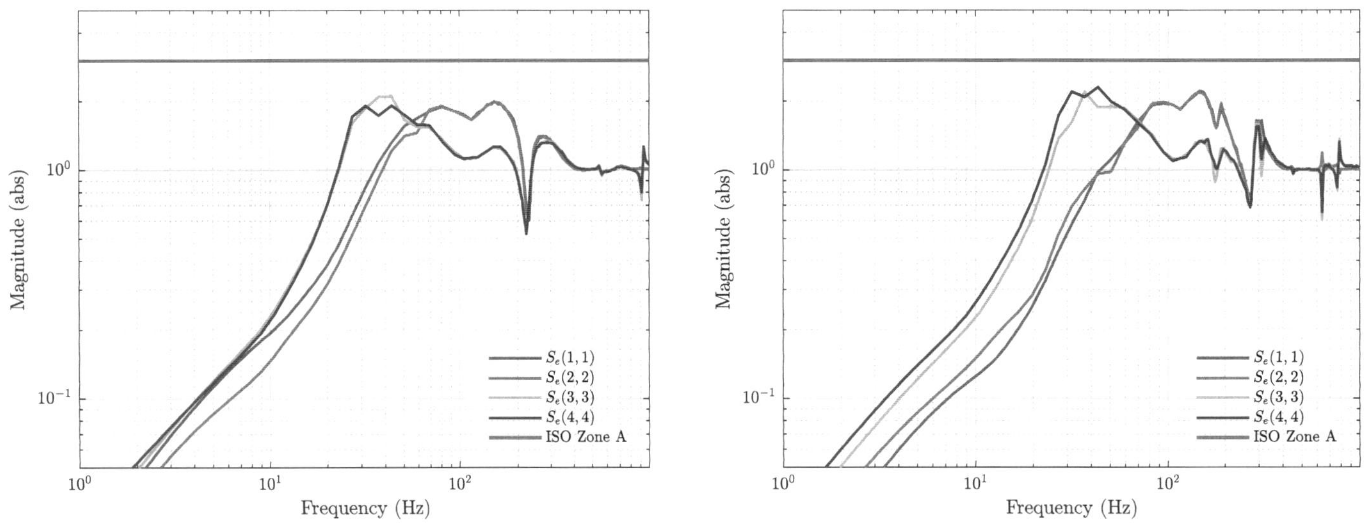

(a) Sensitivity at standstill.

(b) Sensitivity at top speed

Fig. 11: Measured output sensitivities at standstill and top speed according to ISO [9] fall into zone A. The system features substantial robustness and excellent performance.

Linear Superconducting Magnetic Bearing for Urban Transportation

Richard STEPHAN [a], Felipe COSTA [b], Elkin RODRIGUEZ [a], Gabriel MESSER [a]

a Federal University of Rio de Janeiro (UFRJ), P.O.Box 68504,21941-972 Rio de Janeiro, Brazil, richard@poli.ufrj.br

b MagLev Company Brazil (MCB), 22793-237 Rio de Janeiro, Brazil,felipe@maglevcompanybrazil.com

Abstract

This paper describes the evolution and perspectives of the Superconducting Magnetic Levitation (SML) method, based on High Critical Temperature Superconductors (HTS) and Rare-Earth Permanent Magnets, to reach commercial application for urban transportation. All the results showed in this article were obtained by tests made on the Maglev-Cobra track at the Federal University of Rio de Janeiro (UFRJ).

Keywords: *Magnetic Levitation, MagLev, Superconducting Magnetic Levitation, Urban Public Transportation.*

1. Introduction

The same magnetic forces that allow the construction of Magnetic Bearings of rotating machines can be applied for linear movements and transportation systems. But the physical arrangements and propulsion alternatives deserve special attention, justifying a dedicated research area.

There are three levitation methods applied in Magnetically Levitated (MagLev) vehicles: Electrodynamic Levitation (EDL), Electromagnetic Levitation (EML), and Superconducting Magnetic Levitation (SML) (Moon, 1994). The EDL method uses repulsion forces and is well fitted for high-speed operation (> 500 km/h) and not for urban transportation. The EML method is based on the attraction force between an electromagnet and ferromagnetic materials. The resulting system requires feedback control to achieve stable operation. Five urban EML-MagLev systems are already in commercial operation: Nagoya, Japan, since 2005 – Seoul, South Korea, since 2016 – Changsha, China, since 2016 – Beijing, China, since 2017 - Hunan, China, since 2022. But there is no SML Vehicle in commercial use, which opens an opportunity for innovation in Urban Public Transportation.

The present paper describes the efforts to turn the SML-MagLev Technology into a commercial product applied to urban mobility. Since nowadays more than 50% of the world population live in cities, and in many countries of Europe, South and North America this share is higher than 80%, there is a promising potential market.

2. The Superconducting Magnetic Levitation (SML) Vehicle MagLev-Cobra

The SML method depends on High Temperature Superconductors (HTS) and Rare Earth Permanent Magnets. These materials were developed at the end of last century and made commercially available only at the early 2000. The main advantage in comparison with the EML method is the stable operation, which increases the reliability and simplicity, opening great opportunities for urban transportation. The SML MagLev-Cobra project (Stephan and Pereira, 2020) reached Technology Readiness Level TRL07 (European Association of Research and Technology Organizations, 2014) in 2019, but COVID-19 imposed a development gap (Fig. 1).

The evolution of the MagLev-Cobra Project is registered in papers published since year 2000 in the International Conference on Magnetically Levitated Systems and Linear Drives (MAGLEV), a worldly recognized conference devoted to the subject (Nicolsky, et al., 2000, 2002; Motta, et al., 2008; Stephan, et al., 2004, 2006, 2008, 2011, 2016, 2018, 2022).

The step to reach TRL08 in October 2023 and the plan to reach TRL09 by 2025 will be described in the following.

2.1 A Reliable Vehicle

An industrialized vehicle, with reliable operation, including doors, air conditioning, comfortable seats, energy supply, emergency breaking and exits belong to the fundamental components of a commercial product. To reach these demands, a partnership has been established with Aerom, a company that has large experience in the fabrication of wagons for urban mobility. Figure 2 shows some details of the new vehicle.

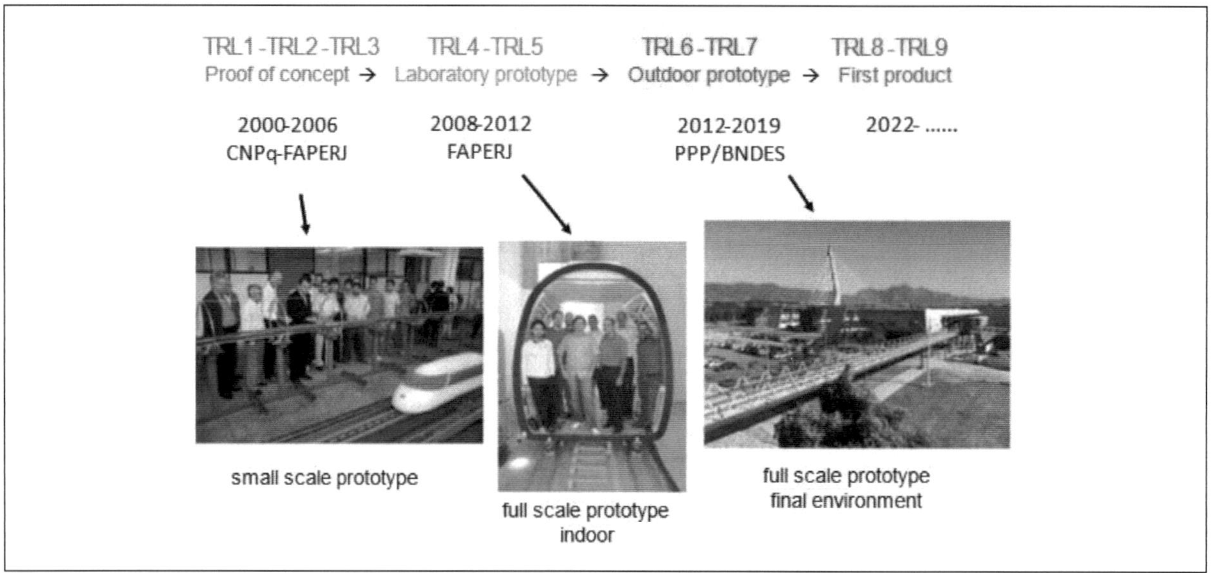

Figure 1 The MagLev-Cobra project measured with the TRL scale.

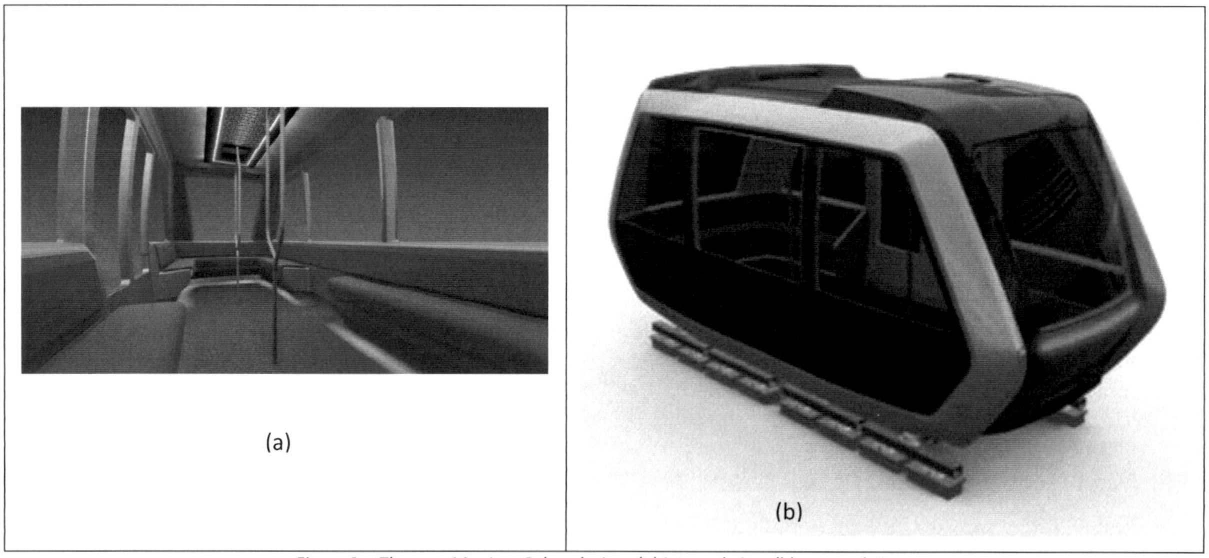

Figure 2 The new MagLev-Cobra design: (a) internal view (b) external view.

The new vehicle proposal is aligned with a modern design and the technological appeal that this technology deserves. The commissioning of the new MagLev Cobra is scheduled to the beginning of October 2023. After that, a fully automated operation will be delivered to the University (UFRJ). This will be the first in the world to offer regularly this kind of transport (SML-MagLev) for employees, students and visitors.

2.2 An Adequate Traction System (Linear Motor)

The linear motor was also improved following Oliveira's (2020 a, 2020 b) Ph.D. research thesis. It is a short primary linear induction machine with aluminium conducting sheets and back iron secondary. The primary is installed under the wagon while the secondary lays along the track (Figure 3).

The new configuration has turned the motor easier to produce and cheaper than the first one, overcoming the upper limit on the vertical movement imposed by the C configuration of the former linear motor (Oliveira 2020 a, 2020 b).

2.3 A Protected Magnetic Rail Infrastructure

One of the main problems observed during 5 years of external operation (TRL 06 – 07) was the oxidation of the magnetic rails. As the magnets weren't completely protected against rain and sun light exposition, micro cracks allowed the penetration of water inside the rails, giving environment conditions for oxidation. To overcome this difficulty, a protective painting based on Zn was successfully applied, but some magnets had already suffered a magnetization loss from circa 0.4 T. This reflected on the levitation height, as will be seen with the measurement results. A new magnetic rail is under development, this time fully protected against water.

2.4 An Automated Operation

The vehicle operation must be autonomous. This is practically the standard of urban transportation that plays a crucial role in modern cities, providing means for people to move around efficiently and effectively. The transportation system must be reliable and able to handle the volume of people using it, without causing significant delays or overcrowding. This can be achieved by optimizing routes and schedules, ensuring that buses, trains, and other modes of transportation arrive and depart on time. An autonomous system is under development and planned to be fully tested by the end of this year.

Figure 3 The new Induction Motor 200 meters long secondary made with Aluminium sheets and back iron.

3. Measurements

For the reliable operation of MagLev vehicles, the levitation height is an important information to ensure some safety level and reliability once the vehicle does not touch the ground. In the case of SML, since there is no closed loop control of the levitation height, and the system regulates itself as result of the interaction between the refrigerated superconductor materials and the magnet rails, this information plays an important role. Therefore, for each wagon, four ultrasonic type position sensors were installed, as shown in Figure 4.

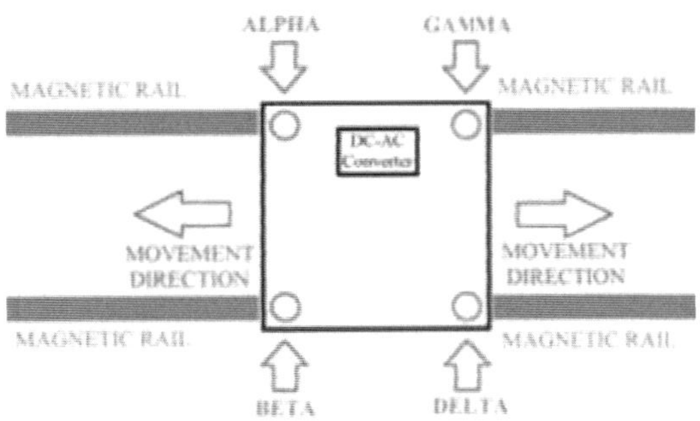

Figure 4 The four levitation height position sensors (alpha, beta, gamma, delta).

The height sensors installed on the wagon are ultrasonic sensors developed by the manufacturer Balluff with a range between 30 and 400 mm. They have the facility of restricting the acquisition range for more precise measurements. In this case, the range has been reduced to 30 up to 80 mm. As an example, Figure 5 shows the behavior of the beta and gamma sensors, which are located at opposite vertices of the wagon's diagonal.

The test trips were carried out in free fall along the test line. This is possible because one end of the test track is higher than the other. Since there is no friction due to levitation, only the force of gravity moves the wagon. In this way, the two trips present the same travelling time, except for the moment when the inverter is activated to brake the vehicle. The figures show the variation in levitation height over two trips. As can be seen, the profile is similar throughout the entire time (for the same sensor) except when the inverter is activated, which causes noise in the signal and alters the final travel time. In addition, the beta sensor had average height variations between 44 and 53 mm, while the gamma ranged between 47 and 56 mm.

Figure 6 presents a velocity test run. The data were obtained with a retro-reflective photoelectric sensor that allows the velocity measurement without contact. Based on this information, by integration, it will be possible to determine the vehicle position. This will be an input for the automation system.

4. Conclusion

This paper was prepared during the first semester of 2023, in the middle of the efforts that will allow the MagLev Cobra Vehicle to start daily operation, in October, connecting two stations (200 meters apart) in the Campus of the Federal University of Rio de Janeiro (UFRJ). If we succeed, it will be the first SML-MagLev System in the world to reach TRL08. The next step, the TRL09 level, will require a line, at least 1 km long, to test operation in ramps, curves and at speeds of 50 to 70 km/h. For that, a technical and economic feasibility study is on the verge to be concluded. This proposal will still be in UFRJ Campus, but, this time, connecting the University Technological Center, mainly devoted to the education of engineers, to the Technological Park, where companies like Siemens, Ambev, FMC, GE, Vallourec, Halliburton, Schlumberger, Tenaris, Suez, have installed research centers. With this second step, we will contribute to turn the UFRJ Campus in an example of a city of the future.

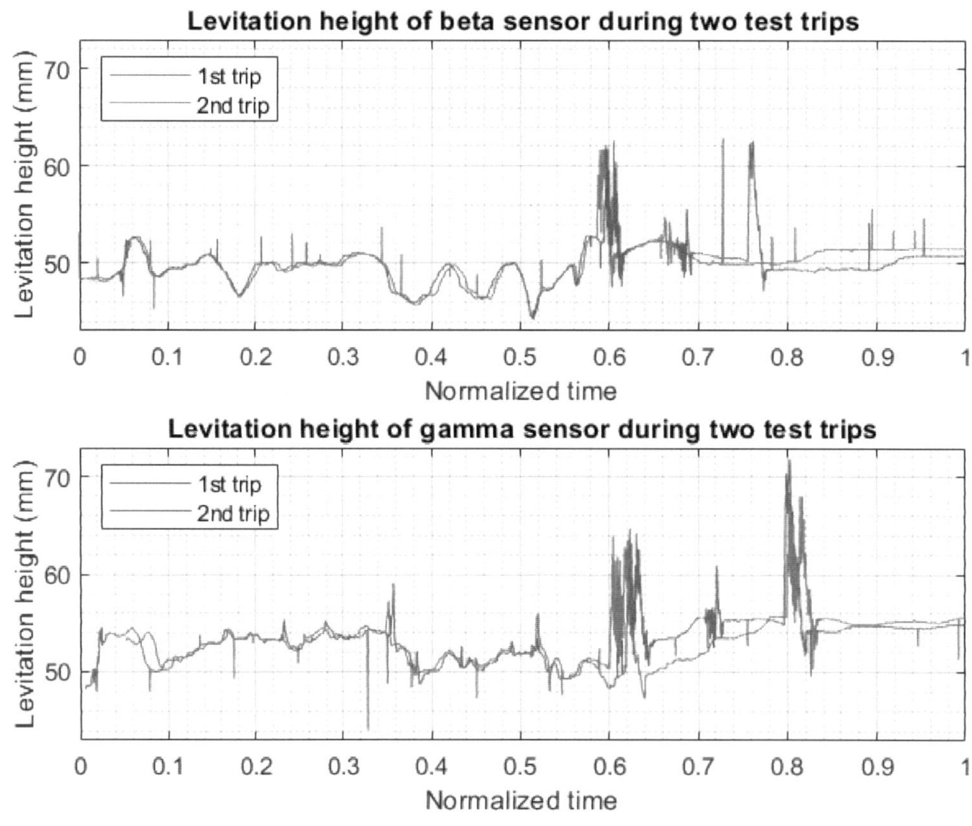

Figure 5 Levitation height measurements during two test ride.

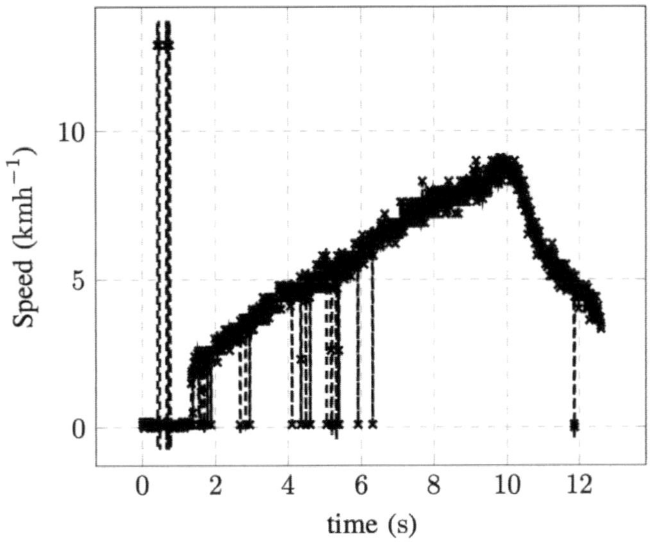

Figure 6 A velocity test run.

References

European Association of Research and Technology Organisations (2014). The TRL Scale as a Research and Innovation Policy Tool; EARTO: Brussels, Belgium..

Moon, F.C. (1994) Superconducting Levitation. John Wiley and Sons.

Motta, E. S.; Stephan, R. M.; Andrade Jr, R. ;Norman, J. H.; Kuhn, L. (2008) "Optimization of a Magnetic Rail for the HTS-Superconducting Vehicle Maglev-Cobra", 20th MAGLEV, San Diego.

Nicolsky, R., Stephan, R. M., Ferreira, A.C., Andrade Jr, R., Moreira, M.A., Rolim, L. G. B., Neves, M. A., Rosário, M. A. P. (2002) "The Current State of the Brazilian Project for a Superconducting Magnetic Levitation Train", In: 17th MAGLEV, Lausanne.

Nicolsky, R., Stephan, R. M., Andrade Jr, R., Ferreira, A.C. (2000) "The Brazilian Project for a Superconducting Magnetic Levitation Train", In: 16th MAGLEV, Rio.

Oliveira RH, Stephan RM, Ferreira AC, Pina J. (2020 a) Design and innovative test of a linear induction motor for urban maglev vehicles. IEEE Transactions on Industry Applications; 56(6):6949-6956.

Oliveira RH, Stephan RM, Ferreira AC. (2020 b)Optimized linear motor for urban superconducting magnetic levitation vehicles. IEEE Transactions on Applied Superconductivity; 30(5):1-8.

Stephan, R.M.; Pereira, A. (2020) The Vital Contribution of MagLev Vehicles for the Mobility in Smart Cities. MDPI-ELECTRONICS, v.9, p.978 - 990.

Stephan, R.M.; Deng, Z. (2022) Past, Present and Future of Superconducting Magnetic Levitation (SML), In: 25th MAGLEV, Changsha (Webinar).

Stephan, R.M.; Costa, F.; Rodriguez, E.; Deng, Z. (2018) Retrospective and perspectives of the superconducting magnetic levitation (SML) technology applied to urban transportation. In: 24th MAGLEV, St. Petersburg.

Stephan, R.M.; Costa, F.; de Andrade, R.; Ferreira, A. C.; Rodriguez, E. F. (2016) "The experimental line of the MagLev-Cobra project for urban transportation". In: 23rd MAGLEV, Berlin.

Stephan, R.M.; Andrade Jr, Rubens ;Ferreira, A. C.; Machado, O.; Santos, M. D. A.; Sotelo, G.G.; Dias, D.H.N. (2011) "The First Full Scale Functional Prototype of the Superconducting Magnetically Levitated Vehicle MagLev-Cobra". In: 21st MAGLEV, Daejeon.

Stephan, R. M.; David, E. G.; Andrade Jr, R.; Sotelo, G.G.; Machado, O.; Haas, O; Werfel, F. (2008) "A Full-Scale Module of the MagLev-Cobra HTS-Superconducting Vehicle". In: 20th MAGLEV, San Diego.

Stephan, R. M.; David, E. G.; Haas, O. (2008) "Maglev-Cobra: An Urban Transportation Solution Using HTS-Superconductors and Permanent Magnets," In: 20th MAGLEV, San Diego.

Stephan, R. M., Costa, G. C., Andrade Jr, Rubens, Nicolsky, R. (2006) "Feasibility Study of an HTS-Maglev Line at the Federal University of Rio de Janeiro", In: 19th MAGLEV, Dresden.

Stephan, R. M., Ferreira, A. C., Andrade Jr, R., Moreira, M. A., Rolim, L. G. B., Neves, M. A., Rosário, M. A. P., Machado, O. J., Nicolsky, R., (2004) "A Superconducting Magnetic Levitation Train Prototype in Closed Loop Track" , In: 18th MAGLEV, Shanghai.

Acknowledgment

To FAPERJ (Fundação de Amparo à Pesquisa do Rio de Janeiro) that supports financially this project. To Aeron and its team, partners in the development of the industrialized MagLev-Cobra Vehicle. To Paulo Roberto Costa, CEO of SeaHorse, for his invaluable help in infrastructure improvements.

Comparing Magnetic Bearings with Symmetry of 3

Domingos F. B. DAVID [a], Afonso Celso DEL NERO GOMES [b]
a Universidade Federal Fluminense, Rua Passo da Pátria 156, 24210-240 Niterói, Brazil
b Universidade Federal do Rio de Janeiro, CP 68504, 21945-970 Rio de Janeiro, Brazil, nero@coep.ufrj.br

Abstract

The stators of Active Magnetic Bearings (AMBs) usually show a symmetry of 4, because the magnetic forces applied to the rotor are generated in 4 points in a plane, two in the canonical *x* direction and two in the *y* direction. AMBs with symmetry of 3 are harder to find, either in real-world applications or in the specific literature. Their stators generate forces in 3 coplanar points; one of these points, at most, can lie in one of the canonical *x,y* directions. In this paper, basic aspects of the symmetry of 3 will be explained; they will help to develop the mathematical models of symmetry of 3 AMBs and to compare some situations in this world with the corresponding facts in the symmetry of 4 devices.

Keywords: *Symmetry of 4, Symmetry of 3, Coupled fluxes, Uncoupled fluxes.*

1. Introduction

The most usual AMBs [9, 2, 10] show the structure in the left side of Figure 1. There are two "U-shaped electromagnets" in the *x* or horizontal direction and two in the *y* or vertical direction, resulting in four independent magnetic flux loops. A different configuration for AMBs is possible [6, 5, 4, 3] with only four windings that lead to interconnected magnetic loops, as depicted in the right side of the same Figure. These structures are sometimes called 08 poles and 04 poles based on the number of stumps that leave the stator and not because of any magnetic property.

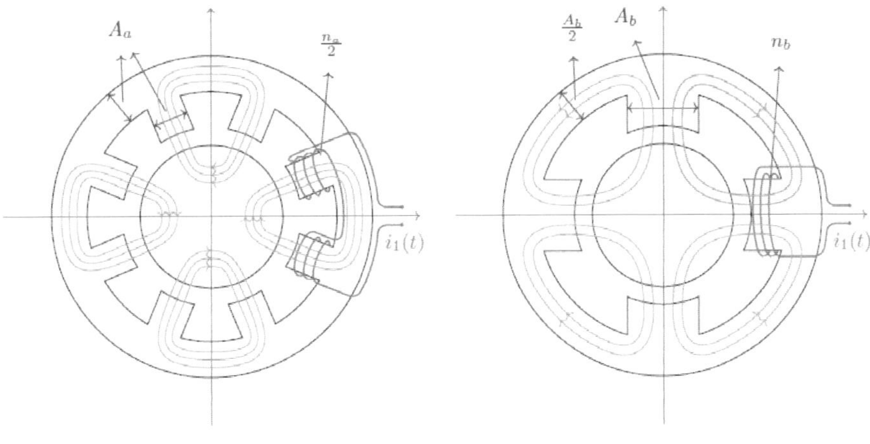

Figure 1: Windings, above, are shown for the positive *x* direction only; opposing pairs of windings along the *x*(*y*) direction control that position. In the traditional AMB, at the left, there are no connections among the flux paths; in the 04 poles case the flux paths are interconnected.

In both cases, the windings in the *x,y* directions are fed with currents $i_0 \pm i_{x,y}(t)$; the $i_{x,y}$ will control the rotor position. The resultant forces $f_{x,y}$ can be expressed in terms of these currents, the air magnetic permeability μ_0, the number of coils *n*, the cross-section areas *A* and the nominal length *h* of the air gaps [9, 2, 10]. After a standard linearization procedure [9] around the operating point $x = y = i_x = i_y = 0$, the forces generated by the traditional AMB are of the form $f_{x,y} = k_p x(y) + k_i i_x(i_y)$ where k_p and k_i are the position and current constants. Notice that the unconnected nature of the magnetic fluxes leads to uncoupled forces.

For the 04 poles stator the linearized expressions are the same, $f_{x,y} = k_p x(y) + k_i i_x(i_y)$, even though the fluxes are coupled; the important issue is that k_p and k_i have higher values than in the previous case and this allows AMBs to provide stiffer suspensions to rotors [6]. It seems safe to state that the 04 poles structure shows a cleaner and more compact design, which will probably result in more cost-effective manufacturing situations. It is also easy to accept that it offers more space for heat dissipation and that the flux losses in its coils are smaller. If all of these theoretical considerations turn out to be true, then the four poles geometry must be seen as a valid alternative to, if not as a better choice than, traditional AMBs [6].

As already mentioned, AMBs with 06 or 03 poles are harder to find, and the specific literature is not so vast [7, 1, 11, 8]. The main goal of this paper is to investigate whether the intriguing relations shown to exist between 08 and 04 poles AMBs do also appear in the symmetry of 3 geometries, between 06 and 03 poles AMBs. Is one of them "better" or stiffer than the other one?

2 Positioning a Body in a Plane with MDs

It is desired to drive a metallic particle to a point in the plane and to keep it there, with forces generated by Magnetic Devices (MDs). Such devices transform electric currents in forces that are applied to some mechanical load.

Planar positioning with 08 or 04 MDs is well known and provides basic examples of AMBs with symmetry of 4. Figure 2 depicts a particle at the origin of a coordinate system with 03 MDs symmetrically placed. An easy way to represent MDs is used in this figure; in real life, most MDs are either cylindrically or U-shaped pieces of a ferromagnetic material surrounded by coils of wire.

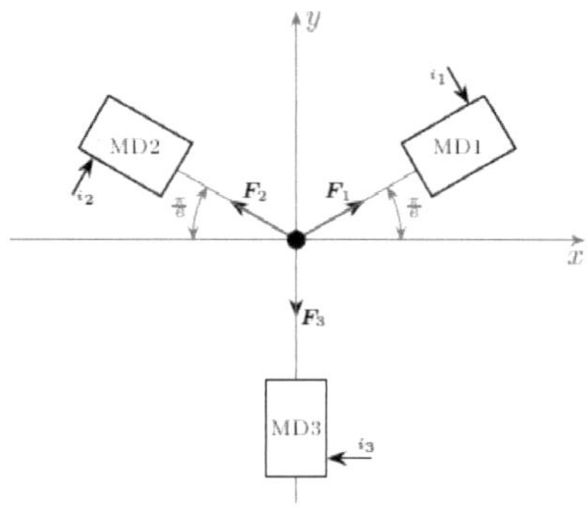

Figure 2: Three MDs are symmetrically placed around the centered particle.

The coordinates $x(t)$ and $y(t)$ (or the rotor position, in the AMBs case) are controlled by the forces $\boldsymbol{F_k}(t)$ for $k = 1,2,3$. The directions of these forces, see figure 2, can be considered fixed, for small displacements; projecting their absolute values in the x and y directions leads to

$$\sum_{k=1}^{3} F_k^x(t) = F_1(t)\cos\frac{\pi}{6} - F_2(t)\cos\frac{\pi}{6} + F_3(t)\cos\frac{3\pi}{2} = F^x(t) \tag{1}$$

$$\sum_{k=1}^{3} F_k^y(t) = F_1(t)\sin\frac{\pi}{6} + F_2(t)\sin\frac{\pi}{6} + F_3(t)\sin\frac{3\pi}{2} = F^y(t). \tag{2}$$

The dynamic characteristics of the particle (and also of the rotor) movements depend, basically, on the modules of the $\boldsymbol{F_k}(t)$. It is well known [9, 2, 10] that forces produced in MDs depend on the square of the magnetic flux and the area occupied by it, as in

$$F_k(t) = \frac{\phi_k^2(t)}{2\mu_0 A_k} \quad \text{for} \quad k = 1, 2, 3, \tag{3}$$

where μ_0 is the air magnetic permeability. For equal areas A_k each

force depends only on the magnetic flux $\phi_k(t)$. This flux is squared in equation (3) and thus MDs generate attractive forces only. In some cases, $\phi_k(t)$ is affected only by the current $i_k(t)$ and by the distance d_k between the body and a point in MDk, usually the closest and most internal one. This case is fairly easy to study but in other situations fluxes in each channel depend on the parameters of all MDs:

$$\phi_k(t) = \varphi_k\left(i_1(t), i_2(t), i_3(t), d_1(t), d_2(t), d_3(t)\right) \quad \forall k = 1, 2, 3.$$

Symmetry assures that the distance *h* between the origin and each actuator is the same. Let *ek* measure the body's displacement in the *k* axis, considered positive when pointing towards MDk. Figure 3 shows the situation for channel 1. When the load dimensions are significant, as in AMBs, *h* is the nominal gap width. In the general case, it is clear that

$$d_k(t) = h - e_k(t) \quad \text{for} \quad k = 1, 2, 3. \tag{5}$$

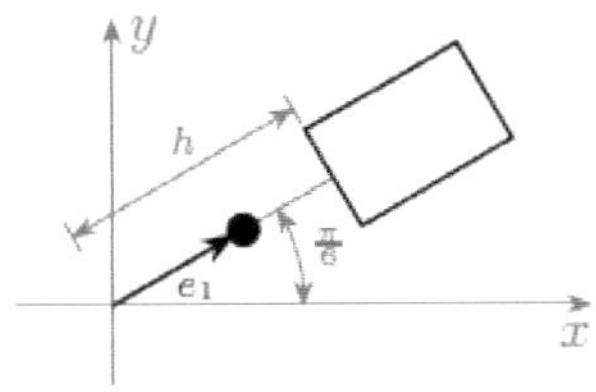

Figure 3: Particle's displacement towards MD1; if the real position is not as above, it should be projected on the line from the origin to the i actuator.

It is easier to measure the *ek*(*t*), the body's displacements from the origin than its distances *dk*(*t*) to the MDs. Therefore, equation (4) can be updated.

$$\phi_k(t) = \varphi_k\left(i_1(t), i_2(t), i_3(t), e_1(t), e_2(t), e_3(t)\right) \quad \forall k = 1, 2, 3. \tag{6}$$

The body's motion depends on 03 injected currents in the MDs, and on 03 positions to be measured. The displacements *e*1,*e*2,*e*3 are redundant to determine the position in the plane, only two numbers are required for this. It is natural to assume that only two displacement sensors are installed in practical implementations, covering the canonical and orthogonal directions *x* and *y*.

When the particle is not located at a particular position in the plane, as in figure 3, the vectors ***x***(*t*) and ***y***(*t*) must be projected on the symmetry of 3 directions as shown in figure 4.

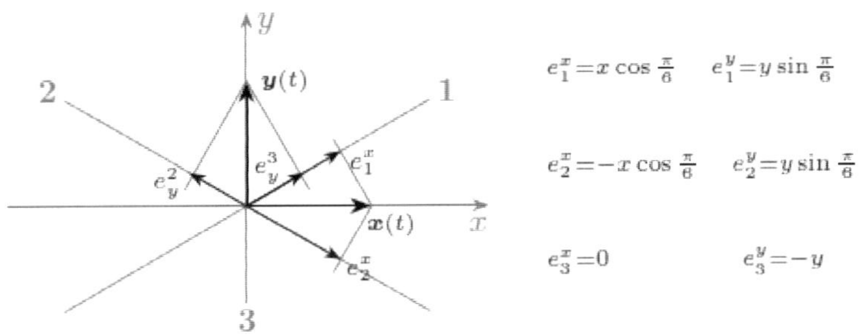

Figure 4: How displacements measured in the *x* and *y* directions are felt in the symmetrical axes 1, 2, 3 of the Mds.

The *ek*(*t*) in figure 4 are related to the measurable quantities *x*(*t*) and *y*(*t*) with the help of

$$e_1(t) = x(t)\cos\frac{\pi}{6} + y(t)\sin\frac{\pi}{6} \tag{7}$$

$$e_2(t) = -x(t)\cos\frac{\pi}{6} + y(t)\sin\frac{\pi}{6} \tag{8}$$

$$e_3(t) = x(t)\cos\frac{3\pi}{2} + y(t)\sin\frac{3\pi}{2} = -y(t) \tag{9}$$

The quantities *dk* = *h* - *ek* using these values are approximations to the real distances between the body and the MDs. Since these displacements are small, the approximations are accepted and used without any care. There are, now, only 05 variables affecting the behavior of fluxes in MDs and the obtained resultant forces. Equations (4) and (6) can be further improved

$$\phi_k(t) = \varphi_k\left(i_1(t), i_2(t), i_3(t), x(t), y(t)\right) \quad \forall k = 1, 2, 3. \tag{10}$$

To analyze the action of the currents $i_k(t)$ and whether it is possible to position the particle with less than 3 control variables, particular cases will be considered in the next sections.

3 AMBs with uncoupled fluxes

A typical illustration is shown in figure 5. The stator is a cylindrical piece of ferromagnetic material in which three pairs of poles can be recognized. The word pole used here has no magnetic meaning, it indicates parts of the stator that point towards the rotor.

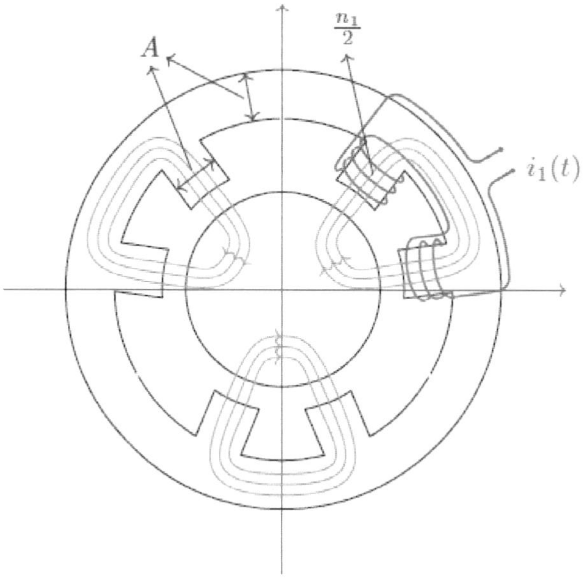

Figure 5: 06 poles configuration for AMBs; windings are shown for channel 1 only; there are no connections among the flux paths. All pairs of windings control the *x* and *y* positions.

Each pair of poles is surrounded by coils carrying currents and performs as an independent U-shaped electromagnet. The three MDs generating forces are autonomous because all the flux injected in the rotor by one of the poles is absorbed by the other pole in the same channel. Each of the fluxes ϕ_k follows a path in the plane that does not touch or intercept the fluxes by the other channels: they are said to be uncoupled, decoupled, or unconnected. Basic results on reluctance forces [9, 2, 10] can be used to find the magnitude of the force generated at each MD:

$$F_k(t) = \frac{\mu_0 A_k n_k^2}{4}\left(\frac{i_k(t)}{d_k(t)}\right)^2 \quad \text{for} \quad k = 1, 2, 3. \tag{11}$$

where μ_0 is the air magnetic permeability. Assuming a uniformly built stator, $A_1 = A_2 = A_3 = A$ and $n_1 = n_2 = n_3 = n$; using equation (5):

$$F_k(t) = K\left(\frac{i_k(t)}{h - e_k(t)}\right)^2 \quad \text{with} \quad K = \frac{\mu_0 A n^2}{4} \quad \text{for} \quad k = 1, 2, 3. \tag{12}$$

The forces on the rotor, in the *x* and *y* directions, come from equations (1) and (2); the result, omitting the (*t*) in the time-varying signals, is

$$F^x = (F_1 - F_2)\cos\frac{\pi}{6} = \frac{K\sqrt{3}}{2}\left[\left(\frac{i_1}{d_1}\right)^2 - \left(\frac{i_2}{d_2}\right)^2\right] \tag{13}$$

$$F^y = (F_1 + F_2)\sin\frac{\pi}{6} - F_3 = \frac{K}{2}\left[\left(\frac{i_1}{d_1}\right)^2 + \left(\frac{i_2}{d_2}\right)^2 - 2\left(\frac{i_3}{d_3}\right)^2\right]. \tag{14}$$

Recalling equations (7), (8) and (9) and since $d_k = h - e_k$, the three distances in the above formulas depend on the displacements x and y only:

$$d_1 = h - \frac{\sqrt{3}}{2}x - \frac{1}{2}y; \quad d_2 = h + \frac{\sqrt{3}}{2}x - \frac{1}{2}y; \quad d_3 = h + y. \tag{15}$$

AMBs operate close to a situation of no displacements and zero external inputs and so it is natural to linearize (13) and (14) around $(i_1^0, i_2^0, i_3^0, x^0, y^0) = (0,0,0,0,0)$. The result would be an uncontrollable model for the AMB. This suggests the use of procedures that work successfully in the symmetry of 04 world: base and differential currents. It is reasonable to consider a fixed base current i_0 and three differential currents $v_k(t)$ to be rigged as

$$i_1(t) = i_0 + v_1(t); \qquad i_2(t) = i_0 + v_2(t); \qquad i_3(t) = i_0 + v_3(t). \tag{16}$$

Equations (15) and (16) have similar structures; the first one shows how the 03 parameters d_k can be expressed in terms of x and y. What if the same thing could be done to the 03 $v_k(t)$ in the last equation? Let's postulate two differential currents $i_x(t)$ and $i_y(t)$ and, guided by (15), choose $v_1 = i_x + i_y$, $v_2 = -i_x + i_y$ and $v_3 = -i_y$. This leads to

$$i_1(t) = i_0 + i_x(t) + i_y(t), \quad i_2(t) = i_0 - i_x(t) + i_y(t), \quad i_3(t) = i_0 - i_y(t). \tag{17}$$

The actions of three differential currents would be done by only two... Will this scheme work? will i_x and i_y control the x and y directions? The (positive) answers to these questions depend on a detailed study of equation (16) that will not be done here.

3.1 Base and differential currents

A mathematical model for the force generation in 06 poles AMBs can be obtained by combining equations (13), (14), (15) and (16). The net result is

$$F^x = f(v_1, v_2, v_3, x, y) \qquad \text{and} \qquad F^y = g(v_1, v_2, v_3, x, y). \tag{18}$$

The traditional procedure for linearizing around the operating point OP = $(v_1^o, v_2^o, v_3^o, x^o, y^o) = (0,0,0,0,0)$ leads to

$$F^x - [F^x]_o = \sum_{k=1}^{3}\left[\frac{\partial f}{\partial v_k}\right]_o (v_k - v_k^o) + \left[\frac{\partial f}{\partial x}\right]_o (x - x^o) + \left[\frac{\partial f}{\partial y}\right]_o (y - y^o) \tag{19}$$

$$F^y - [F^y]_o = \sum_{k=1}^{3}\left[\frac{\partial g}{\partial v_k}\right]_o (v_k - v_k^o) + \left[\frac{\partial g}{\partial x}\right]_o (x - x^o) + \left[\frac{\partial g}{\partial y}\right]_o (y - y^o) \tag{20}$$

A thorough calculation of the parameters and derivatives results in

$$F^x = k_p x + \sqrt{3}k_v(v_1 - v_2) \quad \text{and} \quad F^y = k_p y + k_v(v_1 + v_2) - 2k_v v_3 \quad \text{where} \tag{21}$$

$$k_p = \frac{3Ki_0^2}{h^3} = 3\frac{\mu_0 A n^2 i_0^2}{4h^3} \qquad \text{and} \qquad k_v = \frac{Ki_0}{h^2} = \frac{\mu_0 A n^2 i_0}{4h^2}. \tag{22}$$

The wild guess of a previous paragraph, $v_1 = i_x + i_y$, $v_2 = -i_x + i_y$ and $v_3 = -i_y$ transforms the equations above to

$$F^x = k_p x + 2\sqrt{3}k_v i_x \quad \text{and} \quad F^y = k_p y + 4k_v i_y \tag{23}$$

showing a completely decoupled structure: this is a nice and desired fact. To understand how these magnetic forces depending on 03 (or 02) differential currents act on the rotor dynamics, a detailed mathematical model is needed; this will be not covered in this article. In the next section one of the possible symmetry of 3 correspondents of a 04 poles structure will be studied.

4 AMBs with 03 poles and coupled fluxes

A typical illustration is shown in figure 6. The stator is a cylindrical piece of ferromagnetic material from which three single poles point inwards, in the direction of the rotor. The same considerations made previously about the word pole are valid here and in the rest of the article.

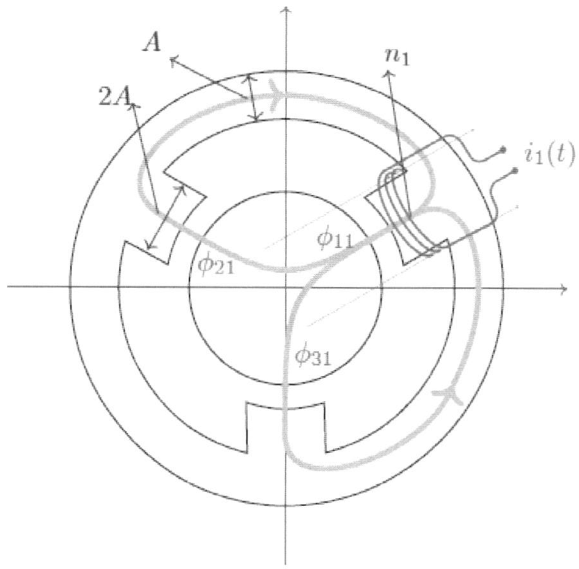

Figure 6: 03 poles configuration for AMBs. Windings are shown for channel 1 only; the flux generated in this channel enters the other poles. All windings control the *x* and *y* positions.

The ferromagnetic connections in a 03 poles structure allow a current injected in any winding to cause fluxes in all three air gaps; figure 6 illustrates the effects of i_1. If ϕ_{jk} denotes the flux in air gap *j* caused by a current in winding *k*, the total magnetic flux ϕ_1 in pole 1 depends on the fluxes ϕ_{11},ϕ_{12},ϕ_{13}. Assuming no air or ferromagnetic losses and positive signs for fluxes headed to the rotating center, the total magnetic fluxes in the poles are:

$$\phi_1 = \phi_{11} - \phi_{12} - \phi_{13}, \quad \phi_2 = -\phi_{21} + \phi_{22} - \phi_{23}, \quad \phi_3 = -\phi_{31} - \phi_{32} + \phi_{33}. \tag{24}$$

For the determination of the ϕ_{jk}, consider the circuit in figure 7 that models the magnetic flux situation. The magneto-motive force generated by i_1 is denoted by $\mathcal{F}_1$ and the reluctance of the air gaps by $\mathcal{R}_1$, $\mathcal{R}_2$, and $\mathcal{R}_3$.

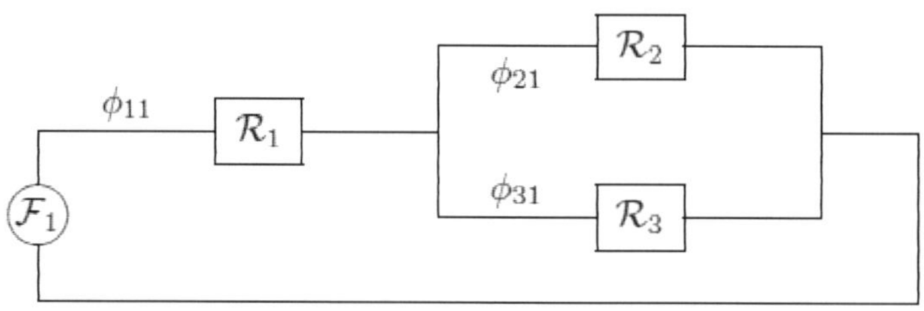

Figure 7: Magnetic flux equivalent circuit associated with current $i_1(t)$.

Recalling that 2*A* is the cross-section area of the poles and that the displacements $d_k = h - e_k$ are explained in figure 3, the reluctances are:

$$\mathcal{R}_1 = \frac{d_1}{\mu_0 2A}, \qquad \mathcal{R}_2 = \frac{d_2}{\mu_0 2A}, \qquad \mathcal{R}_3 = \frac{d_3}{\mu_0 2A}. \tag{25}$$

A simple way to study magnetic circuits is by using the passive electric circuit analogy: fluxes are treated as currents, reluctances act as resistances and the usual operations allowed in Kirchoff's rules are valid. Reluctance $\mathcal{R}_1$

above is in series with the parallel combination of $\mathcal{R}_2$ and $\mathcal{R}_3$. If $\mathcal{R}_*$ denotes the equivalent reluctance of the parallel combination, then the magnetic circuit in figure 7 can be replaced by a single reluctance $\mathcal{R}_1^e$ given by:

$$\mathcal{R}_1^e = \mathcal{R}_1 + \mathcal{R}_* = \mathcal{R}_1 + \frac{\mathcal{R}_2\mathcal{R}_3}{\mathcal{R}_2 + \mathcal{R}_3} = \frac{\mathcal{R}_1\mathcal{R}_2 + \mathcal{R}_1\mathcal{R}_3 + \mathcal{R}_2\mathcal{R}_3}{\mathcal{R}_2 + \mathcal{R}_3} \tag{26}$$

To avoid cumbersome formulas, some auxiliary variables are defined:

$$N = \mathcal{R}_1\mathcal{R}_2 + \mathcal{R}_1\mathcal{R}_3 + \mathcal{R}_2\mathcal{R}_3, \qquad D = d_1 d_2 + d_1 d_3 + d_2 d_3, \tag{27}$$

$$D_1 = \mathcal{R}_2 + \mathcal{R}_3, \qquad D_2 = \mathcal{R}_1 + \mathcal{R}_3, \qquad D_3 = \mathcal{R}_1 + \mathcal{R}_2. \tag{28}$$

Assuming the same number of coils *n* in each pole and using again the imposed currents as in (16), algebraic operations lead to expressions for the fluxes generated by Pole 1 of figure 6:

$$\phi_{11} = ni_1\frac{D_1}{N}, \qquad \phi_{21} = ni_1\frac{\mathcal{R}_3}{N}, \qquad \phi_{31} = ni_1\frac{\mathcal{R}_2}{N}. \tag{29}$$

The same procedure, repeated for currents i_2, i_3 imposed at the windings in Poles 2, and 3 in figure 6 results in:

$$\phi_{12} = ni_2\frac{\mathcal{R}_3}{N}, \qquad \phi_{22} = ni_2\frac{D_2}{N}, \qquad \phi_{32} = ni_2\frac{\mathcal{R}_1}{N},$$

$$\phi_{13} = ni_3\frac{\mathcal{R}_2}{N}, \qquad \phi_{23} = ni_3\frac{\mathcal{R}_1}{N}, \qquad \phi_{33} = ni_3\frac{D_3}{N}.$$

Using these values of the partial fluxes ϕ_{jk} in equation (24) the total fluxes ϕ_k for $k = 1,2,3$ can be determined. Then, with the help of (3), the total reluctance forces generated in a 03 poles AMB can be expressed as:

$$F_k = \mu_0 A n^2 \left(\frac{N_k}{D}\right)^2 \quad \text{for} \quad k = 1, 2, 3. \tag{30}$$

where D comes from (27) and, using $I_{jk} = i_j - i_k$,

$$N_1 = d_2 I_{13} + d_3 I_{12} \qquad N_2 = d_1 I_{23} + d_3 I_{21} \qquad N_3 = d_1 I_{32} + d_2 I_{31}. \tag{31}$$

These last results together with equations (1) and (2) can be used to find the resultant forces in the *x* and *y* directions:

$$F_x = \frac{\sqrt{3}}{2}(F_1 - F_2) = \cdots = K_x\frac{N_1^2 - N_2^2}{D^2} \tag{32}$$

$$F_y = \frac{1}{2}(F_1 + F_2) - F_3 = \cdots = K_y\frac{N_1^2 + N_2^2 - 2N_3^2}{D^2} \tag{33}$$

where $K = \mu_0 A n^2/2$, K_x = Ksqrt{3}/2 and $K_y = K/2$. *It is clear that the forces in the canonical directions*

$$F_x = K_x q_x(N_1, N_2, N_3, D) \qquad \text{and} \qquad F_y = K_y q_y(N_1, N_2, N_3, D) \tag{34}$$

depend on geometric and excitation parameters defined on the symmetry of 3 axes: the positions d_1, d_2, d_3 and the currents i_1, i_2, i_3. Equations (7), (8), (9) together with (5) lead to

$$d_1 = h - \frac{\sqrt{3}}{2}x - \frac{1}{2}y, \qquad d_2 = h + \frac{\sqrt{3}}{2}x - \frac{1}{2}y, \qquad d_3 = h + y. \tag{35}$$

This means that $D = d_1d_2 + d_1d_3 + d_2d_3$ and N_k depend on the canonical parameters *x* and *y*: $D = D(x,y)$ and $N_k = N_k(i_1,i_2,i_3,x,y)$. The resultant forces can be expressed as

$$F_x = K_x q_x(i_1, i_2, i_3, x, y) = f(i_1, i_2, i_3, x, y) \tag{36}$$

$$F_y = K_y q_y(i_1, i_2, i_3, x, y) = g(i_1, i_2, i_3, x, y). \tag{37}$$

The next step, by the script used in the 06 poles case, would be to use the traditional base and differential currents idea, $i_k(t) = i_0 + v_k(t)$, and linearize the above expressions. This scheme fails, unfortunately, so a more detailed study of the problem will be made in the next section.

4.1 Generalized base and differential currents

The injected currents will be

$$i_1(t) = b_1 + v_i(t), \qquad i_2(t) = b_2 + v_2(t), \qquad i_3(t) = b_3 + v_3(t). \tag{38}$$

The base currents are still constant but not the same anymore. Different base currents can be used in horizontal rotors AMBs [1] but the idea here is to study how the 03 poles case linearization behaves in this general situation. Equations (32), (33) and (34) are now turned into

$$F_x = f(v_1, v_2, v_3, x, y), \qquad F_y = g(v_1, v_2, v_3, x, y). \tag{39}$$

The linearization around the operating point OP = ($v1^o$,$v2^o$,$v3^o$,x^o,y^o) = (0,0,0,0,0) relies on equations (19) and (20). A careful calculation of the parameters and derivatives, using Bjk to denote bj - bk and fz^o to denote the partial derivative of f with respect to z evaluated at the operating point leads to

$$f^o_{v_1} = \frac{2K_x}{3h^2}B_{12}, \qquad f^o_{v_2} = \frac{-2K_x}{3h^2}B_{23}, \qquad f^o_{v_3} = \frac{2K_x}{3h^2}B_{21}, \tag{40}$$

$$f^o_x = \frac{K_x\sqrt{3}}{9h^3}\left[B^2_{12} + B^2_{13} + B^2_{23}\right], \qquad f^o_y = 0, \tag{41}$$

$$g^o_{v_1} = \frac{2K_y}{3h^2}(B_{12} + B_{32}), \quad g^o_{v_2} = \frac{2K_y}{3h^2}(B_{21} + B_{31}), \quad g^o_{v_3} = \frac{2K_y}{3h^2}(B_{13} + B_{23}), \tag{42}$$

$$g^o_x = 0, \qquad g^o_y = \frac{2K_y}{9h^3}\left[2B^2_{12} + B^2_{13} + B^2_{23} + b_3(b_3 - b_1 - b_2) + b_1b_2\right]. \tag{43}$$

Useful linearizations for AMBs require that Fx0=Fy0=0. The calculations for this case lead to the necessary conditions

$$\left[F_x\right]_o = 0 \iff (b_1 + b_2 - 2b_3)(b_1 - b_2) = 0, \tag{44}$$

$$\left[F_x\right]_o = 0 \iff (b_1 - 2b_2 + b_3)(b_1 - b_3) + (b_2 - 2b_1 + b_3)(b_2 - b_3) = 0. \tag{45}$$

The chosen values for the bk must satisfy the above equations; in order to make (44) true there are two possibilities:

Choice 01: $b1$ - $b2$ = 0 or $b1$ = $b2$. Plugging this in (45) leads to $b3$ = $b1$ and the overall choice is $b1$ = $b2$ = $b3$ = bo the same base current for all injected currents. Using this in equations (40), (41), (42) and (43) shows us that no linearization can be achieved, because

$$f^o_{v_1} = f^o_{v_2} = f^o_{v_3} = f^o_x = f^o_y = g^o_{v_1} = g^o_{v_2} = g^o_{v_3} = g^o_x = g^o_y = 0. \tag{46}$$

Choice 02: $b1$ + $b2$ - 2$b3$ = 0 or $b3$ = ($b1$ + $b2$)/2. Using this in (45) leads, after some manipulations, to ... $b1$ - $b2$ = 0.

The sad conclusion is that no linearization is possible; the fluxes are so coupled and interconnected in the 03 poles geometry that any practical utilization of them is jeopardized. A parallel physical observation of the 03 poles structure can put reasonable doubts about its validity; the above mathematical procedures seems to give a definite answer to this question. The uncoupled/coupled comparisons between 08 and 04 poles can still be made in the symmetry of 3 world, because there is yet another possible structure for AMBs with magnetic forces separated by (2π/3)rd angles; it will be presented and partially studied in the next section.

5 AMBs with 06 poles and coupled fluxes

A typical illustration is shown in figure 8. The stator is a cylindrical piece of ferromagnetic material from which six equally spaced single poles point toward the rotor. Each pair of diametrically opposed poles forms a channel; poles 1 and 4, 2 and 5, 3 and 6 are responsible for channels 1, 2, 3. The poles in channel k are fed with currents $i_0 \pm v_k$.

The ferromagnetic structure in a 06 poles with coupled fluxes allows currents injected in any winding to cause fluxes in all air gaps; figure 8 illustrates the effects of $i_1(t) = i_0 + v_1(t)$. If ϕ_{jk} denotes the flux in air gap *j* caused by a current in winding *k*, the total magnetic flux ϕ_1 in pole 1 depends on the fluxes $\phi_{11}, \phi_{12} \dots \phi_{16}$. Assuming no air or ferromagnetic losses and positive signs for fluxes headed to the rotating center, the total magnetic fluxes in the poles are:

$$\phi_1 = \phi_{11} - \phi_{12} - \phi_{13} - \phi_{14} - \phi_{15} - \phi_{16}, \ \phi_2 = -\phi_{21} + \phi_{22} - \phi_{23} - \phi_{24} - \phi_{25} - \phi_{26}, \tag{47}$$

$$\phi_3 = -\phi_{31} - \phi_{32} + \phi_{33} - \phi_{34} - \phi_{35} - \phi_{36}, \phi_4 = -\phi_{41} - \phi_{42} - \phi_{43} + \phi_{24} - \phi_{45} - \phi_{46}, \tag{48}$$

$$\phi_5 = -\phi_{51} - \phi_{52} - \phi_{53} - \phi_{54} + \phi_{35} - \phi_{56}, \phi_6 = -\phi_{61} - \phi_{62} - \phi_{63} - \phi_{64} - \phi_{65} + \phi_{66}. \tag{49}$$

More about this will be presented in a future article.

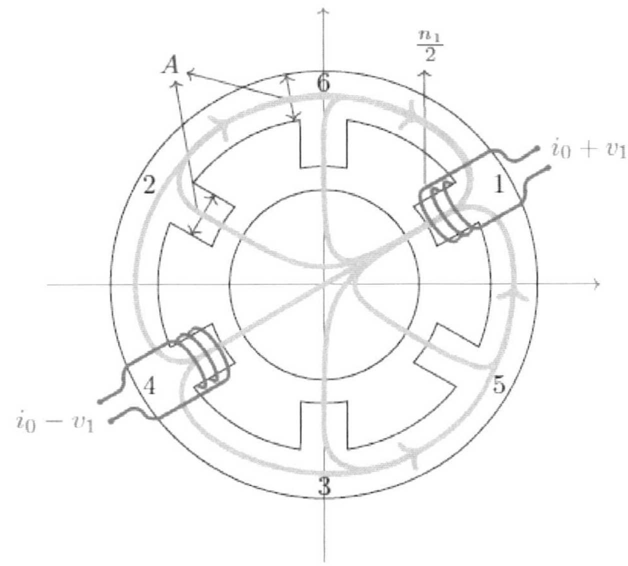

Figure 8: 06 poles and coupled fluxes configuration for AMBs. Windings are shown for channel 1 — poles 1 and 4 — only. The flux generated in any pole enters all the other poles; flux emanating from pole 1 is shown.

6 Conclusions

In the symmetry of 4 world 04 poles devices work as good as, if not better than, 08 poles ones. It was expected that the same would happen in the symmetry of 3 world with 06 and 03 poles, but the above paragraphs show that this is not the case. What remains to be investigated is whether the 03 poles with coupled fluxes structure plays in the symmetry of 3 world a role that corresponds to the 04 poles devices; and, in a case of a positive answer, can AMBs built based on this geometry be of any practical value?

7 Acknowledgements

The authors are very happy to express their deep gratitude to Prof. Richard Magalena Stephan for his continuous presence, friendship and solid technical bits of advice.

References

[1] S. Chen and C. Hsu. Optimal design of a three-pole active magnetic bearing. *IEEE Transactions on Magnetics*, 38:3458–3466, September 2002.

[2] A. Chiba, T. Fukao, O. Ichikawa, M. Oshima, M. Takemoto, and D. Dorrell. *Magnetic Bearings and Bearingless Drives*. Newnes-Elsevier, 2005.

[3] D. F. B. David, J. A. Santisteban, and A. C. Del Nero Gomes. Interconnected four poles magnetic bearings. In *Proceedings of the 1st Brazilian Workshop on Magnetic Bearings*, www.magneticbearings2103.com.br, october 2013.

[4] D. F. B. David, J. A. Santisteban, and A. C. Del Nero Gomes. Interconnected four poles magnetic bearings simulations and testing. In *Proceedings of ISMB14, 14th International Symposium on Magnetic Bearings*, pages 30–35, http://ismb14.magneticbearings.org, august 2014.

[5] D. F. B. David, J. A. Santisteban, and A. C. Del Nero Gomes. Laboratory tests on an interconnected four poles magnetic bearing. In *Proceedings of ISMB15, 15th International Symposium on Magnetic Bearings*, august 2016.

[6] D. F. B. David, J. A. Santisteban, and A. C. Del Nero Gomes. Modeling and testing strategies for an interconnected four-pole magnetic bearing. *Actuators*, 6(21), 2017.

[7] N. R. Hemenway and E. L.Severson. Three-pole magnetic bearing design and actuation. *IEEE Transactions on Industry Applications*, 56(Issue: 6):6348–6359, 2020. https://ieeexplore.ieee.org/document/9154570.

[8] Laura J. M. Mothé, Vinicius R. Vasco, Y. P. Corrêa, D. D. F. Brito, and A. C. D. N. Gomes. Model and control of a six-pole magnetic bearing. In *Proceedings of ISMB17, 17th International Symposium on Magnetic Bearings*, http://www.magneticbearings.org/publications, 2021.

[9] G. Schweitzer, H. Bleuler, and A. Traxler. *Active Magnetic Bearings*. Hochshulverlag AG an der ETH Zürich, 1994.

[10] G. Schweitzer, E. Maslen, H. Bleuler, M. Cole, P. Keogh, R. Larsonneur, R. Nordmann, and Y. Ogada. *Magnetic Bearings: Theory, Design and Applications to Rotating Machinery*. Springer-Verlag, 2009.

[11] W. Zhang and H. Zhu. Radial magnetic bearings: an overview. *Results in Physics*, 7:3756–3766, 2017.

PWM Filter for Active Magnetic Bearings

Alexei Filatov[a] and Larry Hawkins[a]
a Calnetix Technologies
16323 Shoemaker Av., Cerritos, CA, 90703 USA, afilatov@calnetix.com, lhawkins@calnetix.com

Abstract

An example of a PWM filter application in an AMB amplifier is discussed. The design objectives included reduction of both conductive and radiated EMI caused by the PWM pulses, improving system efficiency, reliability, and enabling its use with long AMB cables. The selected filter topology included a "Trap" filter, utilizing a parallel LC resonant circuit to suppress output current harmonics near the circuit resonance, followed by a conventional 2nd-order LCR filter. This choice was dictated by a narrow frequency separation between the 25kHz PWM frequency in the existing amplifier and the required 2kHz bandwidth. Introduction of the filter resulted in the attenuation of the PWM signal at the base frequency ranging from -33.2dB to -52.7dB depending on the output current, which was varying from 0A to 10A. The filter practically eliminated high-frequency resonant ringing on the amplifier output and reduced output voltage dV/dt by more than 800 times (from 8,360V/μs to approximately 10 V/μs). Both factors are known to cause EMI and shorten the lives of the actuator magnet wires and the power electronics. The filter enabled the control current delivery to the AMB actuators in a commercial 350kW 15kRPM air compressor over 1.2 km (4,000 ft) cable. The compressor was subsequently levitated and spun to full speed.

Keywords: *Active Magnetic Bearings, PWM Filter, Switching Amplifier, EMI, Long Cable.*

1. Introduction

Active Magnetic Bearings (AMBs) typically use amplifiers with Pulse Width Modulation (PWM), which is the most common type of switching amplifiers that can be found anywhere from audio equipment to motor drives. While there are no practical alternatives to switching amplifiers because of their superior efficiency, they may also create problems when used without output filters. These include:

1. Electrical ground noises caused by high dV/dt associated with the edges of PWM pulses. These noises may affect the operation of other equipment on the same ground, for example the AMB position sensors (W. Santiago, 2004).
2. Radiated electromagnetic emissions due to
   a. Ripple currents.
   b. High-frequency resonant currents flowing in lightly damped LC-loops formed by the load and/or parasitic inductances in tandem with parasitic capacitances and excited by high-frequency content of PWM signals.
3. Reduced lives of actuator magnet wires and power electronics due to voltage spikes (Essex-Furukawa, 2020).
4. Increased eddy currents and hysteresis losses in the lamination stacks.
5. Difficulties of working with long cables (often this would be impossible without a PWM filter). (Essex-Furukawa, 2020) and (A. von Jouanne, et al., 1996).

2. PWM filter design

The PWM filter presented here is designed to work with an existing Calnetix mid-range amplifier which has a 25kHz switching frequency. Considering that many AMB applications require at least 2kHz control bandwidth, the separation between the upper limit of the control bandwidth and the amplifier switching frequency is only 23kHz. To maximize the attenuation of the PWM frequency without affecting the gains and phases of the system transfer functions within the control bandwidth, a filter topology described in (Y. Sozer et al., 2000) was selected. As shown in Fig.1, it consists of a "Trap filter", using a parallel LC resonant circuit to suppress output current harmonics with frequencies near the circuit resonance, followed by a conventional 2nd-order LCR filter. Because of the manufacturing tolerances on the inductances L_t, the trap filter is tuned by adjusting the filter capacitance C_t to closely match the resonant frequency to the amplifier switching frequency. This is a one-time adjustment that is independent of the load dynamics. In addition to the manufacturing tolerances, the current through the inductor also affects the inductance, causing the filter effectiveness

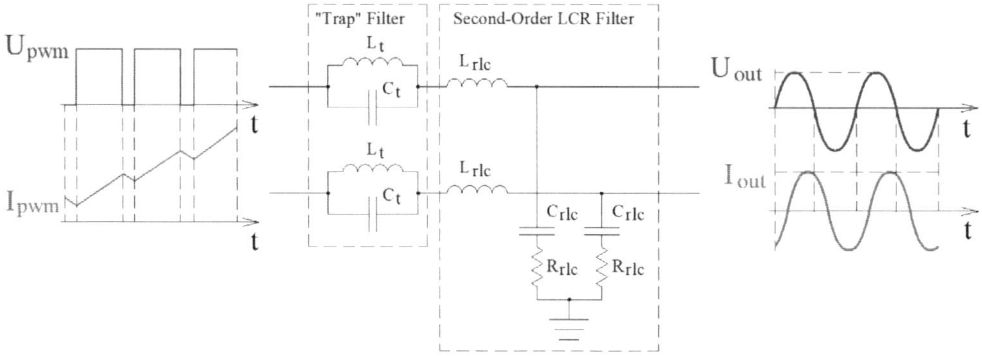

Figure 1. Topology of the PWM Filter.

to be dependent on this current level. Because the amplifier used for testing was rated for 10A continuous current with a typical load consuming less than 5A, the trap filter was tuned for the maximum effectiveness at 5A (-57dB theoretical best-case attenuation at 25kHz). For comparison, a 2nd-order LCR filter alone would yield only -9dB at 25 kHz. However, this simpler "LCR filter only" design would be less sensitive to the current level.

The PWM filter package includes an amplifier current feedback sensor as shown in Fig. 2, which is used in-lieu of the sensor in the standard location at the amplifier output. This allows the output current to follow the commanded current despite the filter dynamics.

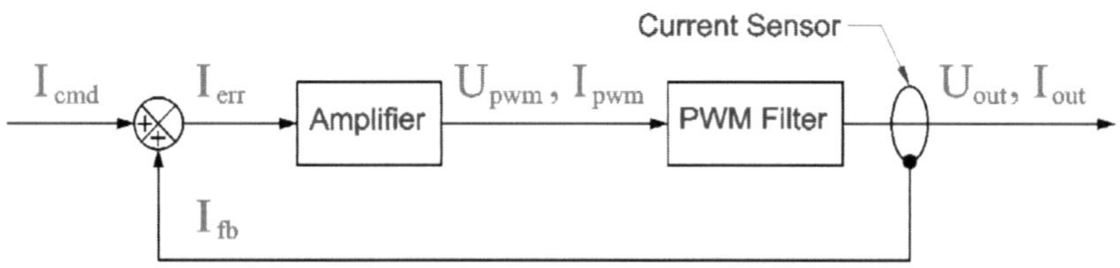

Figure 2. Integration of the PWM Filter into the Amplifier.

Fig. 3 shows a comparison of the voltage transfer functions of a combination "Trap + LCR" filter and a single LCR filter. The main advantage of the "Trap+LCR" filter is a much stronger attenuation of the PWM fundamental frequency, whereas the higher order harmonics, especially the second and the third, are attenuated somewhat less. The difference in the attenuation becomes nearly undistinguishable staring with the fourth harmonic. Stronger attenuation of the fundamental frequency by the "Trap+LCR" filter is very important, because this harmonic is dominant and largely responsible, for example, for the current ripple.

Fig. 4 shows comparison of the first three PWM harmonic amplitudes in the unfiltered voltage (a), after the LCR filter (b) and after the combination "Trap + LCR" filter (c) as functions of PWM duty cycles. When generating the curves shown

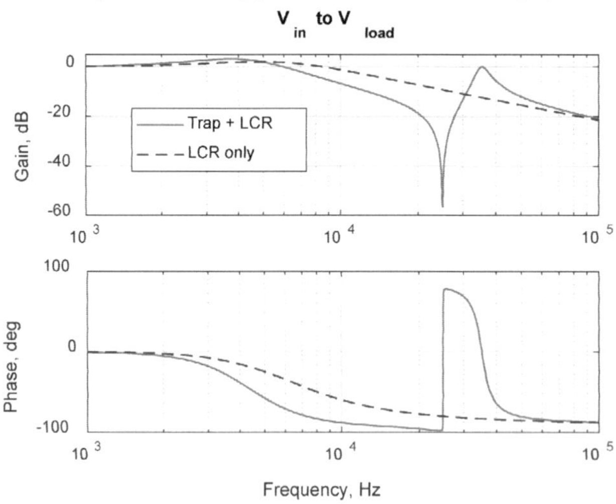

Figure 3. Comparison of the "Trap +LCR" and "LCR only" filter voltage transfer functions.

in Fig. 4, the amplitudes a_n of the PWM harmonics were calculated using equation (1) from (S.W. Smith, 1997):

$$a_n = \frac{2A}{n\pi}\sin(n\pi d) \tag{1}$$

where A is the PWM voltage, n is the harmonic number (n>0), and d is the duty cycle.

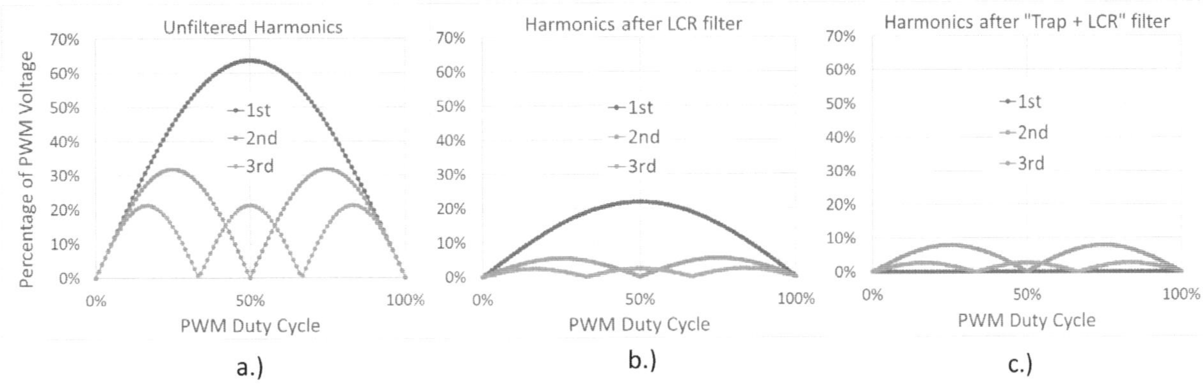

Figure 4. The first three voltage harmonic amplitudes as functions of PWM duty cycles without a filter (a), after the LCR filter (b), after the "Trap+LCR" filter (c).

3. Experimental setup

Fig. 5 shows the experimental setup used to tune the PWM filter and evaluate its performance on a bench. A standard production Magnetic Bearing Control (MBC) box incorporating an amplifier, a control board and a 200V amplifier power supply was used for the testing. The MBC in this setup had only five control channels because it was configured for a machine with all five degrees of freedom controlled by permanent-magnet-biased AMBs (no amplifier bias channels were needed). The PWM filter, however, had six channels, allowing for configurations with an electrically-biased axial AMB.

During tuning and testing, individual amplifier channels were loaded with a 16mH 1.6Ohm test inductor, composed of eight 2mH toroidal inductors connected in series. These inductors had iron powder cores, resulting in much smaller

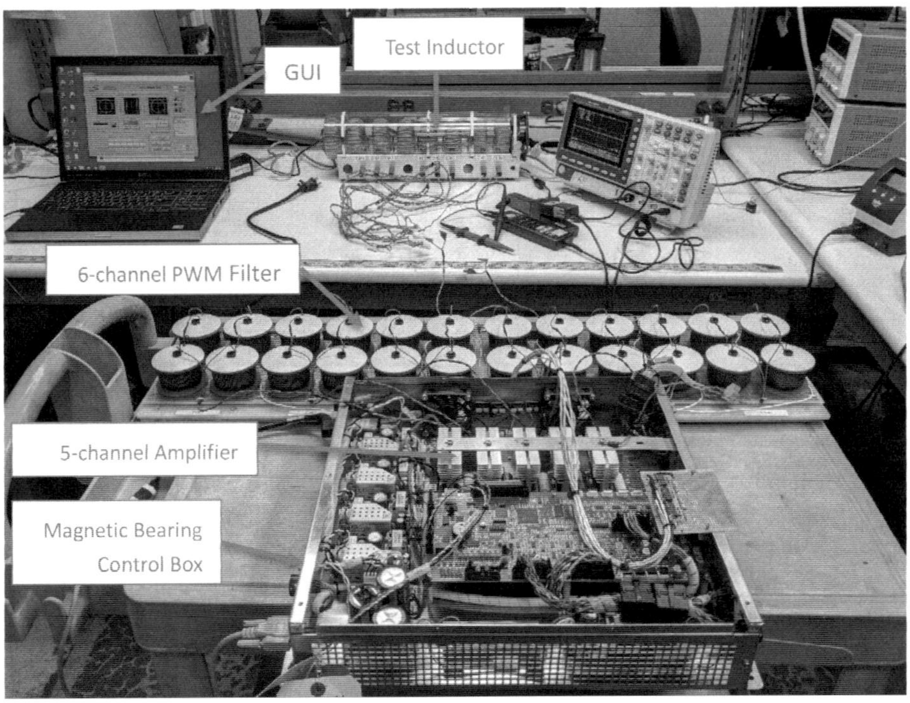

Figure 5. Experimental setup to tune the PWM filter and evaluate its performance on a bench.

eddy-current losses at high frequencies compared to typical AMB actuator coils wound around laminated cores. This represented essentially the worst-case scenario for developing high-frequency ringing on the amplified output induced by harmonics of PWM pulses in the parasitic resonant loops. The ringing would be much smaller if we had used a real AMB actuator with laminated core. A standard AMB control GUI was modified for this test to allow commanding constant duty cycle PWM outputs in addition to the closed-loop operation of the amplifier.

4. Experimental results

4.1. Filter transfer function

Fig. 6 shows overlapped theoretical and measured voltage-to-voltage transfer functions of the proposed filter. The measured transfer function was obtained while outputting 5A DC current, when the maximal PWM harmonic attenuation was expected (-57dB theoretical value at 25kHz). The measured attenuation value at 25kHz in this case was -52.7dB. Because the trap filter inductance values were changing with the output current and the filter was tuned for the resonance to occur close to 25kHz when the output current was 5A, deviations from 5A were leading to lower PWM fundamental harmonic attenuation values. Thus, the attenuation dropped to -36.8dB at 0A and -33.2dB at 10A as shown in Fig. 7 (measured values) – still much higher than the nominal -15dB achievable with the 2nd-order filter only.

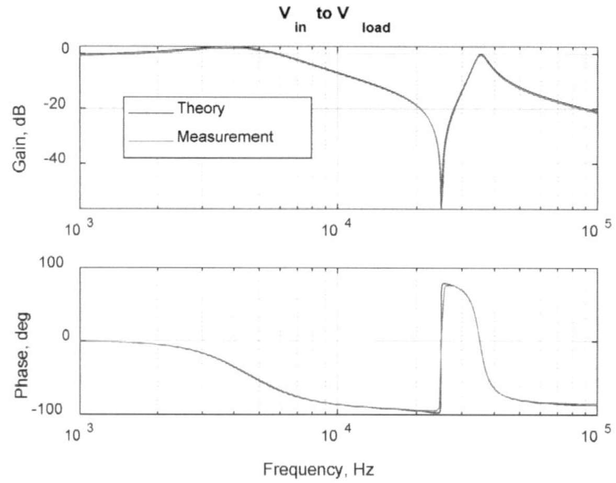

Figure 6. Theoretical and measured with 5A DC output current filter voltage transfer functions.

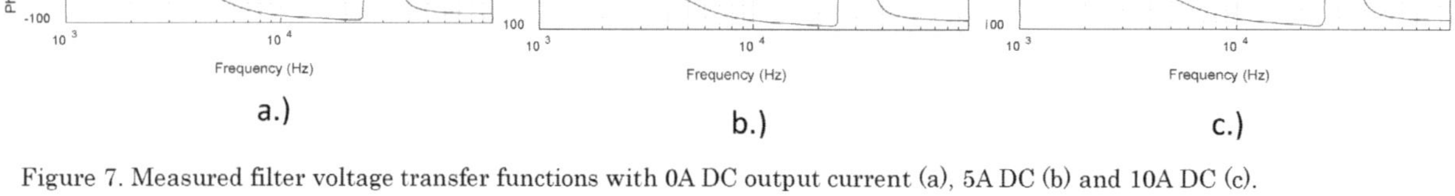

Figure 7. Measured filter voltage transfer functions with 0A DC output current (a), 5A DC (b) and 10A DC (c).

4.2. Open-loop DC-current operation

Figs. 8 and 9 show voltages and currents measured with and without the PWM filter at 5A DC output current level (highest filter effectiveness) and at 10A (lowest filter effectiveness). The amplifier in this measurement was operated in the "open loop" mode, with a DC current level established by manually adjusting the PWM duty cycle. Because the same duty cycle was used with and without the filter, in the latter case the output current was slightly smaller due to the voltage drop on the PWM filter. With both 5A and 10A output currents, the filter effectively suppressed the PWM voltage pulses reducing peak dV/dt by more than 800 times (from 8,360V/μs to approximately 10 V/μs).

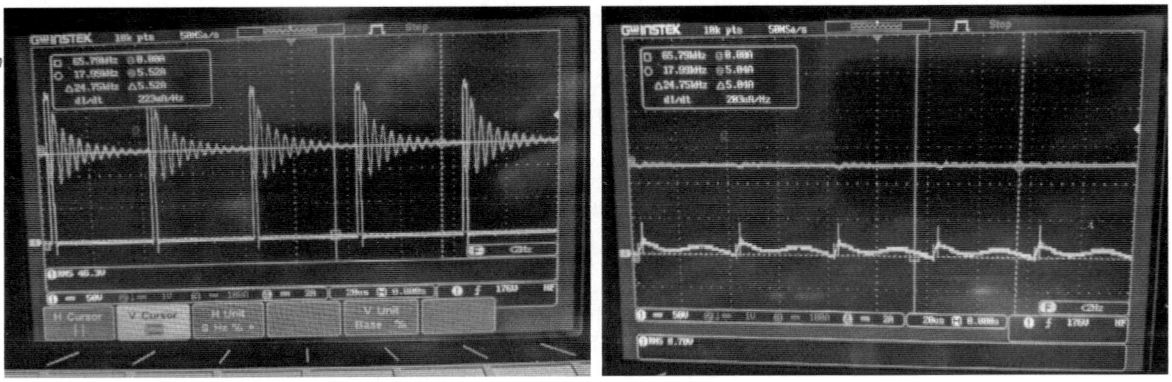

a.) b.)

Figure 8. Measured voltage (yellow) and current (green) on the output of the amplifier without the filter (a) and with the filter (b). The DC output current component was 5.52A without the filter and 5.04A with the filter.

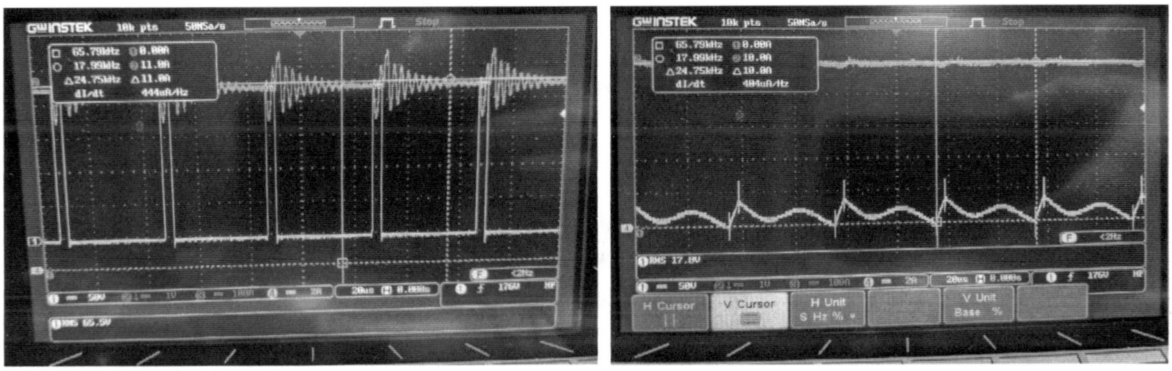

a.) b.)

Figure 9. Measured voltage (yellow) and current (green) on the output of the amplifier without the filter (a) and with the filter (b). The DC output current component was 11.0A without the filter and 10.0A with the filter.

The current waveforms measured without the filter show large oscillations at approximately 250kHz persisting long after the PWM pulses were turned off. Such high-frequency oscillations often occur in parasitic resonant loops on the amplifier output formed by real or parasitic inductances in combination with parasitic capacitances (such as inductor winding capacitances, cable capacitances, amplifier output trace inductances and capacitances, etc.). These oscillations were expected to be unusually large because of low loss level in the test inductors, however the oscillations seen in Figs. 8 and 9 reaching 2A at the peaks are unrealistic, especially if considering that the current probe bandwidth was limited to 100kHz. Most likely these oscillations were electromagnetic noises induced in the scope (or the current probe) by the resonant currents or voltages. Regardless, they disappeared after the introduction of the PWM filter. Fig. 10 shows simulated amplifier output voltage waveforms without the filter (a) and with the filter (b) when outputting 11A DC and

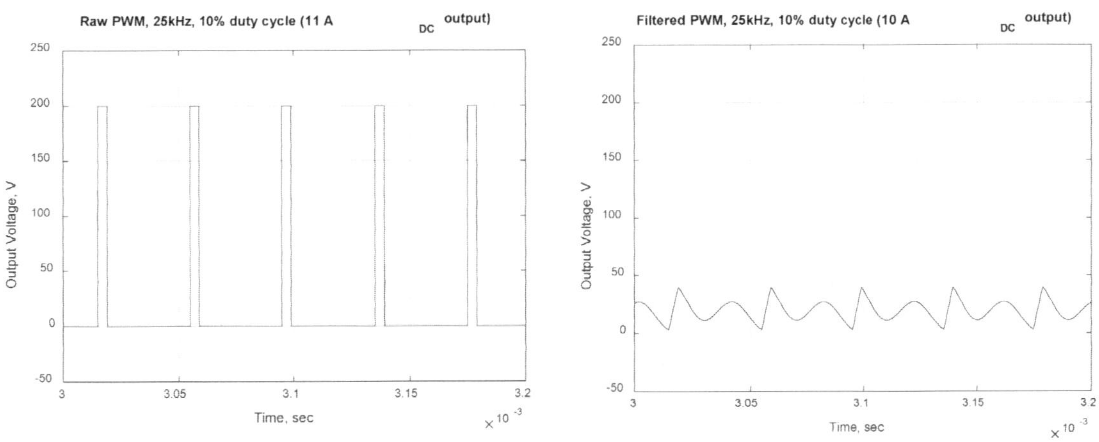

a.) b.)

Figure 10. Simulated amplifier output voltage waveforms without the filter (a) and with the "Trap+LCR" filter (b) when outputting 11A DC and 10A DC respectively. The same 10% duty cycle was used in both cases.

10A DC respectively. The same 10% duty cycle was used in both cases. The simulated filtered voltage waveform Fig. 10b closely resembles the measured ones shown in Fig. 9b.

4.3. Closed-loop AC-current operation

Fig. 11 shows voltages and currents measured after the PWM filter when producing sinusoidal current waveform with 4.4A peak at 100Hz (Fig. 11a) and with 0.8A peak at 1kHz (Fig. 11b) in the closed-loop mode. The current amplitudes at each frequency were limited by the amplifier power supply voltage (200Vdc) combined with the maximum allowable duty cycle (85%). Note that the voltages after the filter are continuous periodical signals rather than trains of PWM pulses and there are no characteristic ripples on the current waveforms. The voltage at 1kHz is nearly sinusoidal, whereas at 100Hz it takes a more complicated shape adjusting to maintain the sinusoidal output current. This is likely because of more pronounced load inductor non-linearity at the larger current achievable at 100Hz.

Fig. 12 shows voltage spectra measured before the PWM filter (a) and after (b) when generating 0.8Apk at 1kHz. The filter does not affect the useful signal at 1kHz, however dramatically reduces PWM sidebands at 13kHz and 16kHz, as well as their harmonics.

Fig. 13 shows simulated amplifier output voltage waveforms when commanding 0.8Apk 1kHz current without the filter (a) with the LCR filter (b) and with the "Trap+LCR" filter (c). Fig. 13c corresponds to the measured waveform shown in Fig. 11b. Clearly the use of the "Trap+LCR" filter (Fig. 13c) results in much cleaner voltage waveform with much smaller ripple compared to the LCR filter alone (Fig. 13b).

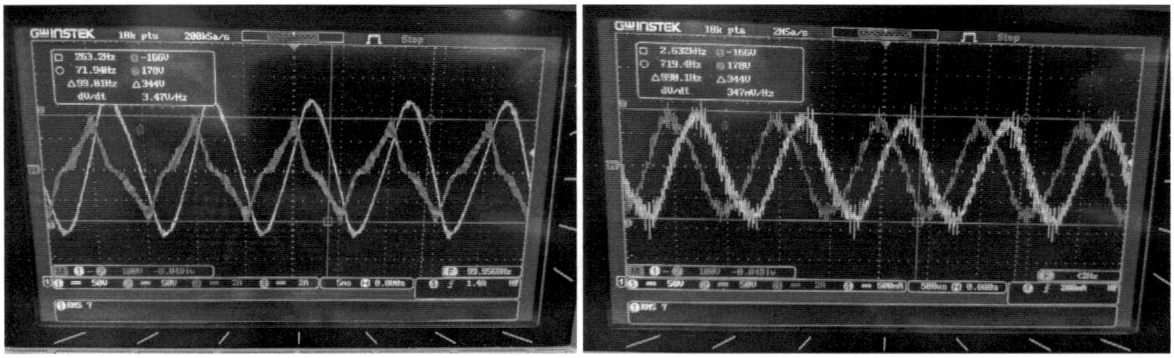

a.) b.)

Figure 11. Measured voltage (red) and current (green) on the output of the amplifier after the filter when producing 4.4Apk 100Hz current (a) and 0.8Apk 1kHz current (b).

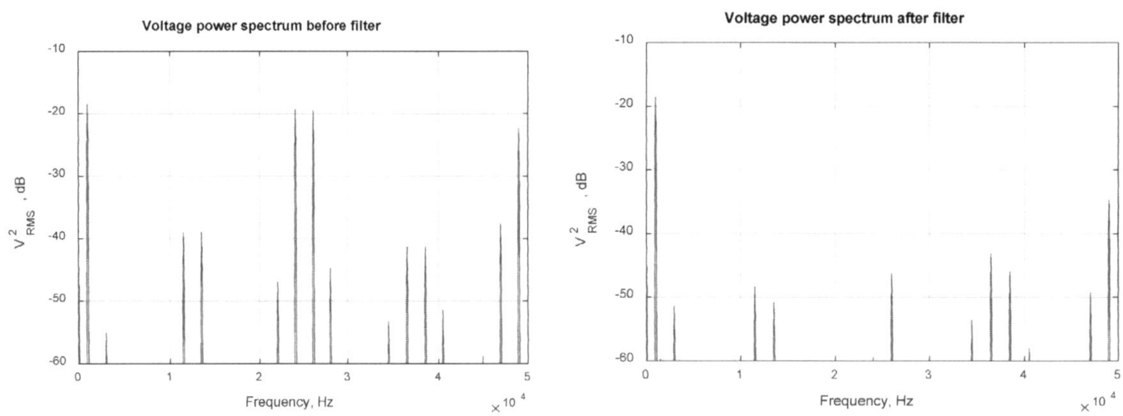

a.) b.)

Figure 12. Measured amplifier output voltage spectra before the filter (a) and after the filter (b) when generating 0.8Apk 1kHz current.

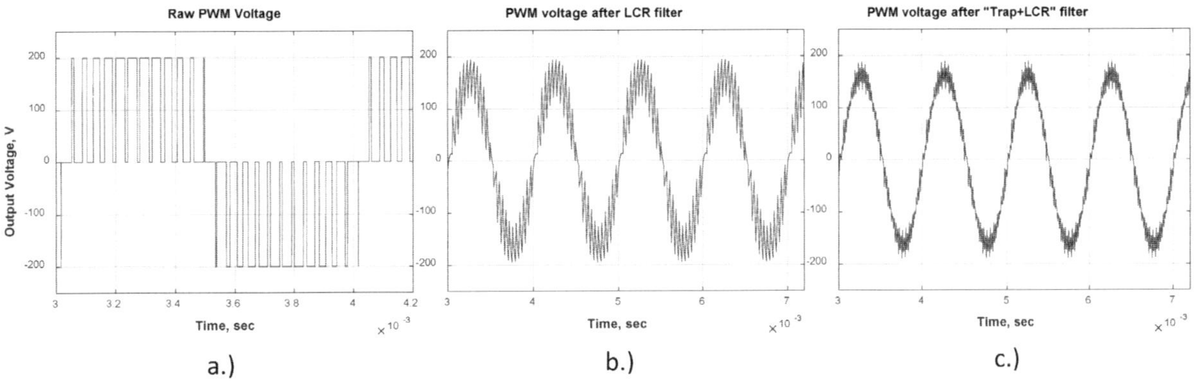

Figure 13. Simulated amplifier output voltage waveforms when commanding 0.8Apk 1kHz current without the filter (a) with the LCR filter (b) and with the "Trap+LCR" filter (c).

4.4. Actuator control over a long cable

Inducing a current in an inductor, such as an AMB actuator winding, using a PWM amplifier becomes challenging when the two are connected by a long cable, which electromagnetic properties cannot be neglected. These properties include specific resistance R_0, inductance L_0 and capacitance C_0 per unit length. For DC voltages and currents, simply more voltage would be needed to drive the same current because of the resistive voltage drop across a cable. The situation becomes much more complicated when high-frequency AC voltages are used, such as PWM pulses. Time-varying electrical signals (voltages and associated currents) propagate along a cable at a specific speed V, which is a fraction of the speed of light in vacuum and is fully defined by the cable parameters L_0 and C_0 (2) (R.K. Moore, 1960).

$$V = \sqrt{\frac{1}{L_0 C_0}} \tag{2}$$

It can be said that all electrical signals travel along a cable as waves with the spatial wavelengths λ inversely dependent on the signal frequencies f as defined by (3). For DC signals $\lambda=\infty$ and this phenomenon can be ignored.

$$\lambda = \frac{V}{f} \tag{3}$$

The level of complications with inducing currents (including DC currents) in an inductor using a PWM amplifier over long cable depends on how close the cable length is to the wavelengths associated with PWM harmonics of significant amplitudes.

1. If the cable length is significantly shorter than the shortest wavelength λ of interest (normally less than ¼ is sufficient), the wave nature of the electrical voltages and currents can be ignored. The cable effect in this case is reduced to adding a lumped cable capacitance in parallel with the inductor. The longer the cable, the bigger the capacitance. When PWM voltage pulses are applied directly to the inductor without a long cable, the current, proportional to an integral of the applied voltage, ends up having a much smoother time waveform than the voltage. In effect, the transfer function from the voltage to the current is an integrator, acting as a low-pass filter. When the cable capacitance is introduced, however, the high-frequency harmonics of the PWM voltage get shorted by this capacitance without reaching the inductor. In addition, the inductor and the cable capacitance form a resonant loop, which may amplify some PWM harmonics.
2. As the cable length approaches the shortest significant spatial wavelength from the PWM spectrum, the voltages and the currents in the cable must be treated as spatial waves and the cable as a "waveguide". Reflected waves may cause the voltages across the amplifier or the inductor to significantly exceed the original amplifier voltage, shortening the life of either the amplifier electronics or the inductor winding, or both. Stationary spatial waves might be also formed with voltage/current peaks and nodes along the cable, which may make AMB control impossible.

In two separate experiments, we used two lines of a 1.2km-long, 3-Phase, 12AWG shielded cable to control currents in either one of the radial channels or the axial channel of a commercial 350kW 15kRPM air compressor. The specific cable inductance and capacitance between the two lines with the shield ungrounded measured at 2kHz (a low frequency expected to be far below the "waveguide" threshold) were found to be 0.578μH/m and 119pF/m respectively. Using (2) and (3), we estimated that the signal frequency at which the wavelength equals four times the cable length is 25.5kHz – almost exactly the 25kHz amplifier switching frequency and lower than many significant higher-order PWM harmonics.

Fig.14 shows the measured transfer functions from the input current into the 1.2km cable to the output current from the cable into the radial actuator control winding (a) or the axial actuator control winding (b). Note that at high frequencies the currents induced in the actuator windings are not equal to the currents injected into the long cable. This,

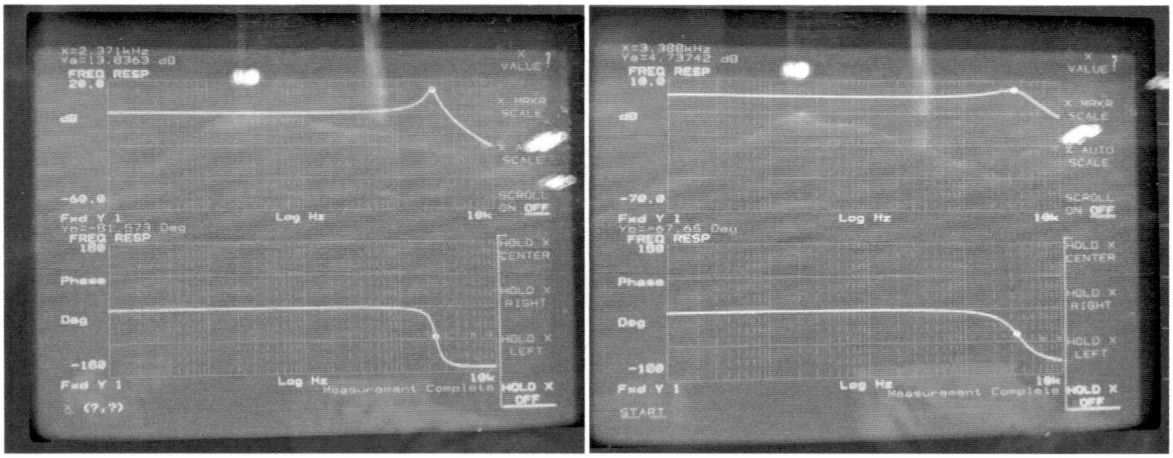

a.) b.)

Figure 14. Measured transfer functions from the input current to the 1.2km cable to the output current from the cable to the radial actuator winding (a) or the axial actuator winding (b).

however, occurs before the "waveguide" threshold and can be explained simply by the presence of the cable capacitance.

To predict the transfer functions such as in Fig. 14 and other effects of a long cable, a finite difference model shown in Fig. 15 was built in LT-Spice. The actuator windings impedances for this model were measured with an LCR meter at several frequencies. Near the resonance at 2.4kHz seen in Fig. 14a, the radial actuator inductance and the effective resistance (accounting for the eddy current losses in the actuator core in addition to the DC resistance) were measured to be 19mH and 43Ohms respectively. For the axial actuator winding near the resonance at 3.4kHz, these parameters were 6.3mH and 80 Ohms. The L/R ratio is much lower for the axial winding because of a non-laminated core.

Fig. 16 shows the current transfer functions as in Fig. 14 predicted using the model shown in Fig. 15 and Table I summarizes measured and predicted resonance frequencies and amplification factors. As can be seen, the analytical predictions match the experimental results reasonably well.

With the PWM filter added, the machine was successfully levitated and spun to full speed with the long cable connected to either the radial or the axial windings.

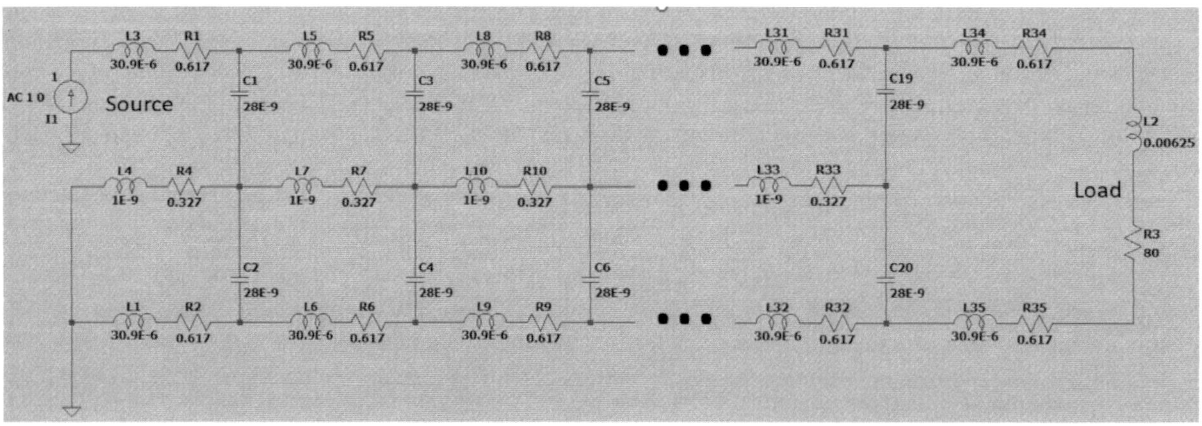

Figure 15. Finite difference LT-Spice model of a long cable with a load.

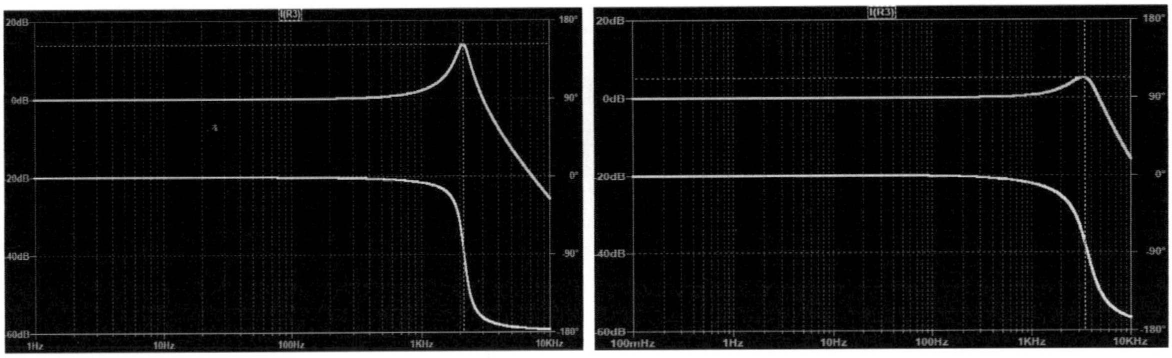

a.) b.)

Figure 16. Transfer functions from the input current to the 1.2km cable to the output current from the cable to the radial actuator winding (a) or the axial actuator winding (b) predicted using the LT-Spice model shown in Fig. 15.

Table I. Comparison of the measured and predicted resonance characteristics on the current over long cable transfer functions.

	f, Hz		Gain, dB	
Bearing	Predicted	Measured	Predicted	Measured
Radial (X2)	2132	2371	13.937	13.836
Axial (Z)	3356	3388	5.116	4.737

5. Conclusions

A combination of a "Trap" filter, utilizing a parallel LC resonant circuit to suppress output current harmonics near the circuit resonance and a conventional 2nd-order LCR filter has been shown very effective in suppressing PWM harmonics in an amplifier with switching frequency close to the upper limit of the required bandwidth. This has been experimentally demonstrated using a PWM amplifier switching at 25kHz with the targeted upper bandwidth limit at 2kHz. The filter has been shown to dramatically reduce switching noises on the output of an amplifier and allowed current control in the radial and axial channels of a commercial 350kW 15kRPM air compressor over 1.2km cable.

References

A. von Jouanne, P. Enjeti and W. Gray (1996) Application issues for PWM adjustable speed AC motor drives. IEEE Industry Applications Magazine, Vol. 2, Issue 5: 10-18.

Essex-Furukawa (2020) Ultrashield Plus and the inverter duty wire market. Available at: https://essexfurukawa.com/wp-content/uploads/2021/10/Essex-Furukawa-White-Paper-UltraShield-Plus-EN.pdf (accessed 11 January 2023).

R.K. Moore (1960) Traveling Wave Engineering. McGraw Hill.

S.W. Smith (1997) The scientist and Engineer's Guide to Digital Signal Processing, 1st edition, California Technical Pub.

Y. Sozer, D. Torrey and S. Reva (2000) New Inverter Output Filter Topology for PWM Motor Drives. IEEE Transactions on Power Electronics, Vol. 15, Issue 6: 911-917.

W. Santiago (2004) Inverter Output Filter Effect on PWM Motor Drives of a Flywheel Energy Storage System. Technical Memorandum TM-2004-213301, NASA, Glenn Research Center, Cleveland, Ohio, USA, September.

Large Diameter and Highly Homogenous Homopolar Active Magnetic Bearings for Energy Storage and Aerospace

Philip SWAN
The Atlantis Project, Redmond, Seattle, USA, philswan@project-atlantis.com

Abstract

High-speed Active Magnetic Bearing (AMB) technology can be used to construct a kinetic energy storage device that achieves significantly lower cost per kWh cycled than better-established technologies such as batteries, Pumped Storage Hydro (PSH), and Compressed Air Energy Storage (CAES). The key is to manufacture a large ring that orbits in a homogeneous magnetic field within a toroidally-shaped vacuum chamber. The supporting magnetic fields prevent hoop stresses within the ring from exceeding the tensile strength of the ring material. This allows a ring of sufficient diameter to store much more kinetic energy per unit of mass than is possible with a flywheel. The Levelized Cost of Energy Cycled is calculated and shown to be inversely proportional to the radius of the ring. Rings with a radius of 30m are cost-competitive with the best Li-Ion battery systems. At a radius of 50m, Ring Energy Storage (RES) is cost-competitive with PSH and CAES. At a radius of 120m, RES paired with scalable but intermittent renewable technologies, such as wind and solar, can supply most of the world's energy grids with reliable power at competitive rates.

If this technology is sufficiently matured and de-risked for commercial energy storage applications, there are more advanced applications of the technology that could be game-changing within the aerospace industry. For example, energy storage on the Moon or Mars where the energy storage capacity per unit of mass shipped between planets is a metric of paramount importance. Non-nuclear energy solutions for off-world use are also valuable for environmental and geopolitical reasons.

Keywords: *Homopolar, Active, Magnetic Bearing, Kinetic Energy Storage, Flywheel*

1. Introduction

High-speed Active Magnetic Bearings (AMBs) that can operate in a vacuum with very low magnetic friction have important applications in the aerospace and renewable energy industries. Areas of active interest include low-cost kinetic energy storage, electromagnetic launch, and hypersonic vehicle testing. Some useful performance metrics for defining application requirements are listed in Table 1.

A new sub-class of homopolar AMBs, called Highly Homogeneous Homopolar AMBs (H3AMBs), is a potential candidate technology for these applications. In a traditional homopolar AMB (an H1AMB) the electromagnets are arranged so that any given point on the rotor's surface will only travel past same-polarity magnetic poles (either all North or all South). An H3AMB is engineered so that, in the absence of field strength changes enacted by control circuits to compensate for perturbations, the nominal pole-to-pole variation in magnetic field strength is minimized. Low magnet friction, and thus a low rate of energy loss, becomes possible because unvarying magnetic fields do not induce eddy currents to flow, and they do suffer from magnetic hysteresis losses.

The technique of magnetic shimming used to achieve magnetic field homogeneities on the order of a few parts-per-million (ppm) in magnetic resonance imaging (MRI) machines, is considered for maximizing magnetic field homogeneity when designing, manufacturing, and tuning H3AMBs. Other techniques include high-tolerance manufacturing, strict process control, mechanical isolation, and the physical separation of electromagnetic motor/generator components from H3AMB components.

The use of a magnetically confined ring for kinetic energy storage has not been reported in recent articles that review the state of the art in Flywheel Energy Storage (Mongird *et al.*, 2020); however, the authors are aware of a relevant study that was published in 1984 (Hull and Iles, 1984). This early study concluded that the idea had promising commercial applications. At the time, potential investors may have disagreed with this assessment as wind and solar were still expensive technologies and climate change was a less pressing concern.

Performance Metrics

Highly disruptive technologies can emerge from heritage technologies that have found niche markets to survive in. A key growth strategy for companies and organizations in niche markets is to, when publishing product performance data, broadly consider all possible growth opportunities, as opposed to only trying to address the needs of existing customers. Those that publish the most comprehensive price and performance data on their products and technology will be offered the best opportunities to enter highly disruptive and potentially very profitable new industries.

Kinetic energy storage is an example of a potentially disruptive new industry for manufacturers of magnetic bearings or magnetically levitated trains. To understand why one must first consider some of the metrics used by the energy storage industry (see Table 1).

Table 1: Energy Storage Performance Metrics

Acronym	***Performance Metric***	***Units***
LCoP	Levelized Cost of Power Input/Output	USD/Watt
LCoS	Levelized Cost of Energy Storage Capacity	USD/Joule
LCoC	Levelized Cost of Energy Cycled	USD/Joule
	Energy Density	W/m^3
	Specific Energy	W/kg
η_{RT}	Round Trip Efficiency	(none)
	Nominal Energy Consumption per Unit of Energy Storage Capacity	W/J
	Nominal Energy Consumption per Unit of Energy Stored	W/J

The term "Levelized" refers to the lifetime cost, including societal costs, which ideally should be comprehensive and include manufacturing, disposal, maintenance, financing, and externalities[1] such as waste disposal, disaster clean up, cost of polluting, etc. Levelized Cost of Energy Cycled is perhaps the most interesting metric in the field of energy storage. It considers how much total energy will be cycled through the device over its lifetime. For example, an energy storage device designed to be coupled with a tidal energy generation facility could cycle energy through storage once every 6 hours and 12 minutes. If the service life of the energy generation system is 50 years, then over its lifetime the system will cycle its storage system 70,694 times. A typical Li-Ion battery can handle only 1200 to 2000 80%-depth-of-discharge charge cycles, so if Li-Ion batteries were paired with tidal energy generation, over the system's lifetime they would need to be recycled and replaced 35 to 59 times. Kinetic energy storage can, in theory, handle an effectively infinite number of charge cycles; therefore, it is a potentially disruptive technology in the grid-scale energy storage market.

Table 2: Energy storage technologies and their performance metrics.

	Li-Ion LFP[2]	Li-Ion NMC[3]	Lead-Acid	Vanadium RFB[4]	CAES[5]	Pumped Storage Hydro	Hydrogen Energy Storage	Ring Energy Storage
Cycle Life	2000	1200	599	5201	10403	13870	10403	13870
Down Time	0.05	0.05	0.05	0.05	0.05	0.05	0.05	0.05
Depth of Discharge	0.8	0.8	0.8	0.9	0.8	0.8	0.8	0.8
Cycles Per Year*	346.75	346.75	346.75	346.75	346.75	346.75	346.75	346.75
Round Trip Efficiency	0.86	0.86	0.85	0.68	0.52	0.80	0.35	
Years of Operation	5.77	3.46	1.73	15	30	40	30	40
Annualized Cost ($ per kWh-year)*	93	140	310	65	29	36	56	(see below)
Levelized Cost of Energy Cycled ($ per kWh-cycled)	0.268	0.404	0.894	0.187	0.084	0.104	0.161	(see below)

*Assumes daily cycling with 5% downtime

Table 2 lists some of the better-known grid-scale energy storage technologies along with their performance metrics obtained from a 2020 report from the Pacific Northwest National Laboratory (Mongird et al., 2020). The rightmost column of Table 2, "Ring Energy Storage" lists the assumptions that will be used here when analyzing the performance of energy storage based on H3AMB technology. It should be noted that the cheapest technologies in the table, such as CAES and Pumped Storage Hydro, need geographically suitable sites to exist close to the population centers that need

[1] In practice, inaccurate estimates of levelized performance metrics are published, for example, when certain societal costs are excluded. Therefore, when publishing, a detailed statement of assumptions is important.

[2] Lithium-ion Iron Phosphate

[3] Lithium-ion Nickel Manganese Cobalt

[4] Redox Flow Battery

[5] Compressed Air Energy Storage

energy storage; therefore, there are many markets where these technologies cannot address the need. Because of the high cost of energy storage, at present intermittent renewable energy generation paired with energy storage does not compete well with fossil fuel energy generation. To transition to a carbon-neutral economy, we will need an energy storage technology that can both:

a) Achieve a 10-fold reduction in cost, and

b) Scale rapidly to meet the huge global demand for energy storage in 2050 (Jorgenson *et al.*, 2050).

3. Comparative Analysis Study

Flywheel energy storage systems so far have not been able to achieve a low enough energy storage cost to be a disruptive technology in the energy storage marketplace. While some of the most advanced flywheel energy storage systems do make use of H1AMBs, they rely entirely on the specific strength of the material in the flywheel's disk to manage the spinning disk's hoop stress. The maximum rim speed, and thus the energy stored per kg of flywheel material, is limited by the specific tensile strength of the material. Increasing the diameter of the disk does not enable the flywheel to store more energy per kilogram of disk material.

However, this limitation can be circumvented if the moving material (that is, the kinetic energy storage medium) is instead confined, and made to travel in a circle, by using magnetic fields. In this case, the amount of kinetic energy that can be stored per kg of material *will* increase with the radius, r, of the circular path. Such a system would not be a flywheel energy storage system (FESS) but rather an energy storge system based on magnetically confining a ring-shaped, hub-less, spoke-less, rotor, or "mass-stream". Technically, the path of the mass-stream does not need to be circular, although a circle is likely the optimal topology for most applications. If we assume a circular, continuous, unbroken, ring-shaped, mass-stream, the total specific energy of a Ring Energy Storage System (RESS), 'E_{Sp}', is given by...

$$E_{Sp} = \frac{1}{2}\frac{m_{rotor}}{m_{system}}\left(\frac{\sigma_t}{\rho} + ra_m\right) \qquad 1$$

Where:
'σ_t' is the maximum tensile stress that rotor material will be required to withstand,
'ρ' is the density of the rotor material,
'r' is the radius of the circular path of motion,
'a_m' is the inward acceleration applied to the rotor by the magnetic field,
'm_{rotor}' is the mass of the rotor, and
'm_{system}' is the mass of the entire energy storage system.

For grid-scale energy storage, it is not difficult in principle to engineer a system where $ra_m \gg \sigma_t/\rho$; therefore, the viability of such a system becomes dependent on engineering considerations (such as energy losses from magnetic friction) instead of physical limitations such as the specific strength of known materials or the fundamental limits of battery chemistry.

The way that the energy density, 'E', scales with 'r' depends on the assumed values for σ_t, ρ, and a_m. For example, if a ring were made from steel, σ_t would be roughly 200 MPa and ρ would be 7840 kg/m^3. However, the key to making a ring energy storage system economically viable is maximizing the 'ra_m' term.

From (Schweitzer G, 2002), in an electromagnet, the force exerted, 'F', is related to the energy, 'W_a', stored in the two airgaps between the electromagnet and the plate that is attracting (note: "airgap" is a term used in the art, but technically it would be a "vacuum gap" in this application).

$$W_a = \frac{1}{2}B_aH_a2A_as \qquad 2$$

Where:
'B_a' is the flux density in the airgaps,
'H_a' is the magnetic field in the airgaps,
'A_a' is the cross-sectional area of each airgap, and
's' is the distance across the airgaps.

If we assume that s is small in relation to A_a, then for small displacements, ds, the magnetic flux, $\mathrm{B_a A_a}$, remains constant. Then...

$$F = \frac{dW_a}{ds} = B_a H_a A_a \tag{3}$$

Because B_a and H_a are related by...

$$\boldsymbol{B} = \mu_0 \mu_{r,H} \boldsymbol{H} \tag{4}$$

...we can substitute for H_a to obtain...

$$F = \frac{B_a^2 A_a}{\mu_0 \mu_{r,H_a}} \tag{5}$$

Where:
'μ_0' is the vacuum magnetic permeability ($4\pi \times 10^{-7}\mathrm{H/m}$),
'μ_{r,H_a}' is the relative permeability of the vacuum in the airgap to μ_0, which in our case is just 1.

The acceleration that can be generated by the magnets is given by $a = F/m$. The mass, 'm', is the rotor material density times its volume. The area of the airgap is...

$$A_a = w_a l = w_a 2\pi r_i \tag{6}$$

Where:
'w_a' is the width of the magnet at the airgap,
'l' is the circumference of the rotor, and
'r_i' is the inner radius of the rotor.

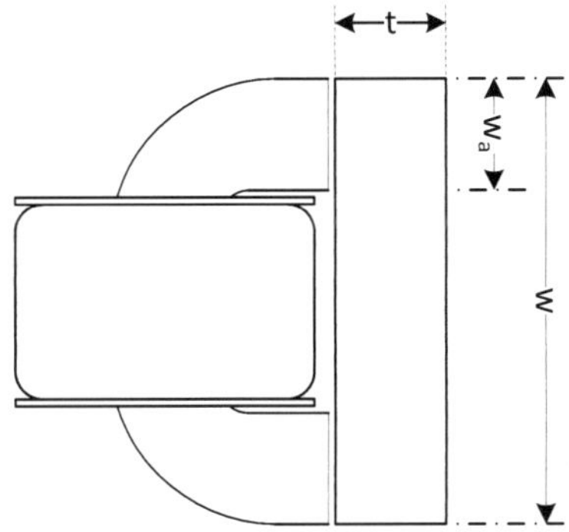

Figure 1: Cross-section of the rotor (right) and primary electromagnets (left)

If the rotor is a circular band with inner radius 'r_i', outer radius 'r_o', and a rectangular cross-section of 't' by 'w' (as shown in Figure 1) the volume of the rotor is ...

$$V = \pi(r_o^2 - r_i^2)w \tag{7}$$

(Note: Figure 1 only illustrates the generation of the primary magnetic field in support of the derivation of equations. Additional magnets that generate levitation and control fields and permanent magnet biasing are not shown.)

The mass of the rotor is...

$$m = \rho V = \rho\pi(r_o^2 - r_i^2)w \tag{8}$$

The inward acceleration of the rotor is thus...

$$a = \frac{F}{m} = \frac{B_a^2 A_a}{\mu_0 \mu_{r,H_a}} \cdot \left(\frac{1}{\rho\pi(r_o^2 - r_i^2)w}\right) = \frac{B_a^2 w_a 2\pi r_i}{\mu_0 \mu_{r,H_a} \rho\pi(r_o^2 - r_i^2)w} = \frac{B_a^2}{\mu_0 \rho} \cdot \left(\frac{w_a 2\pi r_i}{\pi(r_o^2 - r_i^2)w}\right) \tag{9}$$

The storage capacity of the system is...

$$E_{Capacity} = \frac{1}{2} m_{rotor} \left(\frac{\sigma_t}{\rho} + r\frac{B_a^2}{\mu_0 \rho} \cdot \left(\frac{w_a 2\pi r_i}{\pi(r_o^2 - r_i^2)w}\right)\right) \tag{10}$$

This system must be anchored to a foundation. If the mass of this foundation were included in the mass of the system, then the specific energy of the system would be no better than the specific energy of a flywheel. However, in most grid-energy storage applications, the cost of the storage system's foundation is a negligible small portion of the total cost.

In some applications, the mass of the system (not including the mass of the foundations) per unit of energy stored is of critical importance. For example, if we were designing an energy storage system that was to be built on earth and then shipped to the moon to supply power to a lunar base (see Figure 4b), the mass per unit of energy cycled would be extremely important because of the high cost of delivering mass to the surface of the moon.

A figure of merit more appropriate for earthly grid energy storage applications is the aforementioned Levelized Cost of

Energy Cycled (LCoEC). LCoEC is...

$$LCoEC = \frac{LevelizedCost}{n_{Cycles}E_{Capacity}} \quad 11$$

We can estimate the levelized cost from the mass of the system by using cost-over-mass data from similar systems, such as electric motors.

Table 3: Electric motor mass, power, and cost data.

Electric Motor	Model	Mass (kg)	Power (W)	Cost (USD)	Source of Information
RS PRO	238-9715	0.051	5.75	10.42	nz.rs-online.com/
Faulhaber	1727U012CXR	0.028	5.3	64.7	shop.faulhaber.com/1727u012cxr.html
Portescap	Athlonix 17DCT	0.027	4.5	65	portescap.com
Anaheim Automation	23D306S	1.25	1.2	210	anaheimautomation.com
Baldor-Reliance	CEL11303	11	373	812	baldor.com
Oriental Motor	ARM98AK-N5	3.8		1003	orientalmotor.com
Baldor-Reliance	CEBM3546T	20	746	2587	baldor.com
Baldor-Reliance	CEBM3558T	24	2237	3203	baldor.com
Baldor-Reliance	CEM2333T	122	11185	4222	baldor.com
Baldor-Reliance	CEM2514T	88	14914	4490	baldor.com
Baldor-Reliance	EM4102T-G	181	14914	5170	baldor.com
Baldor-Reliance	CEM2539T	171	29828	8769	baldor.com
Baldor-Reliance	CEM2543T	223	37285	9537	baldor.com
Baldor-Reliance	CEM2547T	246	44742	11868	baldor.com
Baldor-Reliance	CEM2551T	207	55927	14217	baldor.com
Baldor-Reliance	CEM2555T	429	74570	16526	baldor.com

Plotting cost versus mass for the data in Table 3 produces Figure 2, which gives us a rough indication of how the cost of an electromagnetic machine is related to its mass.

Figure 2: Electric motor cost versus mass.

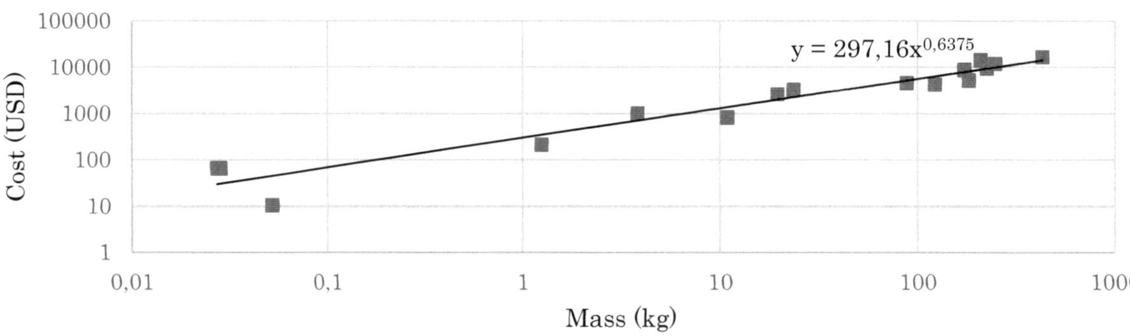

If we use the cost-over-mass relationship of Figure 2, the cycle life and depth of discharge values from Table 2, density and tensile strength values for steel, an engineering factor of 2, and assume $B_a = 1.5T$, $\frac{w}{w_a} = 4$, $t = w_a$, $\frac{m_{system}}{m_{rotor}} = 4$ then we can plot and observe the relationship between LCoEC and the radius of the ring (see Figure 3) for a flywheel and for a magnetically confined ring system.

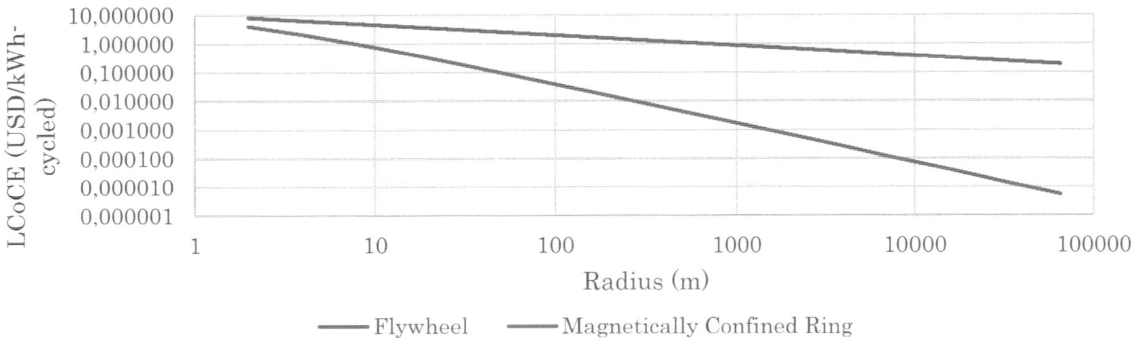

Figure 3: Relationship between Levelized Cost of Energy Cycled (LCoEC) and ring radius.

Without magnetic confinement (that is, operating entirely within the range of speeds that the tensile strength of the rotor's material allows) a flywheel-style kinetic energy storage system does not become cost-competitive with the incumbent technologies on an LCoEC basis - even at the largest scales shown on the chart. However, with magnetic confinement (that is, designing the system so that the energy storing ring orbits within a confining magnetic field), the system becomes cost competitive with incumbent technologies when the radius reaches 30m. A ring with a radius of 120m achieves an LCoEC low enough to make renewable energy generation and storage more economical than fossil-fuel-based energy generation for reliably powering the grid.

Figure 4: Grid-scale energy storage on the Earth and the Moon. Because of high launch costs, high specific energy, operating temperature range, and the potential for in-situ resource utilization are critically important for lunar storage.

While it still may be challenging for energy storage technologies to gain an economic foothold in our fossil-fuel-powered economy, it is generally possible to find microgrid applications where the economics are favorable. These include, for example, remote communities and bases where electric power is generated by using diesel generators, and diesel fuel must be either shipped in or flown in.

Other potential applications include actively supported systems for terrestrial transportation and the development of low-cost space launch infrastructure (Swan, 2023).

The development of this technology from a concept to a minimally viable product may face some engineering challenges. Low levels of magnetic and aerodynamic friction are needed to achieve high round-trip efficiency and to make sure that the self-discharge rate is low. Converting from electric energy to kinetic energy and back must be efficient when the ring is travelling at higher speeds than other types of electric machines will typically operate at.

4. Conclusions

H3AMBs are highly energy efficient because they are engineered to minimize the degree to which metal parts are exposed to time-varying magnetic fields. They may also minimize magnetic friction by laminating magnet iron and employing ferrites. Energy consumption can further be improved by using PM-bias magnets. Significant future applications for such bearings exist beyond the niche markets in the shaft-supporting bearing industry that AMBs currently serve. Some of the more interesting applications are for grid-scale energy storage. If the technology is sufficiently matured and de-risked, additional aerospace applications may emerge.

5. Acknowledgments

The author thanks David Bly (Technical Director at Iron Edison) and Joseph W. Donahue (Consultant) for reviewing this paper and providing their insights. This investigation was funded by The Atlantis Project, an organization that investigates transformative and CapEx intensive technologies for sustainable energy, terrestrial transit, and space launch.

References

Schweitzer G (2002) 'Active magnetic bearings-chances and limitations'.

Hull, J.R. and Iles, M.K. (1984) 'Magnetically confined kinetic-energy storage ring: a new fundamental energy-storage concept'. Available at: https://doi.org/10.2172/6836648.

Jorgenson, J. *et al.* (2050) 'Storage Futures Study: Grid Operational Impacts of Widespread Storage Deployment'. Available at: www.nrel.gov/publications. (Accessed: 5 May 2023).

Mongird, K. *et al.* (2020) '2020 Grid Energy Storage Technology Cost and Performance Assessment'.

Swan, P. (2023) *The Techno-Economic Viability of Actively Supported Structures for Terrestrial Transit and Space Launch*. Big Sky: IEEE Aerospace. Available at: https://ieeexplore.ieee.org/xpl/conhome/1000024/all-proceedings (Accessed: 5 May 2023).

Dynamic Response of Magnetically Suspended Rotor under Base Motions

Yue ZHANG [a], Jin ZHOU [a], Jarir MAHFOUD [b], Yuanping XU [a], Yibo ZHANG [a]

a Nanjing University of Aeronautics and Astronautics, Yudao Street 29, 210016 Nanjing, China, zhangyue08@nuaa.edu.cn

b University of Lyon, INSA-Lyon, CNRS UMR5259, LaMCoS, F-69621 Villeurbanne Cedex, France

Abstract

Active magnetic bearings (AMBs) can provide no-friction support and active vibration control for rotating machinery. The magnetically suspended rotor will inevitably be excited by the base motions when working on the moving carrier, causing the rotor vibration, and even instability. The main purpose of this study is to investigate the vibration response of the magnetically suspended rotor under base motions, especially base rotation, and to evaluate the accuracy of the model. In this work, the model of the magnetically suspended rotor is established, considering the base motions, the closed-loop control system of AMBs and gravity. The vibration response of the magnetically suspended rotor is tested on an AMB-rotor test rig mounted on a 6-DOF robot. Under three different base motion forms, the numerical simulation results based on the model are close to the experimental results, verifying the accuracy of the model. This work can provide guidance for predicting the vibration response of the magnetically suspended rotor under base motions.

Keywords: *Active magnetic bearings, Magnetically Suspended Rotor, Vibration response, Base motions*

1. Introduction

Active magnetic bearings (AMBs) have been used more and more in industrial applications due to their advantages of no friction, no lubrication, high speed, long life, and active control. The rotor supported by AMBs will inevitably be excited by the base motion brought by the moving carrier when applied in vehicle flywheels (Hawkins, et al., 2003), satellite/space-station equipment (Hawkins, et al., 2021), marine thrusters (Shen, et al., 2016), and multi-electric aero-engines (Yi, et al., 2019). There are various base motion forms, including acceleration and deceleration, vibration, sway, shock, and their combination. The results mainly display the rotor vibration. The rotor and stator will not contact when the AMB-rotor system works normally, and the rotor maintains movement between suspension clearances. However, the suspension clearance is often small, and compared with traditional mechanical bearings such as rolling or sliding bearings, the AMBs has low support stiffness and almost no overload capacity. The rotor may exceed the allowable maximum value when excited by the base motions, or even contact and rub with the touchdown bearings, cause instability and damage of the whole system. The most reliable way to investigate the stability of the magnetically suspended rotor under base motions is to conduct various vibration tests. However, it is obviously costly to conduct these tests after the design and manufacture. Therefore, it is necessary to conduct numerical simulation prediction on the vibration response of the magnetically suspended rotor after considering base motions, in order to verify its ability to resist the base excitation in the design process of the magnetically suspended rotor.

In order to evaluate the influence of earthquakes on the turbo-molecular pumps supported by the AMBs, Mural (1989) tested the vibration response of the molecular pump rotors under three seismic waves. The results showed that the AMBs can provide sufficient stiffness to ensure the molecular pump stable under earthquake conditions. Kasarda (2000) designed a single AMB mounted on a shaker to investigate the AMB bearing capacity and system natural frequency under the base sinusoidal excitation of different frequencies and amplitudes. The experimental results show that the stiffness and damping level of the AMB limits the operating performance. Zhang (2012) established a five degrees of freedom rigid rotor coupling model considering the vertical base motion, but did not further analyze its influence. Xu (2020) developed a magnetically suspended rigid rotor model considering the base translation and rotation based on the dual-frame system model, indicating that the translation motion is equivalent to the external force and the rotation motion is equivalent to the torque applied to the rotor. The effects of excitation amplitudes and pulse widths on the rotor response were analyzed by simulation, but no experimental verification was carried out. To control the lateral vibration of the flexible rotor caused by the base motion, Das (2010a, 2010b) installed an electromagnetic actuator (similar to AMB) in the middle of the rotor. The numerical simulation indicated that the rotor vibration is effectively reduced and the stability is improved. Hawkins (2018) conducted a floating shock platform test on an AMB supported chiller compressor. The compressor was subjected to a standard sequence of four different shocks generated from high explosive from different angles and distances. Three of them run normally and one of them suspended again after drop. Jarroux (2018) analyzed the influence of mass unbalance force, base motion, and nonlinear contact force of TDBs on AMB system, and carried out base harmonic excitation test. The test showed that the system could remain stable even if the rotor and TDBs were contacted by the harmonic excitation. Jarroux (2020, 2021) further evaluated the response of the magnetically suspended rotor under strong shock excitation, and the numerical simulation predictions were consistent with the experimental results, verifying the accuracy of the model.

These studies rarely consider base rotation motions. In fact, when magnetically suspended rotating machinery is applied on ships or offshore platforms, it will be affected by waves and winds, resulting in base rolling, pitching, and yawing motions. In this paper, the finite element model of the magnetically suspended rotor under the base motions is established, in which the base motions, the closed-loop control of the AMBs and the rotor gravity load are considered. A 6-DOF robot is used to simulate the base motions of the AMB-rotor test rig, especially the base rotation motions. The dynamic response of the rotor supported by AMBs is studied by numerical simulation and experiments. The response prediction based on the finite element model is compared with the experimental results, verifying the accuracy of the model.

2. Magnetically suspended rotor model under base motions

The model of magnetically suspended rotor under base motions can be expressed as (Zhang, et al., 2023):

$$[\boldsymbol{M}]\{\ddot{\boldsymbol{q}}\} + \left([\boldsymbol{C}]_R - \dot{\phi}[\boldsymbol{G}] + [\boldsymbol{D}]_{bm}\right)\{\dot{\boldsymbol{q}}\} + \left([\boldsymbol{K}] + [\boldsymbol{K}]_{bm}\right)\{\boldsymbol{q}\} = [\boldsymbol{Q}]_{bm} + [\boldsymbol{Q}]_u + [\boldsymbol{F}]_{AMB} + [\boldsymbol{F}]_g, \tag{1}$$

where, $\{\boldsymbol{q}\}$ is the generalized coordinate of the rotor, including translation and rotation of each node. $[\boldsymbol{M}]$, $[\boldsymbol{C}]_R$, $[\boldsymbol{K}]$ are rotor mass matrix, structural damping matrix and stiffness matrix respectively. $[\boldsymbol{G}]$ is the gyroscopic effect matrix. The shrink fit components matching with AMBs and motor on the rotor are considered in modeling. $[\boldsymbol{D}]_{bm}$, $[\boldsymbol{K}]_{bm}$, $[\boldsymbol{Q}]_{bm}$ are respectively the additional damping matrix, additional stiffness matrix and additional generalized force caused by base motions. The base motions are represented by the six rigid body motions of the base. x_b, y_b, z_b, ω_x, ω_y, ω_z represent the translation along the three axes of the base coordinate system and the rotation around the three axes respectively. $[\boldsymbol{Q}]_u$ is the unbalance force. $[\boldsymbol{F}]_g$ is the gravity load. It should be noted that the base rotation will affect the distribution of gravity load in the direction of the coordinate axis. $[\boldsymbol{F}]_{AMB}$ is the electromagnetic force provided by AMBs.

The electromagnetic force is

$$f = \frac{1}{4}\mu_0 A N^2 \cos\alpha_0 \left[\left(\frac{i_0 + i_c}{C_0 - x\cos\alpha_0}\right)^2 - \left(\frac{i_0 - i_c}{C_0 + x\cos\alpha_0}\right)^2\right]. \tag{2}$$

where, μ_0 is the magnetic permeability of the vacuum, A_α is the area of a single magnetic pole, N is the number of turns in one coil, C_0 is the air gap between the rotor and the magnetic poles, x is the rotor position, i_0 is the bias current, i_c is the control current, α_0 is the angle of the magnetic pole, which represents the included angle between the direction of the electromagnetic force and the corresponding magnetic pole. For an eight poles C-type radial AMB, α_0 is usually 22.5 °. The displacement sensor and electromagnetic actuator of AMB are usually not in the same position, which needs to be considered in modeling.

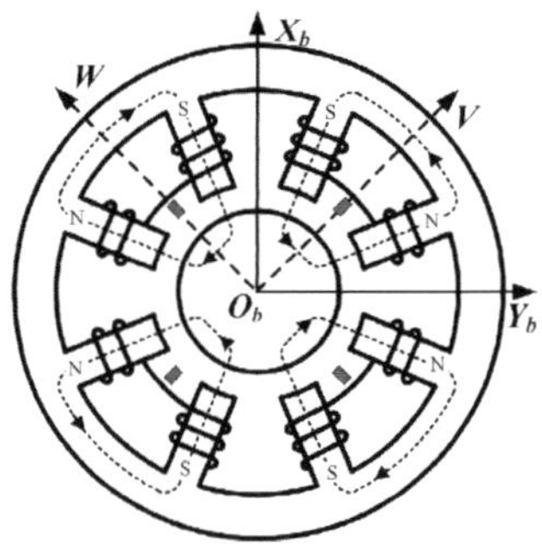

Figure 1 Schematic diagram of radial AMB.

The rotor displacement at the electromagnetic actuator can be expressed as

$$\{\boldsymbol{q}\}_b = [x_{b1} \quad y_{b1} \quad x_{b2} \quad y_{b2}]^{\mathrm{T}} = \boldsymbol{T}_B\{\boldsymbol{q}\}, \tag{3}$$

where, the transformation matrix $\boldsymbol{T}_B$ is to extract the translational displacement at the electromagnetic actuator from all generalized coordinates.

The rotor displacement at the displacement sensor can be expressed as

$$\{\boldsymbol{q}\}_{ss} = [x_{s1} \quad y_{s1} \quad x_{s2} \quad y_{s2}]^{\mathrm{T}} = \boldsymbol{T}_S\{\boldsymbol{q}\}, \tag{4}$$

where, the transformation matrix $\boldsymbol{T}_S$ is to extract the translational displacement at the displacement sensor from all generalized coordinates.

Due to the closed-loop control system of AMBs, the coil control current is obtained from the displacement detected by the displacement sensor

$$\{\boldsymbol{i}\}_c = [i_{cx1} \quad i_{cy1} \quad i_{cx2} \quad i_{cy2}]^{\mathrm{T}} = -G_s(s)G_c(s)G_a(s)\{\boldsymbol{q}\}_{ss}, \tag{5}$$

where, $G_c(s)$, $G_a(s)$, $G_s(s)$ respectively represent the transfer functions of the controller, power amplifier, and displacement sensor in the closed-loop control system. The controller adopts a conventional PID controller. The power amplifier and displacement sensor have low-pass characteristics and their transfer functions are obtained through frequency response from frequency sweep tests.

Therefore, the electromagnetic force of AMBs can be expressed as

$$[\boldsymbol{F}]_{AMB} = [f_{x1} \quad f_{y1} \quad f_{x2} \quad f_{y2}]^{\mathrm{T}}. \tag{6}$$

Based on the above model, the dynamic response of the magnetically suspended rotor under base motions is analyzed, mainly considering the situation of base rotational motion.

3. Numerical simulation and experimental results

3.1 Test rig

The AMB-rotor test rig is shown in the *Figure 2*. The rotor is 840 mm long and 5.25 kg weigh, with a first critical speed of approximately 9000 r/min. The rotor is supported by two radial AMBs (AMB1 and AMB2) and a pair of axial AMB. The main parameters of radial AMBs are shown in the *Table 1*. An induction motor is located in the middle of rotor and is used to drive the rotor to rotate. Touch down bearings (TDBs) are arranged on both sides of AMBs to temporarily support the rotor in case of failure of AMB and prevent the rotor from rubbing with stator. The gyroscopic effect of the rotor is very small. The rotor modal test, system transfer function test and unbalance response test have been carried out to verify the accuracy of the AMB-rotor model without base motions (Zhang, et al., 2023). The origin of the base coordinate system is located at the center of the AMB1.

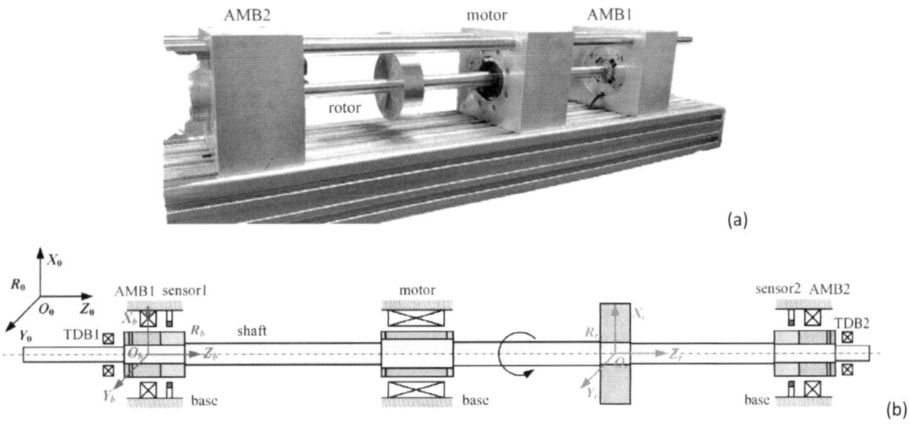

Figure 2 AMB-rotor test rig (a) Photo (b) Structure diagram.

Table 1 Parameters of radial AMBs

	Parameter	*Value*
C_0	air gap between the rotor and the magnetic poles	0.3 mm
C_{TDB}	air gap between the rotor and the TDB	0.125 mm
A_α	pole area	200 mm^2
N	number of turns	120
i_0	bias current	1.3 A

In order to simulate the base motions (such as rolling, yawing and vibration), fasten the above AMB-rotor test rig on the 6-UPS parallel robot, shown in *Figure 3*. The parallel robot can realize the movement of 6-DOF in space. The origin of the upper platform coordinate system is located at its geometric center. The base motions are simulated through the motion of the upper platform. The 6-UPS parallel robot is developed on the basis of the classic 6-DOF Stewart platform, and has the characteristics of large output torque, high motion accuracy, large stiffness, and low inertia. Both ends of each leg are connected to the upper platform and the lower platform respectively through the spherical joint and universal joint. Linear motion is driven by servo motors and screw rods. The upper and lower platforms are hexagonal with an included angle of 120 °. During the experiment, the lower platform is firmly connected to the ground and remains stationary. The motion form of the upper platform is set, and the elongation of each support leg is obtained through dynamic inverse solution. Due to the limitation of leg elongation, while keeping the center of the upper platform stationary, the maximum angle of rotation along each axis is 20 °. Due to the limitation of the driving response speed, the minimum rotation period is 1 s.

The displacement sensors of the AMBs are used to detect the rotor displacement under the base motions. In the experiment, the rotor is only statically suspended without rotation, in order to explore the impact of base motions on the rotor vibration response. The same motions are input into the numerical model for simulation and comparison with experimental results. The test is conducted under three conditions.

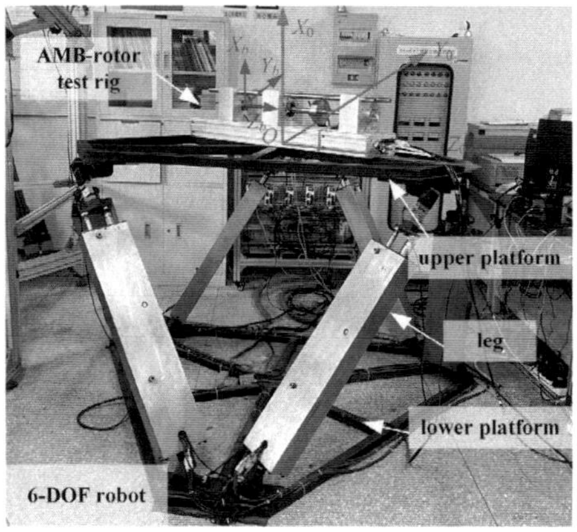

Figure 3 Experimental test rig.

3.2 Simulation and experimental results

Case 1: upper platform rotation around O_0X_0, $\alpha = \alpha_b \sin(\Omega_x t)$.

The relationship between the base and the upper platform coordinate system is shown in *Figure 4*. Then, the base motion of the AMB-rotor test rig can be derived as

$$\omega_x = \dot{\alpha} = \Omega_x \alpha_b \cos(\Omega_x t) \tag{7}$$

$$y_b = l_{0b} \sin \alpha \tag{8}$$

where, α_b and Ω_x are respectively the maximum angle and frequency of the upper platform rotation around O_0X_0, l_{0b} is the distance between the origin of the upper platform system and the origin of the stator system in the z-direction.

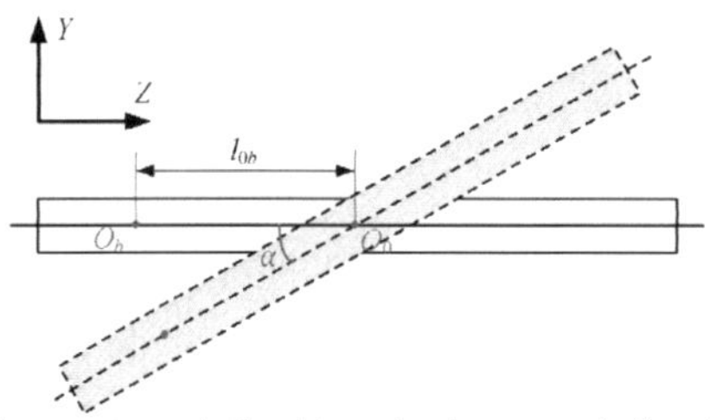

Figure 4 Schematic diagram of coordinate system relationship under the upper platform harmonic rotation around O_0X_0 (top view).

Several sets of numerical simulations and experimental verification of rotor vibration under different base rotation angles and periods were conducted. Due to the limitation of the motion range of the upper platform, the base rotation angle is set at 1-15 ° and the rotation period is 1-10 s. The results are shown in *Figure 5*. The detected rotor vibration signal has high-frequency noise because of the joint friction, signal interference and platform vibration during the operation of the 6-DOF robot. The platform has a very low rotational frequency, so the rotor vibration signal can be filtered and processed. The cut-off frequency of low-pass filter is 5 Hz, and the main frequency information can be retained, which has little impact on the results.

It can be seen that simulation and experiment can well confirm each other. As the base rotation angle and frequency increase, the vibration displacement of the rotor increases. When the rotation amplitude is 15 °, the rotor vibration noise is relatively high, which is caused by the platform. The testing standard for equipment on ships requires a yaw amplitude of 4 ° and a period of 20 s, which is much less severe than the experiments in *Figure 5*. Under this testing condition, the vibration response of the magnetically suspended rotor is very small, indicating that this test has little impact on the operation and stability of the magnetically suspended rotor.

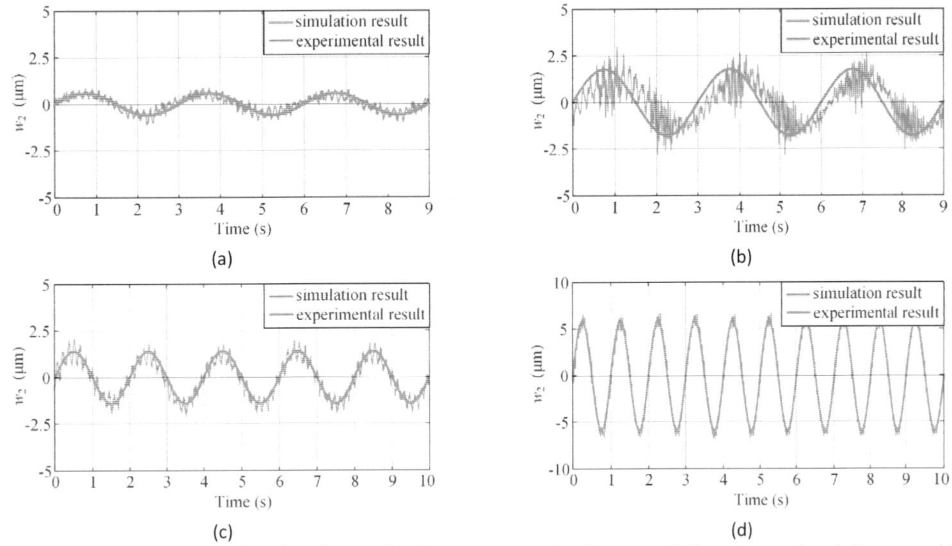

Figure 5 Simulation and experimental results of rotor vibration response under the upper platform harmonic rotation around O_0X_0 (a) 5°-3s (b) 15°-3s (c) 5°-2s (d) 5°-1s.

Case 2: upper platform rotation around O_0Z_0, $\gamma = \gamma_b \sin(\Omega_z t)$.

The relationship between the base and the upper platform coordinate system is shown in *Figure 6*. Then, the base motion of the AMB-rotor test rig can be derived as

$$\omega_z = \dot{\gamma} = \Omega_z \gamma_b \cos(\Omega_z t) \tag{9}$$

$$x_b = h_{0b} \cos \gamma \tag{10}$$

$$y_b = h_{0b} \sin \gamma \tag{11}$$

where, γ_b and Ω_z are respectively the maximum angle and frequency of the upper platform rotation around O_0Z_0, h_{0b} is the distance between the origin of the upper platform system and the origin of the stator system in the x-direction.

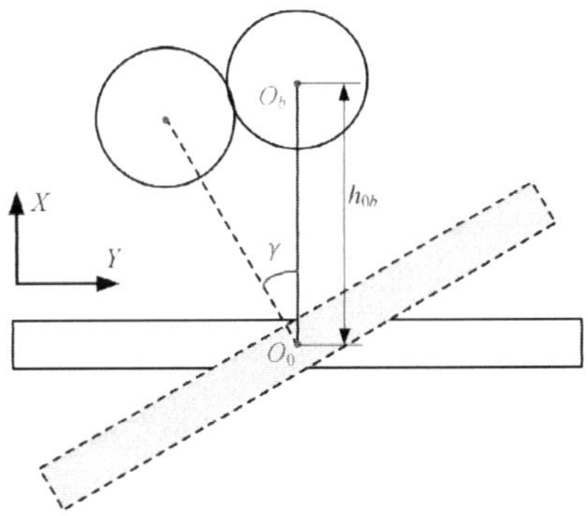

Figure 6 Schematic diagram of coordinate system relationship under the upper platform harmonic rotation around O_0Z_0 (side view).

It should be noted that in this case, the gravity load of the rotor changes. The gravity load in the x and y directions of the stator system varies periodically. When the rotor mass or rotation angle is large, the influence of gravity load distribution cannot be ignored. Taking the W-direction rotor vibration of AMB2 as the research object, several sets of numerical simulations and experimental verifications of rotor vibration under different base rotation angles and periods were conducted. According to the motion range that the platform can achieve, the rotation angle is 1-15 °, and the rotation periods is 1-10 s. The results are shown in *Figure 7*.

The rotor displacement signal in *Figure 7* was also filtered and processed. It can be seen that the simulation and experimental results are close. The experimental results are slightly larger than the simulation results, possibly due to the small rotation angle and experimental platform error, but the influence of rotation angle and frequency on rotor vibration is the same. *Figure 8* shows the rotor orbits at AMB2 when the upper platform has a rotation angle of 15 ° and a rotation period of 5 s. This base motion causes vibration in both the x and y directions of the rotor, resulting in an 8-shaped orbit. The main vibration is in the y-direction, which is about 14 μm, while the vibration in the x-direction is about 2 μm. The asymmetry of the orbit in the experiment is caused by a slight deviation between the AMB-rotor test rig and the upper platform during installation.

The standard for roll testing of equipment on ships requires an amplitude of 45 ° and a period of 5 s. The rotor orbit under this condition can be predicted through simulation, as shown in *Figure 9*. It can be seen that the maximum vibration of the rotor in the y-direction is about 40 μm and the vibration in the x-direction is about 8 μm. If h_{0b} increases, that is, the distance between the magnetically suspended rotor and the motion center of the moving carrier increases, the rotor vibration will further increase, which should be considered in the actual condition.

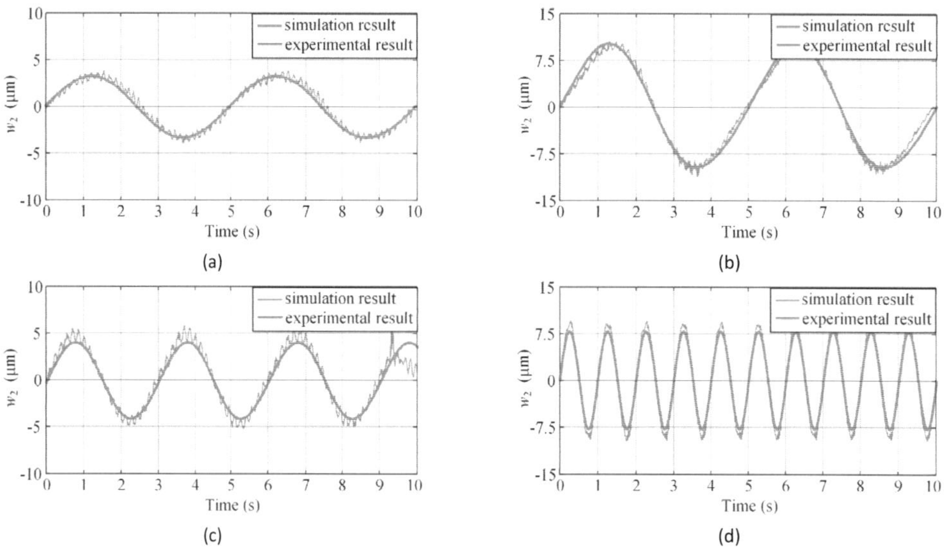

Figure 7 *Simulation and experimental results of rotor vibration response under the upper platform harmonic rotation around O_0Z_0 (a) 5°-5s (b) 15°-5s (c) 5°-3s (d) 5°-1s.*

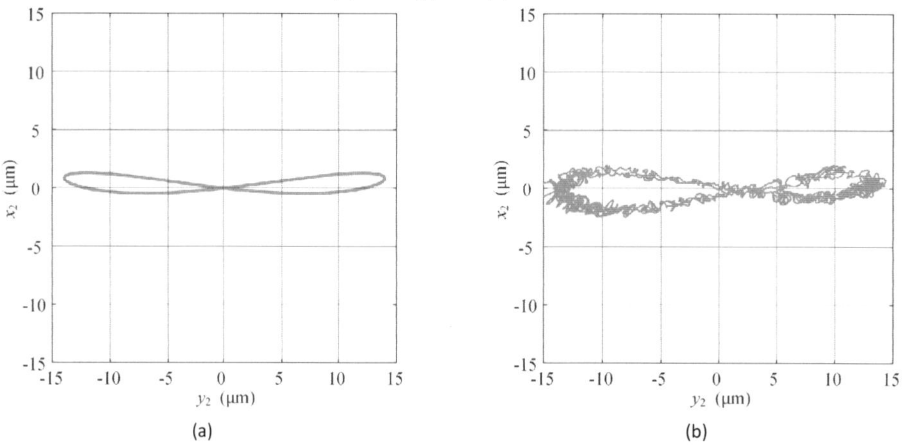

Figure 8 *Results of the rotor orbit under the upper platform harmonic rotation around O_0Z_0 with an angle of 15° and a period of 5s (a) Simulation result (b) Experimental result.*

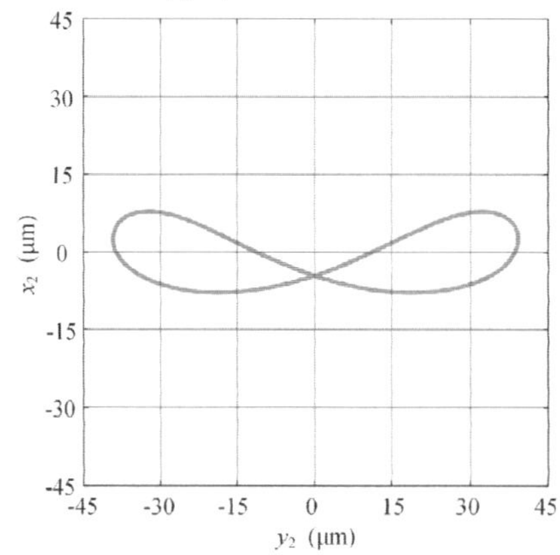

Figure 9 *Simulation results of the rotor orbit under the upper platform harmonic rotation around O_0Z_0 with an angle of 45° and a period of 5s.*

Case 3: upper platform rotation around O_0Z_0, and translation along O_0X_0 at the same time.

Results of the upper platform harmonic rotation around O_0Z_0 with an angle of 10 ° and a period of 5 s, and harmonic translation along O_0X_0 with an amplitude of 50 mm and a frequency of 1 Hz are shown in *Figure 10*. Results of the upper platform rotation around O_0Z_0 with an angle of 5 ° and a period of 3 s, and translation along O_0X_0 with an amplitude of 30 mm and a frequency of 1 Hz are shown in *Figure 11*. The simulation results accorded with the experimental results well. This composite motion makes the vibration of the rotor more complex, including signals with the same frequency as the base rotation and translation, which can be seen as the superposition of two frequency responses.

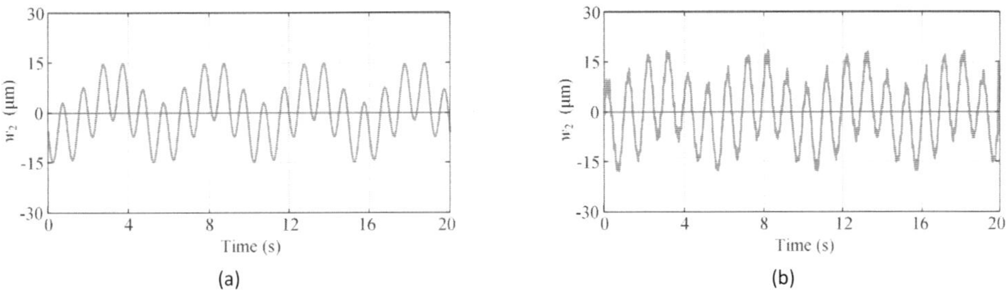

Figure 10 Results under the upper platform rotation around O_0Z_0 with an angle of 10° and a period of 5s, and translation along O_0X_0 with an amplitude of 50mm and a frequency of 1Hz (a) Simulation result (b) Experimental result.

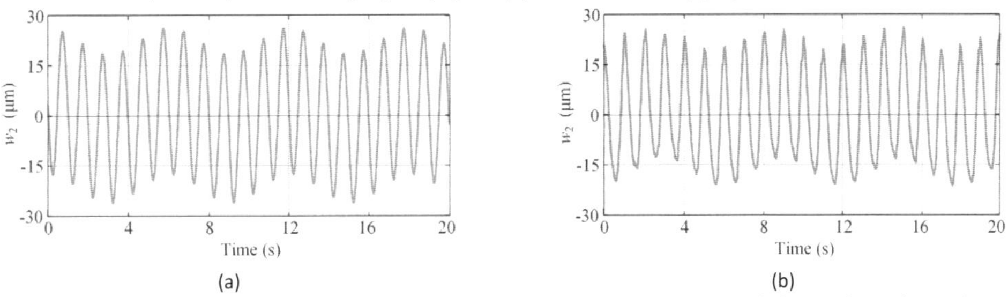

Figure 11 Results under the upper platform rotation around O_0Z_0 with an angle of 5° and a period of 3s, and translation along O_0X_0 with an amplitude of 30mm and a frequency of 1Hz (a) Simulation result (b) Experimental result.

4. Conclusion

The dynamic response of a magnetically suspended rotor considering base motions was investigated. Firstly, a finite element model of the magnetically suspended rotor under base motions was established, taking into account the non-collocated between the displacement sensor and the electromagnetic actuator. Then, an AMB-rotor test rig and a 6-DOF robot are used for the experiments. The 6-DOF robot can realize free motions in space to simulate the base motions of the magnetically suspended rotor, especially the base rotation. The model was verified by comparing the numerical simulation and experimental results of the rotor vibration response. That is, the rotor response under the base motion excitation can be predicted through this model. The results indicate that the rotor vibration is positively related to the maximum angle and frequency of the base rotation. When the frequency of base motion is very low, there is almost no impact on rotor vibration. The base rotation may change the rotor gravity load distribution, which cannot be ignored in the calculation of the rotor vibration response, especially when the rotor mass or the rotation angle is large.

References

Das A, Dutt J, Ray K (2010a) Active vibration control of flexible rotors on maneuvering vehicles. AIAA J. 48: 340-353.

Das A, Dutt J, Ray K (2010b) Active vibration control of unbalanced flexible rotor–shaft systems parametrically excited due to base motion. Appl. Math. Model. 34: 2353-2369.

Hawkins L, Murphy B, Zierer J, et al. (2003) Shock and vibration testing of an AMB supported energy storage flywheel. JSME Int. J. 46: 429-435.

Hawkins L, Wang Z, Nambiar K (2018) Floating shock platform testing of a magnetic bearing supported chiller compressor: Measurements and simulation results. In: Turbomachinery Technical Conference and Exposition, 11-15 June, Oslo, Norway.

Hawkins L, Filatov A, Khatri R, et al. (2021) Design of a compact magnetically levitated blower for space applications. J. Eng. Gas Turbines Power. 143.

Jarroux C, Mahfoud J, Dufour R, et al. (2018) Dynamic behavior of a rotor-AMB system due to strong base motions. In: International Conference on Rotor Dynamics. Springer, Cham.

Jarroux C, Mahfoud J, Defoy B, et al. (2020) Stability of rotating machinery supported on active magnetic bearings subjected to base excitation. J. Vib. Acoust. 142: 031004.

Jarroux C, Mahfoud J, Dufour R, et al. (2021) Investigations on the dynamic behaviour of an on-board rotor-AMB system with touchdown bearing contacts: modelling and experimentation. Mech. Syst. Signal Pr. 159: 107787.

Kasarda M, Clements J, Wicks A, et al. (2000) Effect of sinusoidal base motion on a magnetic bearing. In: IEEE International Conference on Control Applications, Anchorage, AK, USA.

Murai Y, Watanabe K, Kanemitsu Y (1989) Seismic test on turbo-molecular pumps levitated by active magnetic bearing. In: Magnetic Bearings. Springer, Berlin, Heidelberg.

Shen Y, Hu P, Jin S, et al. (2016) Design of novel shaftless pump-jet propulsor for multi-purpose long-range and high-speed autonomous underwater vehicle. IEEE Trans. Magn. 52: 1-4.

Xu Y, Shen Q, Zhang Y, et al (2020) Dynamic modeling of the active magnetic bearing system operating in base motion condition. IEEE Access. 8: 166003-166013.

Yi Y (2019) Review and future of aircraft's propulsion type. In: Journal of Physics: Conference Series. 1345(3): 032075.

Zhang W (2012) Coupled dynamic analysis of magnetic bearing-rotor system under the influences of base motion. Appl. Mech. Mater. 109: 199-203.

Zhang Y, Zhou J, Zhang Y, et al. (2023) Modelling and vibration response of a magnetically suspended flexible rotor considering base motion. Appl. Math. Model. 118: 518-540.

Material Testing Device Using Magnetic Levitation Mechanism

Koichi OKA [a], Mengyi REN [a]

a Kochi University of Technology, Miyanokuchi 185, Tosayamada-cho, Kami, Kochi, Japan (oka.koichi@kochi-tech.ac.jp)

Abstract

This paper describes non-contact material testing devices using magnetic levitation mechanism, in which a specimen can be tested in tension, compression, bending or torsion while the specimen and few parts are being levitated. We mainly focus on the mechanism, control, and experimental results of the tension testing device and bending testing device. The structures of these devices are introduced about material testing methods. For material testing, additional control mechanisms are required except for ordinal levitation control mechanism. Tension testing device and bending testing device have corresponding mechanism respectively. Experimental examinations were performed to examine the feasibility, stability, and robustness under the maxim tension force and bending force. Finally, they demonstrated that the levitation control system has good stability and robustness under the maximum tension force and bending force.

Keywords: *Material testing device, Non-contact testing device, Tension test, Bending test.*

1. Introduction

Magnetic levitation has been widely studied and applied in many industrial applications due to the advantage of non-contact, e.g., high-speed bearings (Supreeth, et al, 2022), (Prrasad, et al, 2021), motors (Kumashiro, at al, 2022), (Wang, at al, 2023), and centrifugal blood pumps (Masuzawa, at al, 2000), (Wu, at al, 2021), etc. As a new application of magnetic suspension system, Tada (2013) has been proposed the magnetic force for the material testing device(MTD). This testing device is easily used for the special liquid environment test. However as this device has an mechanical contact at one end of specimen hold mechanism, if the test in the vacuum environments are required the whole equipment of testing device should be inside a large vacuum chamber. A more sensible approach is to apply a non-contact force from the outside of the vacuum chamber to the specimen inside the vacuum chamber. Magnetic levitation (maglev) technology makes this possible.

In this paper, a mag-lev testing devices, where the specimen can be pulled in a non-contact way, are discussed. We develop a perfect non-contact MTD for compact testing device. As the development of MTD, a tension test device and bending test device are introduced in this paper. This application will bring benefits, because some materials' working environments are special, such as high-purity aluminum (Wan et al,2022), gas-lubricated bearings (Kumar at al, 2022). and S-Glass-Carbon fiber reinforced polymer composites (Elango, et al, 2021). Previously, if we want to test these materials, we have to place a whole testing device including framework, force sensor and actuator, etc. in a container where the special environment can be created. However, some parts of the testing device, e.g., circuits in the force sensor and actuator are sensitive to environmental factors such as temperature, air pressure, humidity, etc. With magnetic levitation, only few parts and specimen need to be put in the container, which greatly increases the feasibility of the tests. The structure of this paper is given as follows. Chapter II presents a structure and experimental results of a tension testing device; Chapter III describes a structure, control method, and experimental results of a 3 point bending testing device working in MLMMTD (magnetic levitation mechanism material testing device).

2. Material tension testing device

2.1 Overall structure

The initial structure of MTD is shown in Figure 1. As shown in the figure, two electromagnets (EM), called top EMs, were fixed at the top of the framework, and one EM, called bottom EM, was fixed at the bottom of the framework by a load cell. The currents of the three EMs were controlled individually by three controllers. The levitated object is in the middle of MTD as shown to the right in the figure, and it includes two top floaters, two top fixtures, two bottom fixtures and one bottom floater. The two ends of the specimen were placed in the slit of the top and bottom fixtures respectively.

The working principle of MTD is, firstly, the two top EMs, whose currents are controlled individually, attract the two top floaters respectively. The two eddy current sensors feed the position of the two top floaters to a controller, which then adjusts the currents of the two top EMs to maintain the top floaters and the fixtures in a constant vertical position. Subsequently, the bottom EM is electrified, and generates a tension force on the bottom floaters, and the current of the bottom EM gradually increases while the top floaters and the fixtures maintain in the original position until the specimen is destroyed. The load cell measures the tension force for recording, and also for control feedback.

The levitation of the two top floaters are controlled by the two top EMs, this may be easy. However, after tension force is loaded, to maintain the levitation, it is necessary to consider suppressing the disturbance caused by the tension force. A material mechanics testing device needs to measure both loading force and specimen elongation to obtain stress-strain curve. In MTD, the load cell can indirectly measure the loading force. The elongation is equal to the amount of change in the bottom air gap. Moreover, the bottom air gap can be calculated according to equation, which describes the magnetic force of an electromagnet.

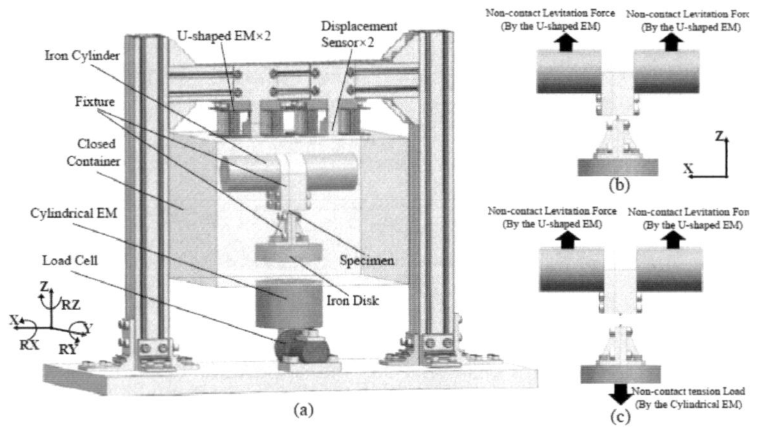

Figure 1 Conceptual illustration of proposed tension test device using magnetic suspension mechanism

2.2 Control system and experimental results

When there is no tension force, the function of the controller is very simple. However, tension force applied to the lower floater will form a disturbance on the system, and it is necessary to compensate. Because the maglev system has large non linearity. In the process of applying tension force, the currents of the top EMs will continuously increase to maintain the levitation, so the plant model will inevitably deviate away from the original plant model. In other words, as the currents of the top EMs increase continuously, the inherent non-linearity of the magnetic levitation system will appear, the constant force feedback gains will no longer fit the plant model after the deviation. Therefore, an improved method shown in Figure 2 was proposed. Dynamic force feedback gains instead of the constant force feedback gains are used for the control system. In the figure, the dynamic force feedback gains consist of a superpositions of the proportion and integral of air gap errors. In addition, the precise tension feedback path in Figure 3 was employed to compensate for the deviation caused by changes in the lower air gap. With such a control system, the tension force was gradually applied at a fixed rate to the bottom EM.

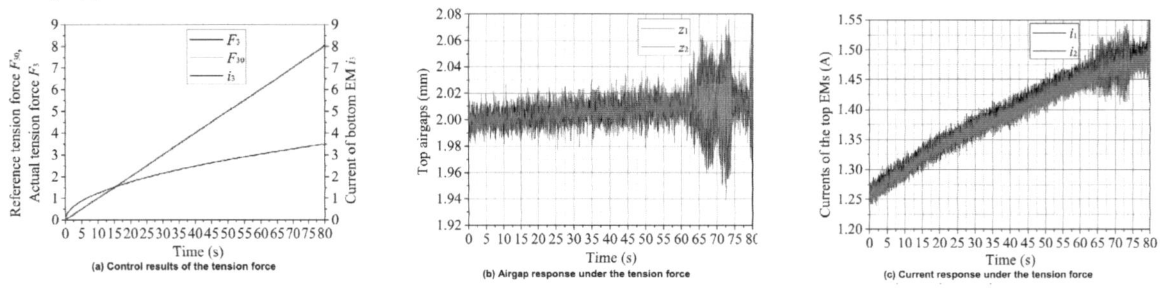

Figure 2 Experimental result of tension test

3. 3 point bending test device

3.1 Overall structure

As shown in Figure 3, the prototype mainly consists of two U-shaped electromagnets, two eddy current sensors, levitated object, a cylindrical electromagnet, and a force sensor. The two U-shaped electromagnets are used to produce levitation force for the levitated object, while the two eddy current sensors are used to determine the position of the two iron cylinders. The cylindrical electromagnet is used to produce a non-contact bending load to bent the specimen in the levitated object, while the force sensor is used to measure the bending load in real time.

The principle is, as shown in Figure 4, firstly, the two U-shaped electromagnets and the two eddy current sensors work to maintain the levitation. The levitation forces will keep the levitated object in a fixed vertical position. As shown in the lower of Figure 4, after the levitation is stable, the cylindrical electromagnet will be energized to produce a non-contact bending load to the iron disk, so that the punch will be pulled down to bend the specimen.

In addition, the applicable maximum deflection of MLBTD is 10mm, which is obtained according to the stroke of the punch component. The applicable maximum bending force of MLBTD is 50N, which is obtained according to the measuring range of the load cell. The applicable specimen stiffness range of MLBTD is more than 430N/m, which is calculated according to the stroke and weight of the punch component, because under the weight of the punch component, too soft specimens will be bent to the maximum deflection before the bending force is applied, so the applicable specimen stiffness has a lower limit. The accuracy of the eddy current sensor is 0.002mm. The accuracy of the load cell is 0.005N.

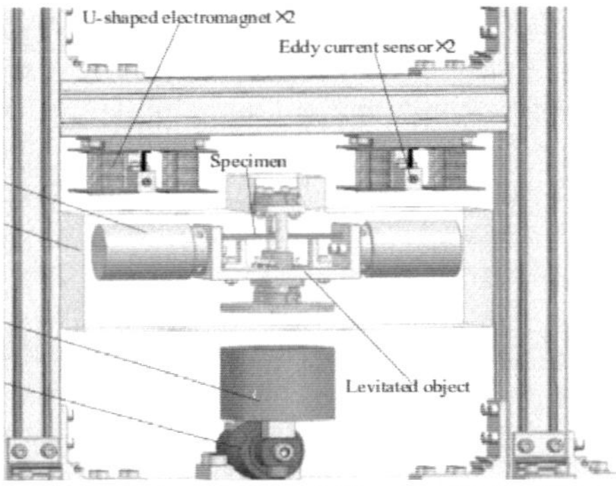

Figure 3 An illustration of 3 point bending testing device

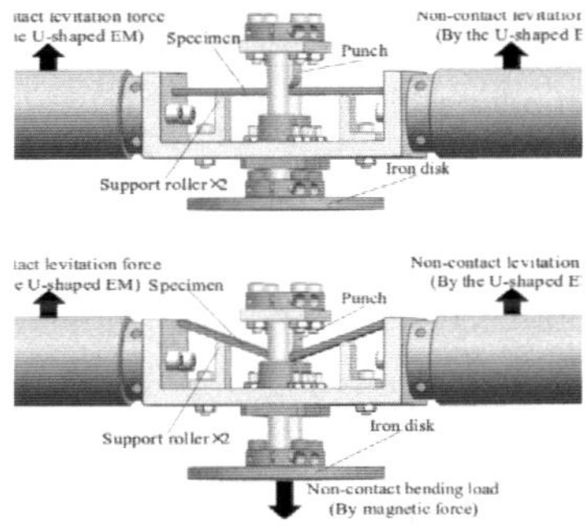

Figure 4 Illustrations of working principle. upper: levitation, lower:applying bending force.

3.2 System modeling

The first step in designing a controller is to get a plant model. Therefore, a state space model for the levitation plant was built as follows. Figure 5 shows an illustration of the levitation plant model. As shown in Figure, i_1 and i_2 denote the currents of the two U-shaped EMs respectively. z_1 and z_2 denote the air gaps between the two U-shaped EMs and the two iron cylinders respectively. F_1 and F_2 denote the attractive force of the two U-shaped EMs to the two iron cylinders. F_b denotes the bending force, which comes from the cylindrical EM and acts to the iron disk. mg denotes the weight of the levitated object. In this model, Both F_b and mg were assumed to always act to the center of mass (COM) of the levitated object. z_M denotes the vertical distance from the COM to the lower surface of the U-shaped EM. θ denotes the tilting angle of the iron cylinders. It was assumed that both the position of the COM and the moment of inertia (MOI) of the levitated object around the COM were invariant even if the punch move.

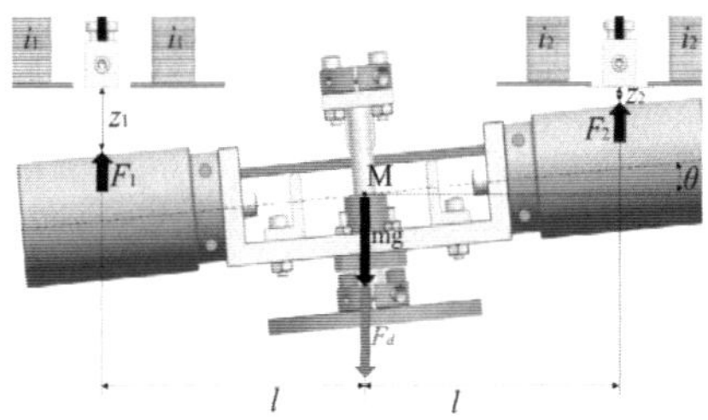

Figure 5 An illustration for levitation plant model

To get a mathematical model, firstly, both F_1 and F_2 were linearized at a working point (i_0, z_0) as follows.

$$\begin{cases} F_1 = F_{1(i_0,z_0)} + k_i(i_1 - i_0) - k_z(z_1 - z_0) \\ F_2 = F_{2(i_0,z_0)} + k_i(i_2 - i_0) - k_z(z_2 - z_0) \end{cases} \tag{1}$$

Where, $F_{1(i0,z0)}=F_{2(i0,z0)}=(mg+F_b)/2$, k_i and K_z are the current coefficient and displacement coefficient of a single U-shaped EM respectively. The dynamics of the levitated object is described as follows.

$$mg - (F_1 + F_2) = m\ddot{z}_M,\ (F_2 - F_1)l = J\ddot{\theta} \tag{2}$$

where, m is the mass of the levitated object; J is the moment of inertia of the levitated object around the COM. According to the geometric relationship, z_M is equal to $(z_1+z_2)/2$. In addition, assuming that θ is extremely small, θ can be considered to be equal to $(z_1-z_2)/2l$. Furthermore, defining $\Delta z_+=z_1+z_2$, $\Delta z_-=z_1-z_2$, $\Delta i_+=i_1+i_2$, $\Delta i_-=i_1-i_2$, and substituting Eq (1) into Eq (2), Two differential equations can be obtained as follows.

$$m\Delta\ddot{z}_+ = -2k_i\Delta i_+ + 2k_z\Delta z_+ ,\quad J\Delta\ddot{z}_- = -2k_i l^2\Delta i_- + 2k_z l^2\Delta z_- \tag{3}$$

Then, two transfer functions can be derived from Eq (3) as follows.

$$\frac{\Delta z_+}{\Delta i_+} = \frac{-2k_i}{ms^2 - 2k_z},\quad \frac{\Delta z_-}{\Delta i_-} = \frac{-2k_i l^2}{Js^2 - 2k_z l^2} \tag{4}$$

where, Δi_+ and Δi_- are the inputs of the systems, Δz_{1+} and Δz_- are the outputs of the systems. The transfer functions reflect the dynamic characteristics of Z DOF and RY DOF respectively.

3.3 Fixed stiffness-damping control of levitation

When the bending force is applied, the levitated object will be in a state of alternating pulling up and down, which poses high requirement on the levitation stiffness. Therefore, the higher the levitation stiffness, the better. However, for a certain hardware, the levitation stiffness that it can hold is limited due to noise and time delay. If we use a traditional PID-controller, with the increase of the bending force, the integrator will increase current to compensate for the bending force, as a result, the levitation stiffness will change. No matter the levitation stiffness increases or decreases, it cannot remain at a fixed value, which means the highest levitation stiffness can only exist for a moment. This will bring some negative effect, for example, if we tune the PD-gains to get the highest levitation stiffness at the beginning, after the bending force increases, the levitation stiffness may exceed the highest value, which will beyond hardware's tolerance and cause instability. On the contrary, the levitation stiffness may also decrease with the increasing of the bending force, which may result in insufficient levitation stiffness to resist the bending force. Therefore, a better solution is keeping the levitation stiffness at the highest value all the time. In addition, the role of levitation damping is to suppress oscillation caused by levitation stiffness, therefore, it is better to keep the levitation damping constant to match the constant levitation stiffness. Given the above discussion, a fixed stiffness-damping (FSD) control was employed, which avoids the risk of insufficient levitation stiffness and levitation stiffness exceeding the upper limit. FSD is also realized based on PD-controller, the difference is, the PD-gains of FSD are scheduled according to a certain rule. The block diagram of the FSD control is shown in Figure 6.

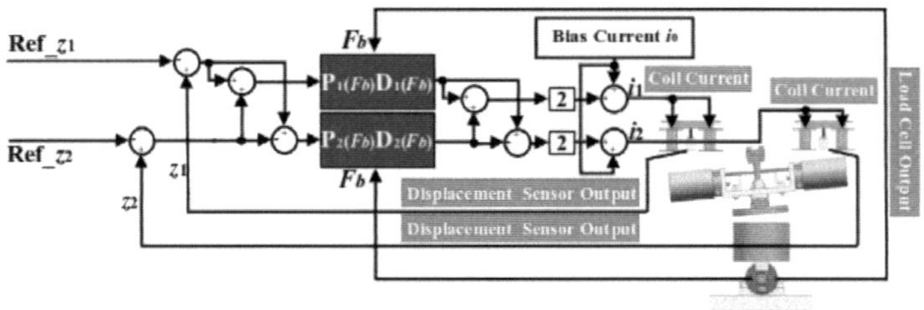

Figure 6 Block diagram of fixed stiffness and damping control system

3.4 Control of the bending force

As shown in Figure 7(a) F_b is the bending force, it is the attractive force of the cylindrical EM to the iron disk. As shown in Figure 7(b), the dynamics of a three-point bending specimen can be regarded as a mass-spring-damper system, where K_S and C_S are the stiffness and viscous damping of the specimen respectively. As l_S is the deflection of the specimen and z_3 is the bottom air gap, the value of z_3+l_S is identified. Therefore, the linearized dynamics model is

$$m_b\ddot{l}_S + C_S\dot{l}_S + (K_S - k_{bz})l_s = k_{bi}i_b \tag{5}$$

where, k_{bz} and k_{bi} is the displacement and current coefficients of the electromagnet.

Our goal is to enable the device to apply ramp forces smoothly over a wide range of materials. For this purpose, as shown in Figure 8, a force control system was designed, and a PI-controller was employed in the force control system because there is a nearly identical case(Ishizaki, at el, 2010) that shown that it is feasible to use PI-controller.

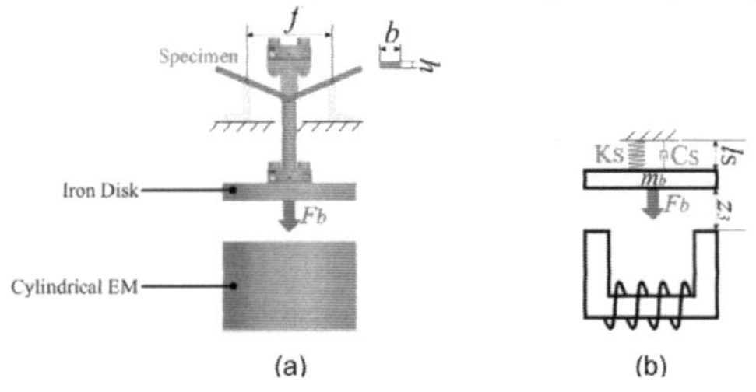

Figure 7 An illustration of bending model

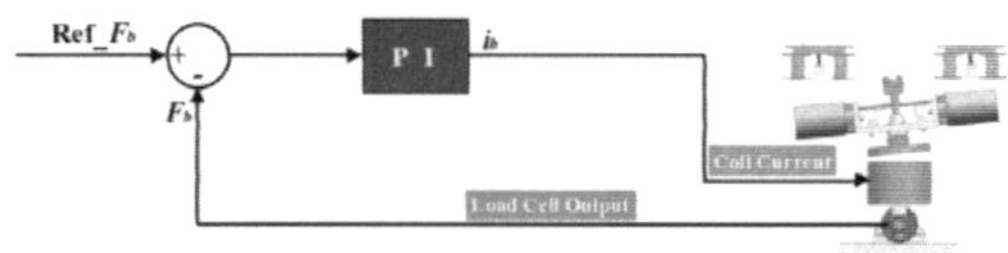

Figure 8 Bending force control system

3.5 Experiment

An experiment was performed to verify the feasibility of the special configuration and FSD control method and to verify the feasibility of the force control system. Figure 9 presents a photograph of the experiment platform.

The experiment was conducted with the identical FSD-controllers and a low-stiffness specimen to test the force control system shown in Figure 8. This is because soft materials will have high deformation and pose high requirement on the force control system, which can test the force control system for good or bad. The specification of the specimen is shown in Table. 1.

Table 1 Specification of material of specimen

Material	*Length*	*Width*	*Height*	*Elastic Modulus*	*Yield Strength*
Z-ABS	86 mm	8 mm	3.7 mm	1080 MPa	30.3 MPa

As shown in Figure 10(a), the air gaps were relatively stable over the 24s. As shown in Figure 10(c), the loading of F_b was relatively stable and smooth over the 24s though there is a little error between the reference input and the actual force in the first 2s; but there had a little oscillation on the F_b curve between 21s and 24s, this is because the iron disk almost had contact with the cylindrical EM due to the large deflection of the specimen; At about 24s, the levitated object fell down, because the iron disk was completely sucked by the cylindrical EM. Overall, the loading process was fairly smooth, the F_b curve can still maintain a slope shape even when the iron disk was extremely close to the cylindrical EM. Therefore, it can be concluded that the force control system is feasible under large deflection.

Figure 9 Experimental setup for examination

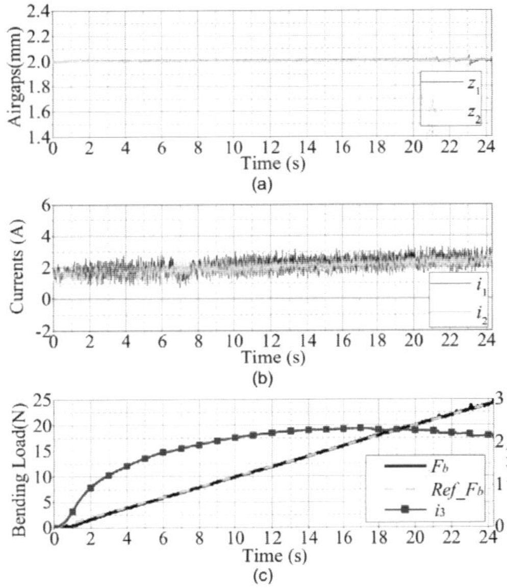

Figure 10 Experimental result with force control system
(a) air gap response, (b) current, (c) force control result

4. Conclusion

This paper presents a non-contact tension and bending testing device using magnetic levitation technology, which is expected to be applied to material testing in special environments. Experiment demonstrated that this structure can guarantee the passive stability of the four DOFs. In addition, a fixed stiffness-damping (FSD) controller was designed to maintain the levitation under the bending force, and a force control system was designed to control the loading of the bending force. The experiment result with the bending test device demonstrated that the force control system allows the device to load the bending force smoothly even under large deflection. It can be seen from the experiments, the air gap oscillation amplitudes with different materials were different, and air gap oscillation amplitude influences test accuracy. Therefore, the dependence of test accuracy on specimen stiffness is worth to be investigated in our future work.

References

Journal article

D. K. Supreeth, S. I. Bekinal, S. R. Chandranna and M. Doddamani, "A Review of Superconducting Magnetic Bearings and Their Application," IEEE Trans. Appl. Supercond., vol. 32, no. 3, pp. 1-15, Apr. 2022.

K. N. V. Prasad and G. Narayanan, "Electromagnetic Bearings With Power Electronic Control for High-Speed Rotating Machines: Review, Analysis, and Design Example," IEEE Trans. Ind. Appl., vol. 57, no. 5, pp. 4946-4957, Sept.-Oct. 2021.

X, Wang, et al. "Study on the Effect of Suspension Windings on the Electromagnetic Vibration and Noise of Ultra-high-Speed Bearingless Permanent Magnet Synchronous Motor for Air Compressor," IEEJ Trans. Electr. Electron. Eng., vol. 18, no. 1, pp. 129-138, Sept. 2023.

T. Masuzawa, S. Ezoe, T. Kato, Y. Okada, "Magnetically Suspended Centrifugal Blood Pump with an Axially Levitated Motor", Artificial Organs, vol. 4, issue6, pp. 468-474, June 2000

P. Wu, et al. "On the optimization of a centrifugal maglev blood pump through design variations," Front. Physiol., vol. 12, pp. 699891, June 2021

H. Wan, et al., "Zone melting under vacuum purification method for high-purity aluminum," Journal of Materials Research and Technology, vol. 17, pp. 802-808, Jan. 2022.

C. Kumar and S. Sarangi, "Dynamic behavior of self-acting gas-lubricated long journal bearing," Mechanics Research Communications, vol. 124, pp. 103950, Aug. 2022.

Natarajan, Elango, et al. "Experimental and numerical analysis on suitability of S-Glass-Carbon fiber reinforced polymer composites for submarine hull," Defense Technology, vol. 19, pp. 1-11, Jan. 2023.

Conference article

A. Kumashiro, A. Chiba, W. Gruber, W. Amrhein and G. Jungmayr, "Novel Reluctance-type Magnetic Geared Motor with Integrated with High-speed Bearingless Motor," 2022 Int. Conf. Power Electron. (IPEC-Himeji 2022- ECCE Asia), Himeji, Japan, 2022, pp. 1762-1768.

N. Tada and H. Masago, "Remotely-controlled tensile test of coppercored lead-free solder joint in liquid using permanent magnet," 2013 8th International Microsystems, Packaging, Assembly and Circuits Technology Conference (IMPACT), 2013, pp. 186-189.

N. Tada and T. Uemori, "Time-dependent fracture of copper-cored lead-free solder ball and nickel rod joints in air, distilled water, and NaCl solution," in Proc. International Microsystems, Packaging, Assembly and Circuits Technology Conference, Oct. 2017, pp. 168-171.

Ishizaki, Takayuki, et al., "PI control system design for Electromagnetic Molding Machine based on Linear Programing," in Proc. IEEE International Conference on Control Applications, Sep. 2010, pp. 2415-2420

Proposal of a One-DOF Actively Controlled Bearingless Motor Using Zero-Sequence Current

Yusuke FUJII [a]

a Tokyo Institute of Technology, 2-12-1 Ookayama, Meguro, Tokyo, Japan, fujii@ee.e.titech.ac.jp

Abstract

This paper proposes a one-DOF actively controlled bearingless motor with 1 kW output power using a three-phase four-wire inverter system equivalent to a three-phase inverter in terms of the number of power-switching devices. In the proposed system, the thrust suspension winding is connected between the neutral point of the Y-connected motor windings and the middle point of the two capacitors in the DC link. The zero-sequence current flowing into the thrust suspension winding generates active thrust force, while the motor is driven by dq-axis vector control. The flux of bias PMs for the thrust magnetic suspension is not flowing into the motor stator core, and the proposed motor design is superior to the conventional design. Additionally, the flux, which is generated by the zero-sequence current flowing into the motor winding, contributes to increasing the thrust force. To demonstrate the performance of both the motor and magnetic suspension, FEA was conducted. It was found that the proposed bearingless motor system has a potential to achieve both stable magnetic suspension and 1 kW-output.

Keywords: *Bearingless motor, Magnetic bearing, Magnetic suspension, Permanent magnet motor, Zero-sequence current*

1. Introduction

Bearingless motors generate rotational torque while the rotor is suspended in a non-contact manner, therefore overcoming many problems related to mechanical bearings. To achieve magnetic suspension in non-contact manner, three translational (x, y, z) and two tilting (θx, θy) motions of the rotor must be magnetically stabilized. Although five-degree-of-freedom (DOF) controlled bearingless motors which actively regulate all of the rotor motions have been developed (Asama, et al., 2010) and (Baumgartner, et al., 2014), they require numerous permanent magnets (PMs), electromagnets, and inverters. Therefore, five-DOF bearingless systems have higher cost and larger size.

A reasonable solution for simplification of the bearingless system is to minimize the number of active magnetic suspensions. One-DOF actively controlled bearingless motor has been developed (Bauer, et al., 2014).The axial force was generated by regulating a suspension current superimposed onto three-phase motor currents. Although the machine is effective for downsizing and simplification of the system, the system requires six half-bridges for independently regulating three types of currents. Another one-DOF actively controlled bearingless system utilizing a single-phase motor has been developed (Gruber, et al., 2019). The axial force was generated by regulating a suspension current superimposed onto single-phase motor currents. The system requires four half-bridges for independently regulating two types of currents. General vector control with dq-axis currents cannot be performed because of the single-phase motor. For further system simplification, single-drive bearingless motors with one-axis active positioning have been developed: axial-gap type (Asama, et al., 2021) and radial-gap type (Sugimoto, et al., 2017). The axial force and torque are generated by d- and q-axis currents, respectively. The remaining radial and tilting positions (x, y, θx, and θy) are passively stabilized by repulsive passive magnetic bearings (RPMB) consisting of the stator PMs and rotor PMs. The system is driven by only one three-phase voltage source inverter. However, flux weakening control with dq-axis vector control, which is able to extend the rotational region with high efficiency, cannot be applied to the single-drive bearingless motor because the d-axis current is dedicated to the axial positioning. Moreover, as far as the authors know, one-DOF controlled bearingless motor over 1 kW has never been reported.

This paper proposes a one-DOF actively controlled bearingless motor with 1 kW using a three-phase four-wire inverter system equivalent to three-phase inverter in terms of the number of power switching devices. In the proposed system, the thrust suspension winding is connected between the neutral-point of the Y-connected motor windings and the middle point of the two capacitors in the DC link. The zero-sequence current flowing into the thrust suspension winding generates active thrust force, while the motor is driven by dq-axis vector control. This paper describes a proposed system including bearingless structure, power electronics circuit, and suspension principle, and then FEA results show the performance of the proposed bearingless motor.

2. Proposed system

2.1 Three-phase four-wire inverter using zero-sequence current

Fig. 1 shows a proposed circuit using a zero-sequence current, i_z (Fujii, et al., 2021). The DC link is divided by two capacitors. The thrust suspension winding is connected between the neutral-point of the Y-connected motor windings and the middle point of the two capacitors. The current flowing into this line is defined as the zero-sequence current i_z, and consequently it generates active thrust suspension force while the three-phase motor is driven with dq-axis vector control, without additional power switching device.

When motor drive and magnetic suspension are performed, the phase currents (i_u, i_v, and i_w) and the terminal voltages (v_u, v_v, and v_w) are expressed as follows:

$$\begin{aligned} i_u(t) &= I_m \cos(\omega_m t + \beta) + i_z(t)/3 \\ i_v(t) &= I_m \cos(\omega_m t + \beta - 2\pi/3) + i_z(t)/3 \\ i_w(t) &= I_m \cos(\omega_m t + \beta - 4\pi/3) + i_z(t)/3 \end{aligned} \quad (1)$$

$$\begin{aligned} v_u(t) &= V_m \cos(\omega_m t + \phi) + v_z(t) \\ v_v(t) &= V_m \cos(\omega_m t + \phi - 2\pi/3) + v_z(t) \\ v_w(t) &= V_m \cos(\omega_m t + \phi - 4\pi/3) + v_z(t) \end{aligned} \quad (2)$$

where V_m represents the amplitude of the motor voltage, v_z represents the zero-sequence voltage, ω_m represents the angular frequency of the motor, and ϕ represents the initial phase. As shown in (2), the zero-sequence voltage v_z in phase is superimposed onto each terminal voltage. Substituting $V_m = 0$ into (2) yields,

$$v_u(t) = v_v(t) = v_w(t) = v_z(t) \quad (3)$$

Fig. 2 shows a zero-sequence equivalent circuit based on (3). The three-phase four-wire inverter system offers a half-bridge function in addition to the three-phase voltage inverter. It should be noted that the zero-sequence current must not have a DC component because the zero-sequence DC current unbalances the two capacitors. Therefore, the zero-power control is required in thrust magnetic suspension so that the zero-sequence DC current is eliminated.

2.2 Structure

Fig. 3 shows the structure of the proposed one-DOF actively controlled bearingless motor, which consists of the repulsive passive magnetic bearing (RPMB), thrust magnetic suspension, and motor units. The RPMBs are placed at the top and bottom ends of the machine, the radial and tilting positions (x, y, θx, and θy) are passively stabilized by the RPMBs. The inner and outer diameters of the rotor repulsive PMs are 37 mm and 53 mm, respectively, with a height of 7.5 mm. The inner and outer diameters of the stator repulsive PMs are 59 mm and 74 mm, respectively, with a height of 7.5 mm. The thrust magnetic suspension unit is configured with the yoke, thrust core, thrust suspension windings with ring shape, and upper/lower bias PMs. The motor unit placed in the center is configured with laminated stator core that has 12-slots, three-phase 8-pole motor concentrated windings, rotor core and motor PMs. The inner and outer diameters of the stator are 60 mm and 110 mm, respectively. The outer diameter of the upper/lower bias PMs and motor PMs in the rotor is 53 mm. The air gaps in the radial direction of the RPMBs and the motor units are set to 3.0 mm and 3.5 mm, respectively, as comparatively wide gap is needed to avoid the touch-down of rotor in the radial direction. The grades of the repulsive PMs and rotor PMs (upper/lower bias and motor) are N52 and N48, respectively.

Fig. 4 shows the magnetization direction of the rotor PMs. Each upper and lower bias PM is magnetized in a unipolar direction and generates a bias flux for the thrust magnetic suspension. Additionally, the flux of bias PMs is not flowing into the motor stator core, thus it has no significant influence on the motor performance.

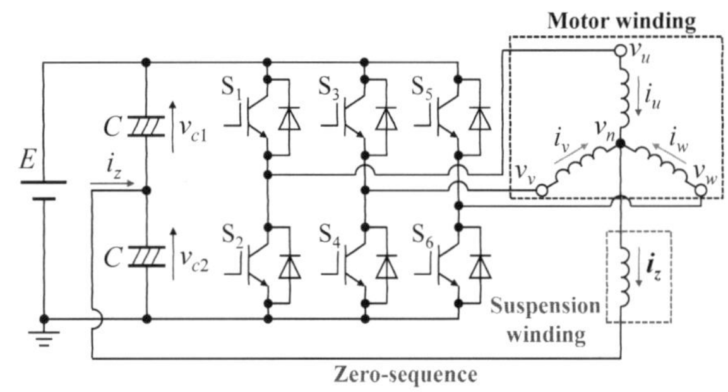

Figure 1 Proposed circuit using zero-sequence current, i_z.

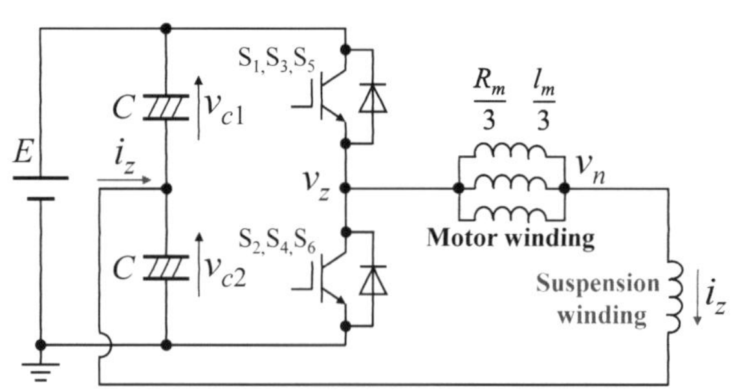

Figure 2 Zero-sequence equivalent circuit.

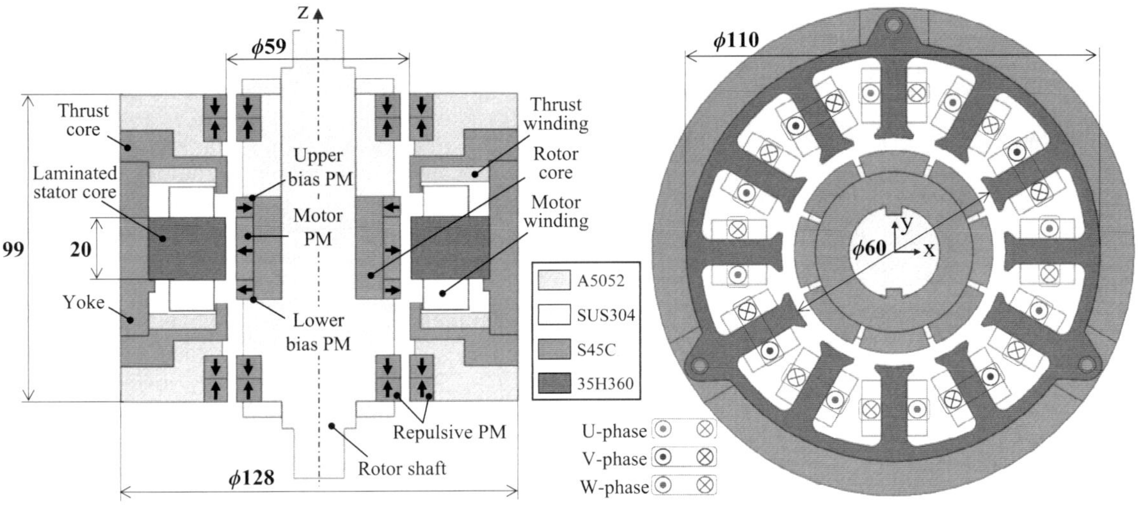

Figure 3 Structure of the Proposed bearingless motor: cross-sectional view of entire model (left) and motor unit (right), respectively.

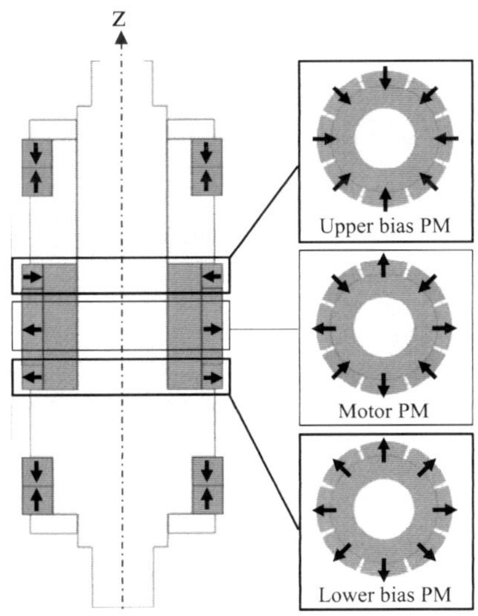

Figure 4 Magnetization direction of rotor PMs.

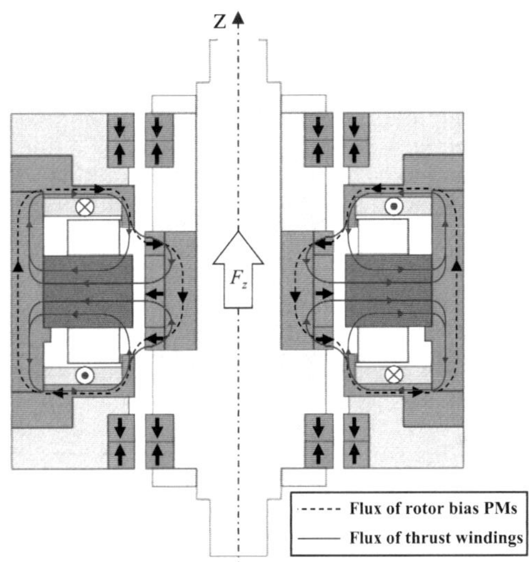

Figure 5 Principle of thrust magnetic suspension.

2.3 Principle of magnetic suspension

Fig. 5 shows the principle of the thrust magnetic suspension, where the positive zero-sequence current is provided into the thrust suspension windings. As indicated by the dotted line in Fig. 5, the flux of the bias PMs passes and loops through the upper bias PMs, rotor core, lower bias PMs, thrust core, and yoke. As indicated by the solid line in Fig. 5, the suspension flux, which is generated by providing the thrust suspension winding with zero-sequence current, passes and loops through the thrust core, rotor core, stator core, and yoke. The flux in the axial direction between the thrust core and rotor core in the upper side becomes denser than that in the lower side, and consequently positive thrust force, F_z, is generated in the rotor. Likewise, the negative thrust force is generated by the negative zero-sequence current.

As indicated in (1), the zero-sequence current for the thrust magnetic suspension flows in the motor windings in addition to the thrust suspension windings. The zero-sequence current $i_z(t)/3$ flowing in the motor windings is denoted as "motor zero-sequence current" herein in order to distinguish it from the zero-sequence current flowing in the thrust suspension windings. The flux generated by the motor zero-sequence current passes and loops through the same place as the thrust suspension flux of Fig. 5. Therefore, it is worth noting that the motor zero-sequence current contributes to increasing the thrust force.

3. FEA analysis

To demonstrate the performance of the proposed one-DOF controlled bearingless motor, three-dimensional (3D) finite element analysis (FEA) was performed using the model shown in the second section.

3.1 Motor torque

Fig. 6 shows the relationship between the torque T and the motor current density J_m. The torque reached 2.87 Nm at a motor current density of 8 A_{rms}/mm^2 (equal to a phase current of 20.1 A_{rms}), which is the rated current. Therefore, the motor is able to achieve 1 kW under the rated current at 3328 rpm.

Fig. 7 shows the torque waveforms at a motor current density of J_m = 8 A_{rms}/mm^2 when providing zero-sequence current. J_z represents the current density of the thrust suspension winding. The torque waveform at J_z = 0 A/mm^2 was identical with that at J_z = 4 A/mm^2. This result implies that the zero-sequence current has no influence on the torque.

3.2 Passive suspension

Fig. 8 shows the force in the x-axis, F_x, acting on the rotor when the rotor is displaced in the x-axis direction, where the average of F_x in one cycle of electrical rotation is plotted. The restoring force was generated by the RPMBs and the rotor positions in the passive directions are stabilized. The total stiffness in the x-axis direction reached 60 N/mm at an x-position of 1.0 mm.

Fig. 9 shows the unbalanced magnetic pull force (UMF) acting on the rotor when the rotor is displaced in the z-axis direction, where the average of the UMF in one cycle of electrical rotation is plotted. The UMF in the z-axis direction was caused by the RPMBs, and was equal to 128 N at the z-position of 0.4 mm in the start-up of the magnetic suspension. Hence, to achieve the thrust magnetic suspension, the active thrust force needs at least 128 N.

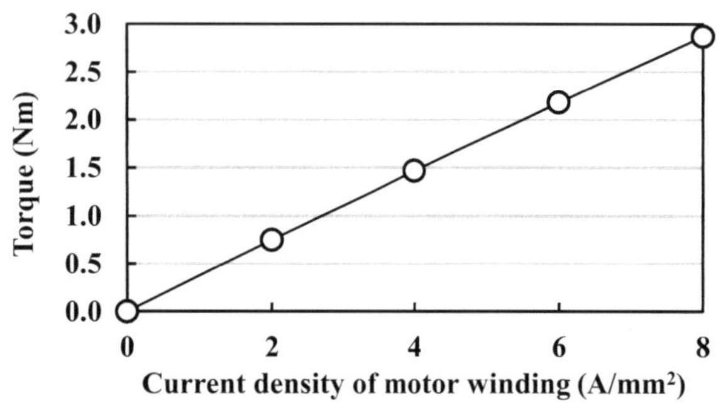

Figure 6 Relationship between torque and motor current density.

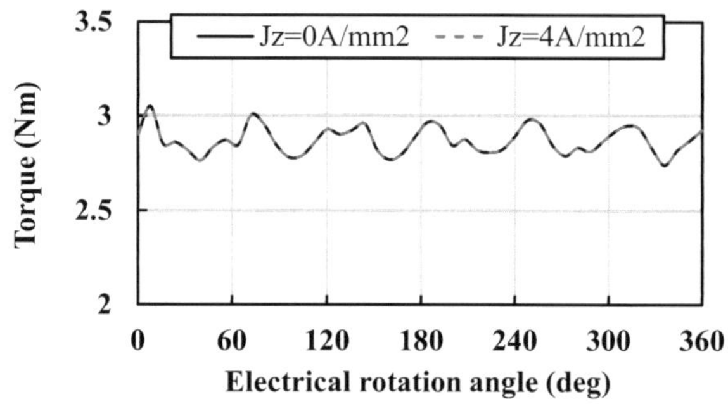

Figure 7 Torque when providing zero-sequence current.

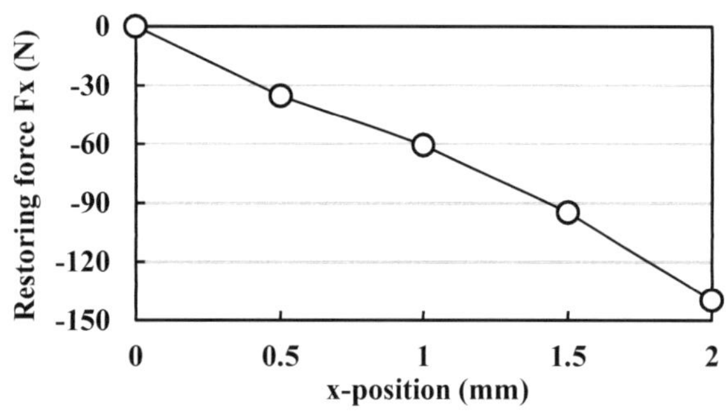

Figure 8 Restoring force in the x-axis direction.

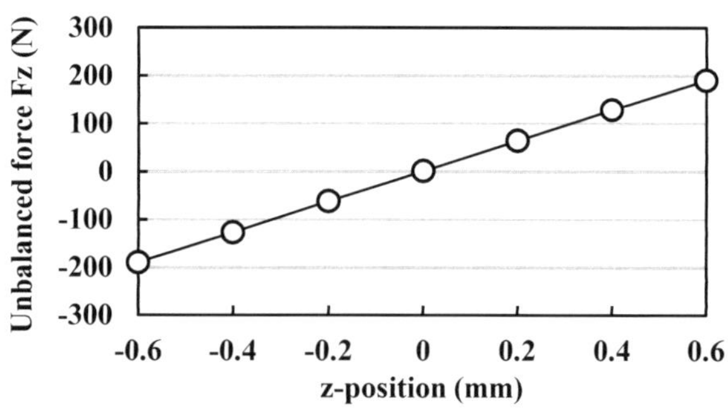

Figure 9 Unbalanced magnetic pull force in the z-axis direction.

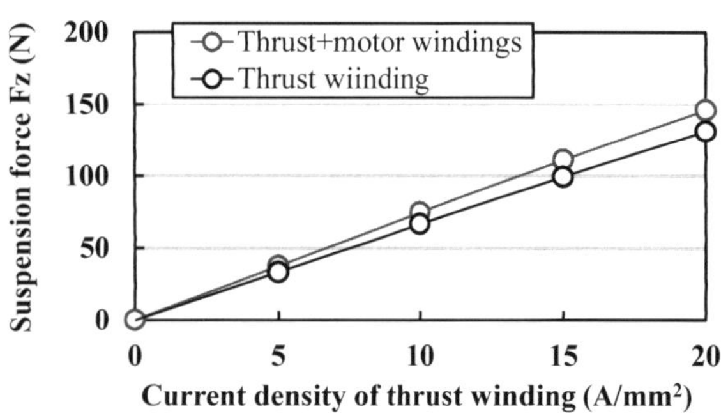

Figure 10 Relationship between thrust force and zero-sequence current.

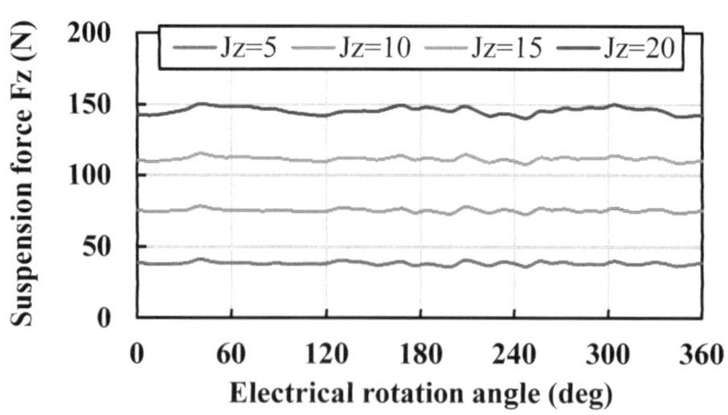

Figure 11 Thrust force waveforms.

3.3 Active thrust magnetic suspension

Fig. 10 shows the relationship between the active thrust suspension force F_z and the thrust current density J_z. The thrust force was increased by the zero-sequence current flowing into not only the thrust suspension winding, but also into the motor winding, and reached 146 N at $J_z = 20$ A/mm^2, which exceeds the UMF in the z-axis direction at the start-up.

Fig. 11 shows the waveforms of the thrust suspension force, Fz. The ripple of the thrust force depending on the rotational angle was comparatively small.

4. Conclusion

This paper proposed a one-DOF actively controlled bearingless motor with 1 kW output power, using a three-phase four-wire inverter system. In the proposed system, the thrust suspension winding is connected between the neutral-point of the Y-connected motor windings and the middle point of the two capacitors in the DC link. The zero-sequence current flowing into the thrust suspension winding generates active thrust force, while the motor is driven by dq-axis vector control. The flux of bias PMs for the thrust magnetic suspension is not flowing into the motor stator core, and the proposed motor design is superior to the conventional design. The 3D-FEA results demonstrated that the proposed motor has a potential to achieve both stable magnetic suspension in a non-contact manner and a 1 kW-output.

References

Asama J, Amada M, Tanabe N, Miyamoto N, Chiba A, Iwasaki S, and Takemoto M (2010) Evaluation of bearingless PM motor with wide magnetic gaps. IEEE Trans. Energy Convers., vol. 25, no. 4, pp. 957–964.

Asama J, Tai T, and Chiba A (2021) Development of axial-flux single-drive bearingless motor with one-axis active positoning. IEEE Trans. Ind. Appl., vol. 57, no. 6, pp. 6792–6799.

Baumgartner T, Burkart R, and Kolar J (2014) Analysis and design of a 300-W 500 000-r/min slotless self-bearing permanent-magnet motor. IEEE Tran. Ind. Electron, vol. 61, no. 8, pp. 4326–4336.

Baur W, Freudenthaler P, and Amrhein W (2014) Experimental characterization of a bearingless rotating field axial-force/torque motor. In: 14th International Symposium Magnetic Bearings, 2014, pp. 132–137.

Fujii Y, Asama J, and Chiba A (2021) Voltage sensorless control of split capacitor for three-phase four-wire motor system with zero-sequence suspension winding. IEEE Trans. Ind. Appl., vol. 57, no. 6, pp. 6823-6832.

Gruber W, Baur W, Wetsch D, Klammer B, and Kurita N (2019) Implementation of a bearingless axial-force/torque motor fan with flex-PCB windings. In: 2019 International Electric Machines and Drives Conference, 2019, pp. 179–184.

Sugimoto H, Shimura I, and Chiba A (2017) A novel stator structure for active axial force improvement in a one-axis actively positioned single-drive be aringless motor. IEEE Trans. Ind. Appl., vol. 53, no. 5, pp. 4414-4421.

A Comparison of GaN Transistors to MOSFETs for Active Magnetic Levitation

Alexander H. PESCH [a], Nyandieka H. ABOKI [a], Jairo E. VILLAMIZAR-VASQUEZ [b], Frank X. LI [b], and Vamsi BORRA [b]

a Mechanical Engineering Program, Rayen School of Engineering, Youngstown State University, 1 University Plaza, Youngstown, OH, USA, ahpesch@ysu.edu

b Electrical and Computer Engineering Program, Rayen School of Engineering, Youngstown State University, 1 University Plaza, Youngstown, OH, USA,

Abstract

Active magnetic bearings have higher efficiency than other bearings because there is no physical contact. However, this benefit is mitigated by the addition of electrical power consumption. It is therefore important for magnetic bearings to have efficient power electronics. Gallium nitride based transistors are a relatively new form of transistor which have shown to be more efficient than MOSFETs. There is insufficient body of literature in the area of application of these transistors for magnetic levitation. This work presents a simple 1 degree-of-freedom voltage controlled levitation experiment in which levitation is achieved with both a gallium nitride transistor and a MOSFET. Voltage losses and current consumption are measured when using each transistor during levitation. In particular, transients during open to closed and closed to open states are measured for PWM pulsing. It is found that the gallium nitride transistor is superior in both switching time and efficiency in situ for magnetic levitation.

Keywords: *Magnetic Bearing, Gallium Nitride, GaN, Voltage Control*

1. Introduction

Active magnetic bearings (AMBs) utilize a magnetic field to suspend a rotor whose relative position is actively monitored and corrected by a controller. AMBs offer a competitive advantage in high-speed rotating machines, vibration isolation and high precision machining applications due to their contactless nature and ability to control and monitor the rotor position (Schweitzer and Maslen, 2009). Recent research has focused on the actuation and control subsystems of AMBs. Force generated by AMBs has a nonlinear relationship with the supplied coil current. AMBs are inherently unstable due to the attractive nature of electromagnets. Therefore, there is a need for a controller to stabilize the system.

Most industrial AMBs rely on a current control strategy which requires amplification of the command signal from the controller to a control current which is fed to the coils. Switching power amplifiers (SPAs) are the most preferred in the current control scheme because of their improved efficiency over the linear power amplifiers (LPAs) (Tonoli, et al., 2012). The application of metal-oxide-semiconductor field effect transistors (MOSFETs) based switching power amplifiers for current control has been well studied over time with suggested improvements from different researchers. Gallium nitride transistors (GaNT) based SPA have been suggested for a wide bandwidth and high switching frequency for the current control scheme (Hu, et al., 2023). GaN are a new type of power electronics component which tend to have lower switching thresholds and better on-state resistance than traditional MOSFETs (Hu, et al., 2023). Utilization of the wide bandwidth and high switching frequency lowers the current ripple and improves performance. For a better current response and reduced current ripple simultaneously, researchers suggested a dual mode power drive system, 2-level pulse width modulation and 3-level pulse width modulation techniques (Chiu and Tsai, 2019), (Yu, et al., 2020), and (Rao, et al., 2023).

However, many suggested improvements on the current control SPAs increase the complexity of the electronic circuitry of the AMBs (Hu, et al., 2023). To eliminate the limitations of MOSFET bandwidth, and its low switching frequency together with the additional circuit subsystem of a current control scheme, a voltage control scheme utilizing the superior characteristics of GaN is suggested. To evaluate the performance of GaN in a voltage control scheme for AMBs, an experimental comparison between GaN and MOSFET for voltage controlled active magnetic levitation was performed.

This paper presents a comparison of the performance of GaN and MOSFET in a voltage-controlled scheme for active magnetic bearings. An experimental study is carried out using GaN on an AMB and compared with MOSFETs on the same

AMB. The current, voltage, and position of the ferromagnetic target is monitored for each case and the results compared. A position setpoint step response Is performed with the GaN and the MOSFET. The overall system performance is evaluated from the results obtained. With the GaN posing higher switching frequency, with a wider bandwidth and reduced switching losses, a superior performance is expected.

This paper is structured as follows. The next section presents the basic ideas of voltage control for AMB levitation and compares it to the more common current control method. Also, a brief introduction to GaN and how they compare to MOSFETs is covered. Then, Section 3 details the experimental test rig used to compare GaN to MOSFETs for single axis levitation. Experimental levitation results are presented in Section 4 using both types of transistors. Finally, some closing remarks are made in Section 5.

2. Theory

2.1 Voltage Control for Levitation

The basic force function for active magnetic levitation is shown in Eq (1). The first term is positive and is for upward force of a top electromagnet. The second term is negative for the downward force of the opposing electromagnet.

$$f = \frac{\mu_0 A N^2}{4}\left(\left(\frac{i_b+i_c}{g-x}\right)^2 - \left(\frac{i_b-i_c}{g+x}\right)^2\right) \tag{1}$$

The parameters μ_0 (permeability of free space), A (pole area), N (number of coil windings), i_b (bias current), and g (nominal air gap) can be considered as constants. x is the position of the ferromagnetic target rotor measured from the center of the bearing. For ease of analysis, the force equation is usually linearized at the operating point in terms of the control variables, current and position.

$$f = k_i i_c + k_x x \tag{2}$$

The levitation problem is inherently unstable; as the rotor rises, the upward force increases, etc. For stable levitation, the control current i_c must be changed in real time to compensate for rotor deflections. This leads to the natural conclusion to control coil current as a feedback of rotor position.

However, current in the coil cannot be dictated directly. It has its own dynamics, even at its simplest, neglecting back emf and other effects:

$$\frac{di}{dt} = \frac{v-iR}{L} \tag{3}$$

Here, applied voltage v induces change in coil current but that is mitigated by resistance R and inductance L. To achieve desired current, and therefore desired levitation force, a servo amplifier is often utilized which has its own current sensor and feedback control law which can automatically account for electrical disturbances or changes in load. The desired current is set by the levitation controller in an outer control loop. That desired current serves as the set point for the inner control loop. This common AMB levitation strategy is illustrated in Fig. 1a.

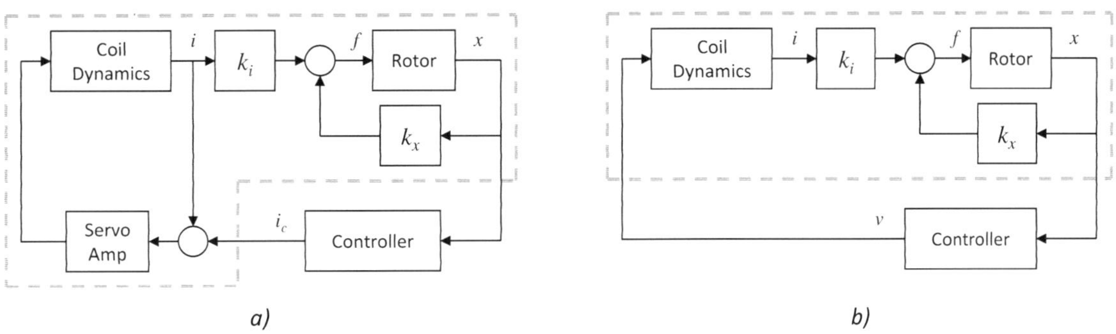

Figure 1 AMB Control Schemes, a) Typical Current Control, and b) Simple Voltage Control used in the Current Study.

The inner current control loop must be faster than the outer levitation control loop, ideally by at least an order of magnitude. Communication from the levitation controller to the current controller, which may be digital or analog, may reduce controller bandwidth. And, the servo amplifier is often expensive. Voltage control for levitation can be implemented without these issues.

A simple form of voltage control for AMB levitation is illustrated in Fig. 1b. Here, the rotordynamics, AMB force function, and coil dynamics are considered as a single plant. The control input to that plant is voltage which can be implemented readily as an equivalent analog with a fast-switching transistor. When the transistor is switched on or off, the suddenly applied voltage induces a transient current in the load. By varying the percentage of time the voltage is switched on, the so-called duty cycle, to that of the total of a fixed switching frequency current can be made to oscillate about a relative level. If the switching frequency is high compared to the plant dynamics, the current signal may be approximated as an analog signal. Because of this strategy, the switching characteristics of the transistor can be significant to the efficiency of the implementation. The next section compares the general switching characteristics of MOSFETs and GaNTs.

2.2 GaNs vs MOSFETs

Nowadays, power electronics require more power, more efficiency, and greater capacity for miniaturization. Also, new devices require operation at higher frequencies. However, operation at higher frequencies will mean greater losses on switching elements (Boutros, et al. 2013).

Currently, machines using power electronics rely almost exclusively on silicon-based power devices which is a well-established technology. However, every day, the power requirements keep growing at a high rate, while power electronic devices make incremental advances to the mature technology. On the other hand, high efficiency applications require low stand-by current, fail-safe devices with high current/voltage capability, and minimum cooling (Ma, 2019).

GaNTs appeared as a disruptive technology that will replace MOSFET in the mid-term future. In fact, GaN switches are projected to have a 100× performance advantage over silicon-based devices and 10× over SiC (Boutros, et al. 2013). Some of the main advantages of GaN over silicon devices are the breakdown voltage it can support, and the specific on-resistance value it can reach (Meneghini, et al., 2017). The construction properties of GaN enable it to have high theoretical value of $V_B^2/R_{on,sp}$ = 5000 MW/cm² (V_B is breakdown voltage and $R_{on,sp}$ is resistance) compared to 40 for Si super-junction MOSFETs and 640 for SiC. These advantages allow GaN to reach higher off voltages without facing temperature issues. Also, it enables the heat sink to be smaller than one used for a MOSFET in an equivalent application. Another advantage of GaN devices is the high saturated drift velocity, which allows operation at more than 10× the switching frequency of Si devices with the same switching loss. Operating at 10× higher frequency allows reduction of the size of passive components, contributing to the miniaturization of power machines (Ma, 2019). In Fig. 2, switching losses of SiC device and GaN are compared. It can be noticed that energy consumption using GaNTs is more efficient at higher frequencies.

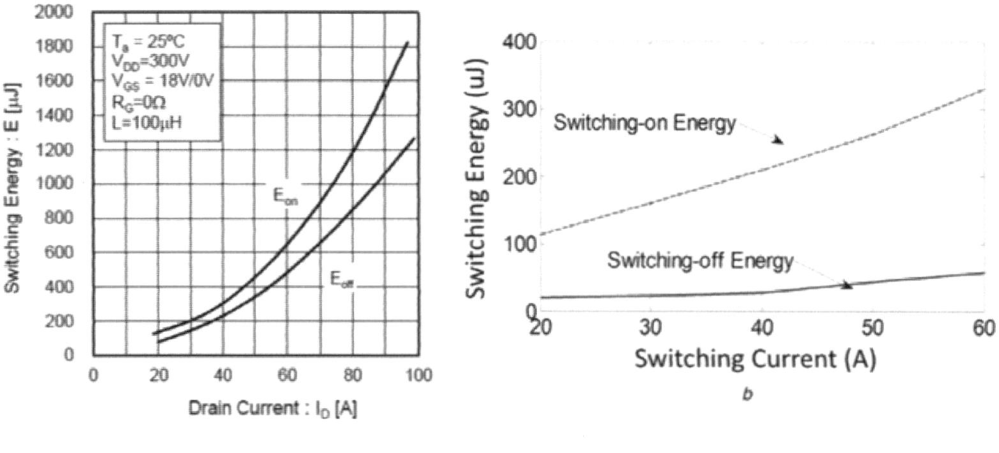

Figure 2. Double-pulse-test results of SiC and GaN a) DPT results of SiC MOSFETs (SCT3022AL), b) DPT results of GaN HEMTs (GS66516T) **(Liu, et al., 2017).**

In (Lui, et al., 2017), a thermal comparison between SiC MOSFET (SCT3022AL) and GaNT (GS66516T) was performed. A DC current was flown through the GaN device. The product of this current and voltage across the transistor represents the power loss. At the same time, two thermistors were attached to the transistor and to an affixed heat sink. The same experiment was performed with the MOSFET device. As a result, Fig. 3a shows the measured temperature.

For the GaNT, the measured temperature reached 0.62°C/W. In contrast, MOSFET devices yielded 0.8°C/W. In terms of gate-drive power, SiC MOSFET consume more than GaNT, given a higher gate-drive voltage and gate charge. Therefore, at higher switching frequencies the MOSFET's power dissipation will be greater (Lui, et al., 2017). In terms of size, GaN devices are much smaller than SiC MOSFET silicon devices. Figure 3b shows the relevant size difference of these devices with similar capabilities (Lui, et al., 2017).

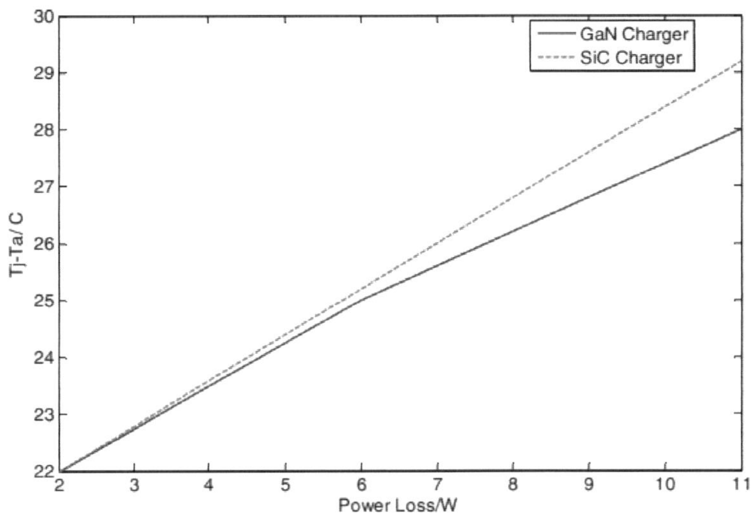

a)

b)

Figure 3. a) Experimental curve of the GaN thermal resistance, **b) Size of Si and GaN counterparts (Liu, et al., 2017).**

3. Experimental Test Rig

The test rig is made up of a 76.2 mm by 76.2 mm by 76.2 mm steel housing. One electromagnet is attached to the bottom of the housing, and another attached to the top. A rotor is suspended between the electromagnets leaving an airgap of 3.175 mm from the top and bottom electromagnet. The two electromagnets used for the construction of the test rig had 480 turns. Each electromagnet was covered with a steel ring around them to control the magnetic fields minimizing the losses of the magnetic force generated. The electromagnets were powered by a 7 VDC supply. To control the airgap, 6.35 mm thick spacers were used. The parameters of the magnetic levitation test rig are shown in Table 1.

Table 1 Test Rig Parameters

parameter	*Description*	*Value*
m	Levitated mass	0.02 kg
k_i	Force –current factor	0.914 N/A
k_x	Force – displacement factor	750.9 N/m
L	Inductance	8.85×10^{-3} H
R	Resistance	3.87 Ω
g	Nominal airgap	3.175×10^{-3} m
N	Number of turns	480
V_s	Supply voltage	9 V

Two infrared sensors are attached to the side of the casing to actively monitor the position of the rotor. The sensors have two emitters on one side and two detectors on the opposite side. They are powered by 3.3V supplied by an Arduino Due. The sensors convert the position of the rotor to an analog voltage signal. The analog voltage signal is input into the 12-bit ADC of the microcontroller for quantization and conversion into a digital signal for processing. The sensors were manually calibrated about the operating point to establish the set point and the relationship between the position and voltage signal which helped implement the control law. Figure 4 shows a photograph and sketch of the experimental test rig.

An ARM Cortex-M3 processor is used to implement the voltage control law. The signal is converted to a digital signal and compared with the calibrated setpoint. The error signal generated from this comparison is fed into a PID controller algorithm for generation of a control signal. A pulse width modulated (PWM) control signal is generated from the controller. The control signal through the MOSFETs or GANT, modulates the voltage supplied to the electromagnetic coils. The different properties of MOSFET and GANT are exploited, and their performance compared.

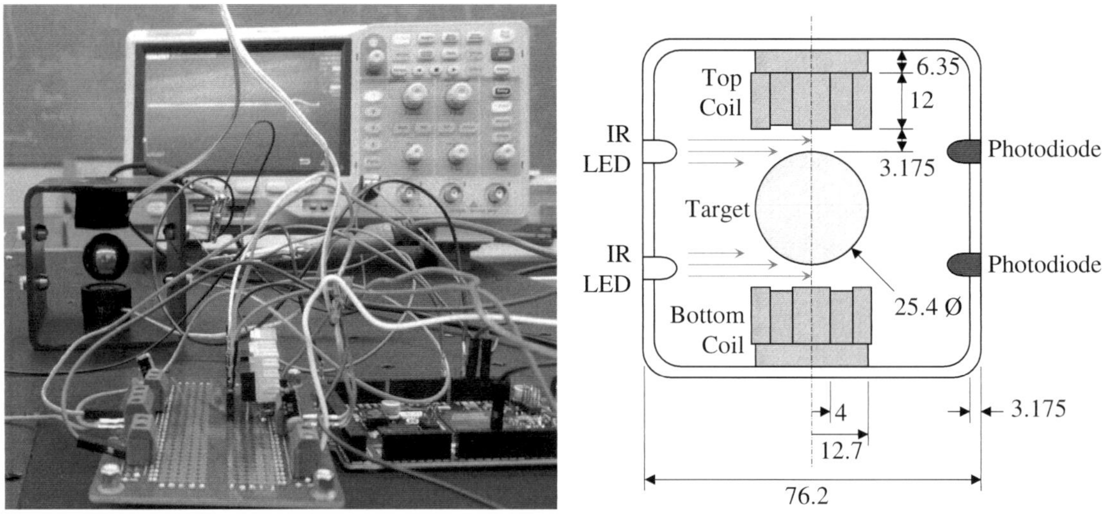

a) *b)*

Figure 4 ***Experimental Test Rig, a) Photograph, and b) Sketch, dimension in mm.***

EPC 2100 GaN and PHP191NQ06LT MOSFET were used for the purpose of this experiment. The first experiment was performed using the GaN transistors and the second one used MOSFET transistors. Both type of transistors receives a PWM signal from the controller and switch to deliver power to the magnets. The specific characteristics of MOSFET and GaN determine the level of noise attenuation, current ripple suppression, and power losses which improves the stability and dynamic performance of the system.

For GaNT implementation, a purchased EPC9059 development board was used. This board contains EPC2100 GaN transistors connected in half-bridge inverter topology. On the other hand, a board for MOSFETs was built to have a similar electrical design. Both transistor boards are shown in Fig. 5. For data acquisition, KEYSIGHT N7026A current probe and KEYSIGHT MSOX4054A Oscilloscope were used.

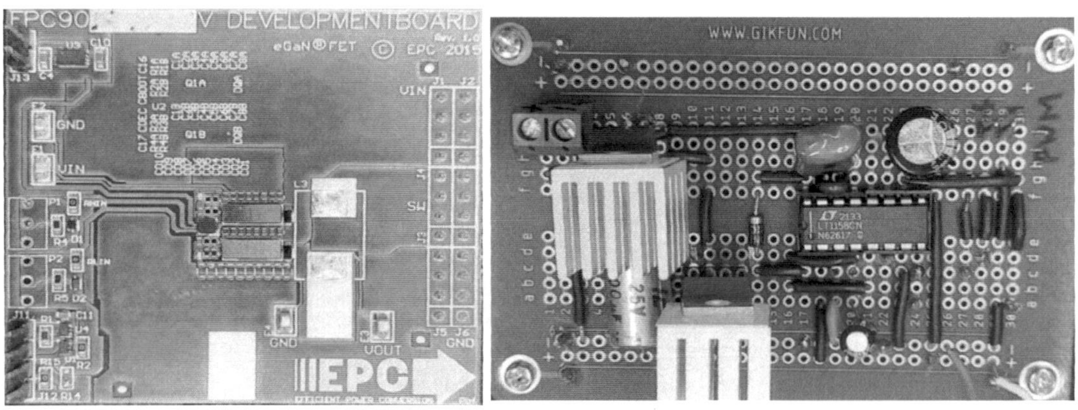

a) *b)*

Figure 5. Power boards, a) EPC9059 GaN development board, and b) MOSFET board based on PHP191NQ06LT.

4. Experimental Results

A levitation experiment was run with the GaN and MOSFET boards. For both tests, drain-source voltage V_{ds} and drain-source current I_{ds} of the top coil transistor were measured. By using this data, switching and conduction losses were estimated. For switching losses analysis, opened to closed and closed to opened transitions were compared for the GaNT and MOSFET. On the other hand, for conduction losses analysis, the dissipated energy is estimated when the transistor is closed.

Switching losses analysis: The energy of a signal is taken as the sum of the power over the switching time. Power is taken as the product of voltage drop across and current through the transistor for the top levitation coil. For one period of the voltage applied to the magnet, there is one closed to opened transition and one opened to closed transition.

Conduction losses analysis: Conduction power losses are estimated by the current squared multiplied by the transistor resistance from the drain to source and the duty cycle. For MOSFET and GaNT, the results will be analyzed for the same duty cycle.

4.1 Levitation using MOSFET

The voltage, current, and power consumption for the MOSFET switching open and closed during levitation are shown in Fig. 6.

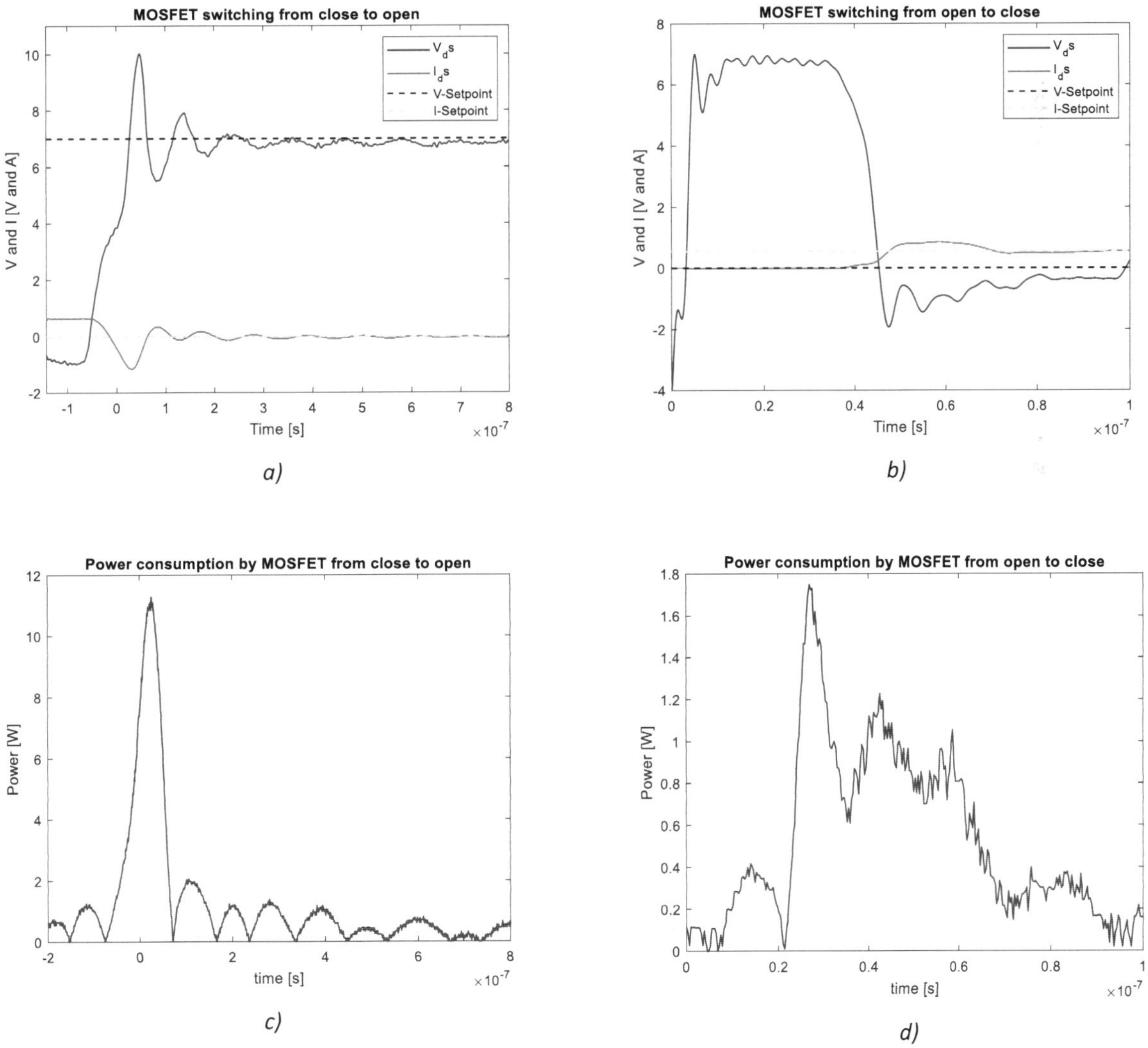

Figure 6. Switching in MOSFET, a) $V_{ds}$and I_{ds} in closed to opened switching, b) V_{ds} and I_{ds} in opened to closed switching, c) power losses in closed to opened switching, and d) power losses in opened to closed switching.

With PWM switching at 1 kHz, the closed to opened transition takes around 0.39 μs to settle. The power losses are observed in Fig. 6. The dissipated energy in the 0.39 μs of switching from closed to opened is 3.21 μJ. For opened to closed transition, it takes 0.48 μs to settle. Also, the dissipated energy in this period is 0.21 μJ. For conductive losses analysis, the drain to source resistance is 75 mΩ. The steady stae current is 0.46 A. Therefore, the power losses with a duty cycle of 60% reach are 6.65 mW.

A position step response is performed during levitation while using the MOSFET resulting in a 48% overshoot with 0.0241 s peak time. The settling time for the step response was 0.12 s with a steady state error of 0.03 V as shown in Fig. 7.

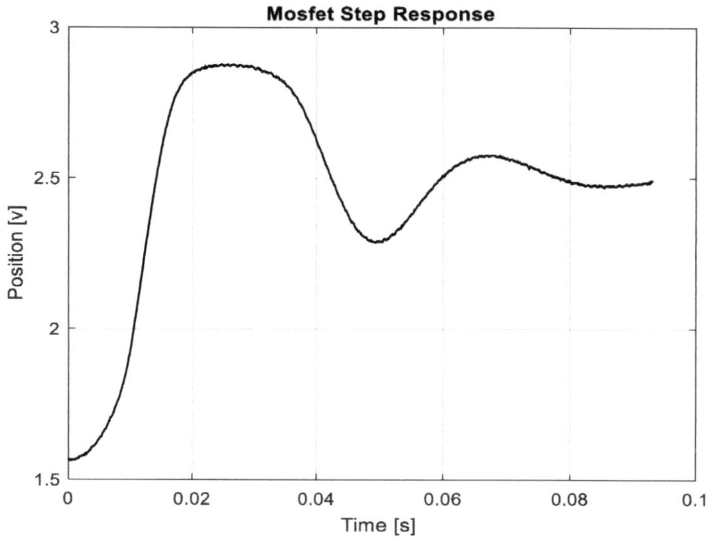

Figure 7. Target Position vs Time using MOSFET.

4.2 Levitation using GaNT

The levitation experiment is repeated using the GaNT, all else being equal. Running at 1 kHz, the closed to opened transition takes around 70.46 ns to settle. The power losses are observed in Fig. 9. The dissipated energy in 70.46 ns of switching from closed to opened is 0.58 μJ. For opened to closed transition, it takes 0.17 μs to stabilize. Also, the dissipated energy in this period is 3.29 nJ. The voltage, current, and power consumption for the MOSFET switching open and closed during levitation are shown in Fig. 8. For conductive losses analysis, the drain to source resistance is 51 mΩ. The steady state current is 0.28 A. Therefore, the power losses with a duty cycle of 41.94% reach are 1.67 mW.

The step response obtained while using the GaN transistors had a 51% overshoot with 0.0212 s peak time. It took 0.0127 s to settle with a 0.02 V steady state error as shown below in Fig. 9. This step response has higher overshoot than that when using the MOSFET. This indicates higher actuator gain and therefore higher closed-loop stiffness leading to lower equivalent damping ratio. Recall that the PID levitation controller is held constant between transistor trials. The higher actuator gain is also leading to a slightly higher natural frequency. Therefore, the use of the GaNT enables the opportunity for renewed controller optimization. Also, the 1 kHz PWM switching frequency is clearly seen in the position signal when levitating with the GaNT whereas it is barely visible when levitating with the MOSFET (See Fig 7). This is also due indirectly to the superior efficiency of the GaNT. Lower total resistance leads to less current smoothing effect of the inductive load. This can be mitigated by faster PWM switching frequency to achieve closer to an approximate analog signal.

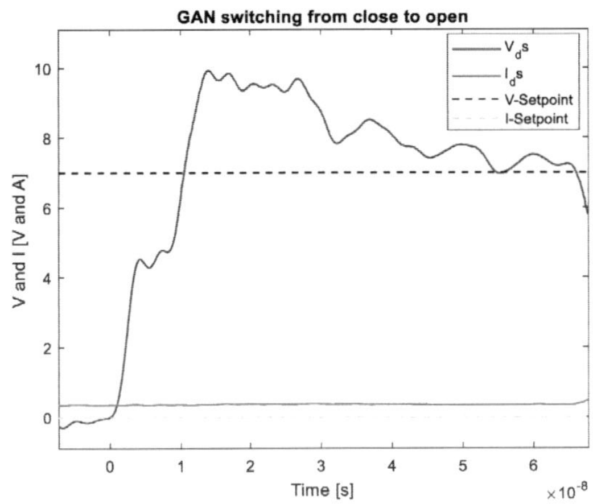

a)

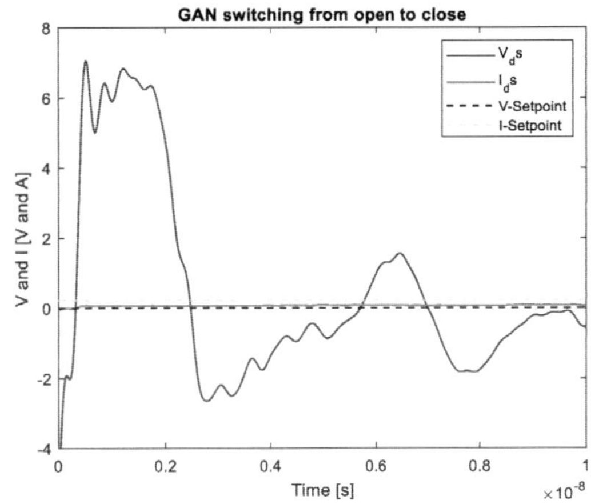

b)

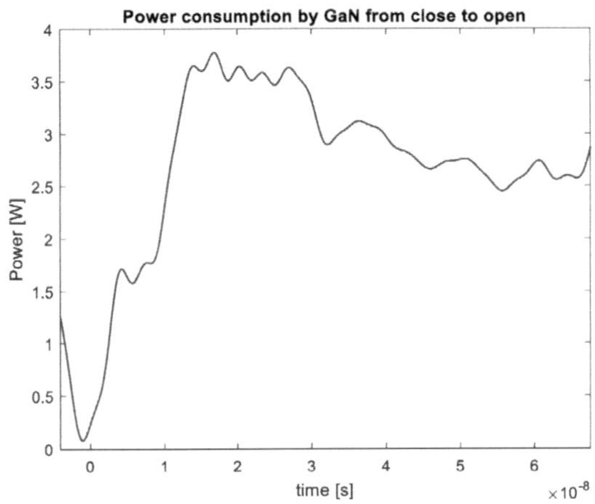

c)

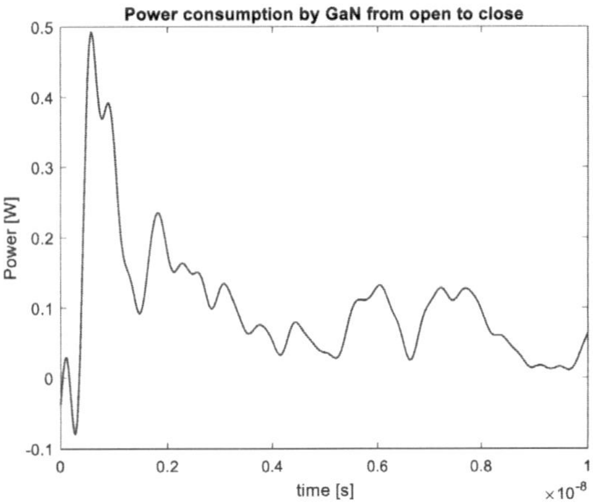

d)

Figure 8. Switching in GaN, a) V_{ds} and I_{ds} in closed to opened switching, b) V_{ds} and I_{ds} in opened to closed switching, c) power losses in closed to opened switching, and d) power losses in opened to closed switching.

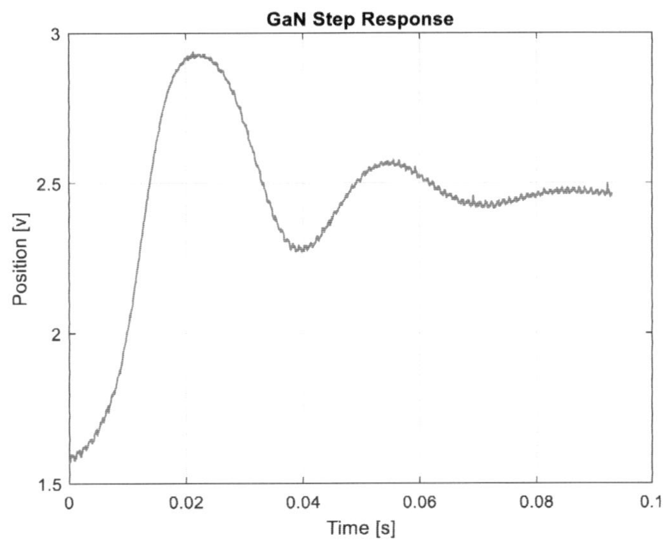

Figure 9. Target position vs Time using GaNT

4.3 Results Summary

Table 2 summarizes the performance of the MOSFET and GaNT during the 1 DoF levitation tests. Overall, the GaNT exhibited lower losses and faster switching time than the MOSFET during levitation.

Table 2 Summary of Transistor Performance

Measure	*MOSFET*	*GaNT*
Switching Losses – Closed to Open	3.21 μJ	0.58 μJ
Switching Losses – Open to Closed	0.213 μJ	3.287 nJ
Switching Time – Closed to Open	0.392 μs	70.47 ns
Switching Time – Open to Closed	0.48 μs	0.17 μs
Conductive Losses	6.65 mW	1.67 mW

5. Conclusions

GaNTs are a relatively new type of power transistor compared to MOSFETs. This work experimentally compared the performance of a GaNT to a MOSFET for a simple 1DoF PWM voltage controlled magnetic levitation test rig. Transistor current, voltage, switching time, and levitation target position were measured for static levitation and a position step response. It was found that the GaNT had faster response time and less losses than the MOSFET.

The more wide spread application of GaNT power electronics to AMBs will lead to AMBs being more efficient and competitive with other types of bearings. Also, the higher actuator gain can enable better AMB performance and the smaller size and less heat generation of GaNTs can lead to more convenient AMB control hardware further increasing the marketability of AMBs.

This study should be repeated for a fully levitated rotor at speed and can be improved by the addition of heat generation measurements.

References

Boutros, K, Chu, R, and Hughes, B (2013) Recent Advances in GaN Power Electronics. In: IEEE 2013 Custom Integrated Circuits Conference, San Jose, CA, USA, 22-25 September 2013.

Chiu, HL, and Tsai, NC (2019) Dual-Mode Power Drive for Active Magnetic Bearings. IET Electr Power Appl. 13(2): 162-167.

Hu, HJ, Liu, K, Want, H, and Wei, JB (2023) A Wide Bandwidth GaN Switching Power Amplifier of Active Magnetic Bearing for a Flywheel Energy Storage System. IEEE Trans Power Electron 38(2): 2589-2605.

Liu, G, Bai, K, McAmmond, M, Brown, A, and Johnson, P (2017) Comparison of SiC MOSFET-based and GaN HEMT-based high-efficiency high-power density 7.2 kW EV Battery Chargers. In: IEEE 5th Workshop on Wide Bandgap Power Devices and Applications, Albuquerque, NM, USA, 11 December 2017.

Ma, B (2019) Driving GaN Power Transistors. In: 31st International Symposium on Power Semiconductor Devices and ICs, Shanghai, China, 19-23 May 2019.

Meneghini, M, Rossetto, I, De Santi, C, Rampazzo, F, Tajalli, A, Barbato, A, Ruzzarin, M, Borga, M, Canato, E, Zanoni, E, and Meneghesso, G (2017) Reliablity and Failure Analysis in Power GaN-HEMTs: an Overview. In: IEEE International Reliability Physics Symposium, Monterey, CA, USA, 2-6 April 2017.

Rao, JS, Arun, V, and Soumendu, K (eds) Lecture Notes in Mechanical Engineering. Available at: http://www.springer.com/series/11693 (accessed 6 May 2023).

Schweitzer G and Maslen EH, (eds) (2009) Magnetic Bearings: Theory, Design, and Application to Rotating Machinery. Heidelberg. Springer Verlag Publishers.

Tonoli A, Bonfitto, A, Silvagni, M, and Suarez LD (2012) Rotors on Active Magnetic Bearings: Modeling and Control Techniques. In: Beltran-Carbajal, F (eds) Advances in Vibration Engineering and Structural Dynamics. Rijeka: IntechOpen, pp. 1-28.

Yu, Z, Gong, L, and Zhu, C (2020) Duty Ratio Stability and Average Steady-State Error of Two-Level Current-Mode Wide Bandwidth Switching Power Amplifiers. IEEE Trans Power Electron 35(4): 4094-4104.

Rotordynamic Design and Control of Three-Stage Centrifugal Compressor with Magnetic Bearings

Sena JEONG [a], Solji RYU [a], Namsoo LEE [a], Myounggyu D. NOH [b]
a Air Solution R&D Center, LG Electronics, 51 Gasan digital 1-ro, Geumcheon-gu, Seoul, South Korea, sena.jeong@lge.com
b Dept. Mechatronics Eng., Chungnam National University, 99 Daehak-ro, Yuseong-gu, Daejeon, South Korea

Abstract

Demand for turbo chillers is increasing for cooling facilities such as large buildings and data centers, and the application of oil-free technology, known for its eco-friendly high efficiency, is expanding to large scale projects. In this study, a large capacity magnetic bearings and rotor dynamics design are described for the development of an oil-free turbo chiller with 2400-3000 USRT performance. First, the rotor and magnetic bearings supporting the three-stage impeller of the large capacity compressor were designed and developed. A constant speed motor rotating at 50 Hz was applied, and the shaft was designed to be long and heavy because of the high load aerodynamic design at low rotation speed. Therefore, the focus was placed on designing sufficient separation margin for the first bending mode. Second, it was important to design the stability of the magnetic bearings with notch filters and PID position control. It is necessary to verify the stability of the bearing support over a wide load range, as efficiency must be maintained not only under maximum load but also during load fluctuations. The system identification of magnetic bearings was used to optimize the controller, and the control stabilization was satisfied through the operation data. Based on these studies, the operation of large capacity turbo chillers was stabilized and the product was successfully commercialized. The chiller introduced in this study was installed in the Middle East and proved its performance by applying a large capacity magnetic bearing for the first time in the oil-free turbo chiller industry.

Keywords: *Centrifugal compressor, Oil-free turbo chiller, Active magnetic bearings, Rotor dynamics*

1. Introduction

The magnetic bearing chiller is an efficient, reliable and cost-effective solution for cooling applications. It utilizes a non-contact bearing system which eliminates the need for lubrication, resulting in reduced maintenance costs and improved efficiency. For this reason, in the HVAC market, demand for products with various capacities to which magnetic bearings are applied is increasing as an alternative to existing oil bearing chillers and screw chillers. In particular, in Middle Eastern countries, where it is difficult to utilize cooling towers due to water shortage, there is a high demand for large air-cooled chillers instead of water-cooled chillers. If oil is used in an air-cooled chiller, oil accumulation in the dry condenser is likely to cause failure. Therefore, in order to build a district cooling system in the Middle East, zero water consumption solutions and oil-free technology are required. That is the need for expansion of oil-free magnetic bearing technology, and it will be of great help to the development of the air-cooled turbo chiller industry in the Middle East. LG Electronics participated in the development of a 2400-3000 USRT class magnetic bearing turbo chiller for installation in Middle Eastern countries. Based on the results of commercialization of a 500-1000 USRT class water-cooled turbo chiller (Noh et al., 2018), we challenged a large capacity air-cooled chiller. In general, expensive high voltage inverter motors are used above 1000 USRT, and constant speed systems using gears have the disadvantage of requiring a separate oil system for gear cooling and lubrication. We tried to implement a system driven directly by a constant speed motor, utilizing a new aerodynamic design that could increase the compression capacity at low constant speed. The goal is to build a constant speed system that does not use a high voltage inverter and develop an oil-free compressor equipped with a large capacity magnetic bearings. With this oil-free technology applied, it is expected to expand its technological advantage and market share in the air-cooled chiller market. In addition, because oil is not used, gear loss and heat exchange loss are reduced, so energy will be saved.

The configuration of the 3000 USRT class oil-free turbo chiller is shown in Fig. 1. The chiller size has a length of 9,355mm, a width of 3,912mm and a height of 4,218mm. Two compressors are fixed on the heat exchanger, and the total weight is about 67 tons. Two compressors are connected in series, and it is a system that compresses a total of five-stages from a two-stage compressor to a three-stage compressor. The three-stage compressor has a longer shaft and heavier impeller weight than the 2-stage compressor. The design and control of rotor bearings with three-stage impellers was challenging. Since the first bending mode is closer compared to the rotor speed of 50Hz, we focused on ensuring the separation margin and stabilizing control according to the new design of the 3-stage compressor.

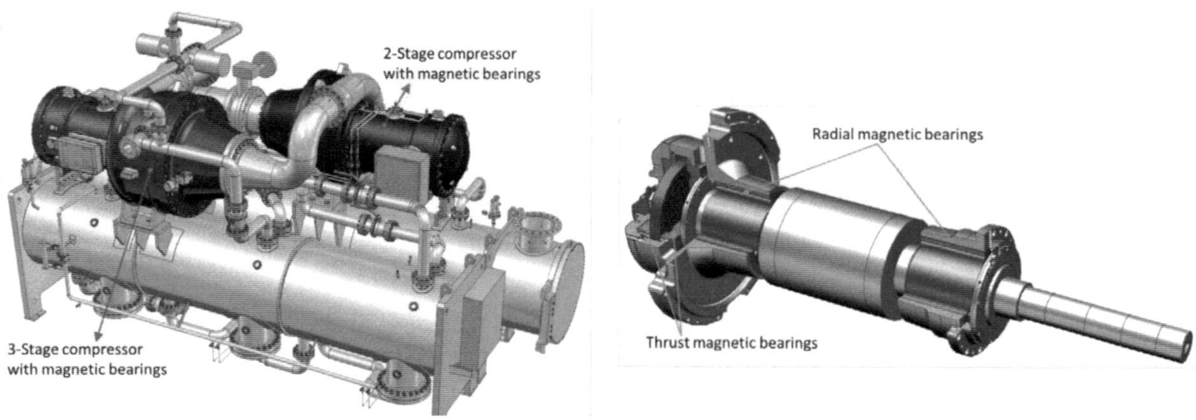

Figure 1 LG Electronics 3000 USRT oil-free turbo chiller with two compressors operated in series (left), and schematic view of rotor-bearing structure for 3-stage compressor (right)

2. Rotor-Bearing design

The shaft has three impellers, and the total length is about 2500mm. The total weight of the rotating parts is 1.3 ton. The mode shape of the rotor is shown in Fig.2. As a result of the rotor stability analysis, the first bending mode was found to be 74 Hz (backward) and 85 Hz (forward), and it was finally confirmed that it had a separation margin of approximately 48% for a constant speed of 50 Hz. In the initial rotor design, the separation margin was not suitable, but the design was modified by reducing the stacking length of the radial magnetic bearing several times. In addition, several modifications were made to reduce weight in the impeller design, while maintaining structural stiffness. As a result, the weight of each impeller was 45 kg, and the magnetic bearing stacking length was designed to be 284 mm. The outer diameter of the magnetic bearing rotor is 248.8 mm, and the bearing load design is 14kN. The load distribution of the radial magnetic bearings supporting both ends of the shaft was calculated as 4900N (NDE) and 7800N (DE).

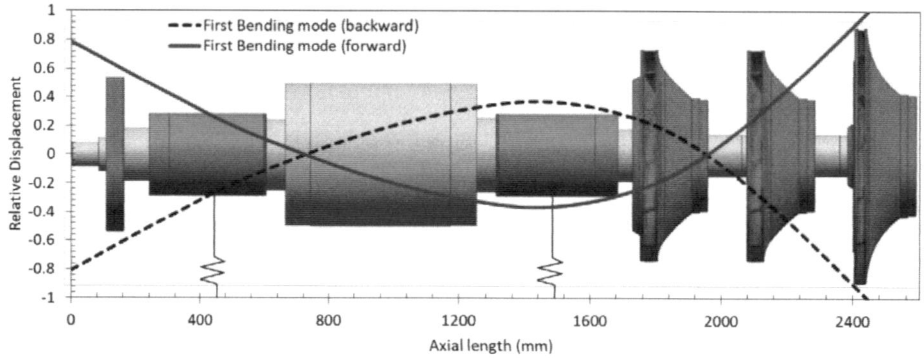

Figure 2 Mode shape of rotor: 1st bending mode 74 Hz (backward), 85 Hz (forward)

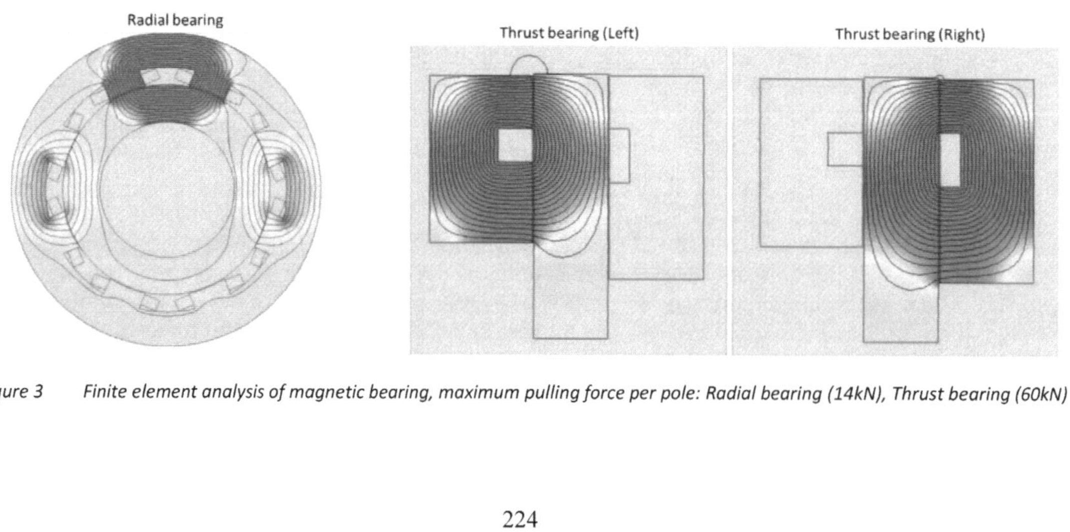

Figure 3 Finite element analysis of magnetic bearing, maximum pulling force per pole: Radial bearing (14kN), Thrust bearing (60kN)

The radial bearings supporting both ends of the shaft were designed to be 14kN, and the safety factor was designed to be more than 1.5 in consideration of the DE Side Bearing to which a lot of load is applied. The bias current is 6A and the maximum current of each pole is 15A. The thrust bearing was designed to support 60kN when 15A was applied, and a safety factor of 2 was applied by predicting the maximum axial load generated from the impeller. Both radial and thrust bearings are designed with an air gap of 1 mm. Fig. 3 shows the finite element analysis results for each bearing.

3. Control design

The magnetic bearing is a system that achieves stabilization through active control, and the performance of the controller as well as the hardware must be evaluated. The design of the controller determines the stiffness through dynamic analysis and designs a controller in the form of a phase lead that obtains stability. Along with this, a notch filter is applied for flexible mode control and a control program is produced. In order to verify the stability of the control design, the system response must be analyzed through the excitation test of the sine wave function. The sine sweep tests were conducted with 900 points from 10 Hz to 1000 Hz. The frequency response results allow identification of the dynamics of the built bearing system. Fig. 4 shows the plant transfer function results for one axis of radial bearing and thrust bearing, respectively. In the plant for radial bearings, the bending modes of the rotor can be identified, the first mode is around 80 Hz and the second mode is around 220 Hz. In the case of the plant for the thrust bearing, no flexible modes appear, but we can see that there are some modes near 150 Hz and 400 Hz. System identification allows more accurate notch filter and PID control design. Fig. 5 shows the sensitivity of each bearing. Based on ISO standard (ISO14839-3, 2006), the sensitivity level satisfies less than 3. The reason that the data before and after controller tuning is almost the same is because the controller was optimized while system identification was repeated several times. Based on this, the stability and reliability of bearing control can be confirmed.

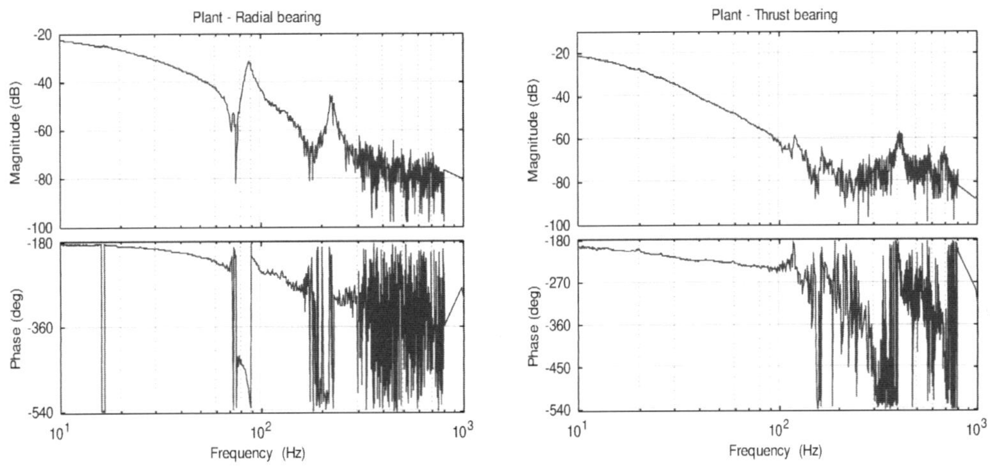

Figure 4 Identified plant transfer function: radial bearing (left), thrust bearing (right).

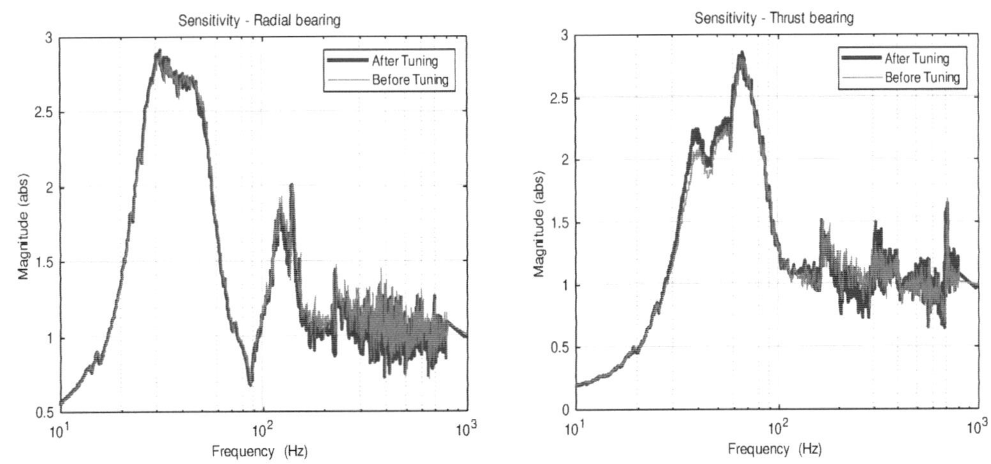

Figure 5 Measured Sensitivity: radial bearing (left), thrust bearing (right).

4. Operation test

The compressor motor operates at constant speed at 50Hz. The refrigerant goes through the two-stage compressor to the three-stage compressor, and the largest load is applied to the three-stage compressor. We observed the vibration of the three-stage compressor in detail, and as a result, it was confirmed that the vibration of the bearing was very good under full load. Fig. 6 shows the raw data and spectrogram of each bearing vibrations. The data sampling rate is 5 kHz, the air gap is 1 mm in the radial direction, and the clearance of the backup bearing is 0.5 mm. In radial bearings, the NDE vibration is slightly larger than the DE side vibration. However, the maximum vibration magnitude does not exceed within 0.1mm. This represents a vibration magnitude of less than 20% of the backup clearance. This result showed a smaller vibration than the criterion of ISO14839, indicating that the magnetic bearings are robust in refrigerant compression systems with load fluctuations or large loads.

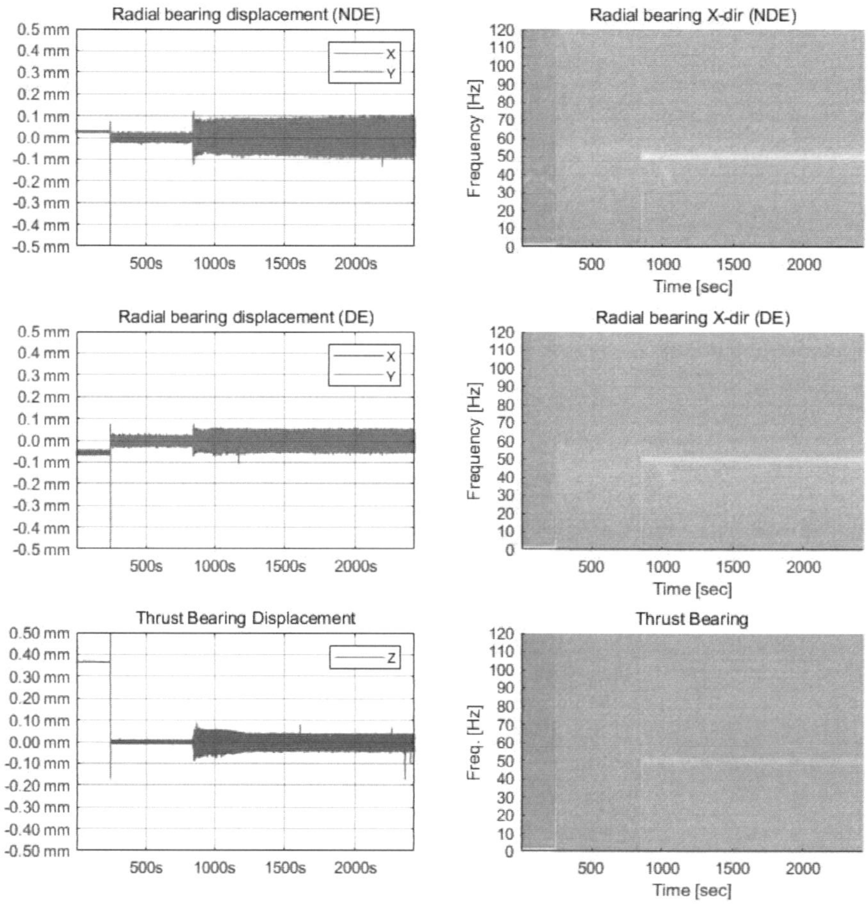

Figure 6 X, Y, Z direction vibrations (raw data) and spectrogram plot of each bearing according to rotation test.

5. Conclusions

In this study, a magnetic bearing system for a three-stage compressor was designed, which required a longer shaft and higher load capacity than a two-stage compressor. Focus on design to stabilize rotor dynamics and verify bearing load capacity. Control optimization is also derived through magnetic bearing design and identification of systems corresponding to high loads. As a result of field operation, the stability of the large-capacity magnetic bearing was confirmed by analyzing the vibration data under high load. With this, LG Electronics achieved its goal of developing an air-cooled oil-free turbo compressor with large capacity magnetic bearings. Based on the results of this study, we can expect to expand the technology of turbo chillers with large capacity magnetic bearings.

References

Noh MD, Jung J, Lee N, Jang S, Lee K and Kim S (2018) Development of Oil-Free Turbo-Chillers Equipped With Magnetically-Levitated Compressors, Proceedings of ISMB 16, Beijing, China.

ISO Standard 14839-3 (2006), Mechanical Vibration – Vibration of Rotating Machinery Equipped with Active Magnetic Bearing: Part 3 – Evaluation of Stability Margin.

Product Carbon Footprint of Cryogenic Turboexpanders Equipped with Active Magnetic Bearings in Hydrocarbon Gas Processing

Louis Mann[1], Gabriel Glynn[1], Dusty Smith[1], Alexandre Kral[2], Ramdane Lateb[2] Antoine Julien[2]
1 Atlas Copco Mafi Trench, Santa Maria, USA
2 SKF Magnetic Mechatronics, St Marcel, France

Abstract

This paper examines the lifecycle greenhouse gas (hereafter "GHG" or "product carbon") footprint of turboexpanders used in an ethylene process, with a focus on comparing those fitted with magnetic bearings versus oil bearings. While the results favor the use of magnetic bearings, the analysis highlights the importance of considering appropriate boundary conditions and working hypotheses, which may introduce biases. For instance, the effectiveness of their sealing systems, as well as the type of gas or simply the simulation models used to evaluate losses, can significantly affect the results. However, this analysis helps quantify the amount of greenhouse gas (GHG) emissions from turboexpanders and indicates reducing emissions would not be the sole factor of bearing technology choice. The choice of magnetic bearings in turboexpanders is mainly driven by other advantages, including reliability, availability, and the elimination of oil contamination, all of which introduce indirect carbon emissions.

Keywords: *Turboexpander, active magnetic bearings, Carbon footprint, greenhouse gas emissions*

1. Introduction – Goal of the Study

Cryogenic turboexpanders are essential components in various oil and gas applications, such as natural gas treatment, gas liquefaction, and petrochemical applications such as ethylene production. Turboexpanders can produce high volume refrigeration by extracting energy from high-pressure fluids and converting it into mechanical shaft power. Detailed descriptions of turboexpanders and their performance can be found in various publications [1]-[3].

The bearings used in turboexpanders play a crucial role in the reliability and safety of the machine. There are two main types of bearings used in turboexpanders: active magnetic bearings (AMB), shown in figure 1, and oil bearings with high-speed journal bearings, tilting pad bearings and sleeve bearings. Turboexpanders with both bearing technologies have proven to have high availability and reliability, with some studies showing AMB having the highest availability of 99.95% [3].

AMBs offer advantages over conventional oil bearings in turbomachinery, such as reduced friction and wear, improved stability and vibration control, the elimination of oil contamination, maintenance cost reduction and finally the reduction of operational cost [4]-[6]. Several studies have investigated the application of AMBs in turboexpanders and their potential benefits in terms of efficiency and reliability [7][8]. Due to the low temperature, high power and zero contamination requirement of many hydrocarbon and ethylene processes, turboexpanders equipped with AMBs are often specified. As a result, the application list for users of turboexpanders with magnetic bearings is quite extensive.

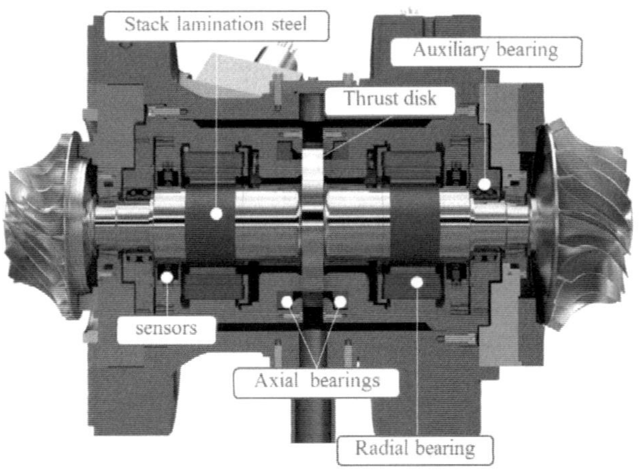

Figure 1 Cross view of a turboexpander with AMBs

The technology of magnetic bearings has become widely prevalent in several fields, making it unnecessary to introduce its principle of operation and control mechanisms [9][10].

Despite the many advantages of turboexpanders and their bearings, several challenges must be overcome to ensure their sustainable and environmentally friendly operation. One of the most critical challenges is the emission of greenhouse gases, primarily methane, during the processing and transportation of natural gas. The energy industry has come under significant pressure to reduce its carbon footprint and adopt cleaner technologies. Therefore, it is crucial to assess the environmental impact of turboexpanders and develop strategies to mitigate their emissions.

The use of magnetic bearings in certain applications such as chillers [11]-[14] and turbo blowers[15][16], has been shown to result in significant energy savings and a reduction in carbon footprint by nearly 30% compared to "oiled compressors". Additionally, the hermetic design of magnetic bearings used in oil and gas applications has been found to reduce gas leakage and thus further minimize the carbon footprint [17]-[18].

This paper aims to compare the greenhouse gas emissions of turboexpanders equipped with magnetic bearings and oil bearings used in ethylene plants. The study will focus on differences in emissions resulting from the use of materials and the operation of turboexpanders during their life cycle considering both replacement portion of new parts and end life stage.

A dedicated section will discuss the results of the study, including the limitations of the analysis. It is important to note that extending the conclusions to other applications beyond ethylene may have limitations, which will be addressed in the discussion section.

2. System Boundaries - Scope

The study evaluates two turboexpander configurations in a representative ethylene process, operating under the same conditions (Table 1). The turboexpander operates in a hermetically sealed environment with all components exposed to a hydrogen-rich process gas. The functional unit is power in kilowatts (kW) of turboexpander refrigeration, evaluated over its 25-year lifespan, including GHG emissions related to material inventory, lifetime maintenance including spares and consumables, in-use utility consumption, energy recovery losses, and end of life disposal. A 25-year lifespan was chosen based on typical machinery service life, which is 25% greater than the minimum service life specified in industry standard API 617.

Table 1 Representative Process Conditions for Turboexpander in Ethylene Process

	Expander	*Compressor*
Inlet Pressure (bara)	15.0	5.9
Inlet Temperature (C)	-150	30
Outlet Pressure (bara)	6.8	7.0
Outlet Temperature (C)	-163	49
Molecular Weight	4.5	4.5
Gas Power (kW)	710	
Shaft Speed	29000	

Only the GHG emission that are different between the two configurations are considered. The two configurations have comparable duty, stage geometry and performance. Both machines are a Frame 3 by the manufacturer's convention, which is determined by the expander wheel diameter. To support this duty, a 110mm bearing is required for the AMB system versus a 51mm tilt pad bearing required for the oil machine. Other assumptions are detailed in the corresponding sections.

3. Turboexpander Description

Turboexpanders are radial inflow turbines used in refrigeration applications to extract energy from a fluid. They commonly consist of an expander stage, a compressor stage, and a bearing center section (Figure 1). When a turboexpander is directly coupled to a compressor, they operate as simple variable-speed machines, requiring no external shaft seal or drives. The aerodynamics of the turboexpander and compressor depend on the bearings' ability to support the required speed, power, and thrust loading. Critical speeds are typically avoided in their design, from zero to trip speed.

3.1. Common Configuration for Both Bearing Systems

Turboexpanders require shaft sealing to prevent cold, unfiltered process gas from entering the bearing carrier. Buffered labyrinth seals are commonly used in both magnetic bearing and lube oil-based systems due to their simple, non-contacting design, which enhances the system's robustness. The hermetic design of turboexpanders allows the seal gas to be contained within the process, eliminating any loss of gas to the atmosphere or a flare system. Turboexpander packages are often equipped with a seal gas conditioning system, which filters and regulates seal gas flow to the turboexpander. While these features are similar, their implementation differs based on the bearing technology.

3.2. Difference in Configuration

To operate the oil bearings, lubrication oil at specific pressure and temperature ranges must be delivered to the machinery. The lube oil support system achieves these pressures and temperatures through a system of valves, piping and vessels, including motor driven pumps and cooler fans. The oil delivery system adds significant material inventory to the turboexpander package. Skid layout differences are shown in Figure 2. For a typical frame 3 turboexpander package, a lube oil skid is 6.5x3 meters (LxW) versus 4x3 meters for a magnetic bearing skid. Rotating assemblies for bearings also vary in size, with AMB having less capacity per unit area and requiring greater shaft and thrust disc surface area for equivalent duty. Figure 3 offers a visual comparison.

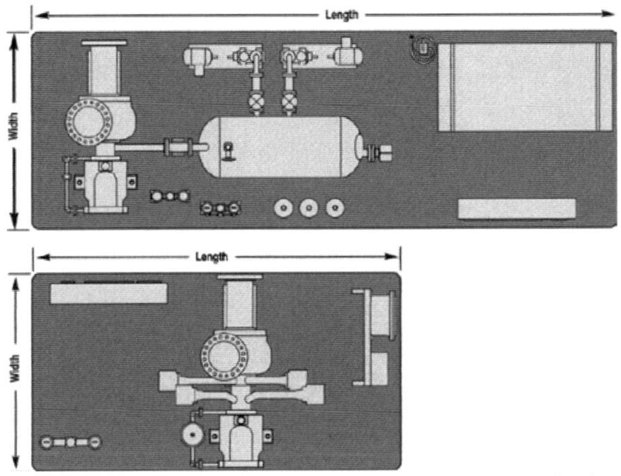

Figure 2 Turboexpander Package Comparison between Oil (Top) and Active Magnetic Bearings (Bottom)

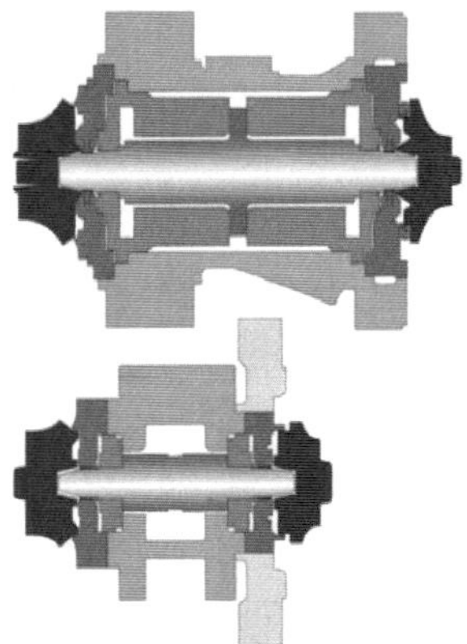

Figure 3 Turboexpander Rotating Assembly Comparison between Active Magnetic (Top) and Oil Bearing (Bottom) Systems

4. Material Inventory and Corresponding GHG Emissions (Supply and End of Life Phases)

Embodied GHG emissions due to material inventory were calculated using emission factors (EF) from ICE database v3 and ecoinvent 3.7.1 databases. The emission factor for raw material has been uplifted by 32% to account for supply chain waste and processing. The emission due to transport and facility operation are comparable for either configuration, therefore are excluded from this study. End of life emissions based on emission factors from Ademe are added to embodied emissions to complete the material life cycle. Maintenance includes capital spares and Oil Bearing maintenance included 3 replacement fills of the oil reservoir. Table 2 shows the results for each respective configuration.

Table 2 Materials Carbon Footprint

	Materials Emissions (kg CO2e)	
Material	**AMB Configuration**	**Oil Bearing Configuration**
Steel*	20952	30099
Aluminum*	4906	4636
Switchboard**	2040	
Epoxy*	2083	1941
Battery**	1012	
Electronics: (chips, boards, etc.)***	853	560
Copper*	732	511
Electrical: Power Supply/Control**	511	1394
Nylon***	243	958
Brass*	188	256
Polypropylene***	24	24
Rubber*	9	9
Electric Motors***	4	6352
Oil****		1680
PVC*		72
Material Total	33556	48493
Maintenance	2182	14167
Grand Total	35738	62660

* ICE Database V3.0 (Inventory of Carbon & Energy), Nov 10, 2019
** Ademe (French Environmental Governmental Agency), 2012, https://base-empreinte.ademe.fr/donnees/jeu-donnees, emission factor "Tableau électrique" and "Pile", retrieved on April 28, 2023
*** ecoinvent v3.7.1 Database, Dec, 2020
****U.S. Environmental Protection Agency, Inventory of U.S. Greenhouse Gas Emissions and Sinks: 1990-2020

Upon analyzing the emissions of each element, several interesting findings have emerged. The lubricated bearings system, for example, requires a larger quantity of steel and motors (pumps, cooling), but fewer electronic and electrical components when compared to the AMB system. When looking at the overall balance, it's worth noting that the AMB system produces 43% fewer GHG emissions due to the materials used.

5. Use Phase GHG Emissions

The carbon footprint of machinery operation depends on power consumption and machine performance. A compressor loaded turboexpander has two functional outputs: refrigeration and gas compression. For simplification, this analysis considers the expander stage refrigeration as the single functional unit used to normalize carbon footprint. Impacts to the compressor output will be converted to a carbon equivalent based on the lost power that requires make-up compression from downstream machinery. For the sake of this analysis, the downstream compressor is assumed to have equivalent efficiency and driven by a gas turbine, which is typical in ethylene applications.

5.1. Machine Performance

5.1.1. Bearing Losses

There are losses due to windage and friction in the bearings that reduces the expander power recovered by the compressor. These are referred to generally as bearing losses. On a machine equipped with magnetic bearing, the only bearing loss incurred is windage as the rotor spins inside the bearing housing. The rotor windage is calculated separately for each surface of the rotor and depends on the geometry, air gap, speed, Reynold's number, and density [19][20].

In addition to windage, machines with oil bearings have a loss associated with the viscous shearing of oil inside the bearing. For a given lube oil supply condition, this loss depends on the bearing geometry, thickness of oil film, the viscosity of the oil and speed.

5.1.2. Aerodynamic Performance

A turboexpander's aerodynamic efficiency has a major influence on its carbon footprint, but bearing selection has little effect on it. With open impellers, the shaft's position affects the clearance between the impeller and non-rotating follower [21]. Oil bearings only engage with the rotor when they are nearly in contact, meaning that oil-bearing rotors operate with minimal endplay. Magnetic axial bearings, on the other hand, keep the shaft centered and force it to straddle the designed endplay without any contact.

5.1.3. Seal Gas

Warm seal gas entering the process can have a detrimental effect on machinery refrigeration. This warming effect is represented in this report as a loss of refrigeration in kW.

Hermetic designs imply that the bearing carrier is submerged in the process fluid. In the case of an oil bearing, trace amounts of oil will become entrained in the seal gas before it is recompressed and sent back to the process. The amount of oil lost is limited by an oil separator. Some processes may not tolerate this trace oil being introduced to the system, and in such cases, the contaminated seal gas must be routed to a flare system. This configuration results in significant process gas loss and carbon footprint impact and is not considered in this study. For this study, the lost oil due to carry over from the seal gas system is considered in the maintenance footprint in Table 2.

In certain situations, it may be necessary to introduce extra cooling gas into the magnetic bearing housing to dissipate heat caused by windage and other losses. This cooling gas, when required, is typically obtained from the same source as the seal gas supply, and accounts for additional flow and system efficiency loss. Hydrogen-rich applications such as ethylene processing tend to have lower power density and higher cooling characteristic from the sealing gas, ultimately resulting in no need for additional cooling streams.

Due to the clearance required between the shaft and the auxiliary bearings, magnetic bearings naturally require a larger clearance at the labyrinth seal when compared to oil bearings. This excess gap can be improved with a floating seal design, where the stationary parts of the seal are allowed to move with the shaft and recenter upon levitation, however seal gas flows are still higher in this configuration when compared to oil bearings. (Figure 4 and Figure 5)

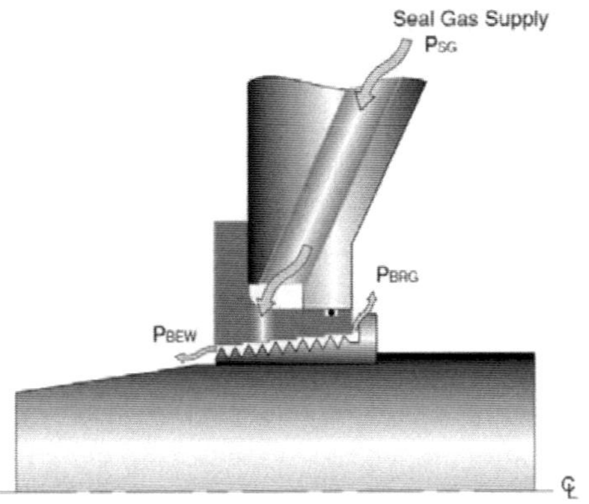

Figure 4: Labyrinth Seal for Oil Lubricated Bearings

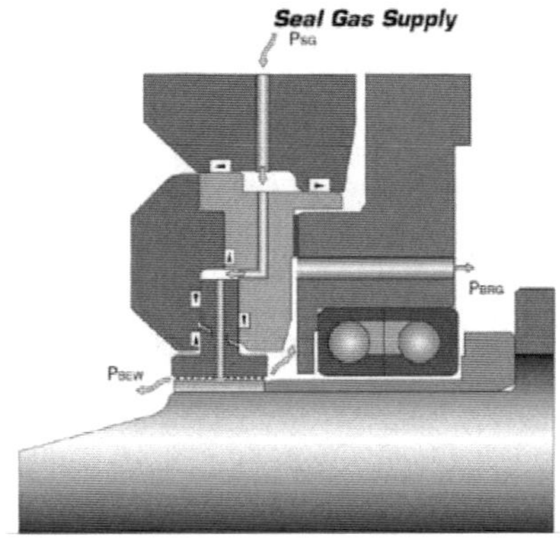

Figure 5: Floating Labyrinth Seal for Active Magnetic Bearings

5.2. Power Consumption

At an ethylene processing plant, the electrical power is typically generated on site from the combustion of hydrocarbon process gas. An emission factor of .44 kg CO_2e per kW-h is used based on the average emission for US natural gas powerplants. [22]

5.2.1. Seal Gas Heater

For ethylene processes, seal gas is typically sourced from the process gas at expander inlet. When the seal gas source is too cold for safe operation of the bearing due to material considerations or oil sludging, it must be heated. The energy requirements of the seal gas heater are determined by the seal gas flow rate, temperature rise, and thermodynamic properties of the seal gas. For oil bearings, a temperature of at least 110°F is required to prevent sludging or freezing of the oil, while magnetic bearings components can tolerate lower temperatures, as low as 10°F. Most commonly this heating comes in the form electric heating element but may also come from steam or other heat source. An electric heater is considered for this study.

5.2.2. Auxiliary Utility Consumption

While the seal gas heater has the largest power consumption of all equipment that supports the turboexpander (further discussed in section 6), other equipment's power requirements must also be considered. The operating power consumption of the magnetic bearing controls and power supply cabinet (Figure 6) is considered as is the power used by the electric motors powering the lube oil system pumps and fans (Figure 7). The power required for instrumentation and controls of the support system is similar for turboexpanders with either bearing selection and therefore has been excluded from this analysis.

Figure 6: Standard AMB control cabinet

Figure 7: Turboexpander with lube oil system

5.2.2.1 AMB Control Cabinet

The AMB control cabinet is comprised of several essential items that work together to ensure the proper functioning of rotating mechanical parts. Firstly, a detection system is included to receive all inputs from the rotating parts. The data is then analyzed by a master controller, which adapts the control outputs for the amplifiers. These amplifiers send currents to the coils of the actuators (AMBs), which adjust the magnetic field and correct the position of the rotor. Additionally, in certain cases, there may be a power supply and associated battery pack included in the cabinet, which have been considered in this study.

6. Results and Comparative Analysis

Table 3. Machine Performance Impact

	AMB Configuration Power (kW)	Oil Bearing Configuration Power (kW)
Tip Clearance (Expander)	18.7	20.3
Seal Gas Heat (Expander)	4.9	3.5
Total	**23.6**	**23.8**
Adjusted Expander Power	**710.0**	**709.8**

Table 4. Machine Power Consumption

Power Consumption	AMB Configuration Power (kW)	Oil Bearing Configuration Power (kW)
Seal Gas Heater (Utilities)	22.2	12.9
AMB Cabinet (Utilities)	2.2	-
Pump (Utilities)	-	4.3
Cooler Fan (Utilities)	-	3.3
Windage Loss (Compressor)	5.6	0.3
Oil Bearing Loss (Compressor)	-	27.0
Total	**30.1**	**47.8**

Table 5. Turboexpander Configuration GHG Emissions Comparison

	Materials GHG Emissions (tCO_2e)	*Use GHG Emissions (tCO_2e)*	*Total Life Cycle GHG Emissions* (tCO_2e)*	*Adjusted Expander Power (kW)*	*Normalized GHG Emissions (tCO_2e/kW)*
AMB Configuration	35.7	2844.8	2880.6	710.0	**4.06**
Oil Bearing Configuration	62.7	4517.9	4580.5	709.8	**6.45**

For the scope considered, where there are significant differences between the two configurations

Reviewing the breakdown of use power consumption in Table 4, note that the bearing losses including both windage and oil shearing are greater in the oil bearing, while the seal gas heater power is greater for the magnetic bearings. Although, the seal gas for magnetic bearings requires less specific heating due to the lower allowable temperature, the increased flow rate results in higher overall seal gas heating duty compared to oil bearings. These two factors, bearing loss for oil bearings and seal gas heating for magnetic bearings, account for the greatest individual carbon footprint for the respective configuration. The other utilities besides seal gas heating are significantly higher for the oil-bearing configuration, suggesting a more favorable performance for magnetic bearings when seal gas heating is not required.

Table 5 shows the materials and use GHG Emissions of machines equipped with magnetic and oil-bearings, with the operating carbon footprint being considerably higher for both bearing technologies. This is due to the turboexpander's extended operational life and high duty cycle, making even minor differences in power consumption or performance have a significant impact on the overall carbon footprint over its lifespan.

In summary, expander performance impact was minimal for either configuration, oil bearings experience greater bearing losses, while magnetic bearings require higher overall seal gas heating duty. This results in a total power consumption reduction of 37% in the AMB system when compared to oil bearing system. This improvement in performance results in the AMB avoiding 68 tCO_2e per year or 1700 tCO_2e over the 25-year life cycle.

7. Discussion and Limitation

In this example, the primary difference in carbon footprint between magnetic and oil-bearing equipment is due to bearing losses. The oil bearing requires a much smaller shaft, resulting in negligible windage loss. In contrast, magnetic bearing losses are exclusively due to windage and are highly dependent on fluid properties, shaft speed, and size.

Therefore, the variation in carbon footprint depends on the equipment's operating fluid. As demonstrated in this study, magnetic bearings have an obvious advantage in lighter gas applications such as ethylene and hydrogen liquefaction. However, as the fluid's density and viscosity increase, the difference in carbon footprint will decrease. For turboexpander applications with heavier process fluids, like natural gas processing, the decision to select magnetic bearings would not be dictated by the reduction in carbon footprint alone, but would depend primarily on other benefits, such as eliminating the need for oil lubrication systems, reducing the size and weight of equipment, improving its reliability, and reducing operating costs.

The second largest impact in this study is from the seal gas. While seal gas heating is common in ethylene processes, it's typically not required in other turboexpander applications such as natural gas processing or liquefaction. In these applications, the increased seal gas consumption experienced by the AMB system has less of an impact on the overall carbon footprint. In this way, the carbon impact from increased windage losses in applications such as natural gas processing may be partially offset by the lack of seal gas heating.

8. Conclusion

The study shed light on several key aspects related to the carbon footprint of various systems. The lubricated bearing system was discovered to have more GHG emissions related to the amount of materials utilized, although this effect was minor compared to the impact during the usage phase. The study showed that the energy consumption of the magnetic bearing system was significantly lower, with a difference of more than 35% compared to the lubricated bearing system. This reduction in power consumption had the greatest impact on carbon footprint and is consistent with similar results reported in other applications cited in the article.

Moreover, it was recognized that the decreased energy usage exhibited by the magnetic bearing system relies heavily on boundary conditions, including the seal gas source and equipment operating fluid. Keeping this in mind, the machinery's carbon footprint should be assessed individually for each application, and a carbon footprint comparison across various applications would be necessary to confirm whether the findings of this study are universally applicable.

9. References

[1] J. Jumonville, "Tutorial on Cryogenic Turboexpanders", *Texas A&M University Turbomachinery Laboratories, Turbomachinery and Pump Symposia, 2010*, [CrossRef]

[2] L.E. Rodriguerz, R.Hoch, « Turboexpander repair and maintenance for sustained performance and profitability", *Gas Processing & LNG magazine, Turbomachinery session*, 2021, [CrossRef]

[3] H.Bloch, C. Soares, "Turboexpanders and process applications", Book, *Gulf Professional Publishing Edition*, 2001, [CrossRef]

[4] J.Sears, S. Uptigrove, "Magnetic Bearing Operating Experience", *Proceedings of the 23rd Turbomachinery Symposium*, pp 235-242, 1994, [CrossRef]

[5] J. Schmied, "Rotordynamic Aspects of a New Hermetically Sealed Pipeline Compressor", *Proceedings of the 20th Turbomachinery Symposium, pp 11-18, 1991*, [CrossRef]

[6] E.G. Foster, V.Kulle, R.A. Peterson, "The Application of Active Magnetic Bearings to a Natural Gas Pipeline Compressor", *proceedings paper, ASME 1986 International gas turbine conference and exhibit*, June 1986, [CrossRef].

[7] J. Jumonville, C M. Ramsey, F. Andrews, " Specifying, Manufacturing, and Testing a Cryogenic Turboexpander Magnetic Bearing System", *Proceedings of the 20th Turbomachinery Symposium*, pp 3-9,1991 , [CrossRef]

[8] F.Davis, R. Agahi, R. Wu, " Full Load, Full Speed Test of Turboexpander-Compressor with Active Magnetic Bearing", *Proceedings of the 35th Turbomachinery Symposium*, pp 81-88, 2006[CrossRef]

[9] G. Schweitzeir, et al, "Magnetic Bearings: Theory, Design, and Application to Rotating Machinery", Book, *Springer-Verlag Edition*, 2009, [CrossRef]

[10] E E . Swanson, E H. Maslen, G. Li, C H. Cloud, « Rotordynamic Design Audits of AMB Supported Machinery", *Proceedings of the 27th Turbomachinery Symposium*, pp 133-158, 2008, [CrossRef]

[11] R. Conry, L. Whelan, J. Ostman, "Magnetic Bearings, Variable Speed Centrifugal Compression and Digital Controls Applied in a Small Tonnage Refrigerant Compressor Design", *International Compressor Engineering Conference*, paper 1500, 2002, [CrossRef]

[12] B. Erpelding, A. Moman, "Small Oil-less Centrifugal Compressors: Bringing Energy Efficiency and Reduced Costs to Chiller Plants", *ACEE Summer Study on Energy Efficiency in Industry*, 2005, [CrossRef]

[13] S.A. Parker, J. Blanchard, " Variable-Speed Oil-Free Centrifugal Chiller with Magnetic Bearings Assessment", *report for the General Service Administration*, Nov 2012, [CrossRef]

[14] E. Byrd, B. Netzel, D.B. Adams, H. Zhang, "Life Cycle GHG Assessment of Magnetic Bearing and Oil Lubricated Bearing Water Cooled Chillers", *Journal of Industrial Ecology*, 2021, [CrossRef]

[15] Kaeser sponsored content, "Turbo blowers with magnetic bearing technology", *WWD magazine*, Aug 2021, [CrossRef]

[16] Y. Kinoshita, T. Aota, "Kawasaki Mag Turbo Single-Stage Sewage Aeration Blower with High-Speed Motor & Magnetic Bearing", *On line paper rubrique Energy and Environment*, April 2017, [CrossRef]

[17] J.L. Gilarranz R, M. Dave, T. Jamison, "Non-Hermetic, Oil-Free Compression Solutions-A Reliable Approach to Reduce Life Cycle Costs for Compressor Applications", *Abu Dhabi International Petroleum Exhibition & Conference*, Nov 2016, [CrossRef]

[18] Baker Hughes online presentation, "Oil-free Turbomachinery with active magnetic bearing technology", June 2021, [CrossRef]

[19] J.W. Daily & R.E. Nece, " Chamber Dimension Effects on Induced Flow and Frictional Resistance of Enclosed Rotating Disks", Journal Basic Eng, pp 217-230, March 1960, [CrossRef]

[20] J.E. Vancik, "Prediction of Windage Power Loss in Alternators", NASA Technical Note D-4849, October 1968, [CrossRef]

[21] R. Dambach, H.P. Hodson, I.Huntsman, "An Experimental Study of Tip Clearance Flow in a Radial Inlfow Turbine", ASME 1998 International Gas Turbine and Aeroengine Congress and Exhibition, June 2-5 1998, [CrossRef]

[22] U.S. Energy Information Administration, State Electricity Profiles, U.S. Profile, 2021, [CrossRef]

Variable Speed Direct Drive Induction Motors Levitated by Active Magnetic Bearings for Oil&Gas Compression Services

Lionel DURANTAY [a], Clément BIRI [a], Alain GELIN [b], Alexandre KRAL [c]

a General Electric Power Conversion, 442 Rue de la Rompure, 54250 Champigneulles, France, lionel.durantay@ge.com

b TotalEnergies, Avenue Larribau, 64018 Pau, France

c SKF S2M, 2 Rue des Champs, 27950 Saint-Marcel, France

Abstract

The purpose of this paper is to present on one hand a state of the art of innovations relating to new electrical architectures for driving high power compressors using induction motor and shaft line levitation with active magnetic bearings, and the other hand, to share return of experience on how to conduct validation tests of magnetically levitated induction rotors in high-power machines.

Keywords: *Factory Acceptance Tests, High-speed Induction Motor, Compression service, Large Magnetic bearings.*

1. Introduction about high-speed induction motors

Over the past three decades, the development of high-speed induction motors, is an alternative solution to conventional compression trains using turbines, for both onshore and offshore applications and has enabled the electrification of various compression services. The solution consists of electric motors, controlled by variable speed drive, directly driving the compressors. The entire drive train is suspended by magnetic bearings, removing the need for the oil system and associated auxiliaries. This system architecture brings with it substantial reduction in weight and footprint.

The advantages of the induction motor compared to the wound synchronous motor include no excitation system and simplified rotor construction, leading to longer run-time with fewer maintenance issues. The main components of an induction motor are (Fig. 1):

- the three-phase stator creating a rotating magnetic field, made of laminations and coils impregnated with resin with a vacuum pressure impregnation,
- the squirrel cage rotor,
- the frame supporting the stator,
- the bearings (oil lubricated or active magnetic),
- the cooling system (water cooled, or gas cooled) with axial path or bilateral paths.

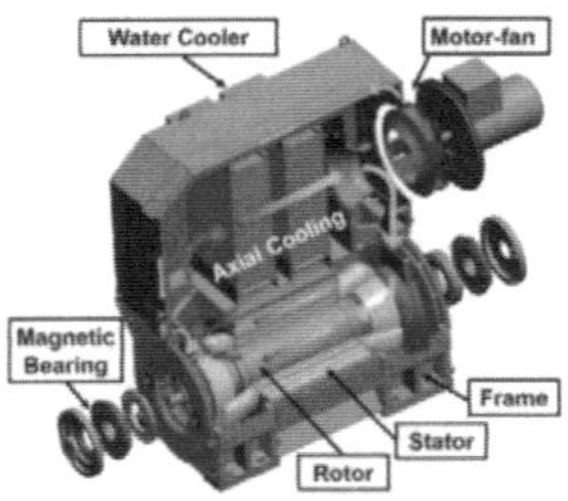

Figure 1 Induction motor design for easy maintenance.

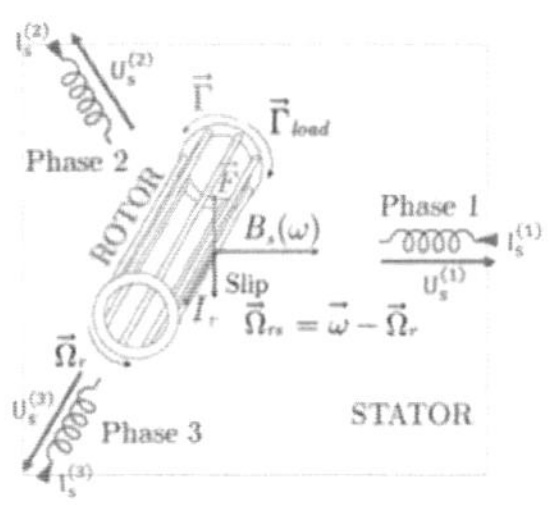

Figure 2 Torque generation with induction rotor

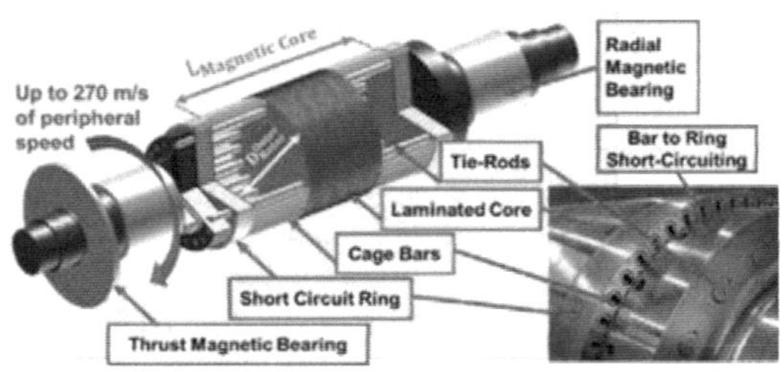

Figure 3 High-speed laminated squirrel cage rotor technology with active magnetic bearings.

The rotor consists of copper bars not insulated, short-circuited at their end by two conductive rings and constitute a "squirrel cage". The cage is swept by the rotating magnetic field. When a resistant torque $(\vec{\Gamma}_{load})$ is applied on the shaft by the load, it induces a rotor slippage (Ω_{rs}) or slip (*slip*) with a rotor pulsation (Ω) a bit lower than the stator pulsation (ω_s) which is fixed by the control of the inverter (Fig.2):

$$\Omega_{rs} = \omega_s - \Omega \tag{1}$$

$$slip = \Omega_{rs} / \omega_s \tag{2}$$

The Lorentz Force induces the rotor current (I_r) into the cage and the Laplace Force induces the torque (Γ) generation in opposite to the resistant (load) torque, where (R_{Rotor}^{Outer}) is the outer diameter of the rotor, and (K_1) and (K_2) are constant:

$$\overrightarrow{I_r} = K_1 . q . \overrightarrow{\Omega_{rs}} \wedge \overrightarrow{B_s} \tag{3}$$

$$\vec{\Gamma} = \vec{F} . R_{Rotor}^{Outer} = \left(K_2 . \overrightarrow{I_r} \wedge \overrightarrow{B_s}\right) . R_{Rotor}^{Outer} \tag{4}$$

2. High variable speed electric systems architectures

In most oil and gas use cases of electrical systems driving compressors, the standalone electric systems with gearbox are the baseline (Fig.4), combining a Direct-On-Line-fed 4-pole electric motor on oil bearings, a gearbox, and a compressor. For the TotalEnergies standpoint, this traditional architecture still presents several main concerns: large and heavy conventional system package, lubricating oil system, long commission time due to the complexity of the package, continuous gas leakage to the flare from complex dry gas seal and oil vapor from bearings (Durantay et Al., 2022). To simplify the system in terms of weight and dimension, standalone direct drive systems can be used for all compression service including gas or steam turbines replacements. The gearbox is eliminated, and the compressor is direct drive coupled to a high-speed motor. It is therefore interesting to use magnetic levitation for the whole shaft line (Fig.5). Below 30MW, standalone electric arrangements can be replaced by fully hermetic integrated moto-compressor using induction motor (Durantay et Al., 2019a) (Thibaut et Al. 2022). The process gas handled by the compressor is used to cool both the induction motor and the magnetic bearings making the unit fully hermetic, allowing to remove the gas dry seals and the conditioning system, reducing the footprint by 40% (Fig.6). Due to the large flexibility of speed range without any power derating, direct drive electric systems can be used for offshore compression services with a speed optimization to cope with the volume flow of each compressor minimizing the weight and the footprint and maximizing the efficiency of each compressor.

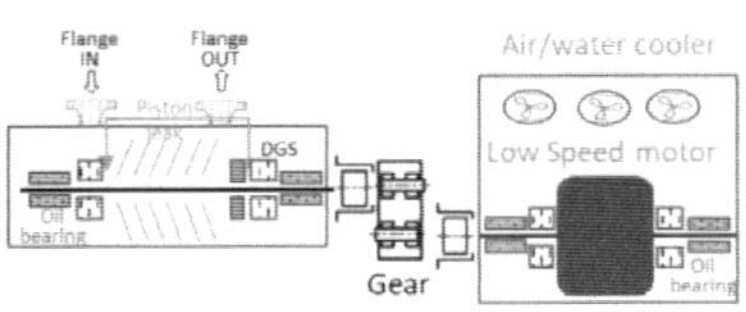

Figure 4 Oil-lubricated gearbox electric architecture.

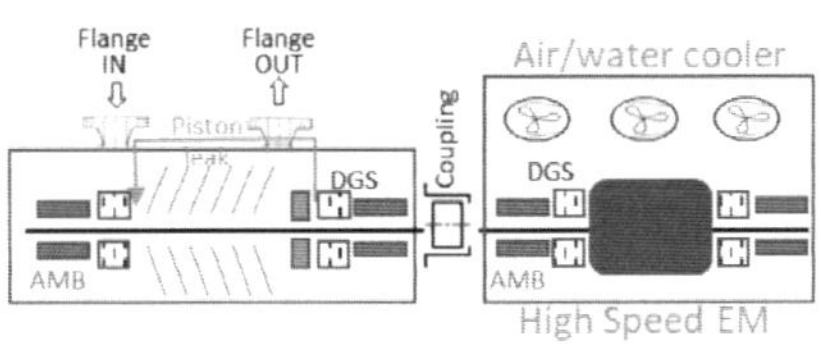

Figure 5 Standalone direct-drive electric architecture.

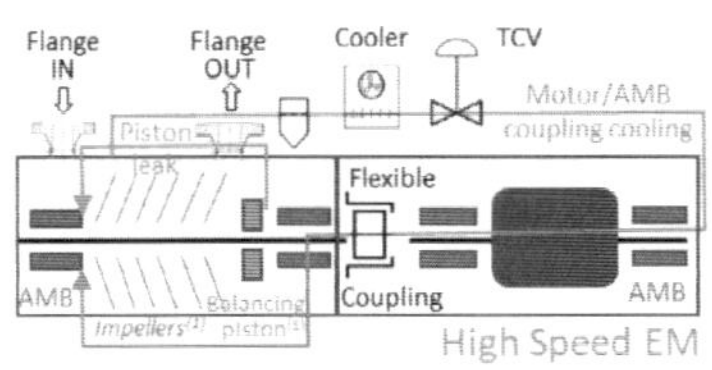

Figure 6 Integrated multi-Stage electric architecture.

An innovative technology of high-speed induction motor has been developed. The induction rotor comprises a steel lamination assembly compressed by tie rods between two end rings and two shaft ends, named shaft-less technology (Fig.3). The cage bars can expand axially through the end ring avoiding any risk of deterioration by thermal fatigue mechanism at the interface of the copper bars and the laminated shaft, respectively having coefficients of thermal expansion of $17.10^{-6}\ K^{-1}$ & $12.10^{-6}\ K^{-1}$. The copper bars are inserted in the slots between the two end rings to form the squirrel cage. The cage can operate reversibly up 200 °C maximizing the torque density. The laminated technology allows high efficiency in reducing the high frequencies stray load losses, and high rotor peripheral speed up to 270 m/s, providing high robustness in terms of aging and stability (Mogenier et Al., 2010) (Mogenier et Al, 2011).

Figure 7 Induction motor 16MW@9000rpm on AMBs.

Figure 8 Active magnetic bearing integration for easy maintenance.

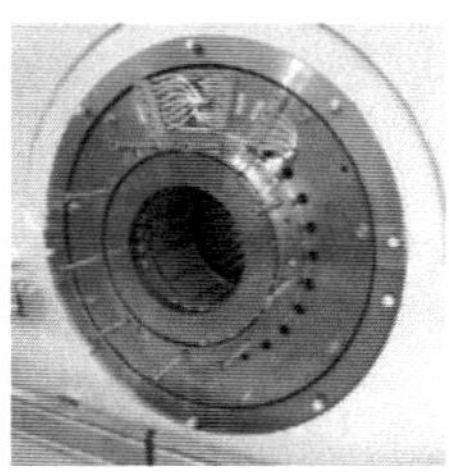

Figure 9 Non driven end active magnetic bearing.

Figure 10 Landing bearing system.

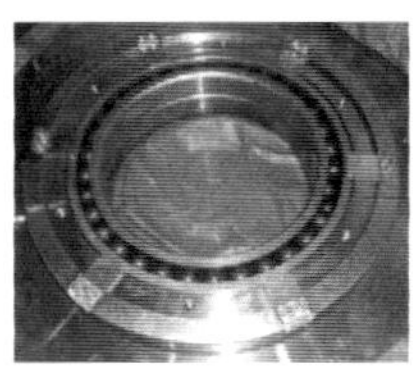

Figure 11 Ceramic landing bearing.

As an example, TotalEnergies has decided to use this technology for an export service sized at 16MW@9000rpm (Fig.7) which is currently the biggest reference in the word of high-speed motors levitated by active magnetic bearings (Fig.8). This high-speed direct drive system architecture using 2-pole induction motor on active magnetic bearings brings substantial reduction in weight and footprint reducing the size of the topside structure and its associated cost for a green field project. A typical cost ratio of 10 to 20 k$ per ton of installed base is the usual assumption for offshore structure (Durantay et al., 2023). As in any bearings, the AMB actuator consists of two mechanical parts, the rotor, and the stator (Fig. 9). The rotor parts are ferromagnetic laminated rings that are shrunk on the shaft. To protect the electromechanical parts in the event of bearing overload or electronic failure, auxiliary bearing set is currently used. The landing system consists of ball bearings located on the stator and having about 50 % magnetic bearing airgap against the rotor (Fig.10). Ball bearings are generally dry lubricated and are made with ceramic balls (Fig.11). The use of active magnetic bearings allows the following advantages for the electric compression: no wear, so no mechanical maintenance and unlimited lifetime, moreover no bearing noise, 200 m/s tip speed achievable, very low friction losses, no process fluid contamination by the bearing, no need for seal, oil lube system and accessories, ability to work in vacuum or hostile environments, permanent control of rotation axis, automatic balancing system because rotor spinning around its inertial axis instead of its geometrical axis, adapting static and dynamic stiffnesses for high accuracy of rotation, vibration free control, permanent monitoring of the system. The electromagnet pulling force of the active magnetic bearing (F_{pull}) is given as follow where (σ_{mag}) is the magnetic pressure and ($Area$) the airgap area:

$$F_{pull} = (\sigma_{mag}).Area = \left(B_{airgap}^2 / 8\pi\, 10^{-7} \right).Area \quad (5)$$

The magnetic induction in the airgap (B_{airgap}) is of the order of 1.5 Tesla because of magnetic saturation in the laminations of the bearing, generating a magnetic pressure around 1 N/mm² and magnetic stiffness around 10^{-3} N/m (Fig.12). The stiffness of AMB is very soft compared to an oil film. Illustrated on a system operating at 6.9MW@13,200rpm (Fig.13), compared to oil bearings, the magnetic bearings, because of their low stiffness control (Fig.10), avoid any modes in the operating speed range of the system (Durantay et al., 2018). The rotor of the high-speed induction motor can operate in subcritical free-free mode when using magnetic bearings, meaning that the first bending mode is above operating speed defining the rotor length (Fig.14). The mode separation margin complies with API 541 even though this standard does not apply to magnetic bearings. The {Power versus Speed} capacity curves of the air-cooled induction motors at 1 and 5 bars cover most of the existing fleet of steam or gas turbines used in mechanical drive in the world (Fig. 15). Currently with existing power amplifiers generating 4 kW and up to 6 kW in peak transient conditions, it is possible to levitate rotors in rigid mode up to 10 tons spinning up to 7000 rpm. By scaling up the power of the existing technology of active magnetic bearing, it will be possible to levitate rotors up to weights of 100 tons with specific arrangements of axial thrust magnetic bearings.

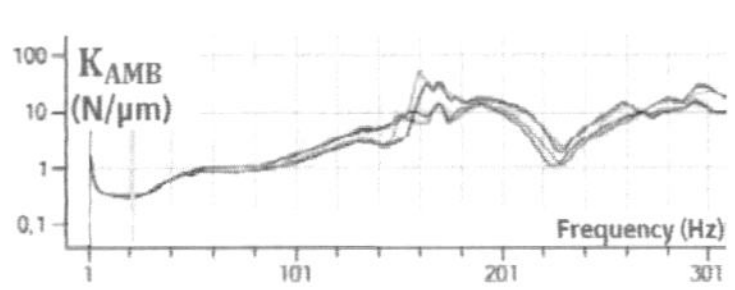

Figure 12 Active magnetic stiffness vs frequency.

Figure 13 6.9MW-13200rpm off-shore export compression Service.

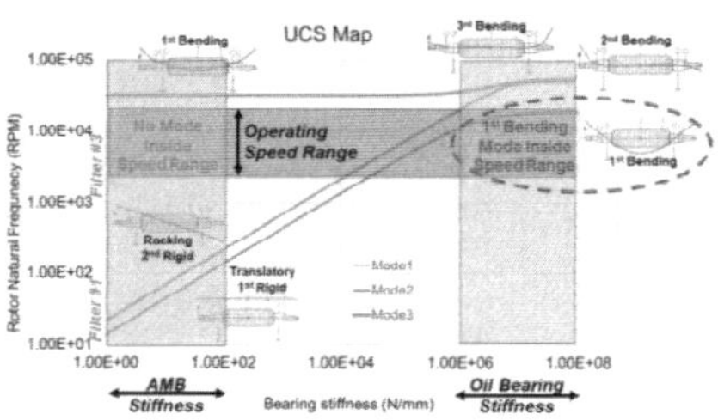

Figure 14 Undamped critical speed map 6.9MW-13,200rpm motor

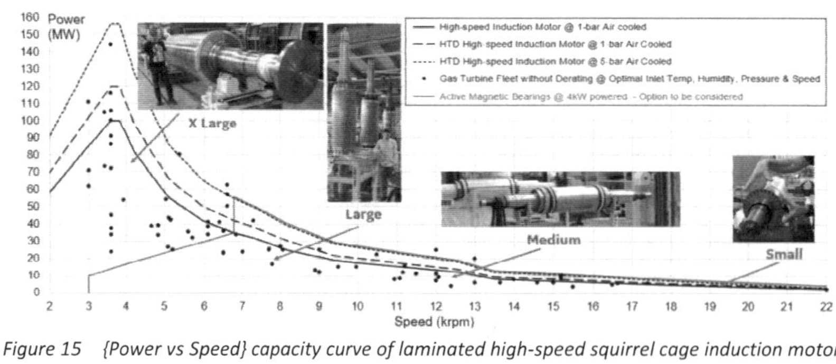

Figure 15 {Power vs Speed} capacity curve of laminated high-speed squirrel cage induction motors.

3. Factory acceptance tests of large high-speed induction motor levitated by active magnetic bearings

3.1. High-speed tests description

High power electric motors, of the order of several megawatts require factory acceptance tests. After an overspeed test useful for relieving the stresses inside the rotor, a rotor heating test is necessary for a high-speed rotor levitated by active magnetic bearings, relieving all residual stresses, reaching stable vibrations over the time in cold and hot conditions, checks residual unbalance, centrifugal and thermal expansion of the different components especially at the interface of the copper bars and the ferromagnetic laminated shaft. The principle of the test is to generate the same joule losses in the squirrel cage as in full load condition and rated speed (Ω), by adjusting the slippage (Ω_{rs}) with the drive feeding the stator at low voltage condition (Durantay et Al, 2019b) (Fig.16) (Fig.17). In this case, the slippage is:

$$\Omega_{rs} = \pm\omega_s - \Omega \tag{6}$$

When the sign of the stator pulsation is identical to the rotor speed, the machine operates in forward rotation mode and counter-rotation mode if the sign is opposite. The cage temperature rise (ΔT_{Cage}) is defined in the following heat transfer equation:

$$\Delta T_{Cage} = R_{thermal}^{rotor} \cdot \frac{3.\, R_{electrical}^{rotor} \cdot I_r^2}{Area_{cage}} \tag{7}$$

where ($R_{thermal}^{rotor}$) is the rotor thermal conductivity, ($R_{electrical}^{rotor}$) the squirrel cage electrical resistance and ($Area_{cage}$) the squirrel cage surface.

Thus, the cage temperature is adjusted with the stator voltage and frequency at rated speed condition imposed by the pony motor (see Fig.5):

$$\Delta T_{Cage} = R_{thermal}^{rotor} \cdot \frac{slip.\, P}{Area_{cage}} = f(slip, U_{stator}, \omega_S) \tag{8}$$

where (P) is the absorbed power and (U_{stator}) the stator voltage.

The campaign of tests is achieved by load tests at variable power and speed conditions:

- in back-to-back arrangement without gearbox, the tested machine in motor mode and one machine in generator mode loading the system (Fig.18) (Fig.19),
- the tested machine in motor mode loaded by a slow speed generator through an intermediate gearbox.

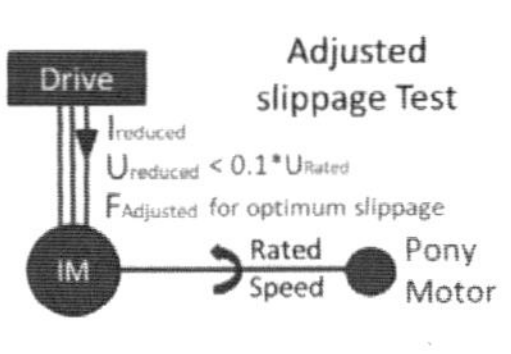

Figure 16 Rotor heating test configuration.

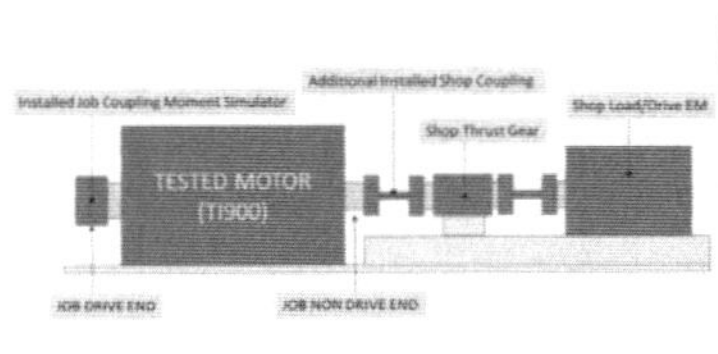

Figure 17 Direct drive coupling arrangement.

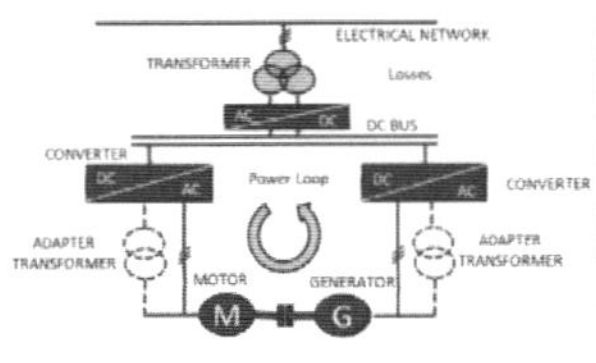

Figure 18 Back-to-back test configuration.

Figure 19 Back-to-back test arrangement.

3.2. High-speed test without axial thrust

The test described is a load test with a gearbox. The high-speed motor tested is an induction motor levitated by active magnetic bearings operating at a nominal speed of 13,840 rpm. The motor is connected to the gearbox with a flexible coupling (Fig.20). The load generator, operating at low speed, is coupled on the other side of the gearbox. This test configuration is conventional. During the increase in speed at no load condition, significant vibrations and shaft displacements are respectively measured on the casing of the gearbox and on the high-speed rotor.

Analysis of vibration signals demonstrates the following cause -effect mechanism. A measurement of the axial displacement of the shaft line highlights an axial resonance at approximately 5Hz self-excited by the slippage modulation of the rotor magnetic field inducing amplitudes of the order of millimeters (Fig.22) and axial vibrations of the gearbox casing (Fig.21) of the order of 9 mm/s rms at 13,000 rpm (Fig.23). This induces a doubling of the amplitude of the high-speed rotor shaft displacements with around 20 umpp of unbalance at 1X the speed of rotation and 20 umpp at 5 Hz (Fig.24). Motor-compressor architectures incorporate the thrust bearing on the compressor side. The electric motors tested do not have an axial thrust bearing. This is not a problem when testing a conventional low-speed oil-bearing motor, but it becomes an issue for high-speed motors levitated by AMB. The problem is solved by inserting between the high-speed coupling and the gearbox a mechanical thrust.

Figure 20 Load test with gearbox (high-speed motor on the right)

Figure 21 Velocimeters instrumentation on gearbox.

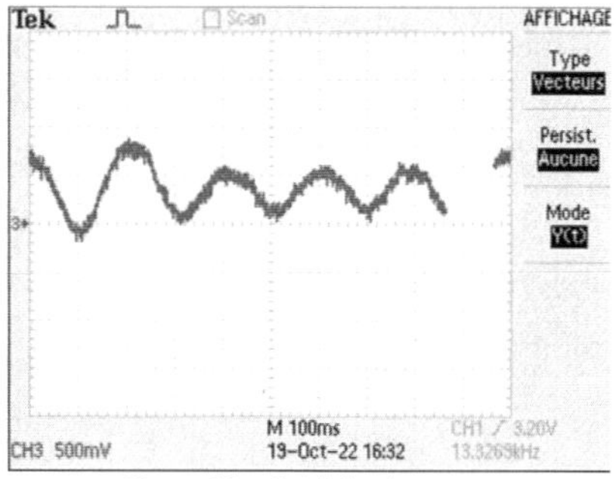

Figure 22 Axial displacement of the induction rotor.

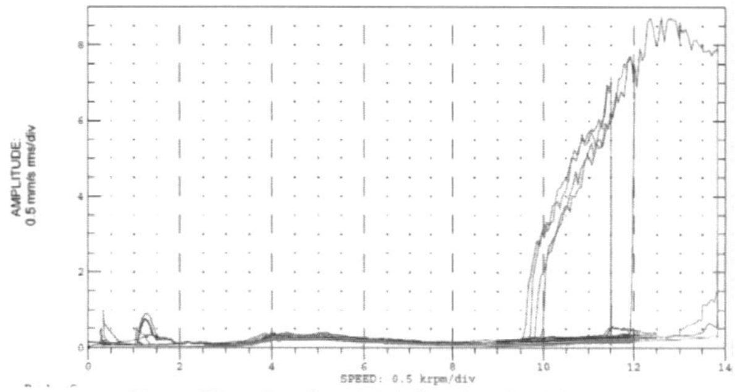

Figure 23 Gearbox Axial vibration @ 13,840rpm.

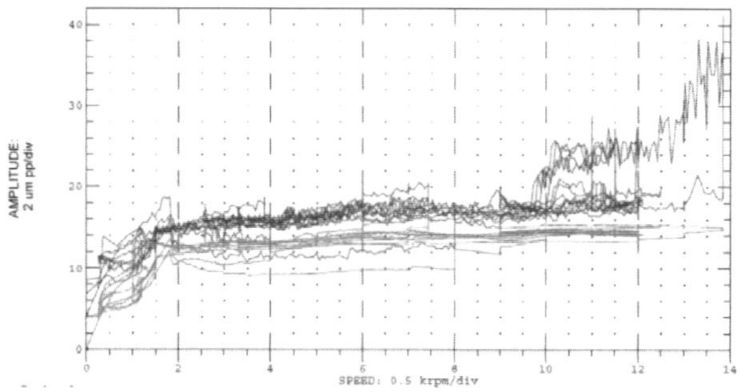

Figure 24 Induction rotor shaft displacements @ 13,840rpm.

3.3. High-speed test with axial thrust

The recommended test setup is to use an axial thrust bearing equipped with ball bearing lubricated by an oil mist. For the rotor heating tests or partial load tests, a high-speed induction motor is used to drive the main motor to be tested. This motor can deliver a power of 6 MW at 15,000 rpm (Fig.25) (Fig.26). It is equipped with customized active magnetic bearings (Fig.27) (Fig.28). This testbench motor has been used since 2012 for high-speed testing.

Tests of the 16MW@9000rpm motor are carried out with this axial thrust bearing (Fig.29) (Fig.30). Vibrations at 8,940 rpm, at no load and at full load at 16 MW are less than 50 umpp (Fig.31) with an alarm and trigger threshold set by mutual agreement with the end user at 100 umpp and 150 umpp.

Figure 25 Testbench motor on AMB (on left) with thrust bearing gear (on right).

Figure 26 Testbench motor magnetic bearing integration.

Figure 27 Testbench motor active magnetic bearing.

Figure 28 Testbench motor active magnetic bearing actuator.

Figure 29 Rotor heat test arrangement (Testbench motor side).

Figure 30 Rotor heat test arrangement (16MW Motor side).

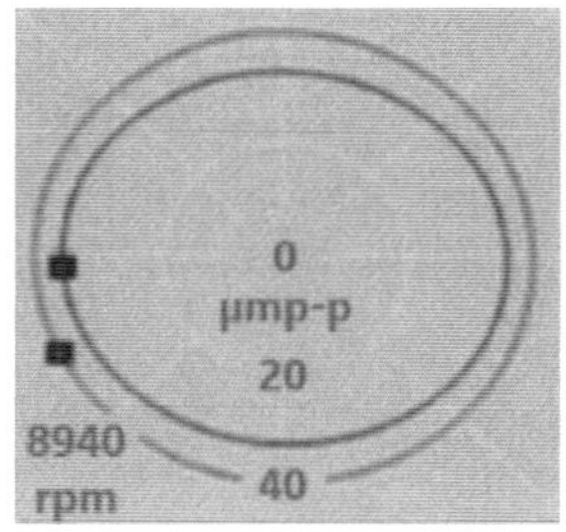

Figure 31 16MW rotor shaft displacement @ 8,940rpm.

4. Conclusion

This publication describes the state of the art in electric compressor drive architectures, in particular the high-speed induction motor technology which makes it possible to cover almost all the demands of compression services in the Oil & Gas market up to 100MW, including gas and steam turbines replacements. Associated with this motor technology, the magnetic bearings allow great flexibility of use for wide ranges of speeds by eliminating the use of oil, with limited risks of interaction with the foundation, and integrated moto-compressor system cooled by the process gas. Unlike a gas turbine or a steam turbine which respectively produce an average of 5,000 tons and 10,000 tons of greenhouse gas per year, the high-speed induction motor produces no polluting emissions if it is powered by carbon-free electricity from a nuclear power plant or a renewable energy source such as offshore wind turbines using direct drive permanent magnet generators (Fig.32 & 33). For these large electrical motors levitated by active magnetic bearings, the recommended test setup is to use an axial thrust bearing equipped with ball bearing lubricated by an oil mist, to reduce the risk of vibrational interactions between the motor to be tested and the testbench machine.

Figure 32 6MW offshore windmill field

Figure 33 6MW@11rpm PM generator

References

API Standard 541:Fifth Edition (2014) Form-wound Squirrel Cage Induction Motors—375 kW (500 Horsepower) and Larger.

Durantay L, Alban T, Siala S and Billaud A (2018) Selection and Tests of Innovative Variable-Speed Motor-Compressor Solutions for a 55-MW Full Electric Offshore Platform Maximizing Availability and Efficiency With Better Environmental Impact. IEEE Transactions on Industry Applications, 55(6).

Durantay L, Gelin A, Thibaut E and Vidalenc Y (2019a) Integrated Motor Compressor versus Conventional Solution. In 16th PCIC Europe Conference, Paris, FR, 7-9 May 2019.

Durantay L, Clark T, Roth L and Galmiche C (2019b) Alternative Package of Tests to Reduce Capex and Delivery Time of Large LNG Train Using Induction Machine Fed by VSI Drive Technology up to 120MW. In 66th IEEE IAS PCIC technical Conference, Vancouver, CA, 8-12 Sep 2019.

Durantay L, Gamiche C, Grosselin C and Verdot P (2022) Innovative Variable Speed Systems Using Induction Motor for Gas Compression. In 8th IEEE-PCIC Middle East Conference, Al Khobar, KSA, 15-16 November 2022.

Durantay L, Gelin A, Thibaut E and Roth L (2023) Rex and Recommandations for Mechanical Integration of Large High-speed Motors. In 19th PCIC Europe Conference, Milano, IT, 6-8 June 2023.

Mogenier G, Dufour R, Ferraris-Besso G, Durantay L and Barras N (2010) Identification of Lamination Stack Properties: Applications to High-speed Induction Motors. IEEE Transactions on Industrial Electronics, 57(1).

Mogenier G, Baranger T, Dufour R, Durantay L and Barras N (2011) Efficient Model Development for an Assembled Rotor of an Induction Motor Using a Condensed Modal Functional. Journal of Computational and Nonlinear Dynamics, 6(2), April, pp. 1–8.

Thibaut E, Pellerin O, Durantay L, Bouyssou S, Bouteille A and Garaudee L (2022) Selection and Return of Experience of Integrated Moto-compressors on Two Oil&Gas Sites. In 18th PCIC Europe Conference, London, UK, 6-8 June 2022.

Offshore gas compression challenges to reduce the carbon footprint in a needed transition.

Massimiliano Ortiz Neri[1], Antonio Cristallo[1], Alexandre Kral[2]

1 *Baker Hughes Nuovo Pignone, Florence, Italy*

2 *SKF Magnetic Mechatronics (S2M), Saint Marcel, France*

Abstract

The Oil and Gas industry must reduce its carbon footprint and transition to a more sustainable operating model while meeting an increase in demand for gas. To address this challenge, offshore operators are seeking new solutions that can reduce carbon emissions, increase reliability, and enable unmanned operations.

This paper proposes oil-free high-speed moto-compressor solutions as a technical alternative to overcome the drawbacks of traditional oil-lubricated compression solutions. High-speed motors offer the possibility to reduce rotor weight and overall dimensions, better match the speed to the compressor's optimal one, and eliminate the need for an intermediate gearbox. Oil-free compression solutions can also result in a reduction in the overall package footprint and weight of up to 50% while providing unmanned capability.

Magnetic bearings have been used in natural gas centrifugal compressors since the late 1980s, providing higher speed capabilities, improved reliability, and reduced friction and wear. Baker Hughes has been a leader in implementing magnetic bearing technology and continues to develop this technology in various applications, including hybrid platform solutions.

The use of stand-alone magnetic bearing compressors is seen as a promising solution to tap offshore wells more effectively and address the challenges of decarbonization, new geopolitical situations, and technology improvements. This paper will compare two type of electric motor possibilities, one based on a Low-speed motor with a gearbox and a stand-alone compressor all on oil bearing solutions and the other with a direct drive high speed electric motor with a stand alone compressor on magnetic bearings (oil free solution).

Keywords: *compressors, active magnetic bearings, carbon footprint, platform*

1. Introduction

The raising need of gas as an energy transition and the Ukraine Crises has pushed the oil and gas industry to expand their production to respond to the needs. Gas recuperation has been a main target and on offshore applications with the new technical developments has allow not only to increase but also to optimize the well utilization and prolong its life. For this centrifugal compressor technology was one of the key factors.

In parallel, the CO_2 emission limitation criteria's have been increased. Meaning that the Gas Turbine driven compressors trains are replaced by more effective and reliable electric motor driven solutions. To supply the needed power, there are different ways of approach:

-On site effective power generators creating the needed energy on the platform

-Implement renewable energy source like offshore wind farmers (possibly combined with gensets) [1]

-Supply the energy from a remote onshore facility [2] [3] [4]

The use of Electric Driven compressors has also allowed the reduction of the needed footprint, a higher availability and to reach towards an unmanned installation.

This paper is focused to compare the different electric driven solutions which are:

-Conventional Electric Motor with gearbox driven compressor

-Oil free High Speed Electric Motor (HSEM) driven compressor

in terms of footprint and different feature that are advantages for the offshore (platform or vessel) applications.

2. Setting system boundary

The base case is the comparison of a compression train from a conventional solution compared to an oil free HSEM solution equipped with Active Magnetic Bearings (AMB). The conventional solution is based on a low-speed motor on oil bearings coupled to a multiplier that accelerates the rotation of the compressor on oil bearings, added to that there is the Dry Gas Seals (DGS) and the lube-oil system. As the conventional solution is better known, we will concentrate on the oil free solution. We have considered a power supplied from shore as both are electrical driven solutions.

3. Oil-Free High speed motor drive compressor description

The Oil-Free High speed motor driven compressor is decomposed of the following components/system:

- Centrifugal compressor equipped with AMB and DGS cartridges
- Electric motor on AMB
- Direct power transmission coupling
- Seal gas panel
- Seal gas conditioning skid with booster and/or heater as required by process gas
- Units control panel

A general description of certain parts will be done to explain the technology.

Fig. 1: Oil-free high-speed motor-compressor

- **The Magnetic bearing technology.**

Active Magnetic bearings are electromagnetic mechatronic systems maintaining the position of a rotor inside the stator part. The bearings are used to levitate the rotating parts and with the help of a closed system of positions, temperature and speed sensor continuously corrects its position as shown in Fig. 2 below.

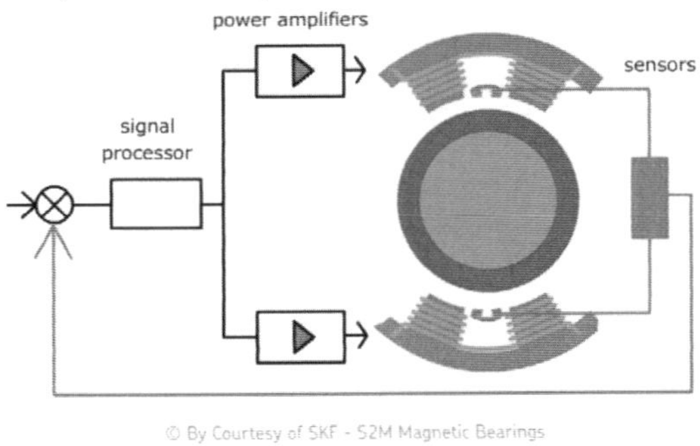

Fig. 2: Bearing and control schematic

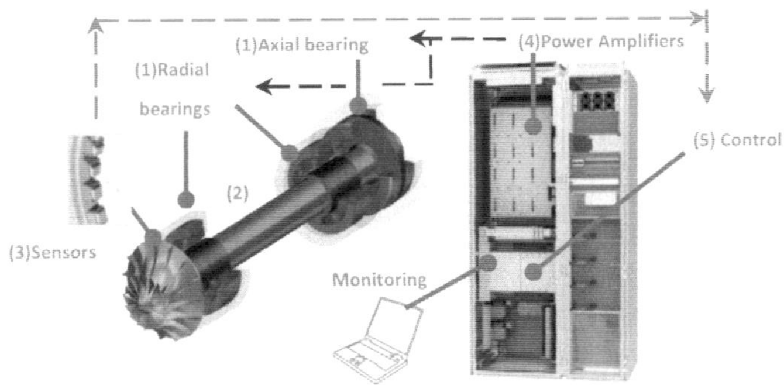

Fig. 3: AMB Components

According to Fig. 3, the magnetic bearing is made of electromagnets (1) that attract the rotor (2) towards the central rotation axis. It is contactless with a radial clearance of 0.3 to 1 mm. The rotor position is measured and controlled in real time by position sensors (3), through a digital control (5) and power current amplifiers (4). To avoid any physical contact between the magnetic bears and the rotor, auxiliary bearings are built in. They allow a safe shut down and protect the system in case of failure or process overload.

Radial Bearing are the bearings that lift the rotor when the machine is horizontal in a static mode and control the shaft in a safe position in the dynamic mode. Axial Bearing is the bearing that controls the forces applied on the thrust disk. Auxiliary bearings, also known as landing bearings, are the bearings on which the compressor rotor is still when the system is in rest condition.

The use of AMB technology versus the traditional oil bearings allow to remove the contact friction and components wear, reduce the mechanical losses, and embed monitoring which allow remote control features .

- **High Speed Electric Motor.**

On HSEM solution inductive motors are typically used with a relevant driver that feeds the motor with higher electric frequency with respect to the grid which in turn causes the rotor to rotate at high speed (typically greater than 3600 rpm) and produce mechanical power.

The high speed introduces additional challenges for the design of the machine that finally bring to technical solution that may differs from traditional machine mainly for what concern the rotor since the higher centrifugal forces limit the usage of shrinking parts on forged shaft, like laminations and require specific smoothed surfaces design to limit losses for friction and turbulence.

High-speed motors have generally smaller footprint and weight compared to a traditional low-speed design (for the same rated power) with the rotor much lighter and smaller in diameter due to the reduced torque. As consequence, the reduction of rotor weight and diameter fit perfectly with the magnetic bearing technology.

- **Centrifugal compressor equipped with AMB and dry gas seal cartridge.**

An AMB centrifugal compressor is like the conventional one: the bearings area is segregated from the process gas by the dry gas seals cartridge allowing the compressor to handle all type of gases (sour, wet, acid etc.). Here below in Fig. 4 the schematic drawing for a typical compressor cross-section.

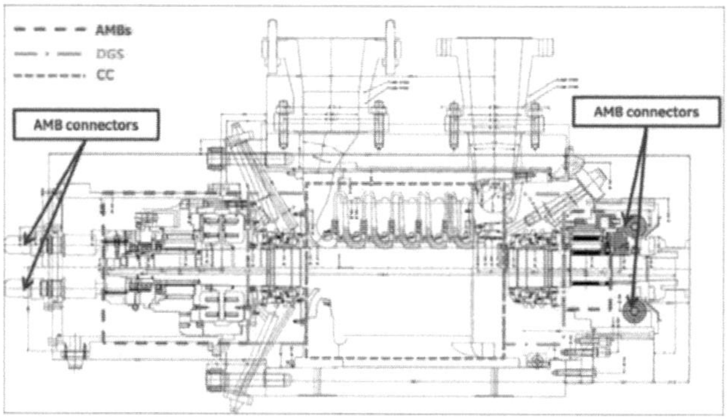

Fig 4: Compressor cross-section

The rotordynamic assessment of a motor-compressor train is typically performed considering each unit alone, being the spacer-type flexible coupling, used to connect the motor to the compressor, capable to segregate the units dynamically.

Differently from oil bearings, the stiffness and damping of the AMB are varying with the frequency and they are dependent by the controller parameters that typically can be defined and adjusted to make the rigid modes and first bending mode well damped allowing the unit to extend the operating speed range.

- **Cooling System.**

Despite the absence of contact, the AMBs are generating power losses (iron, windage and copper losses) so that all active magnetic bearings mounted on centrifugal compressor and electric motor need to be cooled by a dedicated cooling system.

For the centrifugal compressor the AMB cooling circuit is based on a quasi-closed loop using the plant instrument air where air circulation can be forced by a dedicated impeller installed on the compressor shaft as shown in the Fig 5 or through an external blower.

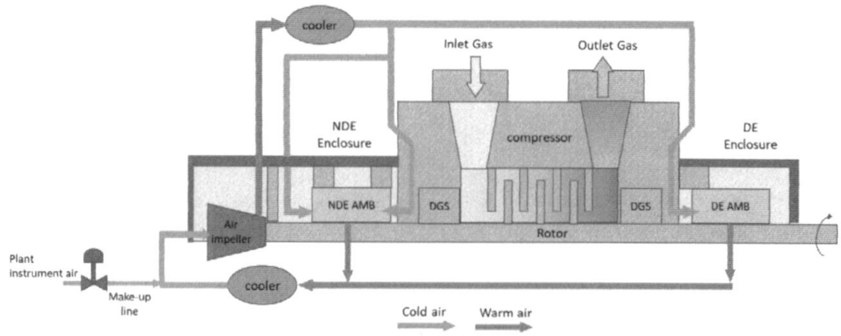

Fig. 5: AMB cooling layout.

The air temperature is regulated by water coolers, and the air leakages on the coupling side are compensated by a make-up line fed by plant instrument air.

For the electric motor, instead, the cooling of the active magnetic bearings is ensured by the internal motor cooling system using the cooled air from ventilation system.

Primary function of air system is to cool the AMB to dissipate the heat generated from air friction and electrical losses.

Secondary function of air system is to prevent any migration of explosive mixture in the AMB enclosure from either external environment or from dry gas seal. A principle of minimum overpressure with respect to those zones is applied.

Same air system is used during purging sequence, at start-up, to avoid any potential explosive mixture in the AMB enclosures before the magnetic bearings energization.

In operation, AMB coil temperature and enclosures pressure are monitored. Below a minimum enclosure pressure threshold and/or above a certain AMB temperature threshold, a general trip of units will occur. Alarms are also displayed to the main control unit screen. We can roughly estimate a power need of 1kW cooling system for 1MW compression power on a AMB equipped compressor.

4. Main benefits for oil-free configurations

When comparing high-speed oil-free compression package with conventional oil-lubricated motor-compressor, the simplification of the architecture is clear as shown in Fig. 6 and the following benefit can be highlighted:

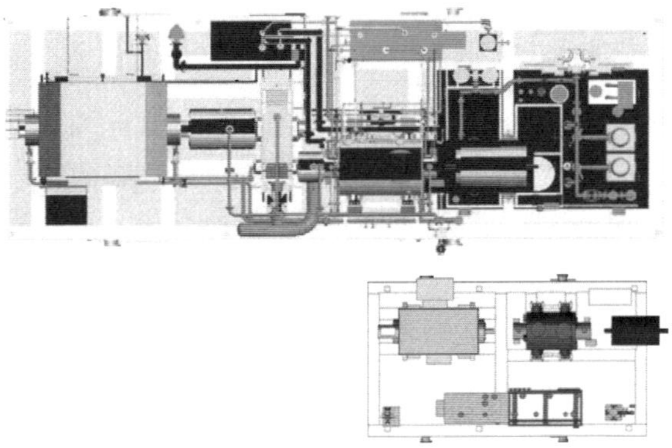

	Dimensions [m]	Footprint [m^2]	Weight [tons]
Conventional train	16.7 x 4.8	80	188
Oil-free train	10.4 x 3.8	39.5	85
Var %	N/A	-50%	-55%

Fig. 6: Oil-free vs conventional

Advantages oil-free high speed motor driven compressor

→ CAPEX:

- Gain of space and weight: The reduction of the weight and the footprint by 50% reduces considerably the needed platform support, cutting cost on the construction.
- Installation and Commissioning: By simplifying the connectivity between machine skid and auxiliaries, avoiding all the oil tube piping work and the flushing system on platform. The installation and commissioning time is reduced by 20%
- Integrated instrumentation and digital technology allow high frequency monitoring, limiting physical additional physical sensor and cabling. This monitoring technology is suited for remote and offshore application and enables easier condition monitoring possibilities.

→ OPEX:

- The all-electric solution allows to avoid all the periodic visual inspections, limit the regular maintenances activities due to the oil-based auxiliaries (oil lude system, oil cooling system, gearbox...).
- The power consumption of the cooling of the AMB is about 2 times lower than the consumption of the full lube-oil system, allowing to lower the energy consumption and the CO_2 emissions of the oil-free compression train.

- By limiting the number of measurement points due less auxiliaries, the cost of maintaining theses points, collecting the data, and analyzing the information are practically divided by 2.

➔ PRODUCTION:

- Higher availability and reliability, through the limitation of auxiliaries and the cleaner technology (no oil pollution) there is less maintenance stoppage needed (every 5 years) and the Dry Gas Seals (DGS) usually the weak point, is less contaminated and increase their reliability. This has been proven by the reliability, availability and maintainability (RAM) analysis.
- Unmanned enabler due to the full electric design and less maintenance needs, little or no crew is needed to survey the installation. This can also affect the living quarters and the spare part storage space required, lowering the general size of the infrastructure.

5. Enhanced configuration

An alternative solution is represented by the semi-sealed compressor on which, similarly to the hermetically sealed compressor concept, the AMBs are cooled by the process gas with only one dry gas seal located on the coupling side. This configuration reduces the gas leakages but an in-depth study on the gas composition needs to be done as it is in contact with the AMB. Today protection technology has improved and new “upstream” projects using more resistant technology have been launched.

The spearhead technology development based on a compressor train directly at the well head, on the seafloor is the proof that everything is possible. Subsea projects are equipped also with SKF S2M bearing have proven their capability since the commissioning in 2015.

6. Conclusion

This study shows the following benefits:

➔ The remote monitoring and control allow to lower the needed crew on the platform and enabling the installation in critical environment locations.

➔ Environment and operational: All electric solution, less construction material to support train...

➔ Total Cost of Ownership: Installation costs, operational costs, maintenance costs, energy costs lowered.

➔ Production: less maintenance stops, higher availability, and reliability.

All these advantages have not been calculated but each of them could be evaluated case by case according to the application and/or specific plant location.

Off-shore project should leverage on these benefits adopting the oil-free high speed motor driven compressors (where applicable) without hesitation.

7. References

[1] Magnus Korpåsa *, Leif Warlanda , Wei Heb , John Olav Giæver Tandea “A Case-Study on Offshore Wind Power Supply to Oil and Gas Rigs” SINTEF Energy Research, N-7465 Trondheim, Norway.

[2] Vishnu Gupta K., Larsen F., Ang D.L., Patel P., “Gas Compressors”, December 2022, IHS Markit, now part of S&P Global.

[3] Basha K., Intieri A., Mantini A., Ermini M., Del Puglia S., “Decarbonizing FPSO: Lifecycle Assessment of State-of-the-Art Rotating Equipment Solutions”, 2023, Offshore Technology Conference, Houston, TX, USA, 1-4 May.

[4] Voldsund M., Reyes-Lua A., Fu C., Ditaranto M, Neksa P., Mazzetti M., Brekke O., Bindigsbo A. U., Grainger D., Pettersen J., “Low carbon power generation for offshore oil and gas production”, 2023, Energy Conversion and Management:X.

[5] Alban T., Guillemin S., Labouré P., and Pellerin O., “Considerations to Optimize Electric Motor Compression Stations”, 2014, 9th Pipeline Technology Conference, Berlin, 12-14 May.

[6] Durantay L., Gelin A., Thibaut E., Vidalenc Y., “Integrated moto-compressor versus conventional solution”, 2019, 16th PCIC Europe, Paris, 7-9 May.

The dynamic characteristics of a diamagnetic levitated electrostatic motor

Yuanping XU [a], Yang WANG [a], Jarir MAHFOUD [b], Riwang LING [a], Jin ZHOU [a]
a Nanjing University of Aeronautics and Astronautics, Yudao Street, Nanjing, China, ypxu@nuaa.edu.cn
b University of Lyon, Lyon, France

Abstract

Diamagnetic Levitation emerges as a promising approach to tackle this friction issue. Due to the challenges of friction in micro motors, diamagnetic levitation emerges as a promising approach to tackle this friction issue and is essential in further developing high-speed, low-energy micro motors. Diamagnetic Levitated Electrostatic Motor has huge potential for development with the advantage of frictionless and passively stable levitation. This paper focuses on the model of the dynamics parameters of diamagnetical levitated rotor in order to accurately calculate the dynamic response of the rotor, which is important for optimising the performance of diamagnetical levitated electrostatic motor.

Therefore, this paper first establishes the kinetic equations of a diamagnetic levitated electrostatic motor and proposes a method for correcting and identifying the magnetic field parameters of permanent magnets based on the equivalent current model. Then, the diamagnetic force on the suspended rotor is modeled, which leads to the investigation of the distribution of the equivalent stiffness in the rotor dynamics system. The electromagnetic forces that impede the motion of the levitated rotor are then modeled based on the principle of electromagnetic induction, thus exploring the distribution characteristics of the damping parameters in the kinematic equations. At last, The radial vibration characteristics of the levitated rotor during operation are investigated by means of numerical simulations, and relevant experiments are designed to verify them.

Keywords: *Diamagnetic Levitation, Electrostatic Motor, Modeling.*

1. Introduction

With the development of MEMS technology, micro motors are the core parts of MEMs for realizing energy conversion, function-driven, and precision control, performing the combined functions of precision actuation and control in critical areas such as aerospace, micro-robotics, and biomedicine.

Diamagnetic levitation refers to the free levitation of diamagnetic object in magnetic field, which is not limited by Earnshaws' theorem[1], to achieve a constant, passive levitation of the diamagnetic object at room temperature without the need for an energy supply. With the inherent merits of frictionless, passive self-stabilization, and increased buoyancy-weight ratio, diamagnetic levitation emerges as a promising application in the micromotor.With the development of drive technology for diamagnetic levitated motors, the miniaturization and increase in speed of diamagnetic levitated motors have been accompanied by the problem of their severe vibration at high-speed operation, which limits further optimization of the performance of diamagnetic levitated electrostatic motors.

This paper investigates the force characteristics of the diamagnetic rotor. It completes modeling dynamic parameters of the diamagnetic rotor and predicting the vibrations in the magnetic levitated electrostatic motor's open-loop rotation.

The first section describes the operating principle of the diamagnetic levitated electrostatic motor and the kinematic characteristics of the levitated rotor. Then the second section develops mathematical models for correcting and identifying magnetic field parameters, diamagnetic forces, and electromagnetic damping forces. Then the third section corresponding experiments are designed to validate the above proposed mathematical model and analyze the dynamic response characteristics of the levitated rotor during the operation of the diamagnetic levitated electrostatic motor. The fourth section concludes with a full summary.

2. Diamagnetic levitated electromagnetic motor

Fig.1 shows a schematic diagram of the stable levitation of the pyrolytic graphite disc. The "concave" magnetic field generated by the permanent toroidal magnets forms a potential trap structure in the horizontal plane, with the minimum value point on the central axis. The magnetic field structure can generate a radial diamagnetic force against the levitated rotor toward the minimum value point. Meanwhile, the "convex" magnetic field generated by the cylindrical permanent magnets gives the levitated rotor a significant magnetic potential energy, which forms a clear potential energy trap structure in the axial direction with the gravitational potential energy. The potential energy trap in the axial direction can generate an axial antimagnetic force against a suspended rotor pointing at a minimum value point. The levitated rotor achieves stable levitation at the intersection of the axial and radial directions.

Fig.2 shows a schematic diagram of the diamagnetic levitated electromagnetic motor, which combines diamagnetic levitated technology with electrostatic drives, using glass as the dielectric rotor[2-3]. The levitated rotor is constituted of a pyrolytic graphite disc (φ10×0.5mm) and a glass disc (φ10×0.15mm), as shown in Figure 2. For the PM array, a concentric ring-shaped magnet (φ18.68×φ7.96×12mm) encircling a cylinder magnet (φ7.94×12mm) with an opposite axial magnetization pattern that provides the radial forces; when the stator is supplied with three-phase AC high voltages, a rotating electric field above the stator will be generated which causes a lateral driven force that propels the glass disc to move based on the dielectric relaxation.

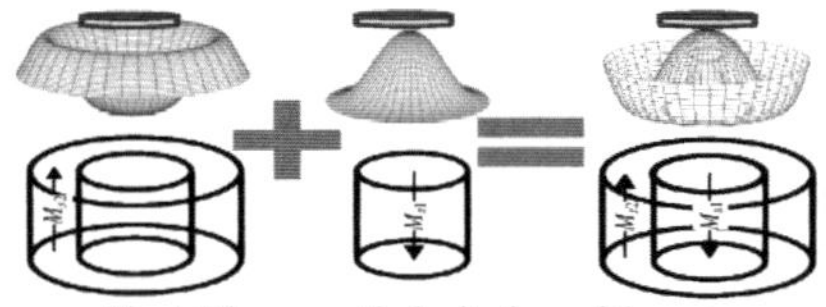

Fig.1 Diamagnetic levitation of the rotor

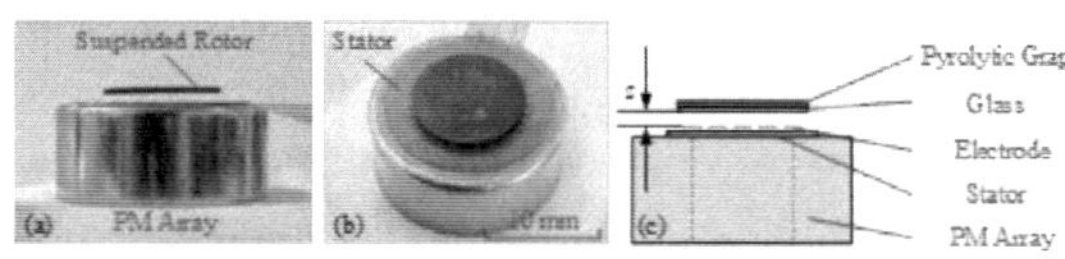

Fig.2 Diamagnetic levitated electromagnetic motor

3. Modelling and calculation of dynamics equation

3.1 Diamagnetic Force

First, the equivalent current model is used to analyze the PM (permanent magnets)[4]. According to this model, under ideal homogeneous magnetization conditions, the magnetic effects of the nuclear currents within the permanent magnet cancel each other out and appear as surface currents on a macroscopic scale. The permanent magnet with a uniform magnetization distribution can be equated to a surface current source distributed on the boundary surface of the magnetizing distribution region.In the column coordinate system, the magnetic flux density generated by a cylindrical permanent magnet at any point in space can be expressed as,

$$\boldsymbol{J}_M = \boldsymbol{M} \times \boldsymbol{e}_n = M_s \sin\theta \cdot \boldsymbol{e}_\phi \tag{1}$$

$$\boldsymbol{B} = \frac{\mu_0}{4\pi} \oiint_S \frac{\boldsymbol{J}_M \times \boldsymbol{e}_R}{|P-P'|^2} ds = \frac{\mu_0}{4\pi} \oiint_S \frac{M_s \cdot \boldsymbol{e}_\phi \times \boldsymbol{e}_R}{|\boldsymbol{P}-\boldsymbol{P}'|^2} \sin\theta ds \tag{2}$$

where $\boldsymbol{J_M}$ is the equivalent surface current density, $\boldsymbol{M_s}$ stands for the magnetization,θ represents the angle between the direction of magnetization and the normal vector of the equivalent current surface, $\mu 0$ is the magnetic permeability of vacuum, P is the source point, P' is the observation point.

The disconnected end face of the permanent magnet causes its magnetization to be in an open circuit, suffering from a demagnetization field[5]. This effect results in an inconsistent magnetic field inside the permanent magnet and does not allow for absolutely uniform magnetization of the permanent magnet. Moreover, according to theoretical proof, only permanent ellipsoidal magnets can meet the requirements for uniform magnetization. The variable magnetization inside the miniature permanent magnets leads to a considerable deviation from the ideal distribution of the external magnetic field, significantly affecting the diamagnetic force calculation. As shown in Figure 3, according to the experiment of magnetic field measurements, this paper equates the magnetized region at the end of the columnar permanent magnet to an ellipsoidal homogeneously magnetized area.

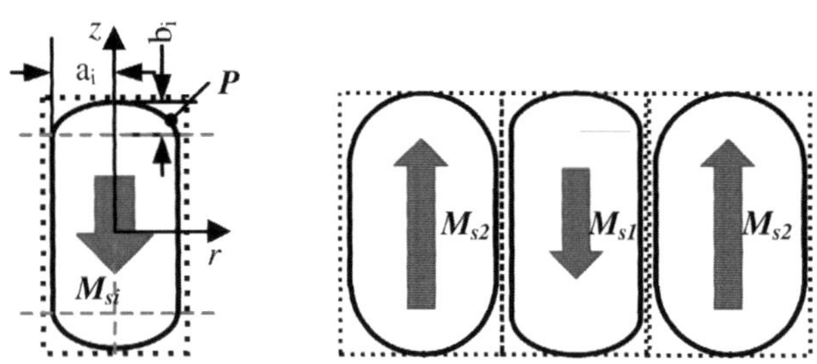

(a) Cylindrical permanent magnet (b) permanent magnets array

Fig.3 Magnetization distribution of PM array

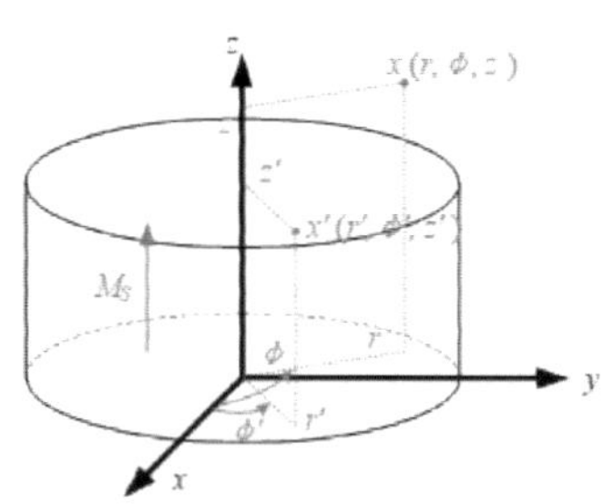

Fig.4 PM Columnar Coordinate System

Based on (2), the equivalent current model for columnar permanent magnets can be obtained as follows

$$\boldsymbol{B} = \frac{\mu_0 M_s}{4\pi} \int_{z1}^{z2} \int_0^{2\pi} \frac{\boldsymbol{e}_{\phi'} \times \boldsymbol{e}_R}{|\boldsymbol{P}-\boldsymbol{P}'|^2} r'(z') d\theta' dz' = \frac{\mu_0 M_s}{4\pi} \oiint_{S_1} \frac{\boldsymbol{e}_{\phi'} \times \boldsymbol{e}_R}{|\boldsymbol{P}_1-\boldsymbol{P}'|^2} \frac{R(z')}{R'} \frac{|\boldsymbol{P}_1-\boldsymbol{P}'|^2}{|\boldsymbol{P}-\boldsymbol{P}'|^2} ds' \tag{3}$$

where S_1 is the boundary surface of the magnetizing distribution region, R' is maximum radius of cross-section, $R(z')$ is radius of the cross-section with z-axis coordinate z'. Based on (5), a new equivalent current model can be obtained as fellows

$$\boldsymbol{J}_{M1}(z') = \frac{R(z')}{R'}\frac{|\boldsymbol{P}_1 - \boldsymbol{P}'|^2}{|\boldsymbol{P} - \boldsymbol{P}'|^2} \tag{4}$$

$$\boldsymbol{B} = \frac{\mu_0}{4\pi}\oiint_{S_1}\frac{\boldsymbol{J}_{M1}(z')\times \boldsymbol{e}_R}{|\boldsymbol{P}_1 - \boldsymbol{P}'|^2}ds' \tag{5}$$

where $\boldsymbol{J}_{M1}$ is the equivalent surface current density. Based on the above model, the equivalent surface current distribution of the permanent magnets is shown in Figure 5.

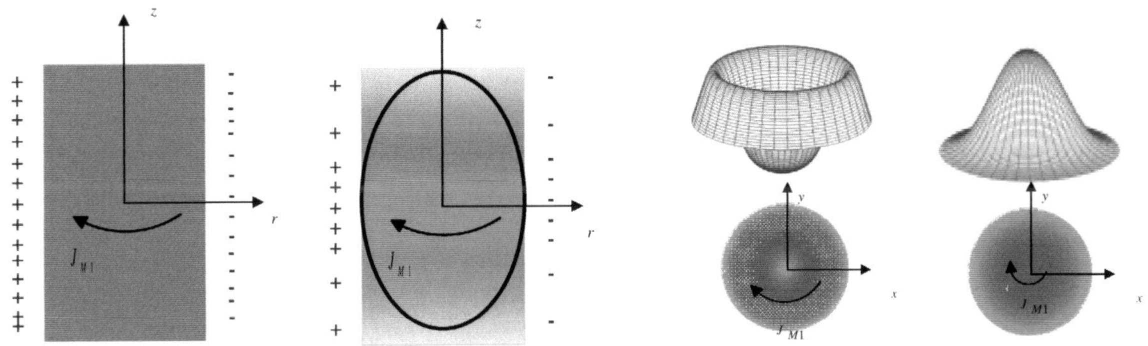

(a) Equivalent surface current axial distribution (b) Equivalent surface current radial distribution

Fig.5 Equivalent surface current axial distribution

Figure 5 shows that when the boundary surface of the magnetizing intensity distribution region is ellipsoidal, the equivalent current magnitude decreases from the middle section of the permanent magnet along the axis toward the ends. In the horizontal direction, the location of the magnetic flux density maximum shrinks from a ring towards the central axis as the radius of the cross-section decreases, forming a clear magnetic potential well structure.

Finally, the magnet array parameters are identified by PSO based on the above-modified model of magnet array parameters.

Table 1 PM array parameters

Symbol	Quantity	Values
a_1	Ellipse 1 long axis radius	3.6 mm
a_2	Ellipse 2 long axis radius	2.9 mm
b_1	Ellipse 1 short axis radius	0.8 mm
b_2	Ellipse 2 short axis radius	1.2 mm
Ms_1	Magnetization for inner cylinder magnet	7.82×10^5 A/m
Ms_2	Magnetization for outer ring magnet	9.21×10^5 A/m

Table 1 lists the parameters for the magnets array. Based on the identified permanent magnet parameters, the magnetic flux density along the radial direction of the permanent magnet array at suspension heights h=1.2 mm、h=1.7 mm、h=2.2 mm was calculated and compared with the experimental data, as shown in Figure 6.

The potential trap structure can generate diamagnetic force against the levitated rotor pointing toward a stable levitated position.

$$\boldsymbol{F} = \frac{x_m}{2u_0}\oiint_{s} B^2\cdot \vec{n}_s dS \tag{6}$$

where x_m is the magnetic susceptibility, and x_m=-450×10^{-6}. From (6), it follows that the diamagnetic force on the levitated rotor at small displacements is proportional to the gradient of the magnetic potential energy, with the direction pointing from the high potential energy to the low potential energy.

The magnetic flux density was calculated based on the PM array parameters. On a horizontal surface with a levitation height h = 0.5 to 0.9 mm, the magnetic flux density is distributed along the radial direction, as shown in Figure 7. At a levitation height of 0.7 mm, the magnetic flux density decays along the axial direction, creating a diamagnetic force on the pyrolytic graphite sheet along the axis, counteracting the effect of gravity, and achieving stable suspension in the axial direction. At a radial distance of 5 mm, the magnetic flux density increases along the radial direction, generating a diamagnetic force on the pyrolytic graphite sheet pointing towards the central axis and achieving a stable radial levitation of the pyrolytic graphite disc. According to (6), the distribution of diamagnetic force is calculated, as shown in Figure 8,

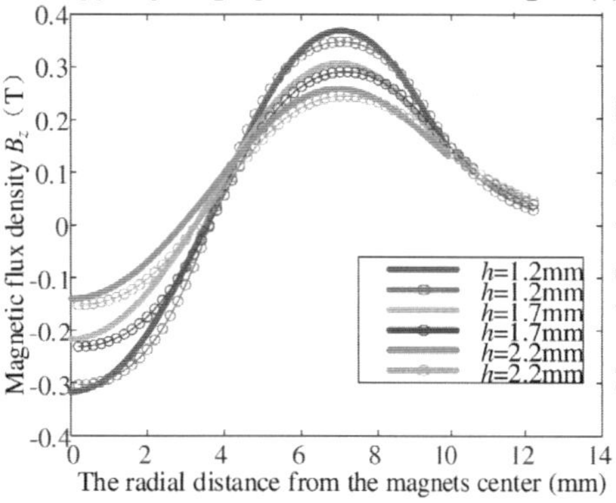

Fig.6 Axial magnetic flux density distribution

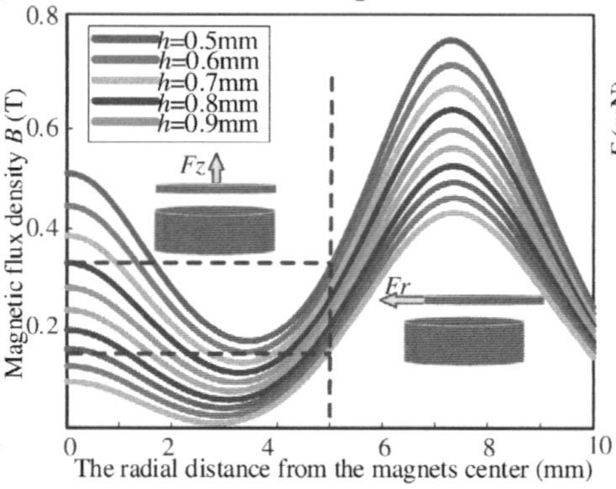

Fig.7 Magnetic flux density radial distribution

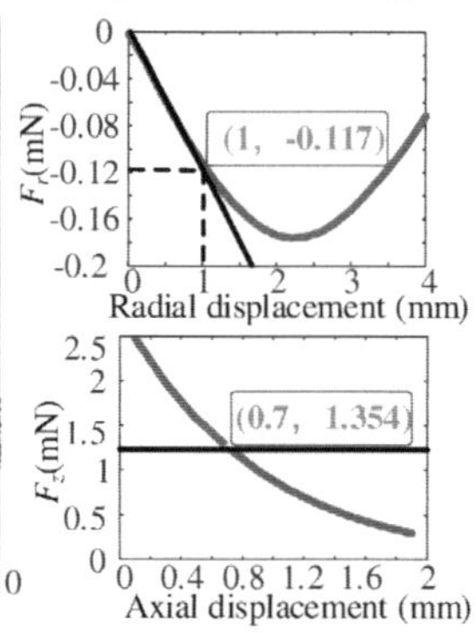

Fig.8 Diamagnetic force

The diamagnetic force on the levitated rotor increases along the radial direction from 0 to 2.5 mm radial displacement, with a maximum radial diamagnetic force of 1.8 mN. And the radial stiffness is equal to almost 0.14 N/m along the radial direction from 0 to 0.2 mm radial displacement.The diamagnetic force on the levitated rotor decreases along the axial direction. At a levitation height of 0.7 mm, the axial diamagnetic force on the pyrolytic graphite sheet is equal to 1.354 mN, counteracting the effect of gravity and achieving stable suspension in the axial direction. And the axial stiffness, greater than radial stiffness, equals almost 1.302 N/m.

3.2 Electronic Damping

The levitated rotor operates under the influence of air damping and electromagnetic damping[6]. At a levitation height of 0.7 mm, electromagnetic damping is the primary source of damping forces on levitation rotors. Pyrolytic graphite moving in a magnetic field generates eddy currents due to electromagnetic induction and is subject to a Lorentz force impeding the motion, which is expressed as follows

$$\boldsymbol{F} = \iint_{S_e} \boldsymbol{I} \times \boldsymbol{B} ds_e \tag{7}$$

When the levitated rotor is vibrating radially in the horizontal direction, as shown in Figure 9, the resulting induced current can be expressed as follows,

$$I_{r'} = -\frac{\partial D'}{r'\partial \theta'} \tag{8}$$

$$I_{\theta'} = -\frac{\partial D'}{\partial r'} \tag{9}$$

$$D = \frac{h}{4\pi\sigma} \oint_{S'} v_r \frac{\partial B_z}{\partial r}(r,\theta) \sqrt{\frac{R^4 + r_Q^2 r'^2 - 2r_Q r' R^2 \cos(\theta_Q - \theta')}{R^2 (r_Q^2 + r'^2 - 2r_Q r' \cos(\theta_Q - \theta'))}} dS_Q \tag{10}$$

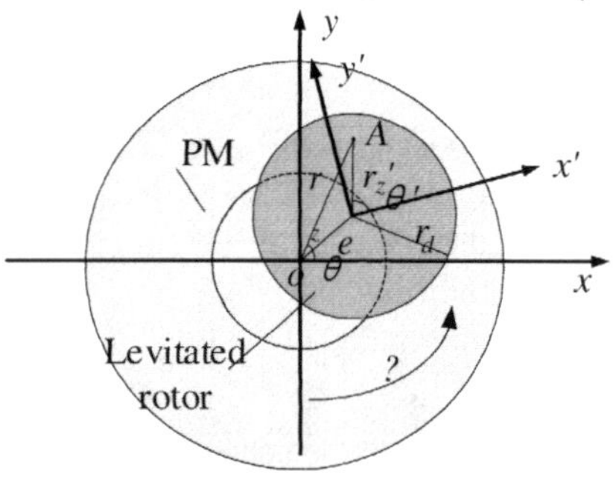

Fig. 9 vibration in the horizontal direction

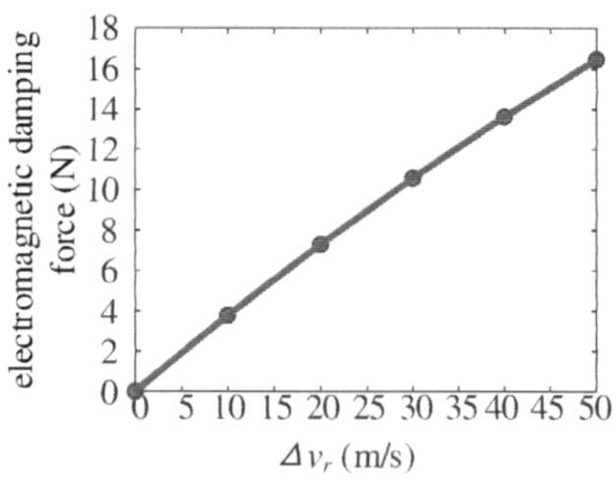

Fig. 10 electromagnetic damping force distrubution

The electromagnetic damping force of the pyrolytic graphite sheet at a distance of 0.2 mm from the center axis of the permanent magnet was calculated using Maxwell simulation and is shown in Figure 10, giving an electromagnetic damping of 3×10^{-4} N·s/m.

3.3 Dynamics Equation

From section 3.1, the rotor is subjected to diamagnetic forces from the permanent magnet stator in all five degrees of freedom, including the *z*-direction, *x* and *y*-directions. In addition, pyrolytic graphite has excellent electrical conductivity and generates eddy currents as well as electromagnetic damping when moving relative to the stator. The kinetic equations can be expressed as follows,

Due to the low speed of the diamagnetic levitated electrostatic motor and the weak influence of the gyroscopic effect on the rotor at low speeds, the rotor's inclination vibration is ignored. The kinetic equations for the rotor can be expressed as follows

$$m\ddot{x}+c_{xe}\dot{x}+k_{\alpha x}x=me\omega^2\cos\omega t \tag{11}$$

$$m\ddot{y}+c_{xe}\dot{y}+k_{\alpha x}y=me\omega^2\sin\omega t \tag{12}$$

$$m\ddot{z}+c_{ze}\dot{z}+k_z z=F_z \tag{13}$$

4. Experiments and Results

4.1 Experiments

Experiments measuring the radial stiffness, damping, and vibration were conducted to verify the dynamics model as shown in Fig. 11, Fig. 12, and Fig. 13.

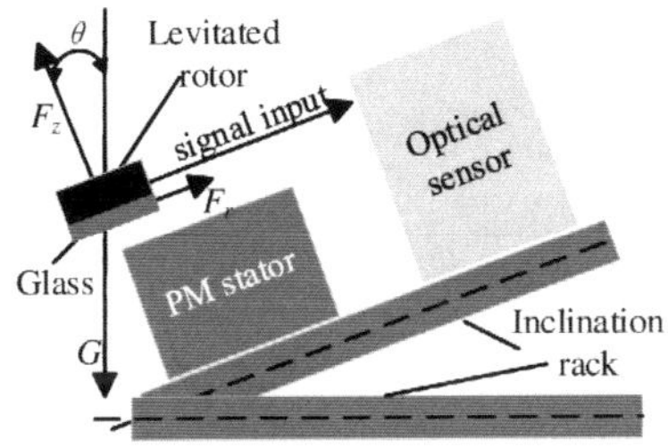

Fig.11 The radial diamagnetic force experiment

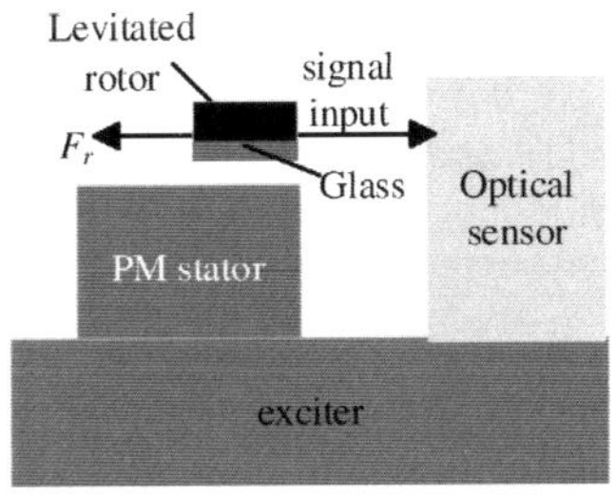

Fig.12 The electromagnetic damping experiment

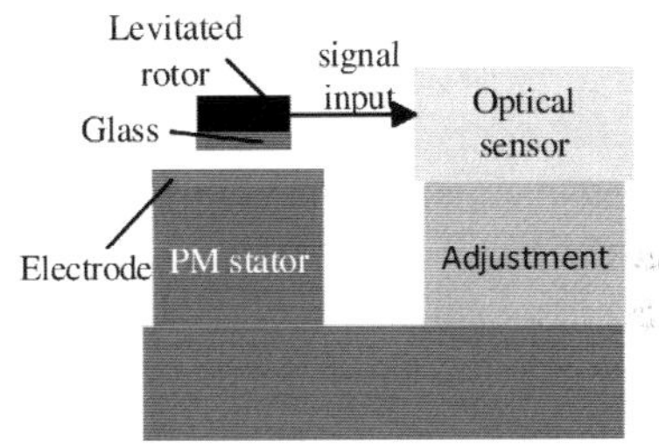

Fig.13 The radial vibration experiment

Fig. 11 shows the structure of the radial diamagnetic force experiment. From top to bottom are the levitated rotor (pyrolytic graphite + glass), the permanent magnets array, and the inclination rack. The inclination rack is used to tilt the permanent magnet. As the axial diamagnetic force cannot balance the gravity of the suspended rotor, the rotor will be displaced radially. Then the rotor is subjected to a certain radial diamagnetic force, which balances the gravity and allows the suspended rotor to return to a stable state of suspension.

Figure 12 shows the structure of the radial electromagnetic damping experiment. From top to bottom are the levitated rotor, the permanent magnets array, and the optical sensor, exciter. The exciter drives the PM array to vibrate horizontally, thereby displacing the levitated rotor relative to the PM array. The levitated rotor will vibrate around the stable suspended position with a gradual decay in amplitude.

Figure 13 shows the structure of. The levitated rotor (pyrolytic graphite + glass), the permanent magnets array, and the adjustment bracket are from top to bottom. The optical sensor is installed on a stand glued to the stator. The levitated rotor is suspended from the stator, with the glass side facing the stator. The laser emitted by the optical sensor illuminates the cylindrical surface of the levitated rotor by adjusting the position of the adjustment bracket.

4.2 Results

Results of the experiments measuring the radial stiffness, damping, and vibration is shown in Fig. 14, Fig. 15, and Fig. 16.

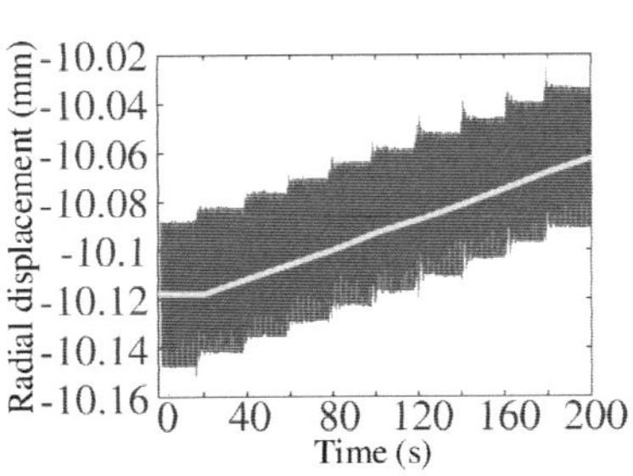

Fig.14 Radial displacement distribution

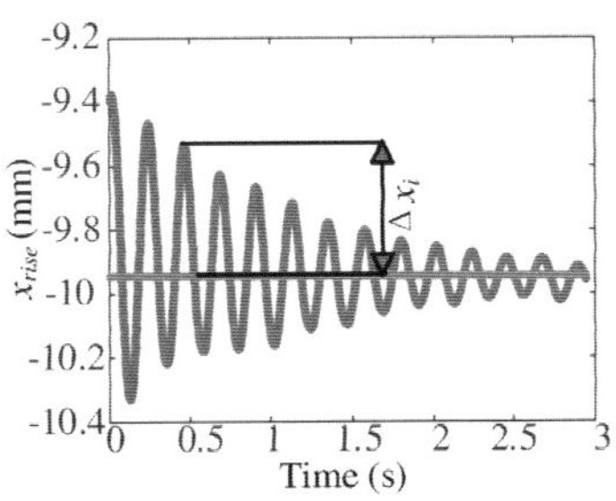

Fig.15 decay of the free response

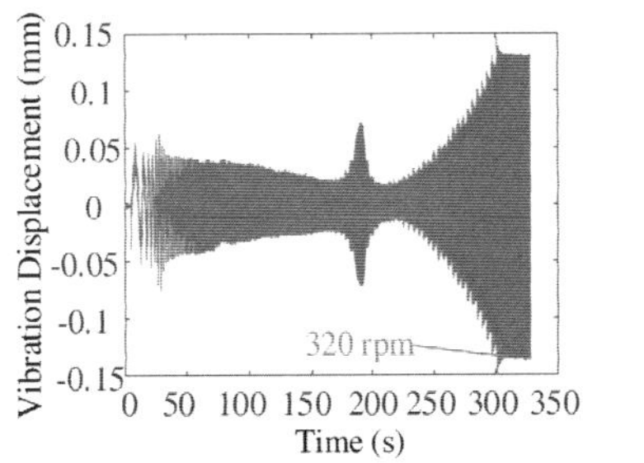

Fig.16 The vibration of the levitated rotor during acceleration

The radial diamagnetic force on the levitated rotor is quantitatively related to the degree of inclination, as shown in Figure 14.

$$F_r = mg\sin\theta = mg\frac{h_{rise}}{\sqrt{{h_{rise}}^2 + l^2}} \tag{14}$$

where l=600mm is horizontal distance between center of the hinge and the micrometer, h_{rise} is the height of the end of the plate being lifted. The radial diamagnetic force is proportional to the radial displacement of the levitated rotor between 0 and 0.06 mm, with a stiffness of approximately 0.135 N/m.

The decay of the radial free response amplitude of the levitated rotor under displacement excitation is shown in Fig. 15, which is related to the electronic damping,

$$\delta = \ln^{\frac{\Delta x_i}{\Delta x_{i+1}}} = \frac{2\pi\zeta}{\sqrt{1-\zeta^2}} \tag{15}$$

$$c = \zeta c_c = 2\sqrt{mk}\zeta \tag{16}$$

where δ is damping coefficient, Δx_i is the amplitude of the free response, ζ is the damping ratio, m=0.1354 g is the mass of the levitated rotor, k=0.135 N/m is the radial stiffness. The radial electromagnetic damping force is proportional to the radial vibration velocity, with average damping of approximately 3.05×10^{-4} N·s/m.

The change of the radial forced response of the levitated rotor with the frequency of input three-phase voltage increasing by 1 rpm at 1 s intervals is shown in Fig. 16. When the speed of the levitated rotor is increased to 300 rpm, the vibration is very intense, and the natural frequency could be calculated of 5.36 Hz. The calculated results are compared with the curve of the rotor vibration amplitude change in the experiment. The eccentricity of the mass of the rotor was identified at 0.02 mm.

5. Conclusion

This paper first establishes the kinetic equations of a diamagnetic levitated electrostatic motor. A magnetic field correction method for permanent magnet arrays is proposed and established based on the equivalent current model. Based on the principle of diamagnetic levitation, the diamagnetic force on the rotor during radial vibration has been modeled. Based on the principle of electromagnetic induction, the electromagnetic damping force on the rotor during radial vibration has been modeled. Meanwhile, the corresponding experiments were designed to verify the properties of the forces on the rotor.

The present study shows that the magnetization of the permanent magnets array decreases along the cross section, which causes the external magnetic field to form a potential well structure. The levitated rotor is subjected to a diamagnetic force proportional to the displacement, equivalent to a stiffness of 0.14 N/m. The levitated rotor is also subjected to an electromagnetic damping force proportional to the speed of motion, equivalent to a damping of 3.1 N/m·s^{-1}. When the speed of the suspended rotor is increased to 300 rpm it is very close to the first order intrinsic frequency and the vibration is very intense, meanwhile the eccentricity of the mass is 0.02 mm. The natural frequency could be calculated of 5.36 Hz.

References

[1] Jianquan L (2020) Magnetic levitation and Earnshaw theorem. Physics Teaching 42(8): 5.
[2] Furlani EP (2001) Permanent Magnet and Electromechanical Devices. San Diego: Academic Press, pp.129-143.
[3] Yuanping X, et al., (2017) Realization of a diamagnetically levitating rotor driven by electrostatic field. IEEE-ASME Transactions on Mechatronics, 22(5): 2387-2391.
[4] Yuanping X, et al., (2020) Modeling and Validation of Diamagnetic Rotor Levitated by Permanent Magnetics. Chinese Journal of Mechanical Engineering.
[5] Yongzhou H (2013) Study of the inhomogeneity of the external magnetic field of permanent magnets. Chinese Journal of Physics (08): 145-151.
[6] Xianfeng C, et al., (2022) Diamagnetic Composites for High-Q Levitating Resonators. Advanced science 9(32).

Observer-based Self-Sensing Techniques for Hybrid Active-Passive Self-Bearing Machines

Joachim VAN VERDEGHEM, Bruno DEHEZ

Mechatronic, Electrical Energy, and Dynamic Systems (MEED),
Université catholique de Louvain (UCLouvain), Louvain-la-Neuve, Belgium,
emails: {joachim.vanverdeghem, bruno.dehez}@uclouvain.be

Abstract

Passive magnetic levitation offers advantages in terms of compactness, reliability and cost thanks to the absence of position sensors, controllers and power electronics dedicated to the rotor suspension. Among these passively levitated systems, those based on a self-bearing machine relying on electrodynamic effects show an operating speed range limited to the high speeds given that the restoring forces are created by induced currents. To address this issue, hybrid active-passive actuation approaches have recently been introduced and consists in actively controlling the axial position of the rotor through the direct-axis component of the currents flowing in the combined winding until reaching the threshold speed beyond which passive levitation can be achieved. However, this active operation requires the addition of an axial position sensor, affecting the benefits related to passive suspension. In this context, this paper proposes self-sensing techniques relying on state observers to estimate the rotor axial position and speed on the basis of the machine electromechanical model. Their performance and robustness are then assessed by means of dynamic simulations.

Keywords: Self-sensing, Sensorless, Observer, Self-bearing, Bearingless

1. Introduction

The demand for high power density magnetically levitated systems has led to the development of self-bearing machines, the latter ensuring both the rotor guidance and drive within a single structure. Generally, the rotor position is actively regulated on the basis of the currents flowing in the armature winding. Although this offers advantages in terms vibration management, sensors and power electronics dedicated to the rotor levitation are required compared to a conventional machine, impacting the cost, compactness and reliability of the system. The current trend is therefore to replace actively controlled degrees of freedom with passive solutions.

In this regard, recent research have investigated passively levitated self-bearing machines relying on a combined armature winding both at the theoretical and experimental levels (Van Verdeghem et al. 2019, Rubio et al. 2023). Nevertheless, the rotor axial levitation arises from induced currents, implying that the restoring force and the underlying stiffness are small at low speeds and even null at rest. Two different approaches have been proposed to allow the rotor suspension in the complete speed range, at the expense of lower drive torque and force capabilities. (Van Verdeghem et al. 2021, Rubio et al. 2021). In both cases, the direct-axis motor current is regulated to control the axial position of the rotor. A position sensor is therefore necessary to operate the machine in this active mode, affecting its reliability, compactness and cost.

Self-sensing techniques can be implemented to prevent the addition of this sensor. Among them, high frequency signal injection is well suited for the start-up and stop phases but observers are preferred at medium and high speeds to prevent interferences between the injected signal and the force and torque currents.

In this context, this paper introduces observer-based self-sensing techniques, derived from the machine electromechanical model, to evaluate the rotor axial position and speed. Dynamic simulations are carried out to validate them and analyse their robustness to parameter errors.

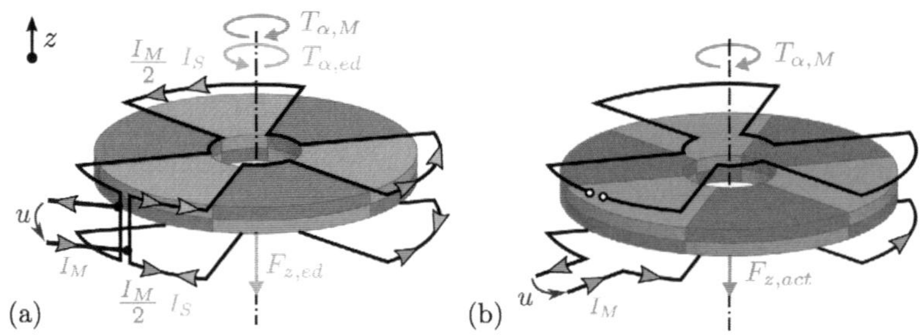

Figure 1: Self-bearing machine (only one phase represented with $p = 3$). (a) Passive operation at high speeds. (b) Active operation at low speeds.

2. Machine Description

This section first describes the structure and the operation principle of the self-bearing machine under study. The electromechanical model predicting its rotor dynamics in active suspension mode is then presented.

2.1 Structure & Principle

As illustrated in Fig. 1, the rotor of the self-bearing machine is composed of axially magnetised permanent magnets creating a magnetic field with p pole pairs. The stator consists of two three-phase windings, placed on both sides of the rotor, whose connection depends on the suspension operation mode.

At high rotor spin speeds, the machine operates in passive mode. As shown in Fig. 1(a), both windings are connected in parallel to the power supply. In case of rotor decentring z, circulating suspension currents I_S are passively induced in the windings due to an unbalance between their permanent magnet flux linkages, creating a restoring force, and add to the conventional motor currents I_M that generate the drive torque.

At low rotor spin speeds, the machine operates in active mode. Only one three-phase winding is supplied whereas the other is left open, as represented in Fig. 1(b). In this way, the rotor axial position can, similarly to the drive torque, be regulated through the motor current I_M thanks to the asymmetric structure. Self-sensing techniques are derived to provide an estimation of the axial position z required to its control while maintaining the advantages of passive suspension.

2.2 Electromechanical Model

The electrical and mechanical equations governing the axial and spin dynamic behaviour of the self-bearing machine when operating in active suspension mode are given by:

$$\begin{aligned} u_d &= Ri_{M,d} + L_c(\dot{i}_{M,d} - p\omega i_{M,q}) + \dot{z}K_z\sqrt{\frac{3}{2}} & m\ddot{z} &= K_z\sqrt{\frac{3}{2}}i_{M,d} - k_z z - C_z\dot{z} + F_e \\ u_q &= Ri_{M,q} + L_c(\dot{i}_{M,q} + p\omega i_{M,d}) + p\omega(K_\alpha + zK_z)\sqrt{\frac{3}{2}} & J_p\dot{\omega} &= p(K_\alpha + zK_z)\sqrt{\frac{3}{2}}i_{M,q} + T_e \end{aligned} \quad (1)$$

where u_d and u_q are the direct and quadrature-axis components of the voltage supplied to the armature, R and L_c are the resistance and the synchronous self inductance coefficient of the windings, K_α is the flux constant, i.e. the amplitude of the permanent magnet flux linked by the windings in centred position, K_z is the flux gradient, namely the proportionality factor between the amplitude of the PM flux linkage and the rotor axial position z, m and J_p are the rotor mass and polar moment of inertia, k_z is the axial external stiffness, e.g. generated by centring bearings ensuring the radial magnetic suspension, C_z is the axial external damping, F_e and T_e are the load force and torque applied on the rotor.

The electrical equations highlight the linear dependence of the direct and quadrature axis back-electromotive force terms to the rotor axial speed $\dot{z}$ and position z respectively. The latter can therefore be extracted from the measurements of the currents and the voltage supplied to the armature winding.

3. State Observers

This section describes the state observers developed based on the machine electromechanical model to estimate the rotor axial position z and speed $\dot{z}$ at low and medium spin speeds without any dedicated sensor.

3.1 Axial Position

The proposed position observer structure relies on the electrical equations of a three-phase RL load in the Park reference frame. Hence, compared to the model (1) governing the self-bearing machine electrical behaviour, the direct and quadrature electromotive force terms are not taken in consideration:

$$\begin{aligned} u_d &= R i_{M,d} + L_c(\dot{i_{M,d}} - p\omega i_{M,q}) \\ u_q &= R i_{M,q} + L_c(\dot{i_{M,q}} + p\omega i_{M,d}) \end{aligned} \tag{2}$$

These equations can be rewritten in a linear time-variant state-space form as follows:

$$\begin{aligned} \dot{\mathbf{x}} &= \mathbf{A}\mathbf{x} + \mathbf{B}\mathbf{u} \\ \mathbf{y} &= \mathbf{C}\mathbf{x} \end{aligned} \tag{3}$$

where the input vector $\mathbf{u}$ includes the direct and quadrature voltages, u_d and u_q, and the state vector $\mathbf{x}$ comprises the corresponding current components, $i_{M,d}$ and $i_{M,q}$. The latter are measured and thus also constitutes the output vector $\mathbf{y}$. The matrices $\mathbf{A}$, $\mathbf{B}$ and $\mathbf{C}$ are given by:

$$\mathbf{A} = \begin{bmatrix} -\frac{R}{L_c} & p\omega \\ -p\omega & -\frac{R}{L_c} \end{bmatrix}, \quad \mathbf{B} = \begin{bmatrix} \frac{1}{L_c} & 0 \\ 0 & \frac{1}{L_c} \end{bmatrix}, \quad \mathbf{C} = \begin{bmatrix} 1 & 0 \\ 0 & 1 \end{bmatrix}. \tag{4}$$

A conventional Luenberger observer can then be established based on this simplified electrical model to provide estimates of the state and output vectors, denoted $\hat{\mathbf{x}}$ and $\hat{\mathbf{y}}$ respectively:

$$\begin{aligned} \dot{\hat{\mathbf{x}}} &= \mathbf{A}\hat{\mathbf{x}} + \mathbf{B}\mathbf{u} + \mathbf{f}(\mathbf{y} - \hat{\mathbf{y}}) \\ \hat{\mathbf{y}} &= \mathbf{C}\hat{\mathbf{x}} \end{aligned} \tag{5}$$

The observer equations include an additional feedback term that involves a function $\mathbf{f}$ of the error between the measured $\mathbf{y}$ and estimated $\hat{\mathbf{y}}$ output vectors. This function usually applies a significant gain $\mathbf{G}$, the stability of the corresponding observer being analysed through the analysis of the eigenvalues of the matrix $\mathbf{A} - \mathbf{GC}$. Other forms can also be used, adding e.g. an integral term or including a sign function (sliding mode). In all cases, this feedback term aims to compensate for errors on parameters of the observed system, on the one hand, but also for unmodelled phenomena and disturbances, on the other hand. Regarding the machine under study and assuming that the winding impedance is properly determined, this term therefore provides estimates for both electromotive forces, denoted $\hat{e}_d$ and $\hat{e}_q$. The rotor axial position z and speed $\dot{z}$ can then be evaluated on the basis of these estimates by direct identification with respect to the complete model (1):

$$\hat{z} = \frac{1}{K_z}\left(\frac{\hat{e}_q}{\sqrt{\frac{3}{2}}p\omega} - K_\alpha\right), \qquad \hat{\dot{z}} = \frac{\hat{e}_d}{\sqrt{\frac{3}{2}}K_z}. \tag{6}$$

This method requires an accurate identification of the machine parameters, notably the flux constant K_α and gradient K_z, to produce precise estimations. In addition, the direct-axis electromotive force e_d, that results from the rotor axial speed $\dot{z}$, is expected to be small due to the low frequency of the axial mechanical oscillations, leading to an unfavourable signal-to-noise ratio. A specific axial speed observer is then developed in Section 3.2

3.2 Axial Speed

The speed observer leans on the following simplified mechanical model predicting the rotor dynamics:

$$m\ddot{z} = K_z\sqrt{\frac{3}{2}}i_{M,d} - k_z z - C_z\dot{z} \tag{7}$$

This equation does not integrate the axial load force F_e, the latter being considered as an unmodelled disturbance. The system under study can be expressed in the state-space form (3) where the state vector $\mathbf{x}$ consists of the rotor axial position z and speed $\dot{z}$ while the input vector $\mathbf{u}$ solely encompasses the measured d-axis motor current $i_{M,d}$. Assuming that the observer (6) properly evaluates the axial position z, its estimate $\hat{z}$ can be seen as an indirect measurement and thus constitutes the output vector $\mathbf{y}$. The matrices $\mathbf{A}$, $\mathbf{B}$ and $\mathbf{C}$ are given by:

$$\mathbf{A} = \begin{bmatrix} 0 & 1 \\ -\frac{k_z}{m} & -\frac{C_z}{m} \end{bmatrix}, \quad \mathbf{B} = \begin{bmatrix} 0 \\ \frac{K_z}{m}\sqrt{\frac{3}{2}} \end{bmatrix}, \quad \mathbf{C} = \begin{bmatrix} 0 & 1 \end{bmatrix}. \tag{8}$$

Pursuing an approach identical to that adopted for the position, an observer can be constructed based on this state-space model. In this case, the rotor axial speed $\dot{z}$ is an internal state of the observer and is consequently directly accessible whereas the feedback term provides an estimate for the unmodelled disturbance force F_e.

4. Case Study

This section aims to validate the operation principle of the proposed observers and to assess their robustness to parameter errors. To that end, dynamic simulations are performed on MATLAB SIMULINK in the Park reference frame on the basis of the electromechanical model (1). A two-level scheme is implemented to control the machine. On the high level, the rotor axial position z and spin speed ω are regulated based on a PID and a PI controller respectively. On the low level, PI controllers allow to regulate the direct and quadrature axis components of the motor current i_M. The case study is conducted on the basis of the reaction wheel demonstrator described and characterised in (Van Verdeghem & Dehez 2021).

4.1 Principle Validation

Assuming an accurate knowledge of the machine parameters, the validation of the observers is carried out based on a specific sequence aiming to verify their tracking capabilities. More precisely, the rotor spin speed ω is first fixed to 1200 (rpm) and then increases up to 1350 (rpm), as shown in Fig. 2(d), while a sinusoidal axial position setpoint characterised by an amplitude of 0.1 (mm) and a frequency of 20 (Hz), equivalent to a synchronous disturbance, is imposed. The rotor is subjected to a step of axial disturbance force F_e at 0.125 (s) as well as to a constant load torque of 50 (mNm).

Analysing first the position observer, Figs. 2(a) and 2(b) illustrate the evolution with time of the direct and quadrature-axis electromotive forces, e_d and e_q, and currents, $i_{M,d}$ and $i_{M,q}$, the dotted and solid lines corresponding to the actual and estimated values respectively. The agreement between them highlights the proper compensation of the machine electromotive forces provided by the feedback term of the observer and confirms its operation principle. It can also be noted that, as expected, the direct component e_d of the electromotive forces is extremely low and would therefore be difficult to exploit in a physical system due to noise. Fig. 2(c) depicts the time evolution of the rotor axial displacement z and its estimate $\hat{z}$, calculated through (6), underlining their accordance and therefore validating the position observer.

Focusing then on the speed observer, Fig. 3(a) represents the time evolution of the axial disturbance force F_e and its estimate $\hat{F}_e$. It can be concluded that the observer feedback term correctly evaluates the axial load force throughout the sequence, even though a small overshoot occurs upon the step. Hence, the estimate $\hat{F}_e$ can be exploited for active disturbance rejection control, through a feedforward action in the position regulator

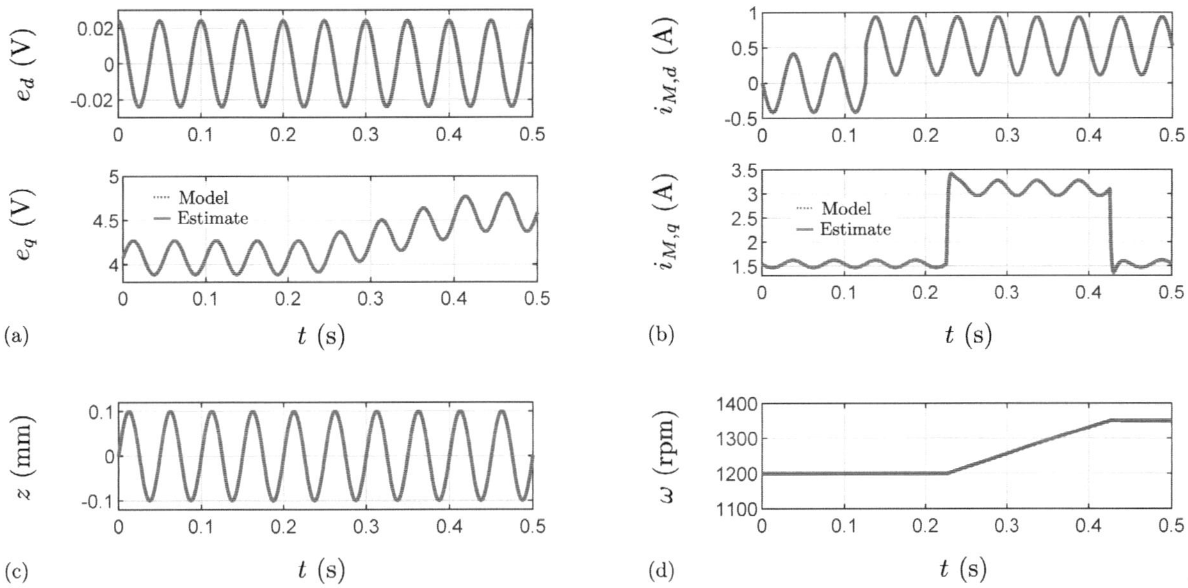

Figure 2: Position observer validation. (a) Direct and quadrature axis electromotives forces, e_d and e_q. (b) Direct and quadrature axis currents, $i_{M,d}$ and $i_{M,q}$. (c) Rotor axial position z. (d) Rotor spin speed ω.

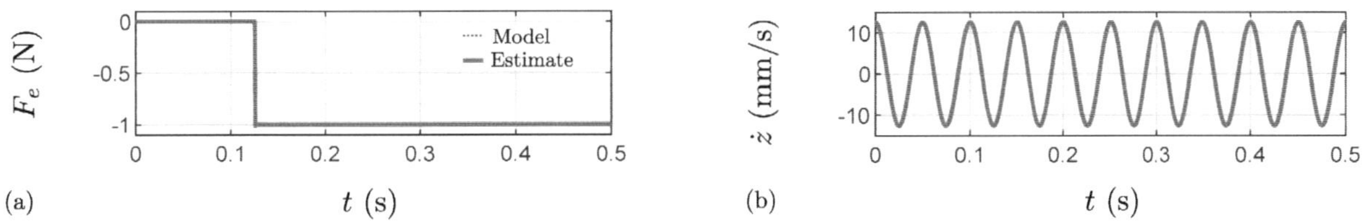

Figure 3: Speed observer validation. (a) Axial load force F_e. (b) Rotor axial speed $\dot{z}$.

(Petersen et al. 2023), but also to adapt to the rotor axial position setpoint in active suspension mode so as to limit oscillations due to the transition to passive operation (Van Verdeghem et al. 2021). Finally, as shown in Fig. 3(b), the rotor axial speed $\dot{z}$ is properly estimated, validating the proposed observer.

4.2 Robustness Analysis

The observer robustness is analysed by studying the impact on the estimates of errors on the self-bearing machine parameters. The spin speed reference is set to 1200 (rpm) and a sinusoidal axial position setpoint is imposed. Constant axial load force F_e and torque T_e, amounting to 1 (N) and 50 (mNm), are exerted on the rotor.

Figs. 4(a) and 4(b) present the time evolution of the estimated axial position $\hat{z}$ and the estimation error $|z-\hat{z}|$ for $\pm 5\%$ variations of the machine parameters. As expected, deviations on the winding impedance directly affect the estimate $\hat{e}_q$ of the quadrature-axis electromotive force and hence conduces to an offset on the position. Specifically, an error on the synchronous inductance coefficient L_c has a low influence on the observer accuracy given that this parameter is small for the ironless structure under study. On the other hand, the resistive term creates a non-negligible estimation discrepancy due to its dependence to the quadrature axis current $i_{M,q}$ and the underlying commanded torque, arising from the load T_e and spin speed reference modifications. The incorrect evaluation of the flux constant K_α also generates a significant offset since the variation with the position of the flux linkages and thus of the resulting electromotive forces, based on which the estimate is established, is of the same order of magnitude as the error. In contrast, as stated in (6), a deviation of the flux gradient K_z

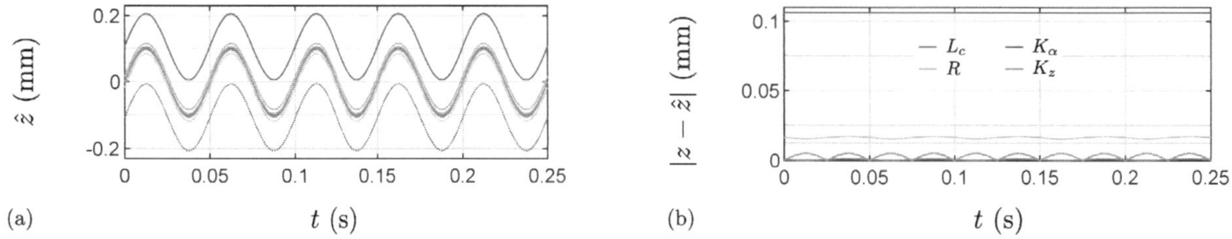

Figure 4: Observer robustness to $\pm 5\%$ parameter variations. (a) Position estimate $\hat{z}$. (b) Estimation error.

solely leads to a scaling error on the estimate $\hat{z}$, thence still allowing to regulate the rotor position in the airgap centre. Regarding the speed observer, inaccuracies in the machine mechanical parameters do not compromise the estimate $\hat{\dot{z}}$ but directly impact that of the disturbance force $\hat{F}_e$. Similarly, the improper estimation of the axial position z due to parameter errors has almost no influence on the speed estimate.

It should be noted that the errors considered in this analysis are substantial when the machine parameters are experimentally identified in the commissioning phase. Moreover, the effect of the temperature on the winding resistance R as well as the flux gradient K_z and constant K_α can be taken into account in the observer.

5. Conclusion

This paper introduces observer-based self-sensing techniques to estimate the rotor axial position and speed of self-bearing machines. The former relies on the motor electrical model and more specifically on the induced electromotive forces whereas the latter takes advantage of the rotor mechanical equation, providing in addition an estimate for the axial load force. Their operation principle and tracking capability are validated through the analysis of a specific sequence including speed, load and position setpoint variations. Furthermore, the study of their robustness reveals in particular that errors on the flux constant and the winding electrical resistance have a significant impact on the position estimate, highlighting the importance of a proper identification of the machine parameters. In contrast, parameter errors have a very small impact on the axial speed estimate.

Future works will include the development of a differential control strategy based on the machine electromotive forces, estimated through the proposed observer for the supplied winding and measured for the winding left open, to regulate the rotor in centred position while suppressing the influence of the flux constant and gradient.

References

Petersen, N., Slininger, T. & Severson, E. L. (2023), 'State observers and run-out reduction for magnetically levitated motor systems', *IEEE Transactions on Industry Applications* **59**(2), 1812–1823.

Rubio, G. C., Fujii, Y. & Chiba, A. (2021), Bearingless motor with asymmetric axial electrodynamic stabilization, *in* 'Joint Technical Meeting on Magnetics/Motor Drive/Linear Drives'.

Rubio, G. C., Hothur Komal, V. C., Fujii, Y. & Chiba, A. (2023), 'Experimental verification of passive axial electrodynamic suspension in a bearingless motor', *IEEE Open Journal of Industry Applications* pp. 1–12.

Van Verdeghem, J. & Dehez, B. (2021), 'Fully passively levitated self-bearing machine implemented within a reaction wheel', *IEEE Transactions on Industry Applications* **57**(6), 5782–5795.

Van Verdeghem, J., Kluyskens, V. & Dehez, B. (2019), 'Experimental investigations on passively levitated electrodynamic thrust self-bearing motors', *IEEE Transaction on Industry Applications* **55**(5), 4743–4753.

Van Verdeghem, J., Severson, E. L. & Dehez, B. (2021), 'Hybrid active-passive actuation approach of passively levitated thrust self-bearing machines', *IEEE Transactions on Industry Applications* **57**(6), 7035–7045.

Modelling and Dynamic Analysis of Rotor-AMBs System with Shrink Fit Assembly

Yang ZHOU [a], Jin ZHOU [a], Jarir MAHFOUD [b], Yuanping XU [a]

a Nanjing University of Aeronautics and Astronautics, Yudao Street 29, 210016 Nanjing, China, zhouyang0216@nuaa.edu.cn

b University of Lyon, INSA-Lyon, CNRS UMR5259, LaMCoS, F-69621 Villeurbanne Cedex, France

Abstract

Rotor-active magnetic bearings (rotor-AMBs) systems nowadays have been widely used in fluid machinery and the shrink fit assembly is widely applied in the rotor. However, when the shrink fit is tight enough, additional external perturbation could be introduced when rotor is levitated. The mechanism of this additional perturbation should be assessed to ensure stable operation of the machine. The effect of the shrink fit interface contact is equated to an additional stiffness matrix by massless spring units, whose certain contact stiffness is uniformly distributed over the interface. Particularly, we consider that there is partial separation in the contact interface due to the AMBs supporting and a novel additional stiffness matrix related to contact status is proposed by calculating the real-time contact area. The contact stiffness is calculated by microscopic contact model based on fractal theory. Finally, numerical simulation shows that the interface contact influences the system robustness and the unstable mode is excited. The increase of interference and contact length both will increase the amplitude of the rotor response. Moreover, the influence and order of the vibration are quantitatively validated by the experiments.

Keywords: *Active Magnetic Bearing, shrink fit, interface contact, flexible mode vibration*

1. Introduction

Active magnetic bearings (AMBs) have been widely used in centrifugal gas compressors and other high-speed fluid machinery applications (Schweitzer and Maslen, 2009). In these rotating machinery, mechanical elements such as disks and impellers are often shrink fitted on the shafts, and the shrink fit assembly forms the interface contact. However, when the connection is tight enough, the flexible mode vibrations will be excited when rotor is levitated. However, when the shrink fit interference decreases, the rotor can be levitated stably. To sum up, the cause of this vibration is the coupling between the assembly interface contact and the AMB supporting, of which few studies have been studied.

Shrink fit assembly is widely applied in rotor – bearing system. Unlike the traditional complete rotor, the presence of the interface contact may bring changes to the dynamic characteristic of rotor. Tan et al. equated the rotor shrink-fit interface to an internal damping force and analyzed its influence on system stability by decomposing the asymmetric stiffness component from the system equations (Tan, et al., 2014). Francesco Sorge studied the internal friction in the rotor – shaft system due to the shrink fit release of the assembly (Sorge, 2013a, 2014b). The internal friction tends to destabilize the over-critical rotor running. The friction works between two rotating parts. The internal friction between rotating parts produces a destabilizing force in the following way (Yukio and Toshio, 2012). When a deflected elastic shaft whirls with an angular velocity that differs from the rotational speed, the fibers of the shaft elongate or contract in the hub. A fiber on the shaft surface contracts while it moves from the outside to the inside of the orbit circle and elongates while it moves from the inside to the outside. Then, dry friction works from the hub to the shaft surface to prevent this motion. However, this vibration mechanism is not able to interpret the flexible mode vibration caused by shrink fit interface contact in AMBs-suspended machinery due to: 1. The rotor with shrink fit assembly is levitated by AMBs; 2. This vibration is excited without rotation.

Yannick Paul et al. mentioned that the poor connection between rotor and the impeller may cause high-frequency vibration during levitation in the expander-compressors with AMBs (Smithanik and Yannick, 2015). Wei and Fang equated the influence of shrinkage fit in a rigid rotor to a disturbance in the rotational direction (Wei and Fang, 2008). Simon and Flowers equated the shrink fit contact effect to an internal damping force and analyzed its influence on the rotor-AMB system stability when rotating (Andras and Flowers, 2007a, 2009b). It can be seen in these results that the interface contact significantly influences the system stability. We also note that these results are all based on excitation of the unbalance force and the internal damping force, but not the contact effect. However, contact effect plays a significant role in amplifying the excitation, which leads to system instability. In AMB suspended rotating machinery, the energy input from the AMB supporting is seen as the excitation and is amplified by the contact effect.

In traditional modeling of the rotor assembled by several components with shrink fit, the Young's modulus and density of lamination stacks are updated based on the experimental identification of natural frequencies (Xu, et al., 2015). This method focuses on the natural frequencies of the rotor and ensures the modeling accuracy. However, this approach should consider the contact characteristics due to the shrink fit. This paper proposed a new modeling of the rotor-AMBs System with shrink fit assembly. The effect of the interface contact is equivalent to an additional stiffness matrix by simplifying the interface contact to massless spring units, whose certain contact stiffness is uniformly distributed over the contact interface. The dynamic response of the rotor supported by AMBs considering the effect of shrink fit interface contact is studied by numerical simulation in this paper.

2. Modelling of rotor-AMBs system with shrink fit

2.1 Description of the rotor – AMBs test rig

The four-degree of freedom rotor-AMBs experimental test rig shown in Fig. 1 for this study consists of two parts: the mechanical system and the electronic control system. The rotor supported by the radial AMBs A and B has a length of 1.004 m and a weight of 10.35 kg. A disk connected to the rotor by shrink fit on the right end to simulate the influence of interface contact. Disks of different inner diameters and axial lengths are manufactured to simulate the effect of interference and contact length on system's response.

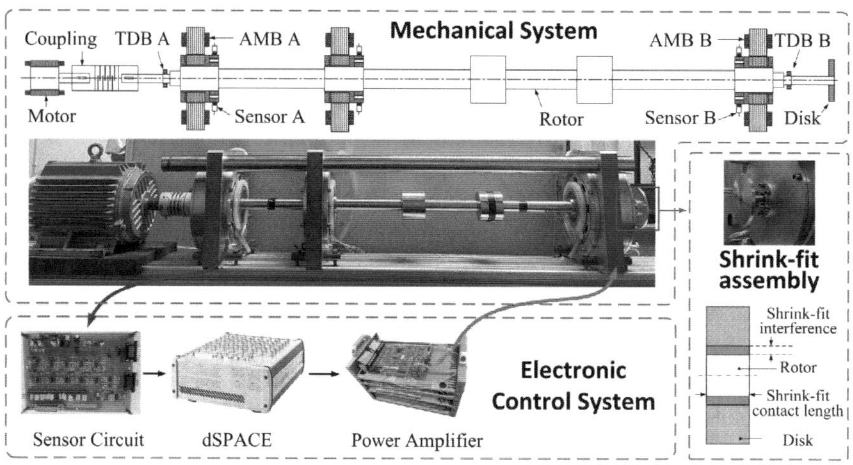

Figure 1 Rotor-AMBs test rig.

2.2 Modelling of the interface contact

This work focuses on the influence of the shrink fit interface on the rotor. The interface contact in the rotor is modeled as a uniformly distributed stiffness over the contact interface as shown in Fig. 2(a), where the disk and the rotor are joined together by shrink fit assembly. The arbitrary node i_s on the outer surface of rotor is connected to its corresponding node j_s by the spring unit. The stiffness of the massless spring unit is referred to as the contact stiffness. The contact stiffness can be subdivided into the normal contact stiffness k_f and the tangential contact stiffness k_q. The directions of k_f and k_q are perpendicular and parallel to the contact interface, respectively. The value of the contact stiffness is related to the interference δ and the contact area A_a (A_a=2πRL). There is relative deformation between the contact interfaces when the rotor vibrates. The energy generated by the spring deformation is calculated to study the influence of the interface contact. According to the energy principle of the spring unit, the energy of the unit spring Δu can be obtained by the linear superposition of the change of the spring potential energy in the *x*, *y* and *z* directions as,

$$\Delta u = \Delta u_x + \Delta u_y + \Delta u_z = 0.5 \cdot k_q \Delta x^2 + 0.5 \cdot k_q \Delta y^2 + 0.5 \cdot k_f \Delta z^2 \tag{1}$$

where Δ*x*, Δ*y* and Δ*z* represent the deformation of each spring unit in the *x*, *y* and *z* directions, respectively.

In the shrink fit assembly, the center node of the rotor is node *i* and the center node of the disk is node *j*. The generalized coordinates of centers *i* and *j* are $(x_i, y_i, \alpha_i, \beta_i)$ and $(x_j, y_j, \alpha_j, \beta_j)$, respectively in absolute coordinates *oxyz*. As shown in Fig. 2(b). By establishing the floating coordinate of node *i* and *j*, the coordinates of node i_s can be expressed as in the absolute coordinate system *oxyz* as

$$\begin{cases} x_{is} = x_i + l\sin\alpha_i\cos\beta_i - R\cos\theta\cos\beta_i \\ y_{is} = y_i + l\sin\beta_i + r\cos\theta\sin\beta_i\sin\alpha_i + r\sin\theta\cos\alpha_i \\ z_{is} = z_i + l\cos\alpha_i\cos\beta_i - r\cos\theta\sin\beta_i\cos\alpha_i + r\sin\theta\sin\alpha_i \end{cases} \tag{2}$$

The coordinates of node j_s can be obtained using this method.

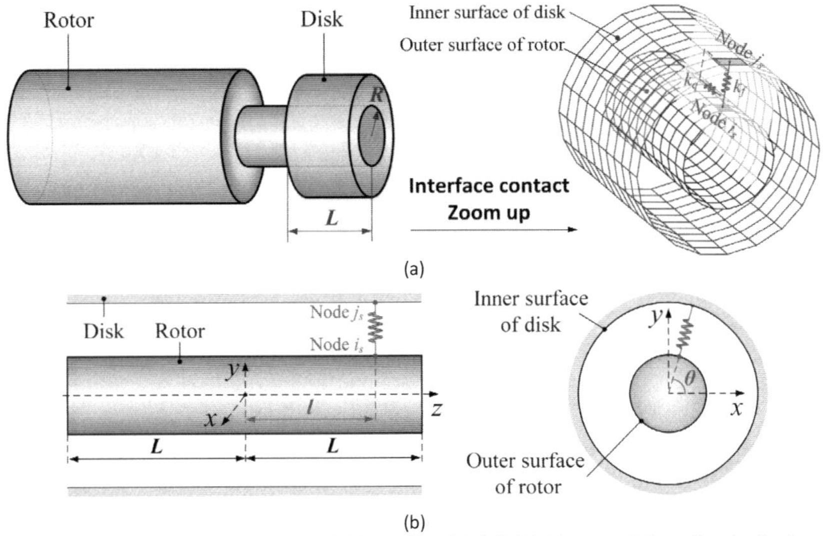

Figure 2 Schematic diagram of the interface contact model formed by shrink fit (a) Diagram of the uniformly distributed spring unit (b) The coordinate of the spring unit.

The deformation of the spring unit *Δx*, *Δy* and *Δz* in the absolute coordinate system is calculated by the coordinates of i_s and j_s as,

$$\begin{cases} \Delta x = x_{js} - x_{is} = x_j - x_i + l\left(\alpha_j - \alpha_i\right) \\ \Delta y = y_{js} - y_{is} = y_j - y_i + l\left(\beta_j - \beta_i\right) \\ \Delta z = z_{js} - z_{is} = R\sin\theta\left(\alpha_j - \alpha_i\right) + R\cos\theta\left(\beta_i - \beta_j\right) \end{cases} \tag{3}$$

The total energy generated by the spring unit is obtained by summing up the unit spring energy *Δu* as follows

$$\begin{aligned} u_s &= \iint_A 0.5\left(k_f \Delta x^2 + k_f \Delta y^2 + k_q \Delta z^2\right)\mathrm{d}A \\ &= \int_{-L}^{L}\int_0^{2\pi} 0.5\left(k_f \Delta x^2 + k_f \Delta y^2 + k_q \Delta z^2\right)\mathrm{d}\theta\mathrm{d}l \end{aligned} \tag{4}$$

where A denotes the contact region and the area of A is given by $A_a = 2\pi RL$ when the two interfaces are fully contacted (Wu, et al., 2021). However, the rotor and disk interfaces are not fully contacted when the rotor is levitated. It is assumed that the contact interface is partly separated as shown in Fig. 3(a). The orange and the blue line are the Intersecting lines formed by the intersection of the outer surface of the rotor and the inner surface of the disk. In the Fig. 3(b), we expand the rotor outer surface along the circumferential direction, it can be seen that the real contact region is enclosed by the blue line $z_1(\theta)$ and the red dotted line $z_2(\theta)$, which is colored.

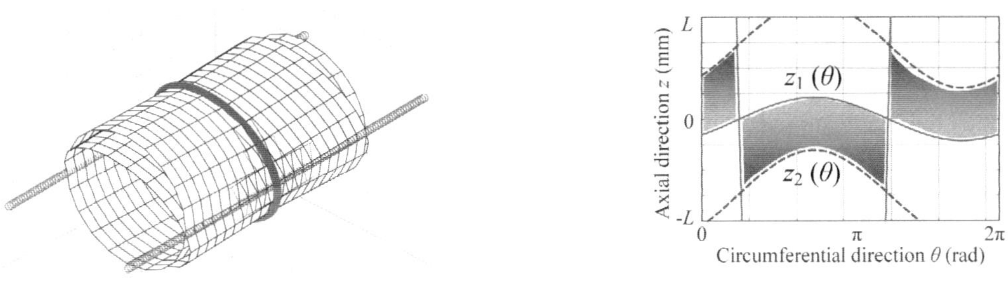

(a) (b)

Figure 3 Schematic diagram of the actual contact status between the rotor and the disk (a) Diagram of the partly separation in 3D view (b) The expansion of the rotor outer surface along the circumferential direction.

For $z_1(\theta)$ and $z_2(\theta)$, the blue and the red line are obtained by deriving the intersecting line equation of the two cylindrical surfaces. The function of $z_1(\theta)$ and $z_2(\theta)$ change with the generalized coordinates of nodes *i* and *j*. Therefore, the total energy is time-variant and can be expressed as

$$u_s = \int_{z_1(\theta)}^{z_2(\theta)}\int_0^{2\pi} 0.5\left(k_f \Delta x^2 + k_f \Delta y^2 + k_q \Delta z^2\right)\mathrm{d}\theta\mathrm{d}l \tag{5}$$

Submitting Eq. (4) into Eq. (6), we can express the total elastic potential energy u_s as

$$u_s=0.5\cdot\begin{bmatrix}x_i & y_i & \alpha_i & \beta_i & x_j & y_j & \alpha_j & \beta_j\end{bmatrix}K_e\begin{bmatrix}x_i & y_i & \alpha_i & \beta_i & x_j & y_j & \alpha_j & \beta_j\end{bmatrix}^{\mathrm{T}} \tag{6}$$

$$K_e=\begin{bmatrix}\mathrm{diag}\left(k_q\iint_A \mathrm{d}A, k_q\iint_A \mathrm{d}A, k_f\iint_A y^2\mathrm{d}A, k_f\iint_A x^2\mathrm{d}A\right) & -\mathrm{diag}\left(k_q\iint_A \mathrm{d}A, k_q\iint_A \mathrm{d}A, k_f\iint_A y^2\mathrm{d}A, k_f\iint_A x^2\mathrm{d}A\right)\\ -\mathrm{diag}\left(k_q\iint_A \mathrm{d}A, k_q\iint_A \mathrm{d}A, k_f\iint_A y^2\mathrm{d}A, k_f\iint_A x^2\mathrm{d}A\right) & \mathrm{diag}\left(k_q\iint_A \mathrm{d}A, k_q\iint_A \mathrm{d}A, k_f\iint_A y^2\mathrm{d}A, k_f\iint_A x^2\mathrm{d}A\right)\end{bmatrix} \tag{7}$$

The time-variant additional stiffness matrix K_e arising from the interface contact will be added to the dynamic equations of the system as the influence of the interface contact. The main contact effect variables are the normal contact stiffness k_f, tangential contact stiffness k_q and the contact length L.

The contact stiffness model of shrink-fit interface is established based on the elastic-plastic contact model and fractal theory (Liu, et al., 2019). The microscopic contact model of the annulus can be simplified to the microscopic contact model of the planes. Considering that the nominal radius R of shrink fit is fixed, the normal contact stiffness and tangential contact stiffness under different interference and contact length L is shown below.

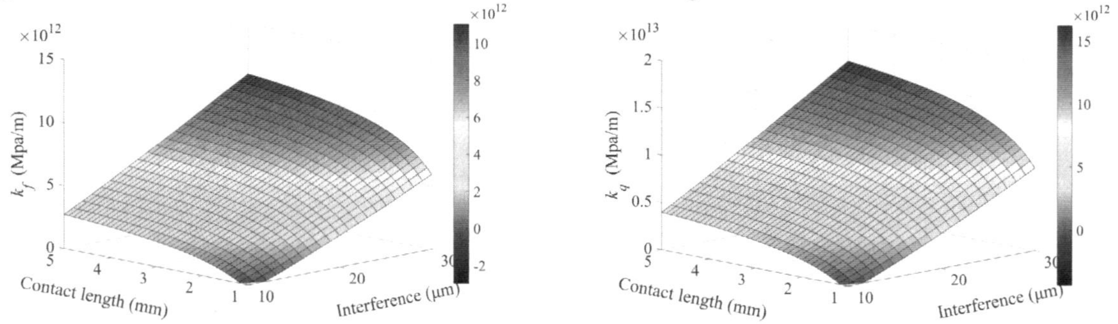

Figure 4 Calculation of the normal stiffness and tangential stiffness of the shrink fit (a) Normal stiffness under different interference δ and different contact length L (b) Tangential stiffness under different interference δ and different contact length L.

2.3 Modelling of the rotor-AMBs system

The theory of Bernoulli-Euler beam is applied to the finite element of the rotor. The rotor is modelled with 66 nodes and 64 elements for a total mesh of 264 DOF for the lateral analysis (see Fig. 5). The equation of motion takes the following form.

$$M_R\ddot{q}+C_R\dot{q}+K_Rq=0 \tag{8}$$

where M_R, C_R and K_R represents the mass matrix, the damping matrix and the stiffness matrix of the system, respectively. F_{AMB} represents the electromagnetic force by the AMBs. Vector q represents the system displacement vector $[x_1,y_1,\alpha_1,\beta_1 \ldots x_{66},y_{66},\alpha_{66},\beta_{66}]^T$, where the generalized coordinates of centers i and j are also included.

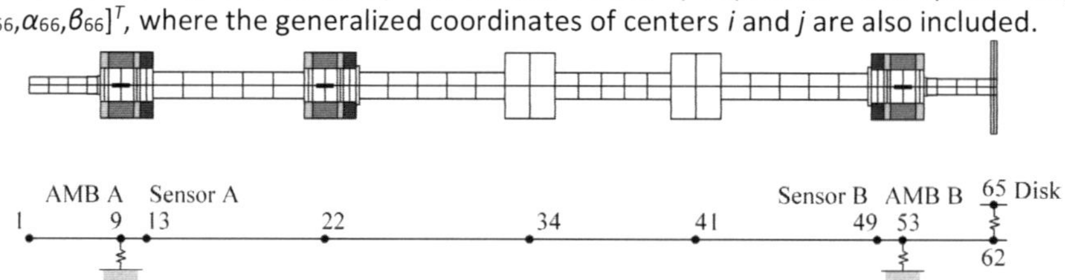

Figure 5 The finite element model for the rotor-AMBs system with shrink fit.

The active magnetic bearing generates the attractive force in differential driving mode, and the electromagnetic force f_{amb} is linearized and expressed as

$$f_{amb}=k_x x+k_i i \tag{9}$$

where k_x and k_i represent the displacement stiffness and the current stiffness, respectively.

Substituting Eq. (8) and Eq. (10) into Eq. (9), the general equation of motion could be written as

$$M_R\ddot{q}+C_R\dot{q}+\left(K_R-T_e^{\mathrm{T}}K_eT_e\right)q=k_xT_a^{\mathrm{T}}T_aq+k_iT_a^{\mathrm{T}}i_a \tag{10}$$

where T_e is the transfer matrix of the interface nodes i and j, T_a is the transfer matrix of the AMB nodes, i_a is the control current of AMBs A and B in the x and y directions.

The state space model of the rotor-AMBs system considering the interface contact can be expressed as

$$\begin{cases}\dot{x}_s=\begin{bmatrix}\dot{q}\\ \ddot{q}\end{bmatrix}=\begin{bmatrix}0 & I\\ -M_R^{-1}\left(K_R-k_xT_a^{\mathrm{T}}T_a-T_e^{\mathrm{T}}K_eT_e\right) & -M_R^{-1}C_R\end{bmatrix}\begin{bmatrix}q\\ \dot{q}\end{bmatrix}+\begin{bmatrix}0\\ k_iM_R^{-1}T_a^{\mathrm{T}}\end{bmatrix}i_a=A_sx_s+B_si_a\\ q_s=\begin{bmatrix}q_{sAx} & q_{sAy} & q_{sBx} & q_{sBy}\end{bmatrix}^{\mathrm{T}}=\begin{bmatrix}T_s & 0\end{bmatrix}\begin{bmatrix}q & \dot{q}\end{bmatrix}^{\mathrm{T}}=C_sx_s\end{cases} \tag{11}$$

where x_s is the state vector, q_s is the displacement of the sensor nodes in the *x* and *y* directions, and T_s is the transfer matrix of the sensor nodes.

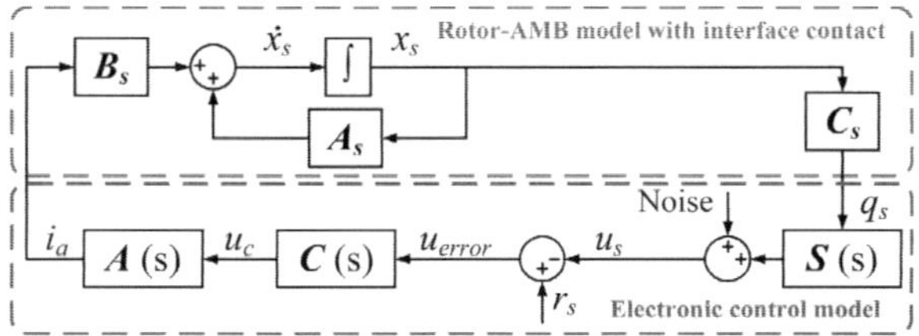

Figure 6 Diagram of the closed-loop mathematical model considering interface contact.

As shown in Fig. 6, beside modelling the rotor and the electromagnetic forces, modelling of the electronic control system, which consists of the eddy-current sensors, the amplifiers and the controllers, is also needed.

The relationship between the control currents and the radial rotor displacement q_s detected by the sensors is given as Eq. (13)

$$i_a = \frac{i_a}{u_c}\frac{u_c}{u_{error}}(r_s - u_s) = A(s)C(s)\left[r_s - S(s)q_s\right] = \frac{k_a(2\pi f_z)}{s+(2\pi f_z)} \cdot \left(K_P + \frac{K_I}{s} + \frac{K_D s}{T_D s+1}\right)(r_s - k_s q_s) \tag{12}$$

where *S*(*s*), *A*(*s*) and *C*(*s*) are the transfer functions of the sensor, the amplifier and the controller, respectively, k_s is the gain of sensor, k_a is the gain of power amplifier, f_z is the cut-off frequency, K_P, K_I and K_D respectively represent the proportional gain, the integral gain, the derivative gain, T_D is the derivative time constant to prevent magnifying the error signal by the controller in the high frequency ranges, and r_s is the reference voltage.

As shown in Fig. 6, the rotor-AMBs mechatronic system considering interface contact is established by combining the rotor-AMBs model considering interface contact and the electronic control model.

3. Numerical simulation

The dynamic response of the rotor supported by AMBs considering the effect of shrink fit interface contact is studied by using numerical simulation in this paper.

In order to study the influence of the interface contact on the system, appropriate control parameters should be tuned to ensure stable levitation of the rotor without interface contact. In order to simplify the model and increase the calculation speed, only the first five bending modes are retained through model reduction. The parameters in the PID controller in simulation are given as: K_P =1.7, K_I =1, K_D =0.0006, T_D =0.0001. In this work, these parameters are kept the same to exclude the influence of the controller.

When *L* is 3 mm, the influence of the shrink fit interference on the rotor displacement is shown in Fig. 7. Varying the value of the interference from 5 μm to 15 μm, the time domain response plotted in Fig. 7(a) shows that, as *L* increases, the amplitude of the response increases, and as *L* increases, the vibration appears faster, which means that the system is more unstable. By comparing the rotor responses at the positions of AMB A and AMB B, the response near the disk (the B_x direction) is greater than the response of the other side (the A_x direction). In the frequency domain, it can be seen in Fig. 7(b) that the rotor vibrates around the 4th bending mode frequency.

When *δ* is 10 μm, the influence of the contact length on the rotor displacement is shown in Fig. 8. Varying the contact length from 2 mm to 4 mm, the time domain response plotted in Fig. 8(a) shows that, as *L* increases, the amplitude of the response also increases. Similarly, response near the disk is greater than the response of the other side. In the frequency domain, it can be seen in Fig. 8(b) that the rotor vibrates around the 4th bending mode frequency.

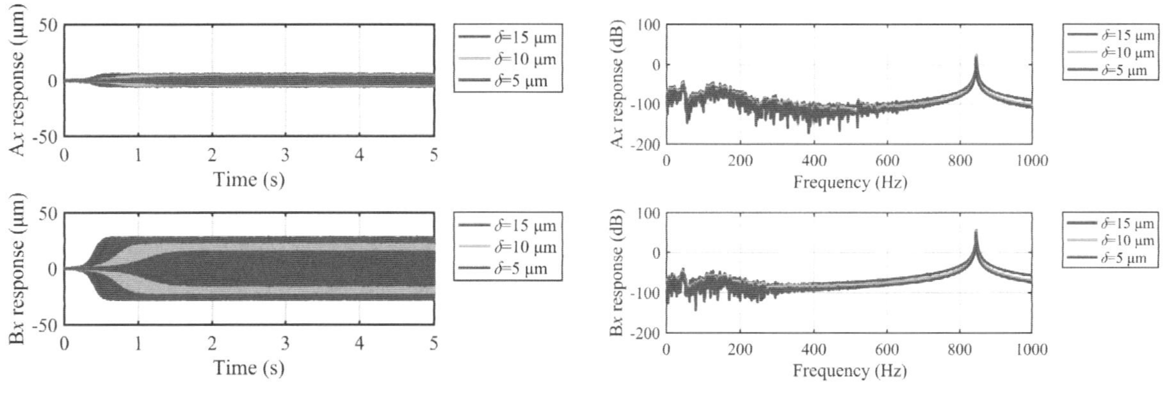

Figure 7 Comparison of dynamic responses in simulation (different interference) (a) Rotor time-domain responses (b) Rotor frequency-domain responses.

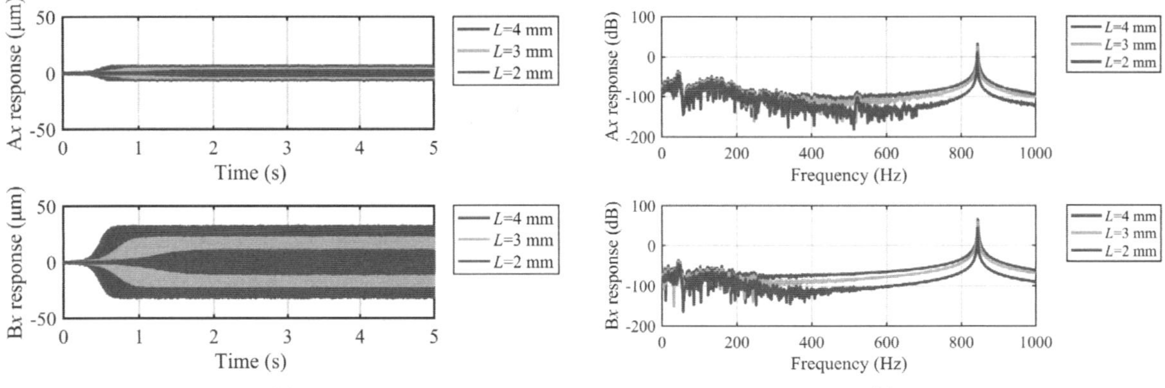

Figure 8 Comparison of dynamic responses in simulation (different contact length) (a) Rotor time-domain responses (b) Rotor frequency-domain responses.

4. Conclusions

This paper studies the perturbation caused by shrink fit interface contact in rotor-AMBs systems when the rotor is levitated without rotating by establishing the mechatronic rotor-AMBs model considering the interface contact. An additional stiffness matrix is introduced in order to characterize the interface contact between the rotor and the disk, and the value of the matrix is related to the contact stiffness and the contact length. Then the contact model for shrink-fit interfaces considering the multi-scale features of the rough contact surface is applied to calculate the contact stiffness under different interference and contact length. Finally, the influence of the interference and contact length on the dynamic characteristics of the system are numerically analyzed. Main conclusions of this study are summarized as follows:

(1) There is fluctuation in the rotor response when the rotor is levitated by AMBs because there is noise in the mechatronic system. When the disk is connected to the rotor by shrink fit, there is also multivariant slight vibration in the interface of shrink fit. The vibration energy caused by interface contact is acted upon the system in way of positive feedback and the resonance will be excited. The frequency of the 4th bending mode is higher than the cut-off frequency of the power amplifier and the controller is unable to control this mode due to the phase lag. As a result, the rotor vibrates at the frequency of the 4th bending mode.

(2) The normal and tangential stiffness increases with the increasing of interference and the increase of contact length.

(3) The steady vibration amplitude near the interface contact (AMB B) is greater than the amplitude on the other end (AMB A). The increase of the interference and the contact length will both increase the vibration amplitude.

References

Andras S and Flowers G (2007a) Suppression of internal damping-induced instability using adaptive techniques. ASME International Design Engineering Technical Conferences/Computers and Information in Engineering Conference, Las Vegas, USA, 4–7 September 2007, pp.1111-1118.

Andras S and Flowers G (2009b) Adaptive disturbance rejection and stabilization for rotor systems with internal damping. ASME International Design Engineering Technical Conferences/Computers and Information in Engineering Conference, San Diego, USA, 30 August–2 September 2009, pp.1063-1072.

Liu J, Ma C, Wang S, et al. (2019) Contact stiffness of spindle-tool holder based on fractal theory and multi-scale contact mechanics model. Mechanical Systems and Signal Processing 119: 363-379.

Schweitzer G and Maslen EH (2009) Magnetic Bearings: Theory, Design, and Application to Rotating Machinery. Heidelberg: Springer Berlin.

Smithanik J and Yannick P (2015) Applying API 617, 8th Edition to Expander-Compressors with Active Magnetic Bearings. In: 44th Turbomachinery & 31st Pump Symposium, Houston, USA, 14–17 September 2015, pp.349–364.

Sorge F (2013a) Approach to rotor-shaft hysteretic whirl using Krylov–Bogoliubov techniques. Journal of Vibration and Control 21(5): 883-895.

Sorge F (2014b) Nonlinear analysis of cylindrical and conical hysteretic whirl motions in rotor-dynamics. Journal of Sound and Vibration 333(20): 5042-5056.

Tan D, Chen J, Liao M, et al. (2014) Instability caused by cylindrical surface fit in rotor system. Mechanical Science and Technology for Aerospace Engineering 33(12): 1786-1790.

Wei T and Fang J (2008) Design of magnetically suspended elastic rotor notch filter based on two-frequency Bode diagram. Optics and Precision Engineering 16(5): 789-796.

Wu X, Jiao Y, Chen Z, et al. (2021) Establishment of a contact stiffness matrix and its effect on the dynamic behavior of rod-fastening rotor bearing system. Archive of Applied Mechanics 91(7): 3247-3271.

Xu Y, Zhou J, Di L, et al. (2015) Active magnetic bearing rotor model updating using resonance and mac error. Shock and Vibration 2015: 1-9.

Yukio I and Toshio Y (2012) Linear and Nonlinear Rotordynamics: A Modern Treatment with Applications. Hoboken: Wiley - VCH Verlag GmbH & Co. KGaA.

Hybrid passive levitation mechanism utilizing thrust force and magnetic force for a pump application

Wataru HIJIKATA [a], Zhaomin ZHENG [a], Ryota MAGARI [a]

a School of Engineering, Tokyo Institute of Technology, I6-8, 2-12-1, Ookayama, Meguro-ku, 1528550 Tokyo, Japan, hijikata.w.aa@m.titech.ac.jp

Abstract

The durability of a blood pump has been enhanced owing to the development of contactless bearings. According to Earnshaw's law, levitation in all six degrees of freedom cannot be realized using a static magnetic field. Therefore, the combination of passive magnetic bearings and active magnetic bearings or passive magnetic bearings and hydrodynamic bearings are commonly adopted in the blood pump. However, the active magnetic bearings exhibit a risk of malfunctioning of their control system, and the hydrodynamic bearings exhibit a risk of blood trauma owing to their narrow gaps. Therefore, the objective of this study is to develop a passive levitation system for an impeller with a large gap without active control. The large gap prevents blood damage, and passive levitation contributes to reduce the risk of device malfunction. In this study, hybrid passive levitation mechanism utilizing thrust force and magnetic force is proposed. The thrust force generated by the rotation of the impeller changes depending on the interaction between the housing geometry and impeller position, even with a large gap. Considering this characteristic, the impeller was levitated by balancing the magnetic force generated by the permanent magnets and the thrust force in the axial direction. The movements in the radial and angular directions are passively supported by the restoring force and torque generated by the permanent magnets. The geometry of the pump and dimensions of the permanent magnets were designed so that the mechanism had positive stiffness in all directions. The levitation of the impeller in the axial direction was confirmed in the experiment by the pump prototype.

Keywords: *Passive bearing, Passive levitation, Thrust force, Magnetic bearing, Blood pump*

1. Introduction

The durability of a blood pump has been enhanced owing to the development of magnetic bearings. In this system, an impeller is levitated by using magnetic force to enhance the durability and hemocompatibility. According to Earnshaw's law(Earnshaw, 1839), levitation in all six degrees of freedom cannot be realized using a static magnetic field. Therefore, all the magnetically levitated (MagLev) blood pumps utilize active magnetic bearings or hybrid bearings with passive magnetic bearings and hydrodynamic bearings(Hoshi, Shinshi and Takatani, 2006). Active magnetic bearings in the field of blood pump have been proposed as a new value in systems that provide additional functions such as flow sensors(Shida, Masuzawa and Osa, 2022)(Hijikata *et al.*, 2015), thrombus detection(Hijikata *et al.*, 2020) and thrombus prevention(Murashige and Hijikata, 2019)(Hatakenaka *et al.*, 2023).

However, the active magnetic bearings exhibit a risk of malfunctioning of their control system, and the hydrodynamic bearings exhibit a risk of blood trauma owing to their narrow gaps. Therefore, in this study, we developed a novel hybrid levitation mechanism with a large gap without any active control system by combining thrust force and passive magnetic bearings.

2. Principle of thrust force hybrid levitation system

The proposed hybrid levitation mechanism installing in an axial blood pump is shown in Fig. 1(Magari and Hijikata, 2022). The mechanism consists of a housing, impeller, flow straightener, diffuser, direct-drive motor, and passive magnetic bearing. The housing used in the proposed mechanism has a unique shape with a larger diameter in the center of the pump. Key feature for levitation of this mechanism is thrust force acting on the impeller as a reaction force against blood flow generation.

The proposed levitation system, which supports five degrees of freedom, except for rotation around the axial direction, is realized by satisfying conditions (i)–(iii) shown below.

(i) Axial direction: As shown in Fig. 2, an equilibrium point exists, and the resultant force of thrust and magnetic force exhibits a positive stiffness around it.

(ii) Radial direction: An equilibrium point exists, and the magnetic force exhibits a positive stiffness around it.

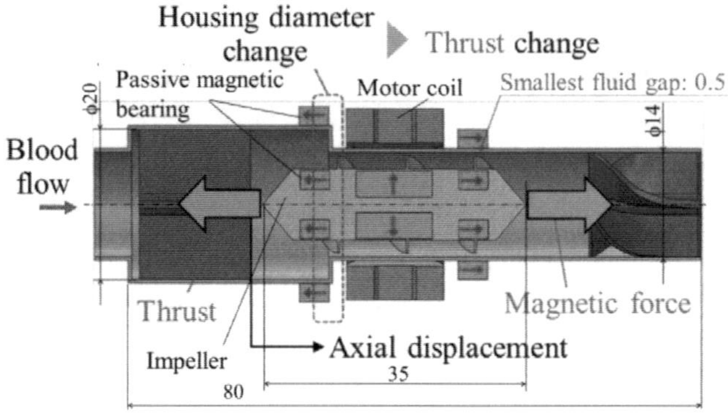

Figure 1 Configuration and dimension of the hybrid levitation mechanism installed in the blood pump.

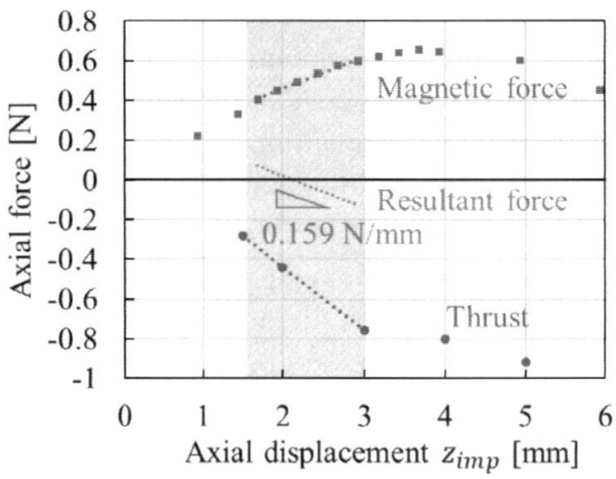

Figure 2 Concept and simulated axial force acting on the impeller.

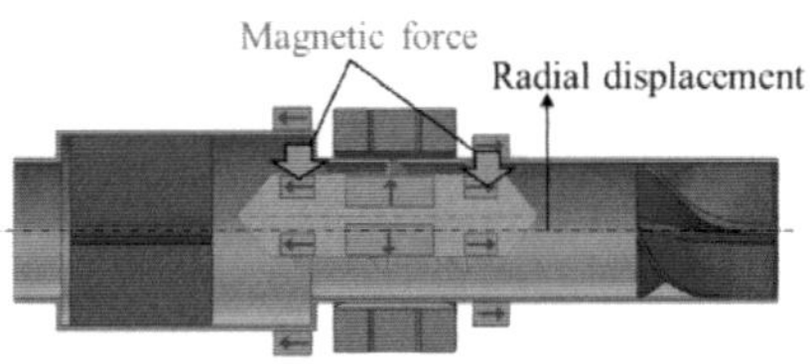

Figure 3 Levitation principle in the radial direction.

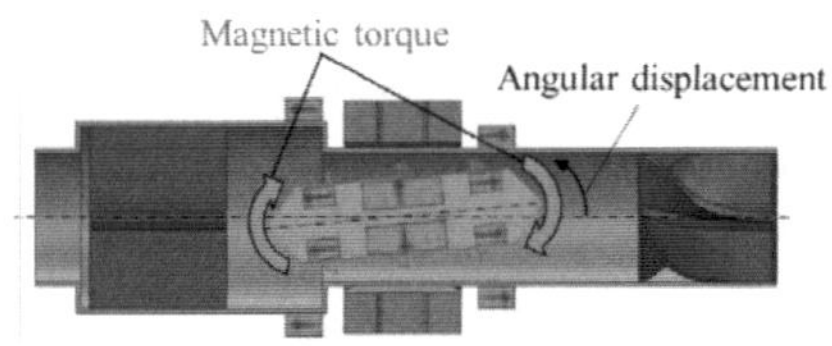

Figure 4 Levitation principle in the angular direction.

(iii) Angular direction: An equilibrium point exists, and the torque generated by the magnetic force exhibits a positive stiffness around it.

Conditions (ii) and (iii) can be satisfied by a normal passive magnetic bearing as shown in Fig. 3 and Fig. 4, respectively. However, under these conditions, axial stiffness becomes negative. In the proposed levitation system, thrust force is utilized to overcome this negative stiffness in the axial direction to levitate the impeller. We found that the thrust force increases as the cross-sectional area of the flow path at the end of the impeller (indicated with red-dashed rectangle in Fig. 1) increases. Owing to this feature, the proposed pump can have a positive stiffness in the thrust force against the displacement of the impeller and hence, the total stiffness in the axial direction can also be positive, that meets the condition (i).

3. Design and Prototyping

By using computational fluid dynamics (CFD) analysis and magnetic field analysis, dimensions of the mechanism and each stiffness were calculated. The designed dimensions were shown in Fig. 1 and stiffness was shown in Fig. 2, respectively. In this calculation, flow rate of the pump and rotational speed of the impeller were considered as 1.5

L/min and 10000 rpm. The equilibrium point was at z_{imp} = 2.1 mm, and the resultant thrust and magnetic force exhibited a positive stiffness of 0.159 N/mm around the equilibrium point. The calculated stiffness in the radial and angular directions by the passive magnetic bearing was 9.5 mN/mm and 0.08 mNm/deg., respectively. These results satisfied the conditions of levitation (i) - (iii).

To validate the CFD and magnetic field analyses, we experimentally measured the axial force as well as observe the axial levitation. Fig. 5 shows the configuration of the pump used for validation and Fig. 6 shows photographs of the experimental prototype. Note that radial motion of the impeller was supported by the shaft and bearings because this experiment aimed to validate the levitation in the axial direction. The thrust force generated on the impeller can be transmitted to load cells via a shaft that supports the impeller. The shaft was supported by linear bearings, and the position of the impeller was adjusted by turning the screws on both ends of the experimental apparatus. As shown in Fig. 7, the prototype pump was connected to the mock circulatory loop consisting of a reservoir, pressure gauges and flow meter. As a working fluid, porcine blood was filled in the loop. The viscosity of the blood was measured with a viscometer (SV-10, A&D Co., Ltd, Tokyo), and it was 5.12 mPa·s (21.9°C).

After measuring the thrust force, axial levitation was demonstrated. In this demonstration, load cells were removed so that the impeller could be displaced in the axial direction and working fluid was replaced to water to measure the axial displacement of the impeller by using a laser displacement sensor. The smallest gap between the impeller and housing was 500 μm.

4. Results

Fig. 8 shows the experimental results of the thrust force measurement. Although the target flow rate was set to 1.5 L/min, the flow rate at 5000 rpm did not reach the target value. Hence, the thrust at maximum flow was recorded, that is, the flow rates were 0.72 L/min, 0.89 L/min, 0.98 L/min, 1.1 L/min, 1.2 L/min, 1.2 L/min, 1.3 L/min, and 1.2 L/min at the axial displacement of the impeller z_{imp} = 1.5, 2.0, 2.5, 3.0, 3.5, 4.0, 4.5, and 5.0 mm, respectively. At other speeds, the flow rate was set as 1.5 L/min. Simulated results of the CFD analysis with the viscosity of 3.6 mPa·s was also indicated in the plots. As expected, the magnitude of the thrust increased with impeller displacement, indicating positive stiffness. The trend was also in good agreement with the CFD analysis.

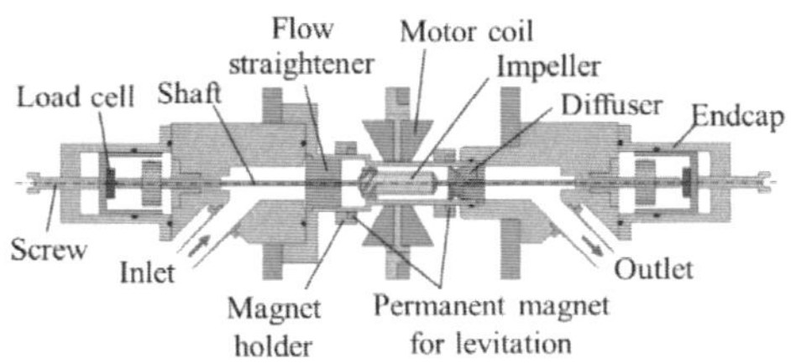

Figure 5 Configuration of the prototype mechanism.

Figure 6 Photograph of the prototype.

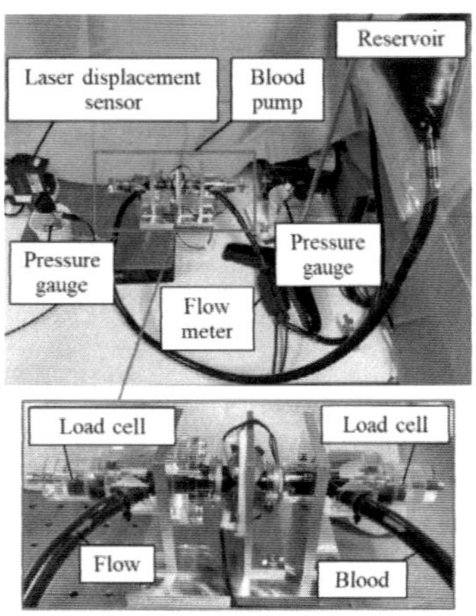

Figure 7 Configuration of the prototype mechanism.

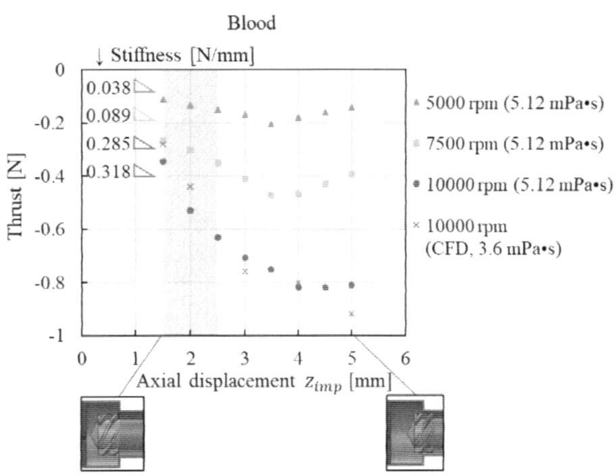

Figure 8 Measured and simulated thrust force.

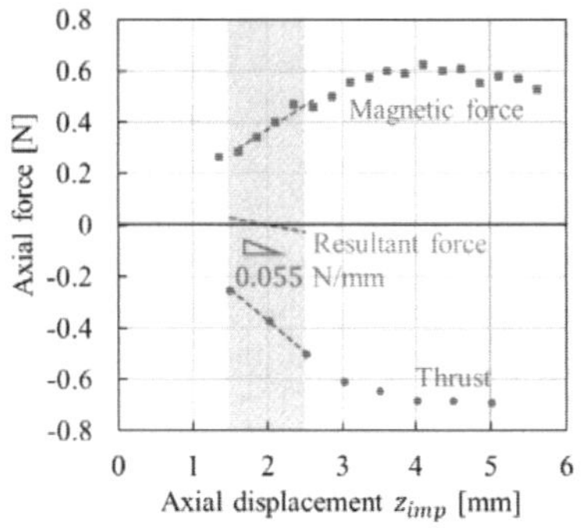

Figure 9 Resultant force measured in the experiments.

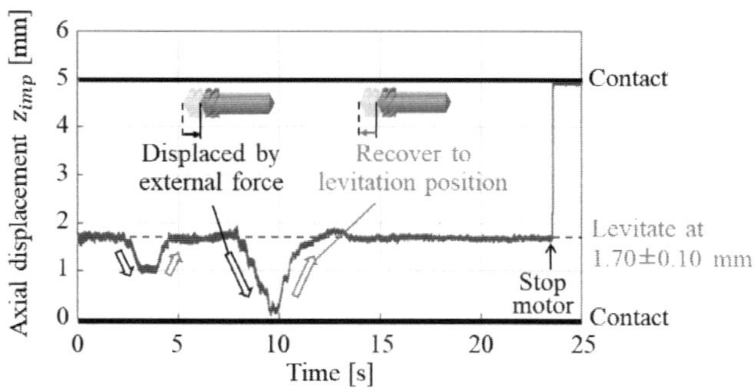

Figure 10 Result of the levitation verification in the axial direction.

Fig. 9 shows the resultant force experimentally obtained from the measured thrust (water, 10000 rpm) and magnetic force. The resultant force had a positive stiffness of 0.055 N/mm, and the equilibrium point existed at z_{imp} = 2.0 mm. Therefore, it is considered to satisfy the axial levitation condition and can levitate in the axial direction. Then, we removed load cells and validate the levitation. As the measured axial displacement shows in Fig. 10, the impeller could levitate at the center position of z_{imp} = 1.70 mm with the rotational speed of 10000 rpm. At approximately 3 and 7 s, the impeller was pushed using a screw installed at the end of the pump to verify the stability of the levitation. The impeller returned to the levitation position after the displacement. These results show that the impeller levitated in the axial direction with the smallest gap of 500 µm without any active control.

5. Conclusion

In this study, hybrid passive levitation mechanism utilizing thrust force and magnetic force was developed to realize blood pumps with large gaps without an active control system. The axial magnetic force and thrust acting on the impeller were experimentally measured, and the results showed that the stiffness of the resultant axial force was positive value of 0.055 N/mm. Furthermore, the demonstration showed that the impeller could levitate in the axial direction with the smallest gap of 500 µm without any active control. The results of this study indicate there is possibility to realize fully passive levitation principle by combining the thrust force and magnetic force. The verification of levitation, including the radial and angular directions, will be examined in a future study.

References

Earnshaw, S. (1839) 'On the Nature of the Molecular Forces which Regulate the Constitution of the Luminferous Ether', *The Nature of the Molecular Forces*, 7(1), pp. 97–112.

Hatakenaka, K. *et al.* (2023) 'Prevention of thrombus formation in blood pump by mechanical circular orbital excitation of impeller in magnetically levitated centrifugal pump', *Artificial Organs*, 47(2), pp. 425–431. doi: 10.1111/aor.14443.

Hijikata, W. *et al.* (2015) 'Estimating flow rate using the motor torque in a rotary blood pump', *Sensors and Materials*, 27(4), pp. 297–308. doi: 10.18494/SAM.2015.1068.

Hijikata, W. *et al.* (2020) 'Detection of thrombosis in a magnetically levitated blood pump by vibrational excitation of the impeller', *Artificial Organs*, 44(6), pp. 594–603. doi: 10.1111/aor.13632.

Hoshi, H., Shinshi, T. and Takatani, S. (2006) 'Third-generation blood pumps with mechanical noncontact magnetic bearings.', *Artificial organs*, 30(5), pp. 324–38. doi: 10.1111/j.1525-1594.2006.00222.x.

Magari, R. and Hijikata, W. (2022) 'Proposition of a passive levitation system utilizing thrust and magnetic force for a ventricular assist device', *Journal of Advanced Mechanical Design, Systems and Manufacturing*, 16(3), pp. 1–12. doi: 10.1299/jamdsm.2022jamdsm0025.

Murashige, T. and Hijikata, W. (2019) 'Mechanical antithrombogenic properties by vibrational excitation of the impeller in a magnetically levitated centrifugal blood pump', *Artificial Organs*, 43(9), pp. 849–859. doi: 10.1111/aor.13541.

Shida, S., Masuzawa, T. and Osa, M. (2022) 'Radial Passive Stability Analysis of Magnetically Levitated Impeller Position for the Development of Flow Rate Estimations of a Ventricular Assist Device', *Advanced Biomedical Engineering*, 11, pp. 194–202. doi: 10.14326/abe.11.194.

Investigation of magnetic field by the Hall sensors embedded into magnetic bearing poles

Adam Krzysztof PIŁAT, Rafal BIESZCZAD, Hubert MILANOWSKI

AGH University of Science and Technology
Faculty of Electrical Engineering, Automatics, Computer Science and Biomedical Engineering
Department of Automatic Control and Robotics
Al. A. Mickiewicza 30, 30-059 Krakow, POLAND
ap@agh.edu.pl, rafbiesz@agh.edu.pl, milan@agh.edu.pl

Abstract

A typical eight poles heteropolar active radial magnetic bearing configuration, consisting of Hall sensors embedded in the poles, is a research target. The active magnetic bearing operates without a safety bearing, and therefore identification was carried out in the maximum range of rotor displacement. The magnetic field was investigated experimentally using digital twin simulations developed in COMSOL Multiphysics software. The conducted research showed nonlinear dependencies of the magnetic flux density measured in the pole pieces as a function of the current, position, and orientation of the rotor. Typically used formulas that describe the relationships between force, magnetic flux density, current, and distance have been shown so that they are not to be used for arbitrary bearing configurations, rotor positions, and orientations due to simplifying assumptions.

Keywords: active magnetic bearing, Hall sensor, nonlinearity, identification, digital twin

1. Introduction

The heteropolar configuration of an active radial magnetic bearing (AMB), comprising eight pole pieces, has emerged as a widely adopted solution for numerous devices. Despite its common utilisation, this configuration remains an active area of research that encompasses both dynamic modelling and control aspects. A different class of control methodologies, known as the flux-based control algorithm, emerges as a viable approach to achieve enhanced positioning precision.Direct flux control requires real-time acquisition of air gap flux density information, which can be accomplished through observer-based techniques or direct measurements. To determine the air-gap flux density in AMBs, Hall sensors must be positioned within the air-gap. However, due to nominal air gap distances of 500 μm or less, it is necessary to use ultrathin sensors or embed them in slots manufactured in the pole surface. Recent research has produced promising results in the development of thin Bismuth Hall sensors Ernst et al. (2016, 2020). Another research avenue pursued involves the fabrication of Hall sensors using a highly thin flexible Kapton foil Mystkowski et al. (2019). An economically viable approach involves the use of standard Hall sensors, which are mounted within slots manufactured on the surface of the pole Voigt & Santos (2012). By embedding the Hall sensors, their fragility is protected, albeit at the expense of a slight decrease in the effective pole area. Consequently, the maximum force achievable from AMBs is reduced by 2%, which might be negligible Kjølhede & Santos (2006), Voigt et al. (2017).

In Chowdhury & Sarjaš (2016) an unscented Kalman filter was used to estimate the position of the levitating body with a permanent magnet attached. UKF is designed based on nonlinear model derived with the Finite Element Method. The filter input is a measurement from the Hall sensor installed directly on the electromagnet surface. The estimated position was utilised as a feedback for nonlinear controller obtained on the basis of back-stepping technique and for a classic linear PID. In Miguel et al. (2018) the position of a levitating ferromagnetic ball is obtained using the measurement of the Hall sensor and the current sensor based on the dependency between the 3 signals identified numerically.

The concept of control based on the Hall sensor embedded in the stator pole was selected by Pilat (2010*a*) in the application of 6 pole radial AMB. Our goal is to realise identification tests to determine the actual characteristics of the magnetic field distribution. These studies are necessary to obtain an analytical mathematical model for the synthesis of direct control of the magnetic field, as well research on state estimators. Particular interest is directed towards the observed nonlinearities shown in magnetic field measurements in a rotor-bearing system.

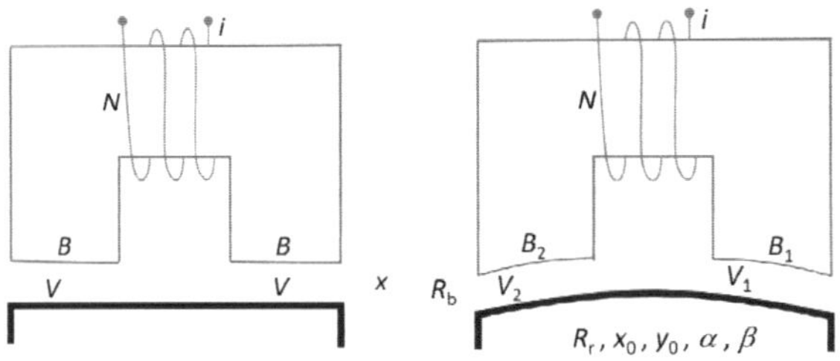

Figure 1: Ideal and real horshoe in active magnetic bearing.

Typically, an ideal configuration of the C-shaped actuator consisting of N coil turns and driven by the current i is considered with a gap volume $V = 2Ax$, where A is a pole area and x is a rotor displacement. In this case, magnetic flux density is calculated as follows: $B = 0.5\mu_0 Nix^{-1}$. In general, the gap volumes V_1 and V_2 vary due to the rotor/bearing radiuses (R_r, R_b), and rotor displacement/orientation (x_0, y_0, α, β). The volume differences are well visible in the case of an elliptic rotor study (see Pilat (2020)). With this research we would like to make a step towards the analytical function describing the magnetic flux density B_1 and B_2 in magnetic bearing poles.

2. Radial Active Magnetic Bearing under study

Without a thin-walled magnetic field sensor, the design of the pole piece was prepared to locate a commercially available magnetic field sensor. A dedicated PCB was developed, 8 sensors were mounted, and then embedded into bearing poles. The magnetic bearing characterised by an internal diameter of 51.8mm was not equipped with a safety bearing. Thanks to which, it was possible to identify the relationship of magnetic flux density $B(\cdot)$ in the full range of 48.5mm diameter rotor displacements and a limited range of currents (0÷2A).

3. Investigation of prototype and digital twin.

Experimental studies were carried out using two actuators located oppositely on the Y axis: EM1 and EM3. The maximum rotor levitation gap is 3.3mm. Each of them was independently controlled by voltage signals U_1 and U_3, resulting in current flow with the intensity of I_1 and I_3, respectively. Thus, measurements of magnetic flux density in each of the pole pieces B_{11}, B_{12} for EM1 and B_{31}, B_{32} for EM3 were obtained, respectively. By linearly controlling the triangular signal with a slope of 2 [V/s], a slight hysteresis was observed in the current signal and changes in magnetic flux density related to the rotor location in terms of the amplitude and width of the hysteresis loop (see Fig. 3a). A step change in the control voltage results in a change in the current that also depends on the location of the rotor, which is then visible in the magnetic flux density measurements (see Fig. 3b). The Bode diagrams obtained by harmonic excitation and parametric optimisation resulted in magnetic flux density to current magnitude M_{ij} and phase shift ϕ_{ij}. The higher values correspond to the rotor short cut in the magnetic circuit (see Fig. 4). The results of these tests showed the properties of the magnetic circuit and the ability to identify the position of the rotor. Precise identification requires the use of a tool that fixes the rotor in the bearing space.

The device was prototyped using the methodology presented by Pilat (2010*b*). The numerical model called a digital twin was developed in the COMSOL Multiphysics and reflects the prototype configuration. The stator is

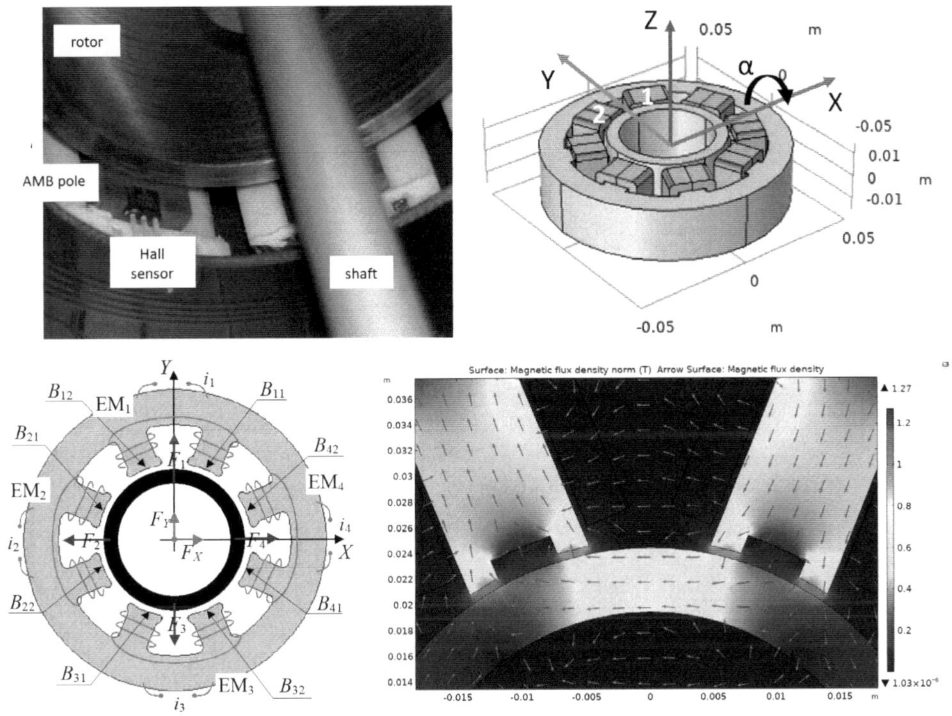

Figure 2: Radial Active Magnetic Bearing prototype with pole emmbeded magnetic sensors, Digital-Tiwn of the considered configuration, schematic bearing diagram, magnetic flux density distribution - cross section at the Hall sensor plane.

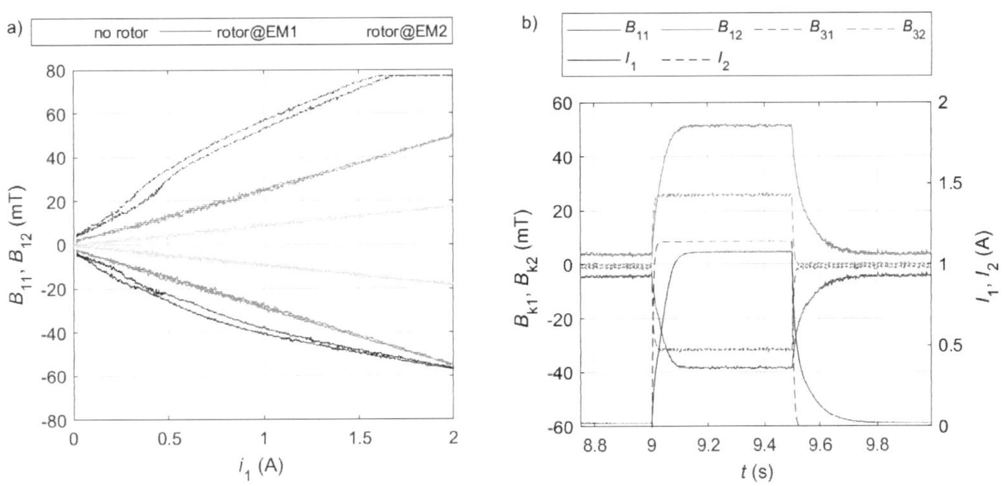

Figure 3: Experimental investigation of magnetic flux density under control: a) triangular excitation, b) square excitation.

modelled as solid, and the coils are modelled as multi-turn with a shape that reflects those produced.

The magnetic flux density measurement points were located in the holes where the Hall sensors are embedded. Using the capabilities of the software, the B_X and B_Y components of the magnetic field at the locations of the real sensor were studied and used in calculations. Simulation tests were carried out under stationary

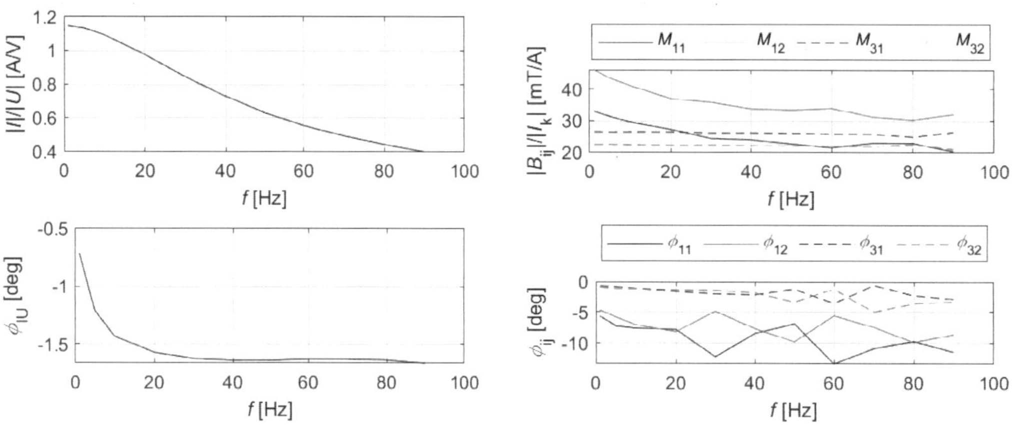

Figure 4: Experimental investigation of magnetic flux density - Bode diagram.

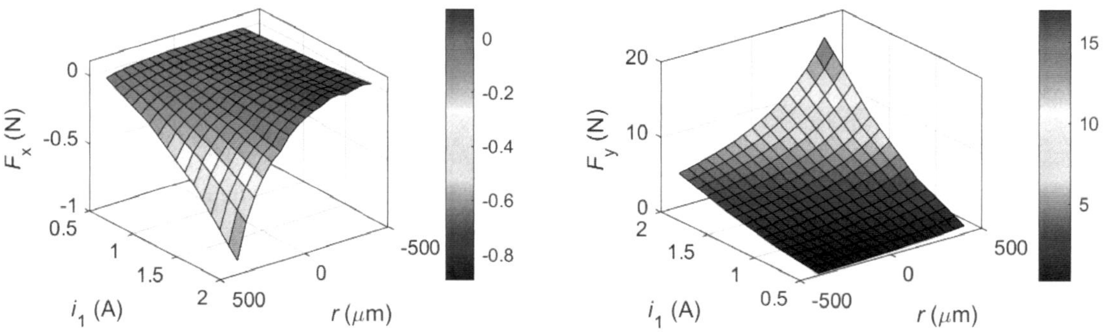

Figure 5: Electromagnetic force components F_x i F_y as a function of radial displacement and coil current.

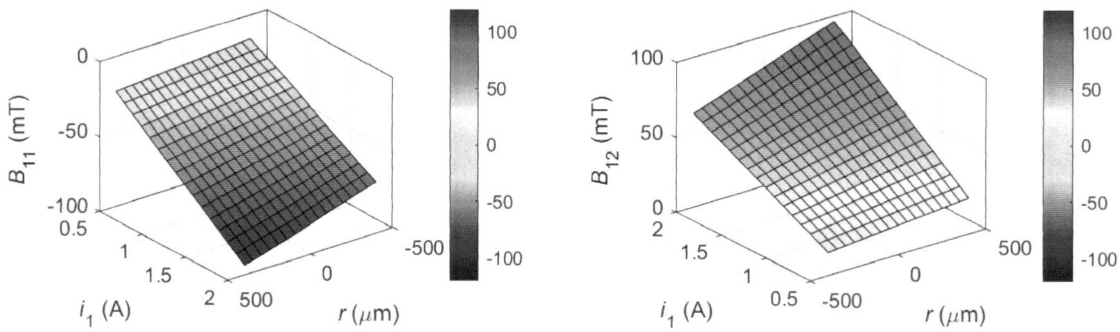

Figure 6: Radial magnetic flux density at AMB poles $B_{11}(r, I_1)$, $B_{12}(r, I_1)$ as a function of radial displacement and coil current.

conditions assuming a defined value of the current in the coil windings.

The digital twin was diagnosed under different conditions, such as axial motion in the X and Y axis, radial motion in the pole axis and rotation with respect to the X axis. Geometric constraints related to the geometry of the rotor were taken into account when determining the displacement area and changing its orientation. Due to the limited space of the ISMB conference article the results of the radial displacement are only presented. The development of the digital twin made it possible to carry out research, the implementation of which under

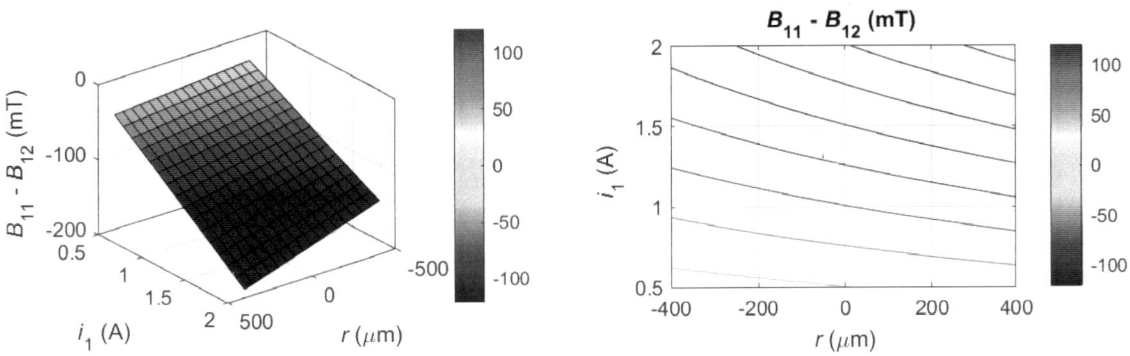

Figure 7: Difference between magnetic flux density measured at both poles $B_{11}(r,I_1)$, $B_{22}(r,I_1)$ at radial rotor displacement.

experimental conditions is difficult or requires the involvement of specialised equipment. It was subjected to simulation tests to obtain the characteristics of the electromagnetic force $F(\cdot)$ and the magnetic flux density $B(\cdot)$ as functions of the current and the position and orientation of the rotor. By moving the rotor along the axis Y - along the axis of the electromagnet, a symmetrical distribution of the magnetic field in both pole pieces and the known characteristics of the electromagnetic force were observed. By moving the rotor radially along the axis passing through pole piece No. 2, the asymmetry of the magnetic field distribution (Fig. 6) and force interactions (Fig. 5) were examined. The Fx component assumes negative values as the gap between the rotor and the pole piece decreases. The asymmetry in the measurement of magnetic flux density is clearly visible in this case and assumes a nonlinear character.

Differences between the values of magnetic flux density recorded by virtual sensors placed in the pole pieces were determined, illustrating the nonlinear nature as a function of the rotor position and current (see Fig. 7).

4. Discussion and Conclusions

The idea of embedding Hall sensors in pole pieces turned out to be important from the research and application point of view. The development of such a solution required an analysis of the limitations resulting from the capabilities of the electromagnetic actuator, the measuring range of the sensor and the appropriate mechanical, electrotechnical and electronic design. The developed research configuration allows for carrying out identification tests of the bearing and the rotor in various configurations of the rotating machine. Despite the availability of a single-axis sensor with specific dimensions, which resulted in the preparation of a pole piece structure, the identification of the magnetic field turned out to be effective and showed structural nonlinearity. Placing the sensor in the pole piece requires a trade-off between the measurement range of the sensor and the performance of the actuator. The developed numerical model allowed one to obtain a number of static characteristics illustrating the electromagnetic force and its individual components, as well as the components of the magnetic field vector in each of the slots of the pole piece. This model allowed us to set the position and spatial orientation of the rotor. The development of the 3D model was necessary due to the use of undercuts for the sensors and thus the lack of representativeness of the 2D model. The observed nonlinearities extended knowledge and will play a crucial role in analytical model, the controller and observer design based on the magnetic-field measurements.

In summary, it was observed that the commonly accepted simplifying assumption concerning the determination of the electromagnetic force and magnetic flux density on the basis of identical values of both gaps between the pole pieces and the rotor is possible only in a very small range of rotor displacements and only in the axial direction relative to the electromagnet. For full analysis of nonlinearity, it is required to determine the volume of gaps at the pole pieces as a function of the position and orientation of the rotor. The development of an analytical mathematical model describing the strength and distribution of magnetic flux density in the gap requires the use of nonlinear functions whose arguments are the position and orientation of the rotor and the value of the current. The development of a control using the measurement of magnetic flux density in the gap requires the use of a nonlinear controller for stabilisation in a wide range of changes in the position and orientation of the rotor. Identification of the position and orientation of the rotor in the levitation area is possible through the use of many magnetic field sensors. Modelling and control using magnetic field measurements requires taking into account the influence of temperature due to the stator to sensor heat transfer. In order to precisely determine the relationship between force and magnetic flux density, the actual volume of the levitation gap between the poles and the rotor should be determined.

References

Chowdhury, A. & Sarjaš, A. (2016), 'Finite element modelling of a field-sensed magnetic suspended system for accurate proximity measurement based on a sensor fusion algorithm with unscented kalman filter', *Sensors (Switzerland)* **16**(9).

Ernst, D., Faghih, M., Liebfried, R., Melzer, M., Karnaushenko, D., Hofmann, W., Schmidt, O. G. & Zerna, T. (2020), 'Packaging of ultrathin flexible magnetic field sensors with polyimide interposer and integration in an active magnetic bearing', *IEEE Transactions on Components, Packaging and Manufacturing Technology* .

Ernst, Mönch, J. I., Bahr, Hofmann, Schmidt & Zerna (2016), Flexible magnetic field sensors with ultra-thin silicon interposer, *in* '2016 6th Electronic System-Integration Technology Conference (ESTC)', pp. 1–4.

Kjølhede, K. & Santos, I. F. (2006), 'Experimental Contribution to High-Precision Characterization of Magnetic Forces in Active Magnetic Bearings', *Journal of Engineering for Gas Turbines and Power* **129**(2), 503–510.
URL: *https://doi.org/10.1115/1.2434345*

Miguel, L., Molina, C., Galluzzi, R., Bonfitto, A., Tonoli, A. & Amati, N. (2018), 'Magnetic Levitation Control Based on Flux Density and Current Measurement', *MDPI Appl. Sci.* .
URL: *www.mdpi.com/journal/applsci*

Mystkowski, A., Kierdelewicz, A., Jastrzebski, R. P., Dragašius, E. & Eidukynas, D. (2019), 'Flux measurement and conditioning system for heteropolar active magnetic bearing using kapton-foil hall sensors', *Mechanical Systems and Signal Processing* **115**, 394–404.

Pilat, A. (2010*a*), 'Active magnetic bearing and control system for active magnetic bearing', *WO* **2011074996**, A2.

Pilat, A. (2010*b*), 'Analytical modeling of active magnetic bearing geometry', *Applied Mathematical Modelling* **34**(12), 3805–3816.

Pilat, A. (2020), '6 pole amb as a drive of elliptic rotor – initial study supported by the virtual prototype', *International Journal of Applied Electromagnetics and Mechanics* **63**(1), 153–170.

Voigt, A. J., Mandrup-Poulsen, C., Nielsen, K. K. & Santos, I. F. (2017), 'Design and Calibration of a Full Scale Active Magnetic Bearing Based Test Facility for Investigating Rotordynamic Properties of Turbomachinery Seals in Multiphase Flow', *Journal of Engineering for Gas Turbines and Power* **139**(5). 052505.
URL: *https://doi.org/10.1115/1.4035176*

Voigt, A. J. & Santos, I. (2012), Theoretical and experimental investigation of force estimation errors using active magnetic bearings with embedded hall sensors, *in* 'Proceedings of ASME Turbo Expo 2012'.

Control and Commissioning of Active Magnetic Bearing (AMB) System for High-speed 1 MW Turbo Blower

Nikita UZHEGOV [a], Alexander SMIRNOV [a], Aleksi SALMI[b], Ville LAHDENSUO[b]

a SpinDrive, Laserkatu 6, 53850 Lappeenranta, Finland, nikita.uzhegov@spindrive.fi

b Runtech Systems, Kastarintie 27, 35990 Kolho, Finland

Abstract

This paper describes the design process, control and commissioning of active magnetic bearings (AMB) for a 1 MW, 11 500 rpm turbo blower used in the pulp and paper industry. Turbo blowers in pulp and paper applications work in a harsh environment. During the operation, the rotor is prone to high unbalance due to dirt buildup on the impeller. Implementing magnetic bearings helped to improve overall system efficiency and allowed stable operation with unbalance forces. AMB topology selection is described and justified from the overall turbo blower unit design perspective. Due to limited observability and controllability, only the robust model-based controller was selected to provide stable levitation throughout the operational range. The model based $\mathcal{H}_\infty$ approach scheduled over the speed with mixed-sensitivity formulation has been selected as a solution. The commissioning was performed by adjusting the plant model to the measured peak values and then by accelerating step by step and capturing identification data at each speed point. It allowed reaching the nominal speed despite working close to the first forward and on top of the first backward mode. Further system tests confirmed stable operation under all load conditions. Stable turbo blower operation has also been validated in situations of high unbalance force tested by installing a dummy impeller with added weights.

Keywords: *AMB, high-speed machine, turbo blower, rotor dynamics, flexible rotor*

1. Introduction

Energy efficiency at various operating points is an essential parameter for industrial equipment. With growing electricity prices, the rotating electrical equipment electricity costs may exceed 70% of the total lifetime equipment costs (De Almeida, Ferreira and Quintino, 2012). For example, a turbo blower is a rotating equipment typically used in industrial and manufacturing settings to provide large volumes of air or gas driven by high-speed electric motors. Turbo blowers can provide high energy efficiency and deliver a constant flow rate, making them a popular choice in many industrial applications, e.g. in pulp and paper. However, high-speed, high-power turbo blowers set challenging requirements for the bearing system needed for an efficient and reliable unit operation. This paper describes active magnetic bearing (AMB) control and commissioning of 1 MW, 11 500 rpm machine. The high output power and high rotational speed of the turbo blower unit resulted in the operation close the first forward mode, and simple control like PID was not able to provide stable operation.

AMBs are used in various industrial applications, where the combination of high rotational speed and high external loads set challenging requirements on the bearings (Uzhegov *et al.*, 2017). Magnetic bearings have several benefits, e.g., improved overall system efficiency due to higher possible rotational speeds and elimination of friction (Smirnov *et al.*, 2017). In addition, AMBs provide virtually maintenance-free and no wear bearing operation due to the absence of physical contact between the rotor and stator. Active magnetic bearings eliminate the risk of oil contamination due to the absence of lubricants. Finally, built-in position sensors and magnetic bearing controllers with significant computational power provide opportunities for condition monitoring of the rotating equipment.

The most common control approach for AMB systems is simple decentralized PID, where each axis is assumed to be independent. The reference for generating such a PID controller is available in fundamental textbooks on magnetic bearings (Chiba *et al.*, 2005; Larsonneur, 2009). The general properties and benefits of PID are summarized by (Knospe, 2006). However, decentralized control has its own limitations, which start to be evident in gyroscopic systems and with flexible rotors. In these cases, the coupling between axes and rotor ends cannot be neglected. To address the challenge, the model-based approaches are utilized such as $\mathcal{H}_\infty$ synthesis. This has also been addressed in the literature for AMB-supported machines (Jastrzebski, Hynynen and Smirnov, 2010; Balini, Scherer and Witte, 2011). Eventually, if a robust model-based approach is not sufficient, the solution is to utilize an adaptive controller. In the AMB system, the most straightforward is adapting to the rotational speed changes. This can be undertaken with the help of the gain-scheduling control theory (Balini, Witte and Scherer, 2012).

This paper describes how limited observability and controllability of the AMB system were solved during the 1 MW turbo blower commissioning. The robust model-based controller takes into account speed dependent model, foundation resonances and external disturbances to achieve the desired performance. During the commission, the model was adjusted, relying on identification results.

Changing turbo blower bearings from roller to active magnetic bearings allowed for optimizing impeller selection. The open impeller was possible to utilize due to the capability of AMBs to monitor and control axial pull. Turbo blower measurements demonstrated an efficiency increase from 62% to 70% by implementing an open impeller and active magnetic bearings. This tested variable speed and variable capacity turbo blower is able to save 30–70% in energy consumption compared with traditional vacuum systems like liquid-ring-pump, depending on the end-user.

2. AMBs for high-speed high power turbo blower

The decision to utilize magnetic bearing in vacuum blower is based on the challenging requirements. These machines are operated in the pulp and paper industry under harsh conditions. There are typically high axial loads with possibilities for surge effects occasionally. In addition, after some time, the dirt accumulates on the impellers resulting in a significant unbalance. The traditional ball bearings under such conditions require frequent maintenance resulting in production downtime, labour work and additional costs.

2.1 AMB Design

The vacuum blower has a standard 16-pole heteropolar radial bearing presented in Figure 1. The primary motivation for this topology is to reduce the journal thickness in order to keep the rotor dynamics acceptable and utilize unified manufacturing for the square-type coils. The system has an identical radial bearing at each end of the rotor.

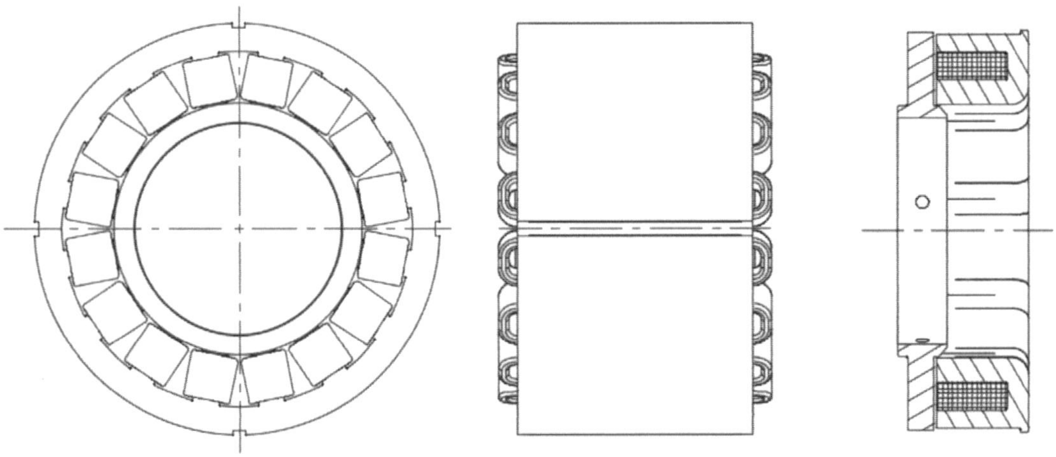

Figure 1 Overview of the radial and axial magnetic bearing actuators

The axial direction is controlled with a classical C-shaped axial bearing. Two axial stators are placed at each rotor end and facing the dedicated disc. The solution allows effectively utilizing the space under end windings of the electrical motor that otherwise will be empty.

The main parameters of the bearings are summarized in Table 1. The magnetic gap for the axial bearing has been increased to accommodate the possibility of uneven thermal expansion between the rotor and the stator.

Table 1 Main parameters of magnetic bearings

Parameter	*Value*
Radial	
Outer diameter	245 mm
Journal outer diameter	150 mm
Magnetic air gap	0.6 mm
Current stiffness	1.07 kN/A
Positions stiffness	30 N/μm
Inductance	40.9 mH

Parameter	Value
Axial	
Magnetic air gap	1.3 mm
Current stiffness	1.81 kN/A
Positions stiffness	9.39 N/μm
Inductance	174 mH

Overall, the bearings can provide the force with enough safety margin to levitate the system and reject disturbance from the process. The axial bearing stator has additional radial slits that break the path for the propagation of eddy currents and thus increase the bandwidth of the axial force.

2.2 Control

To create the linearized radial bearing model for the control purpose the general linearization of the system around its operating point is done. The current and position stiffnesses are given in Table 1. The actuator has been approximated as a first-order low-pass filter with a bandwidth of 350 Hz. The rotor model has been created as 1DOF FEM based on the Timoshenko beam theory. The overall view of that model is presented in **Error! Reference source not found.**, along with free-free modes. It can be seen that the nodes of the first mode go through the actuator locations.

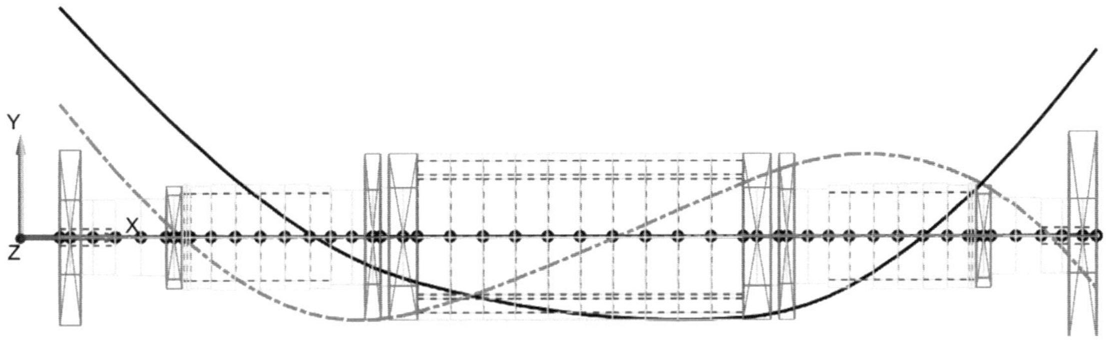

Figure 2 Free-free modes of the rotor model

The summary of the model is presented in

Table 2, where it is seen that 1st model has very limited controllability and observability on the right end of the rotor. Eventually, the resulting rotor model was reduced to keep only information on the two lowest frequency flexible modes to keep the controller order feasible.

Table 2 Main parameters and summary of the rotor model

Parameter	Value
Rotor mass	361 kg
Nominal speed	192 Hz
Polar moment of inertia	3.27 kg m^2
Diametral moment of inertia	62.4 kg m^2
Frequency of the first mode	217 Hz
Frequency of the second mode	412 Hz
Acutator locations from the center of mass (left/right)	490mm/423mm
Sensor locations from the center of mass (left/right)	596mm/529mm
Observability of the 1st mode (left/right)	44%/18.5%
Controllability of the 1st mode (left/right)	16.3%/8.42%

The final model was evaluated in the frequency domain, and the rotor speed was assumed as uncertainty. The corresponding singular value plot is presented in Figure 3. It is seen that the splitting of the first mode is quite considerable and results in roughly 57 Hz spread between forward and backward one at nominal speed. With limited controllability, it provides a challenge for the robust controller to handle. For this reason, the model-based $\mathcal{H}_\infty$ approach scheduled over the speed has been selected as a candidate solution. In particular, the mixed-sensitivity formulation of the problem (Zhou and Doyle, 1998).

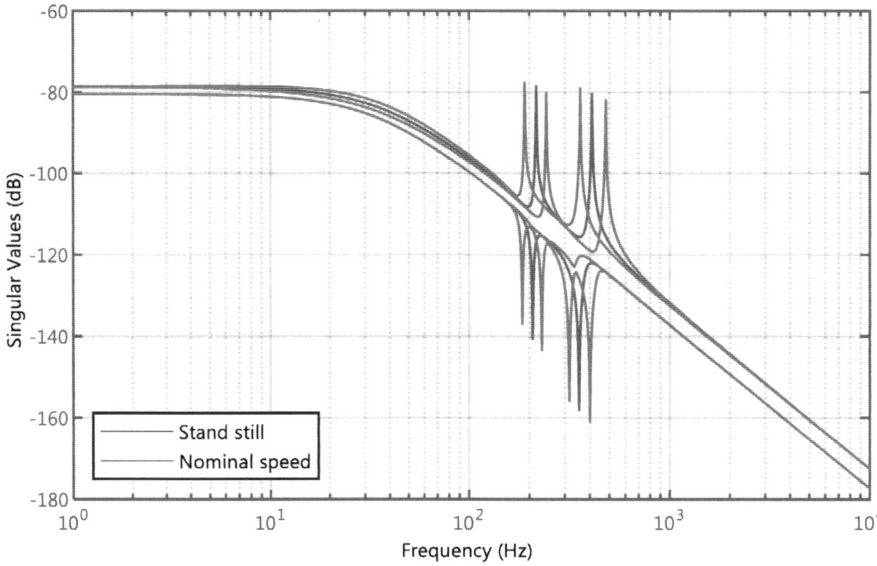

Figure 3 Singular value plot of the model at zero and nominal speeds

The weights for the control effort and sensitivity functions were defined as follows

$$W_{KS} = \frac{10s^3 + 6.87 \cdot 10^4 s^2 + 2.36 \cdot 10^8 s + 4.05 \cdot 10^{11}}{s^3 + 3.44 \cdot 10^4 s^2 + 5.93 \cdot 10^8 s + 5.1 \cdot 10^{12}}$$

$$W_S = 0.45 \frac{s + 150}{s + 1}$$

The control effort weight is rolled off with a 3rd order function at 500 Hz to remove the possibility for exciting higher frequency modes. The sensitivity weight defines the desired DC gain of the controller and limits the possible peak value of the closed loop function. This provides sufficient disturbance rejection.

2.3 Commissioning

The system was assembled in the test facility, where the identification data was obtained after initial levitation. The results are demonstrated in Figure 4, where the predicted model has a reasonable match to the measured results. The

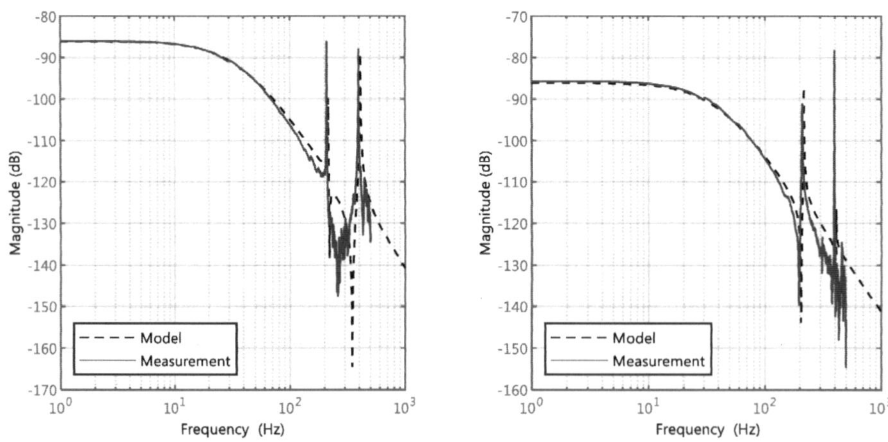

Figure 4 Identification results. Left on left end, right – right end

discrepancy appears in the exact frequencies of the flexible mode. This is expected as the beam model assumes some contact stiffness between the shaft and attached elements, which is difficult to predict. The other discrepancy is in the peak of the first mode for the left end and the second mode for the right end. This is related to the inaccuracy of the mode shape and its damping. These parameters were adjusted in the model with respect to the obtained measurements.

For control purposes, only the first two modes are considered, as their peaks are above the DC gain of the plant, and they require effort from the control system to provide additional dumping. The plant model was adjusted to the actually measured peak values so that the model-based controller could react adequately.

In the next step, the system was accelerated step by step, capturing identification data at each speed point. The resonances are demonstrated as red dots in **Error! Reference source not found.**. They do not perfectly correspond to the linear reduced order model. To overcome the discrepancy, the whole speed range is separated into five points between which the position controller is interpolated. At each point, the frequency of the flexible modes is tuned to the actual measurements to ensure the model match.

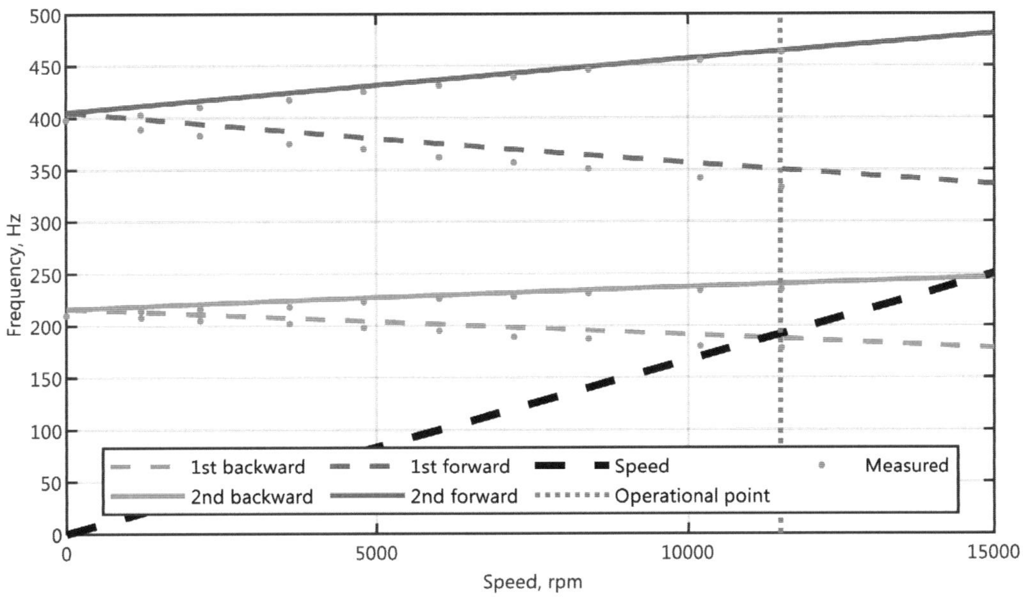

Figure 5 Estimated Campbell diagram and resonance frequencies measured from the setup

This approach allowed for reaching the nominal speed and operating the system under all load conditions. Furthermore, the stability has also been validated in situations of high unbalance force by installing a dummy impeller with specifically added weights. After that, the whole machine has been under test to ensure the overall aerodynamics performance and robustness.

3. Results

The system was thoroughly tested over a 1-year period in all operational conditions. The AMBs demonstrated the expected robustness and maintenance-free operation. In addition, the diagnostics capability of the bearings and, specifically, the possibility to estimate the forces allowed to optimize the impeller. During certain operational conditions, an unexpected axial pull was captured. After analysis of the data, the impeller was updated from closed to open design, which provided a process efficiency increase from 62% to 70%.

The other benefit was obtained by using the dummy impeller with a predefined unbalance level. Based on the experimental data and unbalance orbit estimated by the bearing controller, it was possible to predict the proper intervals for impeller cleaning. On top of that, the AMBs can tolerate high unbalance level without risk of damaging the system.

The high-speed turbo blower with AMBs inside the testing bunked is shown in Figure 6.

4. Conclusions

This paper describes how robust control was implemented for the AMB system used in a 1 MW, 11 500 rpm turbo blower. AMB topology, design parameters and rotor dynamics analysis affecting causing limited controllability and observability were described. The control synthesis procedure is presented, which has taken into account various factors like speed-dependent system behavior, foundation resonances and external disturbances. Finally, the control system was tuned to the actual model of the plant based on the experimental measurements. The gain-scheduling has been applied to provide the necessary performance in the entire operational range because of the flexible modes splitting.

Figure 6 One MW turbo blower under tests

References

De Almeida, A. T., Ferreira, F. J. T. E. and Quintino, A. (2012) 'Technical and economical considerations on super high-efficiency three-phase motors', in *48th IEEE Industrial & Commercial Power Systems Conference*. IEEE, pp. 1–13. doi: 10.1109/ICPS.2012.6229618.

Balini, H. M. N. K., Scherer, C. W. and Witte, J. (2011) 'Performance Enhancement for AMB Systems Using Unstable H∞ Controllers', *IEEE Transactions on Control Systems Technology*, 19(6), pp. 1479–1492. doi: 10.1109/TCST.2010.2097264.

Balini, H. M. N. K., Witte, J. and Scherer, C. W. (2012) 'Synthesis and implementation of gain-scheduling and LPV controllers for an AMB system', *Automatica*. doi: 10.1016/j.automatica.2011.08.061.

Chiba, A. *et al.* (2005) *Magnetic bearings and bearingless drives*. Newnes. Available at: https://www.dawsonera.com/abstract/9780080478975.

Jastrzebski, R. P., Hynynen, K. M. and Smirnov, A. (2010) 'H[infinity] control of active magnetic suspension', *Mechanical Systems and Signal Processing*, 24(4), pp. 995–1006. doi: 10.1016/j.ymssp.2009.10.008.

Knospe, C. R. (2006) 'PID control', *IEEE Control Systems Magazine*, 26(1), pp. 30–31. doi: 10.1109/MCS.2006.1580151.

Larsonneur, R. (2009) 'Magnetic Bearings: Theory, Design, and Application to Rotating Machinery', in Schweitzer, G. and Maslen, E. H. (eds). Springer Berlin Heidelberg, pp. 229–250. doi: 10.1007/978-3-642-00497-1_9.

Smirnov, A., Uzhegov, N., Sillanpää, T., Pyrhönen, J., and Pyrhönen, O. (2017) 'High-speed electrical machine with active magnetic bearing system optimization' *IEEE Transactions on Industrial Electronics*, 64(12), pp. 9876-9885. doi: 10.1109/TIE.2017.2716875.

Uzhegov, N., Smirnov, A., Park, C. H., Ahn, J. H., Heikkinen, J., and Pyrhönen, J. (2017). 'Design aspects of high-speed electrical machines with active magnetic bearings for compressor applications' *IEEE Transactions on Industrial Electronics*, 64(11), pp. 8427-8436. doi: 10.1109/TIE.2017.2698408.

Zhou, K. and Doyle, J. C. (1998) *Essentials of Robust Control*. Upper Saddle River, NJ: Prentice-Hall.

Tuning Methods for Classical SISO PID Magnetic Levitation Controllers via Analytic Models

Nathan P. PETERSEN[a] and Eric L. SEVERSON[a]

[a] *University of Wisconsin–Madison, Madison, WI, USA. Email: nathan.petersen@wisc.edu, eric.severson@wisc.edu*

Abstract

Stable operation of magnetically levitated systems requires feedback control. This paper studies tuning methodologies for the popular stabilizing PID controller which generates force/current references for the inner control loop. The pole placement method and frequency design loop shaping methods are compared in terms of their required tuning rules. Since the control system has five parameters, the tuning process is under constrained since the system only has two parameters: rotor mass and displacement unstable stiffness. Therefore, system performance objective metrics are defined and a case study multi-objective optimization is performed to find Pareto optimal tunings. Six tuning regions are identified and the salient characteristics are given and transformed into tuning rules. For example, it is found that high phase margin does not necessarily translate to robust control, nor a highly damped system response. Future magnetic levitation control designers can use the practical rules herein as a guide for creating robust and high-performance feedback PID controllers for magnetic levitation.

Keywords: magnetic levitation, tuning, PID control, frequency design, pole placement

1 Introduction

Magnetic levitation has been applied to rotating machinery where either active magnetic bearings (AMBs) or bearingless motors (BMs) are used, see Fig. 1. Both AMBs and BMs can be viewed as generic force actuators and made to accurately create a desired force. To achieve stable levitation, motion controllers use displacement measurements to compute a force command which the AMBs/BMs produce. There are potential complexities to the motion control algorithm, e.g., speed-dependent bearing stiffness/damping to handle the dynamics of flexible rotors, co-location issues between the force production and sensor planes, and handling imbalance in the rotor assembly. While complexities may abound for real-world systems, this paper is devoted to understanding the tuning of classical PID controllers for the linear magnetic levitation mechanical plant with mass m and unstable displacement stiffness (positive number) k_δ. While the core theoretic tuning concepts are known in the literature, the goal of this paper is to present a practical and systematic methodology to implement this theory in real systems.

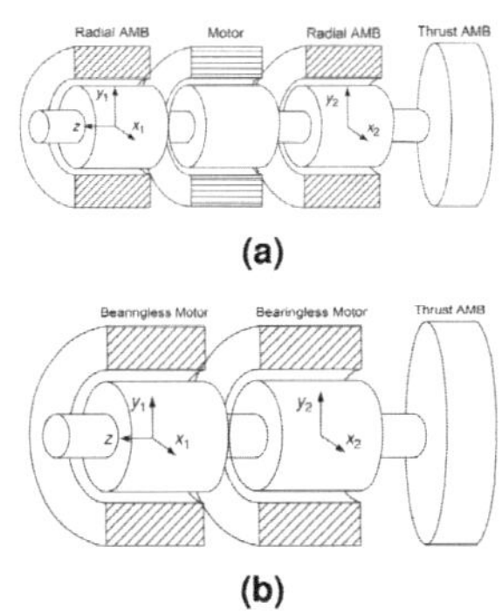

Fig. 1: *Non-contact motor with: (a) AMBs, (b) BMs. (Chiba et al. 2005)*

2 Control Structure

Figure 2 shows control of AMBs/BMs where an inner force (current) regulation loop produces a force command F_c^* with high bandwidth. The outer motion control loop is used to stabilize the system by measuring rotor displacement δ and computing the force reference F_c^* for the inner loop. The starred signals (e.g., δ^*) denote references and i is the current, F_{dis} is the disturbance force, and k_f is the bearing current stiffness constant. The motion controller is denoted $G_c(s)$, the current (force) regulation loop dynamics are modeled as $G_{CR}(s)$, and the magnetic levitation plant is denoted $G_p(s)$. These transfer functions are assumed to be in the form:

$$G_c(s) = K_P + \frac{K_I}{s} + K_D s\left(\frac{\omega_D}{s+\omega_D}\right) \qquad G_{CR}(s) = \frac{\omega_{ci}}{s+\omega_{ci}} \qquad G_p(s) = \frac{1}{ms^2 - k_\delta} \tag{1}$$

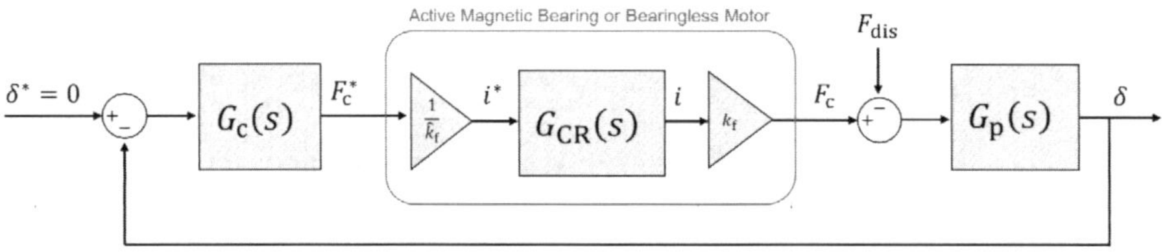

Fig. 2: *Single DOF magnetic levitation control structure showing the AMB or BM as the force actuator.*

The controller $G_c(s)$ is classical PID with an additional low-pass filter (pole) inline with the derivative term to limit the effects of sensor noise. The current regulator $G_{CR}(s)$ is modeled as a first-order low-pass filter (pole) at bandwidth ω_{ci}. The plant $G_p(s)$ is modeled as a mass-spring system with destabilizing spring stiffness; both AMBs and BMs behave according to this model for small displacements from center (Chiba et al. 2005, Maslen et al. 2009). The plant $G_p(s)$ can also be written as two poles where one is in the right-half plane (i.e., unstable) and is located at unstable pole frequency $\omega_0 = \sqrt{k_\delta/m}$. The value of $\omega_0/2\pi$ is typically 20-50 Hz.

3 Tuning Methods

There are several studies published related to controller tuning in magnetic levitation systems. Tuning for PID and PD controllers is investigated in (Bleuler et al. 1994, Anantachaisilp et al. 2012); linear quadratic regulation (LQR) is studied in (Brunet & Rioland 1990, Yoon et al. 2012); automated controller design and tuning using µ-synthesis is investigated in (Sawicki & Maslen 2008). These prior studies give several frameworks for reasoning through controller tuning, however, these studies do not give simple analytical equations and "rules of thumb" that control designers can use when approaching controller gain selection for system bring-up. In this paper, two practical tuning methodologies broadly applicable to SISO magnetic levitation controllers are investigated: (i) pole placement and (ii) frequency domain loop shaping.

Referring to Fig. 2 and (1), a complete controller tuning can then be described as $\boldsymbol{x} \in \mathbb{R}^5$ where $\boldsymbol{x} = [K_P, K_I, K_D, \omega_D, \omega_{ci}]$ and K_P, K_I, and K_D are the PID gains; ω_D is the derivative-term (d-term) filter bandwidth; and ω_{ci} is the inner current regulation bandwidth. In practice, $\omega_{ci} < \omega_{ci,\max}$ and $\omega_D < \omega_{D,\max}$. The filter bandwidths must be high enough so as to not impede stable controller operation.

3.1 Pole Placement Method

It is well known that the closed-loop poles of a system directly impact the system behavior, such as determining bandwidths, settling times, and transient oscillations. Since this system is controllable, linear systems theory states that the closed-loop poles can be placed arbitrarily through gain selection. The task of the control designer is then to select appropriate locations for the poles such that the final system behaves as desired. The following steps are required to algebraically solve the pole-placement problem. First, the open-loop $\mathrm{OL}(s)$ and closed-loop $\mathrm{CL}(s)$ transfer functions of the 1-DOF control system can be derived as:

$$\mathrm{OL}(s) = \frac{k_f}{\hat{k}_f} G_c(s) G_{CR}(s) G_p(s), \qquad \mathrm{CL}(s) = \frac{\delta(s)}{\delta^*(s)} = \frac{\mathrm{OL}(s)}{1+\mathrm{OL}(s)} = \frac{A_2 s^2 + A_1 s + A_0}{s^5 + B_4 s^4 + B_3 s^3 + B_2 s^2 + B_1 s + B_0} \tag{2}$$

A generic polynomial $P(s)$ of equal order to the characteristic polynomial (denominator of $\mathrm{CL}(s)$) is defined:

$$P(s) = (s-a)(s-b)(s-c)(s-d)(s-e) = s^5 + P_4 s^4 + P_3 s^3 + P_2 s^2 + P_1 s + P_0 \tag{3}$$

The coefficients of the characteristic polynomial and $P(s)$ are equated: $P_k = B_k$ for $k = 0 \ldots 4$. This forms five equations with five unknowns. The five unknowns are the controller gains $\boldsymbol{x} = [K_P, K_I, K_D, \omega_D, \omega_{ci}]$ which can be found by solving the system of equations, i.e., $\boldsymbol{x} = f(a,b,c,d,e)$. Due to the fifth-order system, it is recommended to use a symbolic math package to solve for $\boldsymbol{x}$.

3.2 Frequency Design Method

The frequency design method is an alternative way of reasoning about the tuning of the PID controller. By studying the open-loop transfer function $\mathrm{OL}(s)$ and viewing the controller $G_c(s)$ in pole-zero form, Bode plot analysis can be readily used, i.e., metrics like gain/phase margin and gain/phase crossover frequency. The PID controller from (1) can be rewritten in pole-zero form as (4) where there are two zeros at frequencies ω_{z1} and ω_{z2}, one pole at frequency ω_D, one pole at the origin, and gain K_o. To tune the control system using the frequency design method, the control designer must specify five metrics: phase margin ϕ_{PM}, gain crossover frequency (i.e., controller bandwidth) ω_c, one of the two zeros ω_{z1}, pole ω_D, and current regulator bandwidth ω_{ci}. Based on these inputs, the second zero ω_{z2} is computed to achieve the desired phase margin ϕ_{PM} and the gain K_o is computed such that $|\mathrm{OL}(j\omega_c)| = 1$, see the expressions from (5).

$$G_{c,zpk}(s) = K_o \frac{(1+\frac{s}{\omega_{z1}})(1+\frac{s}{\omega_{z2}})}{s(1+\frac{s}{\omega_D})} \text{ where } K_P = \left(\frac{1}{\omega_{z1}} + \frac{1}{\omega_{z2}} - \frac{1}{\omega_D}\right) K_o,\ K_I = K_o,\ K_D = \frac{K_o}{\omega_{z1}\omega_{z2}} - \frac{K_P}{\omega_D} \tag{4}$$

$$\omega_{z2} = \frac{\omega_c}{\tan\left(\phi_{PM} + \frac{\pi}{2} - \mathrm{atan}(\frac{\omega_c}{\omega_{z1}}) + \mathrm{atan}(\frac{\omega_c}{\omega_D}) + \mathrm{atan}(\frac{\omega_c}{\omega_{ci}})\right)},\quad K_o = \frac{1}{\left|G_c(j\omega_c)\right|_{K_o=1}\left|G_{CR}(j\omega_c)\right|\left|G_p(j\omega_c)\right|} \tag{5}$$

3.3 Required Tuning Rules

The process of using either the pole placement method or the frequency design method simply requires evaluating closed-form expressions to solve for the controller gains. However, both methods require the control designer to specify five tuning "rules of thumb" to fully constrain the math. The remainder of this paper investigates how to reason about "good" tuning rules.

4 Tuning Rules for PID Controllers

This section explores trade-offs in system performance between tunings with the goal of presenting guidelines for selecting tuning rules which give optimal controller performance for both the pole placement and frequency design methods. Numerical methods combined with an evolutionary algorithm are used to search the magnetic levitation controller tuning design space to identify Pareto optimal tuning approaches.

4.1 Controller Optimization Formulation

The controller performance is evaluated with three objectives (to maximize):

- O_1: phase margin ϕ_{PM}—generally desired in control systems and helps ensure stability even when unmodeled dynamics are present, for example, eddy currents which limit achievable force bandwidth.
- O_2: allowable parameter error ($\pm$) in m, k_δ, or k_f before closed-loop instability—all real-world systems have some degree of uncertainty in plant parameters which the controller must tolerate.
- O_3: $\frac{1}{\|\delta(s)/F_{dis}(s)\|_\infty}$, i.e., the lowest dynamic stiffness value over all frequencies—since magnetic levitation systems only respond to force disturbances, this term denotes the effective controller effort.

Three constraints are enforced: $O_2 > 10\%$, $\omega_{ci} \leq \omega_{ci,max} = 2\pi \times 1500$ Hz, $\omega_D \leq \omega_{D,max} = 2\pi \times 800$ Hz. The plant studied has $m = 1$ kg and $\omega_0/2\pi = 50$ Hz. The optimization procedure uses a population-based multi-objective evolutionary genetic algorithm. Each individual in the population maps to one set of tuning gains. The optimization free variables are the s-plane location of the five closed-loop poles from Section 3.1. Three optimization cases are performed: (1) all real poles, (2) three real poles with one pair of complex conjugate poles, and (3) one real pole with two pairs of complex conjugate poles. The genetic algorithm was run until convergence, i.e., the Pareto front remained static over multiple generations. Multiple optimization runs were conducted with random initial populations to ensure global optimal solutions.

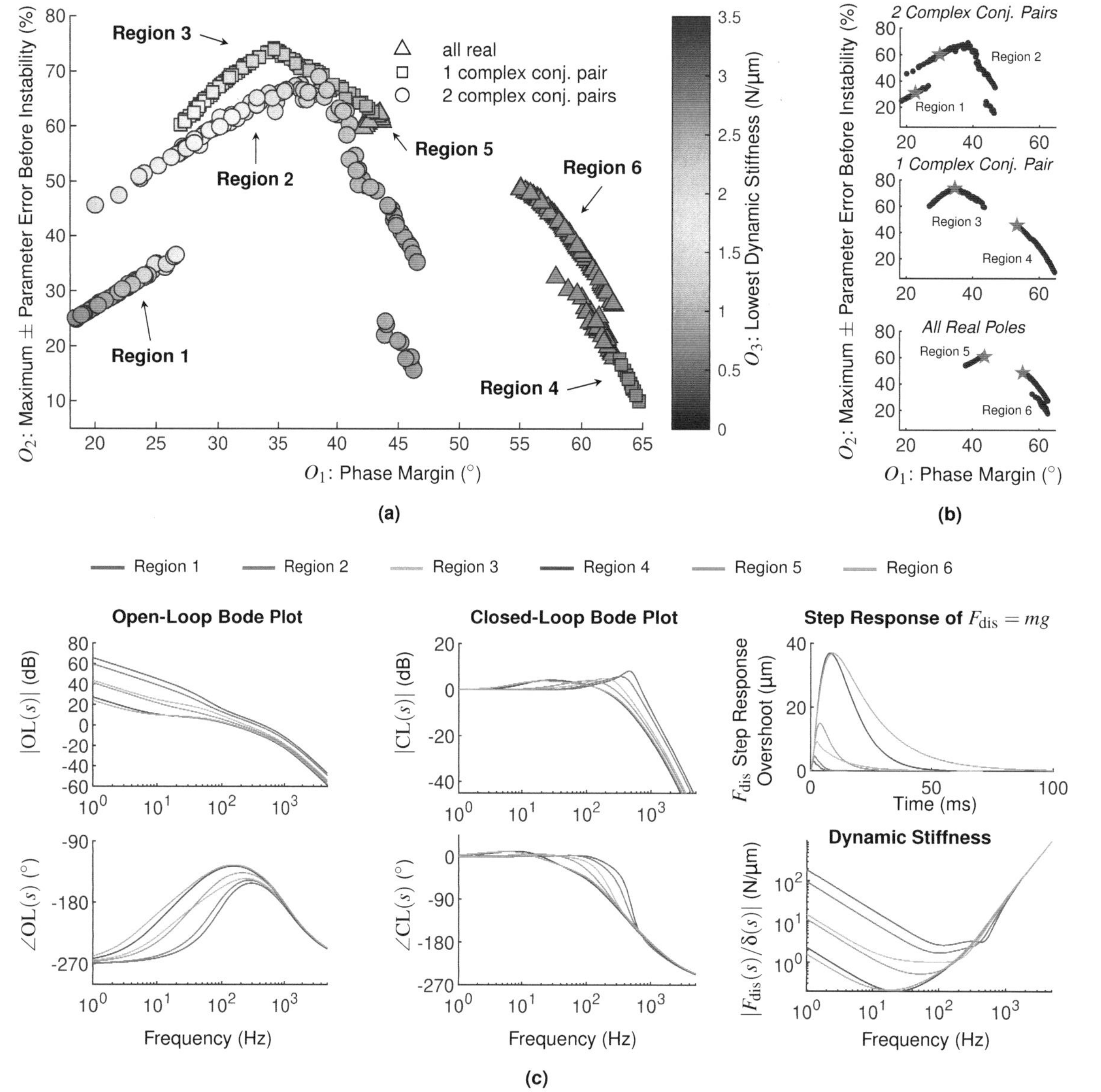

Fig. 3: *Optimization Pareto fronts. Each marker denotes one controller tuning set. (a) Combined global Pareto front; (b) Individual Pareto fronts. (c) System response for the marked tunings from (b).*

Region	$\omega_c/2\pi$ (Hz)	ϕ_{PM} (°)	$\omega_{z1}/2\pi$ (Hz)	$\omega_{z2}/2\pi$ (Hz)	$\omega_D/2\pi$ (Hz)	$\omega_{ci}/2\pi$ (Hz)	$-a/2\pi$ (Hz)	$-b/2\pi$ (Hz)	$-c/2\pi$ (Hz)	$-d/2\pi$ (Hz)	$-e/2\pi$ (Hz)
1	449	23	74.2 + j57.4	$\omega_{z1}^{\dagger}$	769	1386	1714.8	132.1 + j483.1	$b^{\dagger}$	87.6 + j65.1	$e^{\dagger}$
2	358	30	66.6 + j39.0	$\omega_{z1}^{\dagger}$	781	1432	1700.0	172.5 + j382.3	$b^{\dagger}$	84.4 + j43.1	$e^{\dagger}$
3	242	35	13.6	114.6	794	1393	1588.1	299.3	142.2 + j195.7	$c^{\dagger}$	15.0
4	141	53	9.0	41.5	797	1213	1388.9	375.4	200.7	22.4 + j5.9	$d^{\dagger}$
5	192	44	22.4	64.2	800	1412	1584.5	234.2	233.0	131.2	28.8
6	152	55	5.5	42.8	798	1368	1526.1	349.7	247.0	33.6	9.7

Table 1: *Summary of selected Pareto optimal tunings denoted in Fig. 3b.*

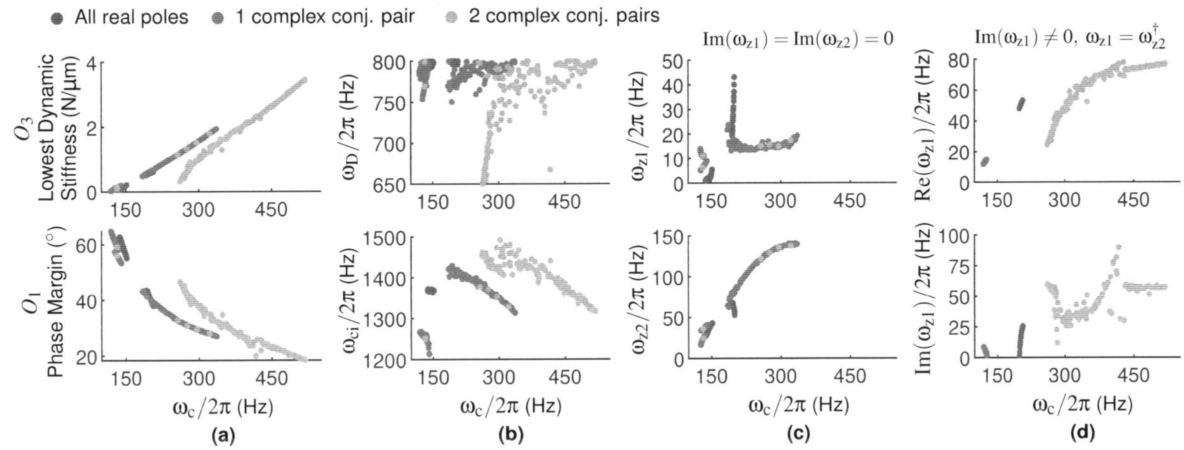

Fig. 4: *Trends in Pareto optimal tunings vs. PID controller bandwidth, i.e., gain crossover frequency* ω_c.

4.2 Optimization Results

Fig. 3 shows the Pareto fronts from the controller tuning optimization studies. Fig. 3a shows the global Pareto front while Fig. 3b shows the Pareto fronts from each case. There are six regions labeled on the Pareto fronts based on the natural segmentation of the results, each resulting in different system performance. The starred points from Fig. 3b are evaluated and the response is shown in Fig. 3c and Table 1. Fig. 4 shows trends in control metrics across Pareto optimal tunings compared to the PID controller bandwidth ω_c. Fig. 4a shows the objectives O_1 and O_3. It is seen that the control bandwidth ω_c is proportional to the lowest dynamic stiffness, and for a given ω_c, the case of two complex conj. pole pairs has worse stiffness compared to only one complex conj. pole pair. However, with two complex conj. pole pairs, the controller can be tuned to bandwidth/stiffness values far beyond what is possible with only one complex conj. pole pair. The phase margin ϕ_{PM} degrades with increasing control bandwidth ω_c. Fig. 4b shows the filter bandwidths for the d-term pole ω_D and current regulator ω_{ci}. The optimization has pushed the d-term filter bandwidth to its limit for many cases, however, this is not true for the current regulation bandwidth. The controller zero (ω_{z1}, ω_{z2}) placement is given where the pure real zeros are shown in Fig. 4c and the complex conj. zeros are shown in Fig. 4d.

4.3 Controller Tuning Regions

The characterization of each of the six tuning regions is compared to give tuning recommendations such as the form of the closed-loop poles and the placement of the PID controller zeros. In this section, the closed-loop poles $[a,b,c,d,e]$ are assumed to be sorted by increasing real part, i.e., $\Re(a) \leq \Re(b) \leq \Re(c)$ etc. For all tunings, the fastest pole a is purely real and is spaced far from the next pole b, i.e., $|\Re(a)| \gg |\Re(b)|$. The placement of a is set according to the current regulation bandwidth limit $\omega_{ci,max}$, however, $|a| \neq \omega_{ci,max}$ since the closed-loop and open-loop poles are not equal.

4.3.1 Regions 1 and 2—two complex conjugate closed-loop pole pairs

Regions 1 and 2 are characterized by the controller zeros—in the pole-zero form from (4)—being complex conjugates, i.e., $\omega_{z1} = \omega_{z1,r} + j\omega_{z1,i}$ and $\omega_{z2} = \omega_{z1}^{\dagger}$. The frequency design method tuning equations presented in Section 3.2 do not handle complex conjugate controller zeros, however, the pole placement method from Section 3.1 inherently does. For all tunings from both Regions 1 and 2, the slower pair of complex conjugate poles (i.e., d and e) are located close to the complex conjugate controller zeros, thus achieving partial pole/zero cancellation. By doing this, the system response has no noticeable oscillations, even though the closed-loop poles consist of two pairs of complex conjugates, and the faster pair has no zero cancellation effect.

4.3.2 Regions 3 and 4—one complex conjugate closed-loop pole pair

Region 3 is characterized by the closed-loop poles c and d being the complex conjugate pair, i.e., the slowest pole e is purely real. In other words, as sorted, $c = c_r + jc_i$ and $d = c^\dagger$. In the pole-zero controller form from (4), all tunings in Region 3 have purely real controller zeros, i.e., $\omega_{z1}, \omega_{z2} \in \mathbb{R}$. Furthermore, the controller zeros are placed on either side of the plant unstable pole ω_0, i.e., $\omega_{z1} < \omega_0 < \omega_{z2}$. Region 4 is characterized by the closed-loop poles d and e being the complex conjugate pair, i.e., the three fastest poles are purely real: $[a, b, c] \in \mathbb{R}$. In other words, as sorted, $d = d_r + jd_i$ and $e = d^\dagger$. Traversing Region 4 from lower to higher phase margin, the poles d and e migrate towards the imaginary axis. In the region with highest phase margin—approximating where $\phi_{PM} > 60°$—the complex conjugate poles dominate and induce excessive oscillations in the system response, even though phase margin is maximized. This counters the common intuition that higher phase margin results in a more damped response and is always better.

4.3.3 Regions 5 and 6—all real closed-loop poles

Region 5 is split in two sub-regions based on complex conjugate or purely real zeros. Traversing Region 5 from higher to lower phase margin, the zeros start as purely real and enclose the slowest closed-loop pole e and the plant unstable pole ω_0, i.e., $\omega_{z1} < |e| < \omega_{z2}$, $\omega_{z1} < \omega_0 < \omega_{z2}$. Traversing from the region limit of approximately $\phi_{PM} = 44°$ to $\phi_{PM} = 41°$, the controller zeros come together and converge at $\omega_{z1} = \omega_{z2} = \omega_0 = |e|$. Then, continuing to traverse from $\phi_{PM} = 41°$ to the region limit of approximately $\phi_{PM} = 38°$, the zeros split off axis. With complex conjugate zeros, the closed-loop poles are all faster than the real part of the zeros, i.e., $|a| > \ldots > |e| > \Re(\omega_{z1}) = \Re(\omega_{z2})$. This is the only region with complex zeros and no pole/zero cancellation occurs. Region 6 is characterized by the controller zeros—in the pole-zero controller form from (4)—being both purely real and both slower than the plant unstable pole ω_0, i.e., $\omega_{z1} < \omega_{z2} < \omega_0$. Furthermore, the controller zeros are set where the slowest two closed-loop poles are sandwiched between the zeros, i.e., $\omega_{z1} < |d| < |e| < \omega_{z2}$.

5 Conclusion

This paper investigates analytical tuning for magnetic levitation PID controllers using either pole placement or frequency design methods. Pareto optimal tunings are identified based on three system performance metrics. Overall, Fig. 3c presents the range of possible system responses given Pareto optimal tunings with the goal of users understanding the tuning design space. Users select the desired Pareto optimal tuning based on Figs. 3 and 4, depending on the desired bandwidth (control effort as dynamic stiffness) and robustness to uncertainty (unmodeled dynamics and/or parameter error). The tuning region is identified from Fig. 3 and Table 1 gives example controller parameters. The evolution of the pole-zero controller formulation is presented in Fig. 4 for how to select the controller zeros, pole frequency limits, and phase margin given the desired control bandwidth.

References

Anantachaisilp, P., Lin, Z. & Allaire, P. (2012), Pid tuning methods for active magnetic bearing systems, *in* '13th International Symposium on Magnetic Bearings'.

Bleuler, H., Jeon, J. & Higuchi, T. (1994), Self-tuning control for magnetic bearings, *in* '4th Int. Symp. on Mag. Bearings'.

Brunet, M. & Rioland, J. (1990), Self-tuning digital state controller for active magnetic bearings, *in* '2nd International Symposium on Magnetic Bearings'.

Chiba, A., Fukao, T., Ichikawa, O., Oshima, M., Takemoto, M. & Dorrell, D. (2005), *Magnetic Bearings and Bearingless Drives*, Newnes.

Maslen, E., Schweitzer, G., Bleuler, H., Cole, M., Keogh, P., Larsonneur, R., Nordmann, R., Okada, Y. & Traxler, A. (2009), *Magnetic Bearings—Theory, Design and Application to Rotating Machinery*, Springer Berlin Heidelberg.

Sawicki, J. & Maslen, E. (2008), Toward automated amb controller tuning: Progress in identification and synthesis, *in* '11th Int. Symposium on Magnetic Bearings'.

Yoon, S. Y., Lin, Z. & Allaire, P. (2012), Iterative tuning of linear quadratic controller for ambs in a high speed compressor, *in* '13th International Symposium on Magnetic Bearings'.

Analysis of Force Capacity in Magnetic Bearings and Bearingless Motors from the Perspective of Airgap Space Harmonic Fields *

Anvar KHAMITOV[a], Eric SEVERSON[a]

a University of Wisconsin-Madison, Engineering Drive 1415, 53706 Madison, USA, khamitov@wisc.edu, eric.severson@wisc.edu

Abstract

This paper studies the force creation capabilities of active magnetic bearings (AMBs) and bearingless motors from the perspective of multiple airgap space harmonics/pole-pairs. This approach is analytic-based and is useful in explaining the underlying physics of the machine and conducting force capacity analysis for different numbers of phases/poles. The presented per unit (p.u.) model makes the force capacity results applicable to any motor dimensions and peak airgap field value. An explanation of the force capacity in bearingless motors is provided when only two harmonics are controlled (which is the typical approach in bearingless motor literature) and the relationship between torque, force, and magnetizing field values is identified. Using this relationship, optimal magnetizing field values for maximum torque-force capability are identified, which is useful to consider when designing a bearingless motor. This paper extends the force capacity analysis to bearingless motors with multiple (more than two) controllable space harmonics and proposes that force enhancement can be achieved through the control of the magnitudes and angles of these harmonics. Results show that potential force enhancement of over 40% in bearingless machines can be achieved when controlling four airgap harmonics as opposed to two harmonics. These results suggest that being able to control multiple harmonics can yield high performance designs.

Keywords: active magnetic bearing, bearingless motor, self-bearing motor, current sequences, force enhancement

1. Introduction

Magnetic levitation technologies have been developed as an alternative to contact bearings to provide contact-free and lubricant-free support of the motor shaft, eliminating any point of wear, bearing friction, and contamination issues. AMBs have been successful in many applications over different power and speed levels (Chen et al. 2019). However, their use has been limited by high design cost and large size, which reduces the overall motor power density and increases the total shaft length. Bearingless motors have the potential to solve the issues of AMBs. However, they have only been successfully deployed in low power applications as they cannot yet meet high power requirements. Most of the bearingless motor prototypes discussed in the literature are aimed at low power ratings and only a small number of them have efficiencies exceeding 90%. Both technologies are known to have a lower specific load capacity (Maslen & Schweitzer 2009) than other bearing types. Specific load capacity is calculated by dividing the rated force (maximum force that can be created at any angle) by the projected rotor area (length × diameter) (Chiba et al. 2005). Typically, AMBs have a specific load capacity of 30-40 N/cm^2, with cobalt-alloys reaching up to 65 N/cm^2 (Maslen & Schweitzer 2009, Jastrzebski et al. 2021, Mushi et al. 2011, Swanson et al. 2008). Bearingless motors are often assumed to have a significantly lower load capacity of approximately 9 N/cm^2 (Chiba et al. 2005). In most publications, force creation and control in AMBs is considered in terms of individual poles/teeth. This paper will present an alternate analysis method from the perspective of airgap space harmonics. The paper will extend force capacity analysis to bearingless motors and show how controlling multiple harmonics can be used to design a bearingless motor with equivalent force capacity to AMBs.

The core contributions of this paper are: 1) explanation of the force capacity of AMBs from the perspective of controllable airgap space harmonic fields, 2) explanation of the force capacity in bearingless motors when only two space harmonics are controlled, and 3) enhancement of the force capacity in bearingless motors by controlling multiple airgap space harmonics and comparison to AMB performance limits. Section 2. reviews pole- and space harmonic-based force models and shows relations between controllable harmonic fields, current sequences, and forces. Section 3. uses the harmonic-based model to explain force capacity in AMBs (contribution 1). Section 4. explains torque-force capability and presents force enhancement potential using multiple harmonics (contributions 2 and 3).

2. Force Creation from the Perspective of Airgap Space Harmonic Fields

This section reviews the force creation in AMBs and bearingless motors based on the airgap space harmonic field interactions presented by the authors in Khamitov & Severson (2022) and identifies the relationship with the conventional method of calculating force in magnetic bearings. Section 2.1 reviews the force vector models and Section 2.2 presents the relationship between airgap harmonics, current sequences, and controllable force vectors. The proposed model (2) is used in subsequent sections to calculate the force capacity in AMBs and bearingless motors.

*This work was supported in part by the USA National Science Foundation under Grant #1942099.

2.1 Pole- and Harmonic-Based Force Vector Models

The conventional method of calculating force in AMBs is based around a magnetic equivalent circuit, where the total force is the vector sum of the forces created by individual poles/teeth Maslen & Schweitzer (2009). This model assumes that the field is normal to the tooth surface and is constant over the tooth span. The cross-section of a general AMB with one coil per pole is depicted in Fig.1a. Each pole i can create an attraction force of $F_i = \frac{A}{2\mu_0}B_i^2$, which depends on the square of flux density B_i in tooth i and the tooth surface area A facing the airgap. Under saturation operation, the pole field has a linear relationship with the pole current. When there are n_p equally spaced poles with the same surface area A as in Fig. 1a, the total created total force vector is (1), where $a^{i-1} = e^{\mathrm{j}(i-1)2\pi/n_p}$ indicates the force direction created by pole i.

$$\vec{F} = Fe^{\mathrm{j}\phi} = \frac{A}{2\mu_0}\left(B_1^2 + aB_2^2 + a^2B_3^2 + \ldots + a^{n_p-1}B_{n_p}^2\right) = \frac{A}{2\mu_0}\sum_{i=1}^{n_p} a^{i-1}B_i^2 \tag{1}$$

An AMB can be equivalently viewed as a multiphase winding with m drive connections (m independent currents), as shown in Fig. 1b. This equivalence allows applying the theories used to analyze multiphase windings in bearingless machines (Generalized Clarke Transformation, winding function theory) and studying force creation in AMBs from the perspective of airgap space harmonic fields. Instead of individually viewing each tooth field, the total airgap field can be studied in terms of its spatial harmonic content along the airgap angle α (see Fig. 1c). Each harmonic of order h is expressed as $B_{\mathrm{n},h}(\alpha) = \hat{B}_{\mathrm{n},h}\cos(h\alpha - \phi_h)$ with a magnitude $\hat{B}_{\mathrm{n},h}$ and an angular location ϕ_h/h, and the total field is the sum of all space harmonics present in the airgap. As an example, Fig. 3b shows a plot of the total airgap field and its space harmonics of orders 1-6 along α. Using harmonic field equations and the Maxwell Stress Tensor, it can be shown that the force is created from the interaction between adjacent harmonics h_i and $h_j = h_i + 1$. This is analogous to the force creation in bearingless machines due to torque p and suspension $p_s = p \pm 1$ pole-pairs. When multiple space harmonics are present, the total created force $\vec{F}$ is the vector sum of the forces due to each pair of adjacent space harmonics:

$$\vec{F} = \sum_{h_i=1}^{n_h-1}\vec{F}_{h_ih_j} = k\sum_{h_i=1}^{n_h-1}C_{h_{ij}}\vec{b}_{h_i}^{*}\vec{b}_{h_j}, \quad \text{where } k = \frac{V_r}{2\mu_0 r} \text{ and } C_{h_{ij}} = \left(1 - h_i\frac{\delta_{\mathrm{eff}}}{r}\right)\left(1 + h_j\frac{\delta_{\mathrm{eff}}}{r}\right) \tag{2}$$

where $\vec{b}_{h_i} = \hat{B}_{\mathrm{n},h_i}e^{\mathrm{j}\phi_{h_i}}$ is a vector that represents the field harmonic h_i, k is a constant in terms of the rotor volume V_r and radius r, and $C_{h_{ij}}$ is a factor considering the effect of the tangential field components (δ_{eff} is the effective airgap length). $C_{h_{ij}} \approx 1$ when $h_{i/j}\delta_{\mathrm{eff}} \ll r$. Some of the harmonics in (2) can be directly manipulated through the control of the equivalent current space vectors/current sequences (more details on this in Section 2.2).

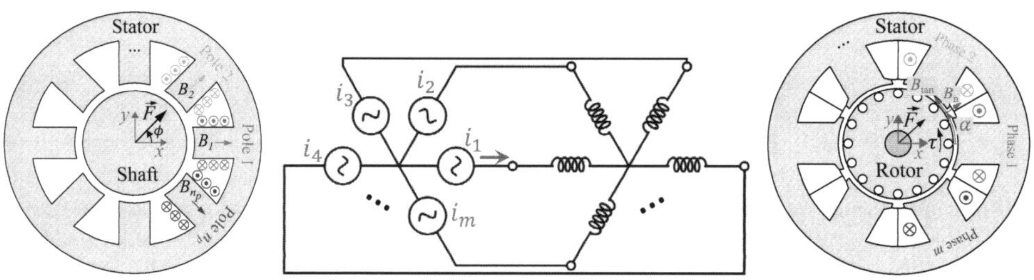

(a) (b) (c)

Figure 1: Cross-sections and definitions: (a) an example AMB (× and • indicate the coil direction into and out of the page, fields are shown for positive currents), (b) multiphase winding circuit diagram, and (c) an example bearingless motor with the definition of axes, fields components, and vectors.

2.2 Relationship to Current Sequences

The concept of current sequences is convenient when studying airgap harmonic fields and forces created from multiphase winding currents. This allows identifying the spatial harmonic content, behavior of created harmonics (rotating or oscillating), and the relationship between the number of independently controllable harmonics and force vector components. This subsection discusses these aspects for AMBs and illustrates using examples. More details on the relationship between harmonics and current sequences are presented in Nishanth et al. (2022) and Khamitov & Severson (2022).

The number of controllable currents m directly determines the number of independently controllable harmonics and their spatial orders. These harmonics can be independently controlled by current sequences. Any set of m-phase currents $\boldsymbol{i} = \begin{bmatrix} i_1 & i_2 & \ldots & i_j & \ldots & i_m \end{bmatrix}^T$ can be decomposed into multiple current sequence components. Phase currents consist of multiple sequences $i_j = i_{j,1} + i_{j,2} + \ldots + i_{j,s} + \ldots$, where a sequence of order s is defined as $i_{j,s} = \hat{I}_s\cos\left(\phi_s - (j-1)s\frac{2\pi}{m}\right)$ with amplitude $\hat{I}_s$, phase angle ϕ_s, and phase separation $s\frac{2\pi}{m}$. Sequence s simultaneously creates harmonics of orders

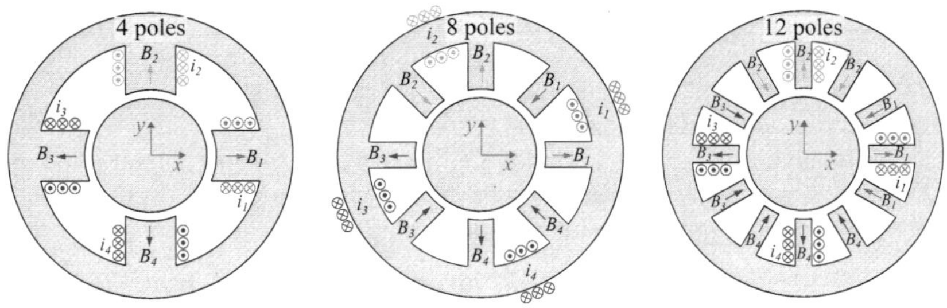

Figure 2: AMBs with 4, 8, and 12 poles (to apply electric motor winding analysis methods, these AMBs can be viewed as each pole wound by a coil).

$h = \pm s + mb$, where b is an integer number. Certain sequences can only control the magnitude of a harmonic (oscillating field), while other sequences can control both the magnitude and angle (rotating field):

$$s = \pm 1,\ \pm 2,\ ...,\ \pm n_{\text{ind}} \text{ control rotating fields, } n_{\text{ind}} = \begin{cases} \frac{m-2}{2} \text{ for } m \in \mathbb{N}_{\text{even}} \\ \frac{m-1}{2} \text{ for } m \in \mathbb{N}_{\text{odd}} \end{cases} \qquad s = 0,\ \frac{m}{2} \text{ control oscillating fields} \tag{3}$$

where n_{ind} is the total number of controllable rotating harmonics. AMBs with $m \in \mathbb{N}_{\text{even}}$ can control two oscillating and $n_{\text{ind}} = \frac{m-2}{2}$ rotating harmonics, while AMBs with $m \in \mathbb{N}_{\text{odd}}$ can control one oscillating and $n_{\text{ind}} = \frac{m-1}{2}$ rotating harmonics.

The force vector model (2) in terms of current quantities can be used to identify the number of controllable force vectors. Using the Generalized Clarke Transformation, current space vector representations of current sequences can be obtained as $\vec{\boldsymbol{i}} = \boldsymbol{C_m i}$ (Wilamowski & Bogdan 2011), where $\vec{\boldsymbol{i}} = \begin{bmatrix} \vec{i}_1 & \vec{i}_2 & ... \end{bmatrix}^T$ is an array of all space vectors. Using the relation between harmonic fields and current sequences, the space vector form of (2) is

$$\vec{F} = \sum_{i=1}^{n_f} \bar{k}_{q,h_{ij}} \vec{i}_i^* \vec{i}_j \tag{4}$$

where $\bar{k}_{q,h_{ij}}$ is the force per ampere squared and n_f is the number of controllable force vectors. This number increases with respect to the number of phases m, which allows for more granular control over the suspension force and potential increase in specific load capacity. To control an individual force vector component $\vec{F}_{h_i h_j} = kC_{h_{ij}} \vec{b}_{h_i}^* \vec{b}_{h_j} = \bar{k}_{q,h_{ij}} \vec{i}_i^* \vec{i}_j$, at least one magnitude and one angle out of four quantities ($\hat{B}_{h_i}$, $\hat{B}_{h_j}$, ϕ_{h_i}, ϕ_{h_j}) must be controllable. This means one harmonic must be able to rotate, while the other harmonic can have a rotating or oscillating behavior. There are $n_{\text{ind}} - 1$ force vectors created independently when controlling all n_{ind} adjacent rotating harmonics. For an even m, an additional force vector can be created when the magnitude of an oscillating harmonic is controlled by sequence $s = 0$ or $\frac{m}{2}$. For an odd m, there is also an additional force vector created by a single sequence of the highest order $s = \frac{m-1}{2}$. This sequence creates two adjacent harmonics of the orders $h_i = s$ and $h_j = -s + m = h_i + 1$, which rotate in opposite directions ($\phi_{h_j} = -\phi_{h_i}$). This explains the way force is created in a three-pole AMB, where harmonics $h_i = 1$ and $h_j = 2$ create force $\vec{F}_{h_i h_j} = kC_{h_{ij}} \hat{B}_{h_i} \hat{B}_{h_j} e^{-2\phi_{h_i}}$. In summary, for any m number of independent winding currents, there are $n_f = n_{\text{ind}}$ independently controllable force vectors.

Consider force creation in 4-, 8-, and 12-pole AMB examples in Fig. 2, where all AMBs have $m = 4$ independent currents. From (3), one rotating harmonic ($n_{\text{ind}} = 1$) from sequence $s = 1$ and one oscillating harmonic from $s = 0$ or $s = m/2$ can be controlled in these AMBs. Controlling one rotating and one oscillating harmonic allows controlling $n_f = 1$ force vector component. The dominant force creating harmonics depend on the winding layout of these AMBs (number of poles per drive connection). Similar to the winding analysis in electric motors, these dominant harmonics can be identified based on their winding factors. The winding factor $\hat{k}_{\text{w},h}$ can take values from zero (harmonic h cannot be created) to one (harmonic h is one of the dominant harmonics). The first dominant harmonic has order $h = n_p/2$ (oscillating), which is created when the subsequent poles in Fig. 2 have alternating magnetic field polarities. In 4- and 12-pole AMBs, this is achieved with $s = 2$ when currents are 180° phase separated (when subsequent currents are equal but opposite). However, in 8-pole AMBs, $h = n_p/2 = 4$ is created by $s = 0$ (all currents are equal). The second dominant harmonic has order of $h = n_p/2 - 1$ (rotating). This harmonic may not be trivial to visualize in 8- and 12-pole AMBs because some of its peaks are not located in front the poles. From the relation $h = \pm s + mb$, this harmonic is created by sequence $s = 1$ in 4- and 12-pole AMBs and by $s = -1$ (negative sequence) in 8-pole AMBs. Since one sequence can create multiple harmonics, sequence $s = \pm 1$ also creates harmonic $h = n_p/2 + 1$ in addition to $h = n_p/2 - 1$, which rotate in opposite directions. If the force equation (2) models one harmonic per sequence (harmonic with the highest magnitude), the interaction between $h = n_p/2 + 1$ and $h = n_p/2$ would be ignored, resulting in force vector error. However, since the control in AMBs is based on individual pole currents instead of particular harmonics, all harmonics (including higher order harmonics) are inherently considered in the force model.

3. Force Capacity of AMBs from the Perspective of Multiple Airgap Harmonic Fields

This section explains the force capacity/design sizing of AMBs from the perspective of multiple airgap space harmonics. This new explanation approach provides insight into identifying and controlling components of the total force vector from each harmonic interaction. The proposed approach to calculate force capacity is explained and the results are presented.

The force vector model in (2) can be used to determine the specific load capacity f_c. The procedure is as follows:

1. Find the maximum force magnitude $F_{\text{max}}(\phi)$ that an AMB can create at every force angle ϕ, while keeping the peak airgap field $\max\left[|B_{\text{n}}(\alpha)|\right] = \max\left[\left|\sum_{h_i=1}^{n_{\text{ind}}} \hat{B}_{h_i}\cos(h_i\alpha - \phi_{h_i})\right|\right]$ below the maximum airgap field B_{max}.
2. Calculate the force capacity as $f_c = F_{\text{rated}}/DL$, where $F_{\text{rated}} = \min\left[F_{\text{max}}(\phi)\right]$ is the maximum force that an AMB can create at any force angle ϕ without exceeding the airgap field limit.

To make the force rating and force capacity results applicable to any DL and B_{max} values, a dimensionless form of the force vector model can be used in the above calculations. By setting the base value to $F_{\text{base}} = kB_{\text{max}}^2$, where k is defined in (2), the normalized form of (2) becomes $\vec{F}' = \sum_{i=1}^{n_h-1} \vec{b}_{h_i}'^{*}\vec{b}_{h_j}'$ (assuming $C_{h_{ij}} \approx 1$). Here, each field quantity is normalized by B_{max}. After calculating the normalized rated force F'_{rated} using this model, the force capacity can be calculated as

$$f_c = \frac{kB_{\text{max}}^2 F'_{\text{rated}}}{DL} = \frac{\pi}{4\mu_0} B_{\text{max}}^2 F'_{\text{rated}} \tag{5}$$

This result shows that the force capacity does not depend on DL, but instead depends on the square of the maximum airgap field B_{max} and the harmonics present in the airgap.

This paper shows results of force capacity calculations for AMBs considering only controllable rotating harmonics n_{ind}, resulting in $2n_{\text{ind}}$ control variables. The calculation of the force capacity for $n_{\text{ind}} \geq 2$ has to be solved as an optimization problem as infinitely many solutions exist. Furthermore, the problem has to be solved numerically due to nonlinear objective and constraints. When controlling two rotating harmonics h_i and h_j, the force magnitude and angle of (2) are calculated as $F = kC_{h_{ij}}\hat{B}_{h_i}\hat{B}_{h_j}$ and $\phi = \phi_{h_j} - \phi_{h_i}$. Following the procedure of calculating the specific load capacity, it is found that the rated force of $F'_{\text{rated}} = 0.32$ p.u can be obtained. Substituting this in (5) gives $f_c = 45$ N/cm^2 for $B_{\text{max}} = 1.5$ T (12.8 N/cm^2 for $B_{\text{max}} = 0.8$ T). This result is expected because the force capacity reported in the AMB literature is approximately 40 N/cm^2. Similarly, the force capacity for more than two independent harmonics can be calculated. Figure 3a summarizes the force capacity results for different numbers of harmonics when $B_{\text{max}} = 1.5$ T. This plot shows that an approximately 31% increase in the force capacity can be obtained when increasing n_{ind} from 2 to 6, which is equivalent to increasing the number of poles from 6 to 14 according to (3). As an example, Fig. 3b shows the optimal distribution of the field harmonics when $n_{\text{ind}} = 6$ ($n_p = 14$) to create maximum force along the x-axis. The blue dots indicate the locations of the pole centers. In this case, all harmonics have their peaks located at $\alpha = \phi_{h_i}/h_i = 90^\circ/h_i$, resulting in the total field (indicated with a solid black line) that has better iron utilization. Interestingly, Fig. 3b shows that poles 1 and 8 that are located along the force direction (indicated with a red dashed line at $\alpha = 0$) have zero fields, while three poles on either sides of the force direction (2-4 and 12-14) have the highest field values.

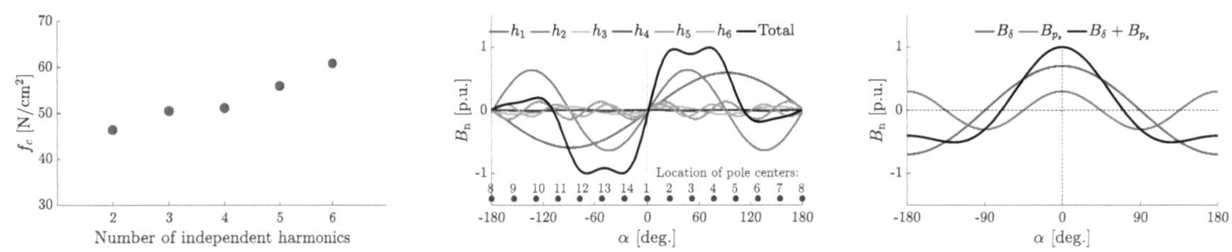

(a) (b) (c)

Figure 3: (a) Results of force capacity analysis in AMBs for different numbers of independent harmonics n_{ind}, (b) optimal airgap field distribution in AMBs for $n_{\text{ind}} = 6$ and $\phi = 0^\circ$, and (c) optimal airgap field distribution in bearingless machines for $n_{\text{ind}} = 2$ and $\phi = 0^\circ$.

4. Force Capacity in Bearingless Motors

This section presents the second and third contributions of the paper. Section 4.1 explains force capacity in bearingless motors when only two space harmonics are controlled, identifies optimal magnetizing field values, and provides an analytical estimation of the torque-force capability using the p.u. model for different magnetizing field values. Section 4.2 proposes potential force capacity enhancement by controlling multiple harmonics. As the primary contribution of this paper, this section finds that, enhancement of force capacity in bearingless machines equivalent to that of AMBs can be achieved through the control of multiple airgap harmonics.

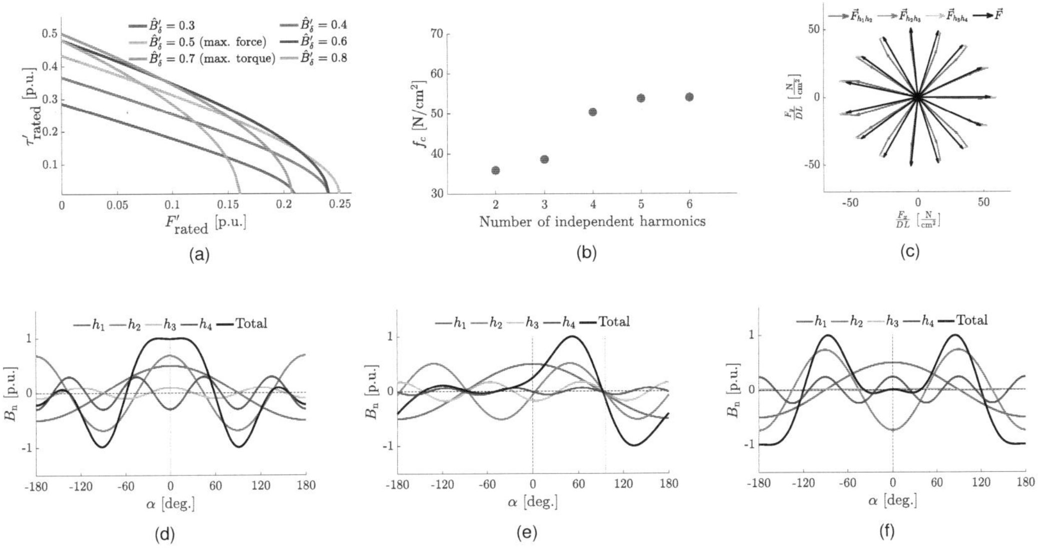

Figure 4: Analysis of force enhancement in bearingless motors: (a) rated torque-force curves for different magnetizing field values when $n_{\text{ind}} = 2$, (b) force capacity vs. number of independent harmonics n_{ind} when $B_{\text{max}} = 1.5$ T and $\hat{B}_\delta = 0.75$ T, (c) optimal force vector components when $n_{\text{ind}} = 4$, and (d)-(f) optimal airgap field distributions for $n_{\text{ind}} = 4$ and $\phi = 0^\circ,\ 90^\circ,\ 180^\circ$.

4.1 Torque-Force Capability from Two Airgap Harmonics

The standard method of creating torque and suspension force in bearingless motors is through controlling p and p_s pole-pair fields. The p pole-pair field consists of a torque creating (q-axis) and a magnetizing field (d-axis) component B_δ, which impacts both torque and force creation. In addition to the force capacity calculation steps described in Section 3. , the magnetizing field harmonic p must have a fixed amplitude at $\hat{B}_\delta$ and an angular location at the rotor rotational angle θ. This subsection identifies the optimal magnetizing field values $\hat{B}_\delta$ when there are two harmonics p and p_s. The optimal magnetizing field is defined as the value creating the maximum force or torque without exceeding the airgap field limit.

First, the optimal $\hat{B}_\delta$ to create the maximum force is identified. With two harmonics in the airgap, the maximum force that can be created at any angle has the magnitude of $F'_{\text{rated}} = \hat{B}'_\delta \hat{B}'_{p_s}$, according to the normalized form of (2). The peak airgap field must not exceed B_{max}, which in p.u. is $\hat{B}'_\delta + \hat{B}'_{p_s} = 1$, as shown in Fig. 3c. Substituting $\hat{B}'_{p_s} = 1 - \hat{B}'_\delta$ gives $F'_{\text{rated}} = \hat{B}'_\delta(1 - \hat{B}'_\delta) = -\hat{B}'^2_\delta + \hat{B}'_\delta$. This is an inverted parabola with the peak $F'_{\text{rated}} = 0.25$ p.u. at $\hat{B}'_\delta = 0.5$ p.u., which gives 35.2 N/cm^2 at $B_{\text{max}} = 1.5$ T and 10 N/cm^2 at $B_{\text{max}} = 0.8$ T. AMBs have approximately 28% higher force capacity at these field values (45 N/cm^2 and 12.8 N/cm^2, see Section 3.) because AMBs do not have the requirement to create the magnetizing field. Now, the optimal $\hat{B}_\delta$ to create the maximum torque is identified. The maximum torque is created when torque creating and magnetizing field components of harmonic p are 90° phase separated electrically, so that $\hat{B}'^2_\delta + \hat{B}'^2_\tau = 1$. The torque is expressed in terms of the normal field components as $\tau_{\text{rated}} = k_\tau \hat{B}_\delta \hat{B}_\tau$, where $k_\tau = V_r p \delta_{\text{eff}}/\mu_0 r$. Defining the base torque value as $\tau_{\text{base}} = k_\tau B^2_{\text{max}}$ (which includes fixed parameters), the normalized form of the torque expression becomes $\tau'_{\text{rated}} = \hat{B}'_\delta \hat{B}'_\tau = \hat{B}'_\delta(1 - \hat{B}'^2_\delta)^{1/2}$. Solving $d\tau'_{\text{rated}}/d\hat{B}'_\delta = 0$, the optimal magnetizing field is found as $\hat{B}'_\delta = 1\sqrt{2} = 0.707$ p.u. with the maximum $\tau'_{\text{rated}} = 0.5$ p.u. The above calculations show that the optimal magnetizing fields for the maximum force ($\hat{B}_\delta = 0.5$ p.u.) and torque ($\hat{B}_\delta = 0.707$ p.u.) creation are different.

The above analysis can be extended to the case when maximum torque and suspension forces are created simultaneously. In this case, the peak airgap field is constrained by $(\hat{B}'^2_\delta + \hat{B}'^2_\tau)^{1/2} + \hat{B}'_{p_s} = 1$. Using this expression, $F'_{\text{rated}} = \hat{B}'_\delta \hat{B}'_{p_s}$, and $\tau'_{\text{rated}} = \hat{B}'_\delta \hat{B}'_\tau$, the relationship between the rated force, torque, and magnetizing field can be derived:

$$\tau'^2_{\text{rated}} = F'^2_{\text{rated}} - 2\hat{B}'_\delta F'_{\text{rated}} + \hat{B}'^2_\delta(1 - \hat{B}'^2_\delta) \tag{6}$$

This equation can be used to determine torque-force capabilities of a bearingless machine for different magnetizing fields, as demonstrated in Fig. 4a. This plot shows that the optimal range of $\hat{B}'_\delta$ is between approximately 0.5-0.707 p.u. depending on the desired range of τ'_{rated} and F'_{rated}. These results can be used to select a magnetizing field value based on the torque and force requirements when designing a bearingless motor.

4.2 Force Enhancement Using Multiple Airgap Harmonics

This section shows that enhancement of the force capacity can be achieved in bearingless motors by controlling more than two rotating space harmonics. To find the force capacity, a similar approach to that described in Section 3. is used with the additional constraints on the magnetizing field harmonic to have fixed magnitude and angle at any force angle ϕ. This reduces the number of degrees of freedom to $2n_{\text{ind}} - 2$ for bearingless motors as opposed to $2n_{\text{ind}}$ in AMBs.

Analysis results are now presented by comparing bearingless motors with different numbers of controllable airgap harmonics n_{ind}. The results are obtained for $p = 1$ and $\hat{B}_\delta = 0.5$ p.u., with the remaining space harmonics $h = 2, 3, 4, ...,$ n_{ind} being used to create force. Figure 4b shows the plot of the force capacity vs. n_{ind} for $B_{\text{max}} = 1.5$ T. This result shows that increasing the number of controllable harmonics from $n_{\text{ind}} = 2$ to $n_{\text{ind}} = 4$ can significantly increase force capacity (by approximately 42%). Figure 4c shows the optimal force vector components from each harmonic interaction required to create F_{rated} at different force angles ϕ for $n_{\text{ind}} = 4$. Figures 4d-4f show the optimal field distributions for this example when $\theta = 0^\circ$ and $\phi = 0^\circ$, 90°, and 180°. An interesting trend is observed from the optimization results: the harmonic peaks are typically aligned or opposite when creating forces at angles 0° and $\pm 180^\circ$, and they all have one intersection point when creating forces at angles $\pm 90^\circ$. Furthermore, at certain angles, additional harmonics are used to improve the total airgap field distribution, but not to contribute to the total force creation. This can be observed in Fig. 4f where the third harmonic is zero and the fourth harmonic is used to reduce the peak airgap field, helping to increase the force capacity.

These results show that, by manipulating the magnitudes and angles of multiple harmonics, force enhancement comparable to AMBs can be achieved, while keeping the peak airgap field below the maximum limit.

5. Conclusion

This paper researches force capacity of AMBs and bearingless motors using multiple airgap space harmonic field interactions. It is found that the key factors affecting the force capacity are the peak allowable airgap field and the distribution of the harmonic fields along the inner bore of the stator. The paper finds that, for the same number of controllable currents, bearingless machines generally have lower force capacity than AMBs due to two main reasons. First, AMBs inherently use more harmonics by individually controlling currents. Second, AMBs do not have the requirement to create the magnetizing field harmonic. This increases the degrees of freedom to find the optimal field distribution in AMBs because harmonic magnitudes and angles do not have to be fixed with respect to the force and rotor angles. It has been identified that the optimal range of the magnetizing field magnitude to have the maximum torque-force capability is between 50% (max force) and 70% (max torque) of the peak airgap field. The paper also showed that, by increasing the number of controllable harmonics and optimally manipulating their magnitudes and angles, enhancement of force capacity in bearingless motors is achievable. The results demonstrate that potential force enhancement of over 40% can be achieved in bearingless machines when controlling four airgap harmonics (10 phases) as opposed to the typical approach of controlling only two harmonics (six phases). For example, with 0.8 T peak field, the force capacity of a bearingless motor can be enhanced from 10 N/cm^2 to 14.3 N/cm^2. Overall, this paper proposes a comprehensive framework for analysis of force creation capabilities in AMBs and bearingless motors, and demonstrates the importance of controlling multiple airgap harmonics for achieving high-performance bearingless motor designs.

References

Chen, J., Zhu, J. & Severson, E. L. (2019), 'Review of bearingless motor technology for significant power applications', *IEEE Transactions on Industry Applications* **56**(2), 1377–1388.

Chiba, A., Fukao, T., Ichikawa, O., Oshima, M., Takemoto, M. & Dorrell, D. G. (2005), *Magnetic bearings and bearingless drives*, Elsevier.

Jastrzebski, R. P., Putkonen, A., Kurvinen, E. & Pyrhönen, O. (2021), 'Design and modeling of 2 mw amb rotor with three radial bearing-sensor planes', *IEEE Transactions on Industry Applications* **57**(6), 6892–6902.

Khamitov, A. & Severson, E. L. (2022), Exact torque and force model of bearingless electric machines, *in* '2022 IEEE Energy Conversion Congress and Exposition (ECCE)', IEEE, pp. 1–8.

Maslen, E. H. & Schweitzer, G. (2009), *Magnetic bearings: theory, design, and application to rotating machinery*, Springer.

Mushi, S. E., Lin, Z. & Allaire, P. E. (2011), 'Design, construction, and modeling of a flexible rotor active magnetic bearing test rig', *IEEE/ASME transactions on mechatronics* **17**(6), 1170–1182.

Nishanth, F., Khamitov, A. & Severson, E. L. (2022), Design of multiphase motor windings for control of multiple airgap fields, *in* '2022 IEEE Energy Conversion Congress and Exposition (ECCE)', IEEE, pp. 1–8.

Swanson, E. E., Maslen, E. H., Li, G., Cloud, C. H. et al. (2008), Rotordynamic design audits of amb supported machinery, *in* 'Proceedings of the 37th Turbomachinery Symposium', Texas A&M University. Turbomachinery Laboratories.

Wilamowski, A. & Bogdan, M. (2011), 'The industrial electronics handbook power electronics', *Taylor and Francis Group* .

2nd Review of Developments in Bearingless Motors

Hubert MITTERHOFER [a], Akira CHIBA[b], Wolfgang GRUBER[c]
a Linz Center of Mechatronics, Altenberger Strasse 69, 4040 Linz, Austria, Hubert.mitterhofer@lcm.at
b Tokyo Institute of Technology, 2-12-1 Ookayama, Meguro, Tokyo, Japan
c Johannes Kepler University Linz, Altenberger Str. 69, 4040 Linz, Austria

Abstract

The present publication gives an overview about bearingless motor research by means of literature study, following the objective of enabling researchers in the field to develop a good overview. To that aim, the literature research methods offered by modern search engines and online repositories are not only applied but also explained. The actual overview about the technological developments is provided indirectly through the review of diverse overview publications with excerpts and references to especially noteworthy findings that contribute to a holistic understanding of the bearingless motor technology. From the more recent literature, observed trends are listed and described. The concluding discussion deals with the question on how to learn from the literature basis and reflects the author's opinions on rewarding future research questions.

Keywords: *Bearingless motors, development review, overview publications, terminology*

1. Introduction

The history of bearingless motors dates back to the 1970's and 1980's when the first ideas have been proposed and first research projects were started. An early publication titled *"A Review of Developments in Bearingless Motors"* (Salazar, et al., 2000) appeared at the International Symposium on Magnetic Bearings (ISMB) in the year 2000. That publication outlined the different types of bearingless motors, clarified technical notations and listed a substantial amount of citations on the – back then – relatively small total amount of bearingless motor literature. As such, the paper has proved to be very useful for many researchers in the field and was, thus, frequently referred to in following research works. With more than two decades having passed, it is not trivial to provide a renewed overview of the developments in the field, as the research activities have increased tremendously. Following the titling of the initial paper, the authors of this paper intend to present an updated review.

It would be scientifically improper to only review the actual publications on bearingless motors themselves, without also reviewing the numerous review and overview publications that have been published on diverse motor types, power ranges, or rotor topologies. After an initial reflection of the publication statistics and terminology in Section 2, Section 3 will, therefore, deal with existing review articles. Some trends that have been identified in the more recent literature are presented in Section 4. A discussion on the observed obstacles on the way to a broader industrial application of bearingless motors on the one hand and future chances and rewarding research questions conclude this article.

2. Literature review

Developments like online search engines and the collection of magnetic bearing related research papers at dedicated web portals, e.g. (magneticbearings.org, 2023) reduce the purpose of a review publication for just *finding* other research papers. However, the now available, powerful search methods can be used to determine interesting perspectives on the literature and its genesis, which will be described in the following paragraphs.

Terminology

Especially in the beginning of research activities, different terminologies have been used for the same idea:

Bearingless motors

The first documented usage of the term "bearingless" can be found in (Bosch, 1988). The genesis of the term can be attributed to the term "brushless motor", where the brushes in DC motors are replaced and their functionality is integrated by field oriented control. The same is true for a "bearingless motor", where the mechanical bearings are replaced and their functionality is integrated by suspension current control or combined current control in separated or combined winding systems, respectively. The same analogy can be drawn to the word "sensorless", where the physical motor angle sensors are replaced and their functionality is integrated by signal processing of the motor's winding currents. Publications using this term have first occurred in 1988 and continue to mark the largest share of publications.

Self-bearing motor

The second most widely used term is "*self-bearing*" motor. It is mostly used by several dedicated research groups with wide activities in the field of blood pumps. The term gained popularity around the turn of the millennium and continues to be used in a smaller amount of papers today. It can be observed that the term is often used by researchers with a strong background in mechanical engineering as there is bearing function provided by electromagnetic effects and control of the mechanical motor device *itself*.

(Combined) motor-bearing

Inversing the idea that a motor has an integrated bearing functionality leads to a bearing with integrated motor functionality which motivates the term "motor-bearing" or "combined motor bearing". Most scientific publications titled with these terms have been published between 1999 and 2006.

Figure 1 shows the results of a literature search study using the Google scholar™ engine: The available database was searched for scientific publications (excluding patents) with the respective search term in the title. Several particularities during the search and the result analysis have been observed that may be helpful for other researchers:

- The term "bearingless" is frequently used, especially up to the year 2010, in publications on the mechanical construction of helicopter rotor hubs. In order not to distort the search results, relevant terms have been excluded and all results have been manually screened.
- The term "self-bearing" is also applied to describe self-supporting structures in the construction industry. It hence can be found on numerous publications on special construction technologies using steel, glass, or concrete. As with the term "bearingless", suitable measures have been taken to not distort the results.
- Publications using the term "motor bearing" have to be searched for with much care: Mostly, the meaning will relate to the (conventional) bearings of a motor, as, e.g., in "Lubrication failures in motor bearings".
- A certain amount of publications are not available online but can only be found as *citation*, i.e. the search result will indicate the citation but will not link an explicit document. On the one hand, this concerns many of the older publications, especially before the year 2000. On the other hand, certain repositories either do not explicitly hold the publication files or are not properly indexed by the used search engine. The most prominent ones are "Proceedings of the Chinese Society of Electrical Engineering", "Chinese Journal on Mechanical Engineering", "Advanced technology of electrical engineering and energy", or "Control Engineering of China". Beginning with 2010, a rising number of bearingless research papers appear with new open-access publishers, sharing the fate of the above-mentioned journals: "Dianli Dianzi Jishu / Power Electronics", "Control Theory & Applications", or "Wei-Te Dianji (Small & Special Electrical Machines). Also, some university journals have published substantial amounts of relevant publications that are not properly indexed, with the most prominent ones being: "Journal of Jiangsu University", "Journal of Sichuan University for Engineering Sciences", or "Dalian Jiaotong University".
- The frequent inability of non-Chinese editors to distinguish Chinese first names from last names leads to a certain confusion and to duplicate search results with identical publication title, e.g. once by authors "J. Gao and S. Huang", once by authors "G. Jian and H. Shoudao", each time referring to the same authors.
- As only the publication titles were searched, a number of publications may be missing in this Figure.

3. Previous review and overview work

Since the initial development review of (Salazar, et al., 2000), not only a lot of publications on bearingless motors themselves have been published but also several review and overview papers that have each selected specific sections of the available literature.

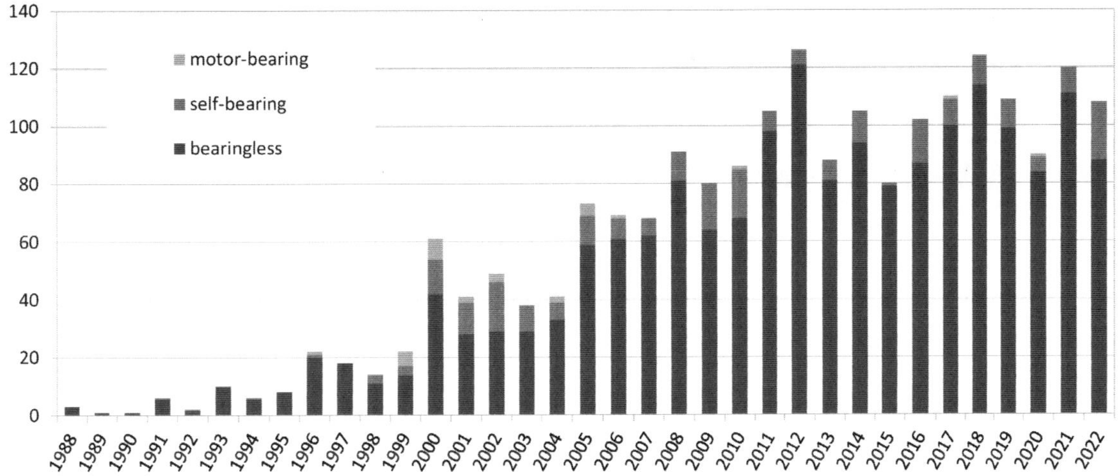

Figure 1 Number of scientific publications with a title holding "bearingless", "self-bearing", or "motor-bearing", respectively. The rising research activity and a certain dent in the Covid19-plagued year 2020 become clearly visible.

Review of bearingless induction motors (BLIM)

In 2014, the authors of (Sun, et al., 2014) describe the advantages and the most typical structures of BLIMs, e.g. the specific rotor short circuit rings which typically differ from induction motor rotors with conventional bearings. Multiple arrangement forms for the stabilization of different numbers of degrees of freedom (DOF) are summarized. The common mathematical models for torque and force calculation with separated winding systems and, subsequently, the most frequent control types such as vector control, direct torque control, and nonlinear decoupling control, are presented. The BLIM review paper closes by comparing the different control types and identifies advanced control for higher force dynamics, sensor integration, and generally increased reliability as the most important research trends.

Review of bearingless switched reluctance motors (BLSRM)

A first collection of BLSRM literature was presented and reviewed in 2012 by the authors of (Ahmed, et al., 2012). Besides a brief description of the working principle of a BLSRM, the authors review different design methods and published designs as well as control methods for BLSRM.

A very recent publication from 2023 (Xiang, et al., 2023) presents the typical coupling problem of standard 12/8 (denominating stator teeth / rotor teeth) BLSRM. In the following, the authors describe different decoupling – or rather, coupling-reduction – strategies. In order of appearance, these include hybrid stator topologies with dedicated windings on separated suspension force and torque teeth, different stator/rotor topologies e.g., 8/10 or 12/14 BLSRMs, asymmetric rotor pole shaping, permanent magnet bias flux injection in the stator, axially aligned dual stator BLSRMs, concentrically positioned dual stator BLSRMs (e.g., torque generation on the outside, force generation on the inside of the rotor) with or without permanent magnet bias flux excitation, wide tooth configurations for an increased angular window of force generation, and, ultimately, present a side-by-side arrangement of two radial active magnetic bearings (AMBs) with a BLSRM in the axial middle, whose motor winding extends onto the AMB stators. Comparisons of the behavior, the main characteristics in terms of torque/force decoupling and torque ripple, and general advantages and disadvantages are listed in an impressive appendix which can be recommended for a broad overview of the BLSRM state of the art. The concluding assessment on the industrial readiness of the BLSRM technology highlights the need for further research in topologies and control since the generically robust and potentially permanent magnet-free BLSRMs have not yet found their way into actual applications.

Review of bearingless synchronous motors (BLSM)

An early overview work on BLSM was presented in 2012 (Sun, et al., 2012) in Chinese language. A later, refined version was published in English (Sun, et al., 2013). Both papers give an overview about the force generation depict the different arrangement options of shafted bearingless motors in an electric machine assembly as shown in *Figure 2*.

The authors also give an overview about the state of the art of BLSM control methods including artificial intelligence-based methods, which seems remarkable, given the early publication year.

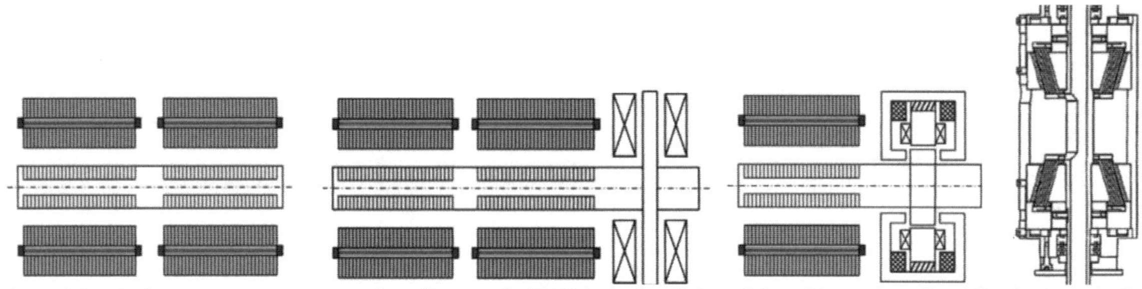

Figure 2 Bearingless motor arrangements from (Sun, et al., 2012) *including (left to right, with number of stabilized DOFs): dual side-by-side BLSM - 4DOF, dual side-by-side BLSM and active axial bearing - 5DOF, single BLSM with combined radial axial active bearing - 5DOF, dual conical BLSM - 5DOF. Copyright courtesy of original publication authors.*

Very recently, the authors of (Pei, et al., 2022) have released a renewed review of BLSM developments. As an interesting means of characterization, the authors divide the different BLSM types according to the nature of the magnetic excitation field (see *Figure 3*, reproducing the image contents presented in (Pei, et al., 2022)):

Passive BLSM do not refer to classic passive permanent magnetic bearings but to bearingless motors with additional electrodynamically stabilized DOFs, i.e. the stabilizing field is generated in a passive manner but still due to the operation of the motor.

Rotational heteropolar active BLSM denotes the major group of BLSM where the suspension force generating magnetic excitation field is heteropolar and rotating.

Stationary heteropolar or homopolar active BLSM includes the more exotic category of BLSM where the excitation field for force creation is non-rotating, either because it is homopolar or because the motor topology allows a stationary heteropolar field.

After introducing this categorization, an impressive literature overview is presented, including numerous figures from representative publications. It is especially these motor cross sections, drawings, and schematics that turn (Pei, et al., 2022) into an excellent summary and orientation point for the BLSM topologies to explain the differences and advantages of the individual concepts. A long list of prominent literature for deepened study is equally provided. The authors also give an interesting comparison of the different concepts, linking the number of stabilized DOFs and the output power and speed. It is noticeable that the 4DOF rotating heteropolar BLSM category (side-by-side BLSM) is dominant in both features.

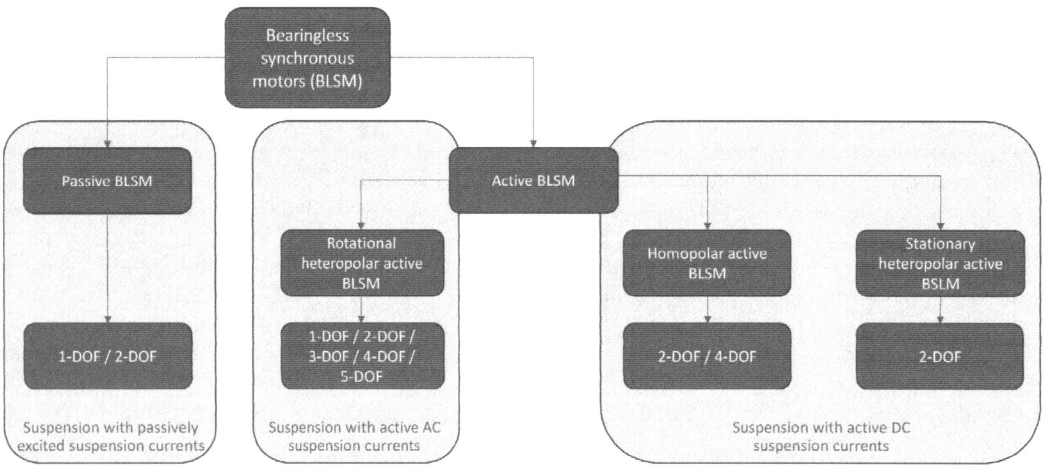

Figure 3 Topologic categorization of BLSM, reproduced from (Pei, et al., 2022). *From left, three groups are used to classify the existing BLSMs: Passive BLSMs, rotational heteropolar active BLSMs, and homopolar active or stationary heteropolar BLSMs.*

Review of bearingless motors for high output power

The authors of (Chen, et al., 2019) have performed an impressive literature review, identifying over 130 selected publications on bearingless motors. The focus for selecting the cited and reviewed publications was placed on presented experimental results, as the aim of the publication itself was to identify the most suitable bearingless designs for realizing increased output power. After listing the known applications of magnetic levitation in the fields of transportation and industry, the provided references for bearingless systems are categorized regarding several different aspects, which makes the paper a valuable piece of information for bearingless motor researchers:

The authors have screened and grouped previous works on all major bearingless motor categories, being: BLIM, BLSM, BLSRM, bearingless consequent pole, bearingless homopolar, bearingless flux switching, and bearingless hysteresis

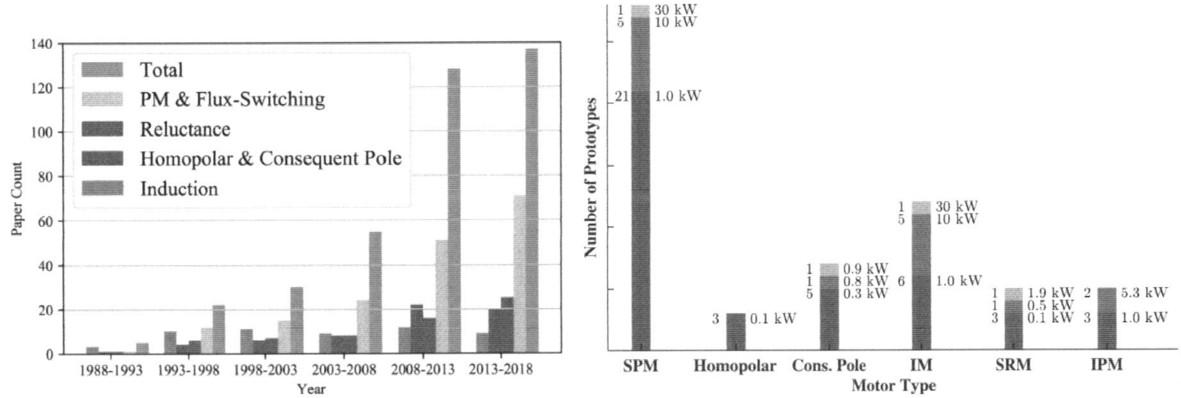

Figure 4 Categorization of existing literature on bearingless motor prototypes from (Chen, et al., 2019)*: Distribution to the main topologies (left), Numbers of realized prototypes including their demonstrated output power per topology (right). Copyright permission obtained from IEEE-CCC.*

motors and have provided an overview about the number of experimentally demonstrated designs for each motor topology (cf. *Figure 4*, left image). The figure clearly shows the dominance of the permanent magnet based BLSM both in absolute numbers as well as in growth rate over time. Lateron, the power of the prototypes found in the literature was also mapped onto the motor topologies, as shown in *Figure 4*, right image. Again, the BLSM dominates (especially when adding up the surface mounted (SPM) and interior permanent magnet (IPM) types), with the BLIM being second-ranked in this quantitative assessment for high-power suitability.

As in (Pei, et al., 2022), also the authors of (Chen, et al., 2019) investigate the number of actively controlled degrees of freedom; They describe the topologies used for 0-DOF (using electrodynamic forces) up to 5-DOF bearingless motors and reference the existing literature, again with a focus on experimentally supported publications. The 2-DOF topology with one bearingless motor and a combined radial-axial AMB and the 4-DOF topology with two bearingless motors and an axial AMB are again found to be the most promising ones for high output power. Consequently, the available data base was used for performance comparisons according to the found efficiencies, output power levels, attained speed regions, rotor diameters, and torque densities per rotor volume. The result of this impressive task again shows that the BLSM (in its SPM, IPM, sliced and shafted variants) is the most performant and hence, the most suitable topology for high output powers. A following parametric geometry optimization for BLSMs and BLIMs with either combined or separated winding systems with the optimization targets of efficiency and torque density per rotor volume confirms that the BLSM with combined windings is most suitable.

Review of bearingless slice motors

A special geometric topology is under review in (Gruber, 2013) and (Gruber & Silber, 2016): Bearingless *disk* or *slice* motors which have been reported in the literature as early as 1995 (Barletta & Schoeb, 1995) with the principle of passively stabilized disk rotors even dating back to 1994 (Bleuler, et al., 1994). Due to their short axial length, these shaftless motors are stabilized passively in axial and tilting directions. With only two necessary DOF left to be stabilized actively, disk motors are amongst those bearingless motors that can be stabilized completely without a second bearingless motor or additional AMBs. The first review publication, (Gruber, 2013), presents an overview about the most common stator and rotor compositions, including a review of motors with separated and combined winding systems. After highlighting the different motor forms, e.g. segment motors for large diameter systems, compact slotless designs for high speed applications or high-pole count variants for torque motors, the industrial applications of bearingless slice motors – mainly in medical and semiconductor industries – are referenced. As the paper was published in 2013, high purity bearingless disk motor fans can be added as newly appeared product (Levitronix, 2023). The last chapter is dedicated to bearingless disk motors with permanent magnet-free rotors, enabling cheap disposables for sensitive applications.

The later review paper (Gruber & Silber, 2016) first presents an overview about the slice motor topologies and depicts a timeline of reported developments with experimental results, structured in designs with separated and combined winding sets. Then, a comprehensive overview is given in the form of comparison tables, listing 50 different prototypes, structured according to performance characteristics such as maximum speed or achieved force and torque coefficients, geometric or magnetic dimensions such as air gap width, rotor length and diameter, and pole count, and electric system parameters such as winding scheme, DC link voltage, or rated current. holds extensive tables, comparing test results presented in the literature. A graphic representation of the achieved surface speeds, suspension force densities, and shear stress densities (as a measure of torque density) conclude the paper.

4. Trends and novelties in bearingless machines

As shown in *Figure 1* and presented in the initial section, the research in bearingless machines has stabilized at a high level. The following Table 1 is dedicated to observed trends and topologic novelties from the past years in and around the bearingless motor research.

Table 1 Selection of trends in bearingless machine research

Observed trend	*Description*	*Literature references (exemplary)*
Non-conventional levitation technology	Permanent magnet Halbach arrays in one stator part, together with a pyrolytic graphite rotor, can be used to passively stabilize 5-DOFs using the diamagnetic effect and an axial flux stator is placed on the opposite rotor face to also provide torque onto the rotor.	(Sugimoto, et al., 2023), (Ozawa, et al., 2022)
Non-conventional torque generation technology	The hysteresis effect can be used for creating torques with very small amplitudes, yet in a synchronous, low torque-ripple affected manner. Selection of suitable suspension field pole number, the torque and radial force creation can be completely decoupled. The rotor structure is simple and motors can be operated without rotor angle sensors.	(Noh, et al., 2017), (Circosta, et al., 2018)
Integration of bearingless motors and magnetic gears	Many different forms of magnetic gears have been presented in the literature as they allow high torque density when high input speeds are provided by the attached motor. The integration of magnetic gears and bearingless motors allows eliminating the losses in the high-speed ball bearing stage and benefitting from all the advantages that magnetic levitation has to offer for high-speed motors.	(Kumashiro, et al., 2020), (Gruber, et al., 2022)
Trend towards higher motor output power	Diverse potential applications require significant output power of beyond 50kW, e.g. blowers for aeration of wastewater treatment facilities or compressors for diverse gaseous media. It becomes clear from published design studies or scaling investigations, that the race towards higher power bearingless motors is ongoing.	Realized prototypes: (Redemann, et al., 2000), (Munteanu, et al., 2011), (Liu, et al., 2020), Motor design targets: (Fu, et al., 2017): >50kW, 60krpm, (Kepsu, et al., 2019): 280kW, 25krpm

5. Discussion and Outlook

Obstacles on the way to broader application

The current state of the art shows a tremendous diversity of bearingless motors in theoretic works and academic prototypes. In the mid-term and long-term perspective however, increased industrial usage is the key factor to maintain and intensify the research interest amongst academics and the willingness to fund this research amongst technology companies and funding bodies. Up to the current point in time, bearingless motors were only able to conquer niche markets where the unique selling proposition of magnetic levitation (abrasion-free, lubrication-free, hermetically sealable operation) was beneficially combined with the more compact arrangement of a bearingless motor when compared to AMB-based solutions. It appears to the authors that there are two types of obstacles, that impede a broader application in industry.

Technical issues

The control of bearingless motors is complex in theory and implementation. Some parts of the extensive literature on bearingless motor control may be too decoupled and not well available to those researchers actually designing, realizing, and testing prototypes for industrial applications. Abstractions in control theory and accessible tools and methods may help to enable better control for more bearingless drives.

Bearingless drives typically require non-conventional power electronics due to unusual switching frequencies, phase count, or programmability. Custom-built electronics negatively influence the bearingless motor development in two ways: First, the hardware costs may overturn potential business cases or prototyping budgets. Second, the person-power and knowledge to design or acquire and to program such custom-made devices may not be present in many research groups.

The necessary non-contact distance sensors in bearingless motors are not available as cheap, reliable mass market products. Additionally, knowledge on EMC issues and signal treatment are usually necessary to achieve the required high position signal quality.

Non-technical issues

Bearingless motor research publications often still present novel topologies or related control methods which is understandable given the huge variety of possibilities. Future research will also have to focus on increasing efficiency, power ratings, and improved designs for better industrialization in terms of winding technology, material usage, or required precisions in manufacturing.

High system complexity and the lack of widespread industrial use cases reduces the willingness of industry to invest in bearingless motor developments and, thus, creates a *chicken and egg* problem.

Bearingless motor R&D work requires thorough understanding of electric motors, mechanical design, manufacturing technology, and control theory, making them truly mechatronic components. Also, practical knowledge concerning prototyping, commissioning, and test bench measurement are necessary. On the academic research side, this may be the central explanation why most research results with experimental results are created in one of the established research groups. On the industrial side, this illustrates the costs and risks associated with bearingless motor product development.

Chances and opportunities in the future

Based on the mentioned shortcomings and on decades of bearingless motor research experience, the authors have identified chances and opportunities and want to express their opinions on possibly rewarding research questions.

- The need for global decarbonization of energy supplies results in a tremendous technological push. Many new needs for highly performant motors will arise in this context and, especially in the field of high power density, high rotational speed, and lubrication and abrasion free operation, bearingless machines can play an important role.
- The concepts of reusability, remanufacturability, and eventually recyclability as well as lifecycle assessment will play an increased role in future products and systems. This imposes new demands on the design of bearingless motors which must be answered in respective research activities.
- The availability of robust and cost-efficient components for, e.g., position sensors, touchdown bearings, or inverters will play an essential role. Research on assessing the suitability of existing off-the-shelf components to this aim and on best-practice on how to implement and – for the example of the position sensors – process the related data would be very rewarding and of high service to the community.
- The technologic development of electric motors has received a strong push due to the electrification trend in the mobility sector. Development methods such as the multi-physics simulations applied in e.g., automotive traction motor design can also be of high value when applied to bearingless motors.
- For low power, high volume applications: The combination active suspension force creation with cost-effective passive stabilization can reduce material costs and electronics needs and is, therefore, essential for cost reduction. Significant research has already been conducted but further activities will be necessary.
- For high power applications: Higher power motors in industrial applications often feature long on-times, which basically combines well with the reduced maintenance intervals and high expected lifetime offered by bearingless motors. However, focus on increased efficiency will be essential to be able to compete with conventional motors with both, conventional bearings or AMBs.
- Research on bearingless motors has come a long way. In contrast to the very first years of scientific activity, today's researchers can rely on well-established terminologies, a wide range of investigated and classified

motor topologies, and a broad range of scientific conferences and journals that are known to host bearingless motor publications. The wide range of available information will help to avoid repeating errors and accelerate research towards a *more bearingless* future.

Acknowledgment

The authors want to thank Prof. Eric Severson and his co-authors for providing the literature list used for the elaboration of (Chen, et al., 2019). This work has been supported by the COMET-K2 Center of the Linz Center of Mechatronics (LCM) funded by the Austrian federal government and the federal state of Upper Austria.

References

Ahmed, F. et al., 2012. Self-bearing switched reluctance motor: A review. *Proc. 1st International Conference on Power and Energy in NERIST (ICPEN),* pp. 1-6.

Barletta, N. & Schoeb, R., 1995. Design of a Bearingless Blood Pump. *Proc. 3rd International Symposium on Magnetic Suspension Technology.*

Baumgartner, T. & Kolar, J., 2013. Multivariable state feedback control of a 500 000 rpm self-bearing motor. *Proc. International Electric Machines and Drives Conf.,* p. 347–353.

Bleuler, H. et al., 1994. Micromachined Active Magnetic Bearings. *Proc. 4th International Symposium on Magnetic Bearings.*

Bosch, R., 1988. Development of a bearingless electrical motor. *Proc. ICEM International Conference on Electrical Machines,* pp. 331-335.

Chen, J., Zhu, J. & Severson, E., 2019. Review of bearingless motor technology for significant power applications. *IEEE Transactions on Industry Applications,* Volume 56, p. 1377–1388.

Chiba, A. et al., 2005. *Magnetic Bearings and Bearingless Drives.* s.l.:Elsevier.

Circosta, S. et al., 2018. Analysis of a Shaftless Semi-Hard Magnetic Material Flywheel on Radial Hysteresis Self-Bearing Drives. *Actuators,* Volume 7.

Farhan, A., Johnson, M., Hanson, K. & Severson, E., 2020. Design of an Ultra-High Speed Bearingless Motor for Significant Rated Power. *Proc. IEEE Energy Conversion Congress and Exposition (ECCE).*

Fu, Y., Takemoto, M., Ogasawara, S. & Orikawa, K., 2017. Investigation of a high speed and high power density bearingless motor with neodymium bonded magnet. *Proc. IEEE International Electric Machines and Drives Conference (IEMDC),* pp. 1-8.

Gruber, W., 2013. Bearingless slice motors: General overview and the special case of novel magnet-free rotors. *Proc. 9. GMM/ETG Symposium on Innovative Small Drives and Micro-Motor Systems.*

Gruber, W., König, T. & Miliker, E.-M., 2022. Design of a Magnetic Geared Bearingless Slice Motor with Combined Windings. *Proc. International Power Electronics Conference,* pp. 1756-1760.

Gruber, W. & Silber, S., 2016. 20 Years Bearingless Slice Motor – its Developments andApplications. *Proc. 15th International Symposium on Magnetic Bearings.*

Jungmayr, G., Marth, E. & Segon, G., 2019. Magnetic-Geared Motor in Side-by-Side Arrangement - Concept and Design. *Proc. IEEE International Electric Machines & Drives Conference (IEMDC),* pp. 847-853.

Kepsu, D., Jastrzebski, R. P., Pvrhönen, O. & Kurvinen, E., 2019. Scalability of SPM Bearingless High - Speed Motor for 180 – 280 kW Applications. *Proc. IEEE International Electric Machines & Drives Conference,* pp. 329-335.

Kumashiro, A. et al., 2020. Proposal of Magnetic Geared Motor with Bearingless High-Speed Rotor. *Proc. 23rd International Conference on Electrical Machines and Systems (ICEMS),* pp. 1744-1747.

Levitronix, 2023. *Levitronix® BFS Chemically Resistant MagLev Fans.* [Online]
Available at: https://www.levitronix.com/products/bearingless-fans/
[Accessed 05 2023].

Liu, Z., Chiba, A., Irino, Y. & Nakazawa, Y., 2020. Optimum Pole Number Combination of a Buried Permanent Magnet Bearingless Motor and Test Results at an Output of 60 kW With a Speed of 37000 r/min. *IEEE Open Journal of Industry*

Applications, Volume 1, pp. 33-41.

magneticbearings.org, 2023. *Paper repository of magneticbearings.org.* [Online]
Available at: https://www.magneticbearings.org/publications/
[Accessed 21 05 2023].

Messager, G. & Binder, A., 2014. Analytical comparison of conventional and modified winding for high speed bearingless permanent magnet synchronous motor applications. *Proc. International Conference on Optimization of Electrical and Electronic Equipment,* p. 330–337.

Mitterhofer, H., 2017. *Towards high speed bearingless disk drives.* s.l.:Dissertation, Johannes Kepler University Linz, Austria.

Mitterhofer, H., Gruber, W. & Amrhein, W., 2014. On the High Speed Capacity of Bearingless Drives. *IEEE Transactions on Industrial Electronics,* Volume 61, pp. 3119-3126.

Munteanu, G., Binder, A. & Schneider, T., 2011. Loss measurement of a 40 kW high-speed bearingless PM synchronous motor. *Proc. IEEE Energy Conversion Congress and Exposition,* pp. 722-729.

Noh, M., Gruber, W. & Trumper, D. L., 2017. Hysteresis Bearingless Slice Motors With Homopolar Flux-Biasing. *IEEE/ASME Transactions on Mechatronics,* Volume 22, pp. 2308-2318.

Nussbaumer, T., Karutz, P., Zuercher, F. & Kolar, J., 2011. Magnetically Levitated Slice Motors—An Overview. *IEEE Transactions on Industry Applications,* Volume 47, p. 754–766.

Ozawa, Y., Fujii, Y., Chiba, A. & Suzuki, H., 2022. Novel Axial-Gap Bearingless PM Motor with Full-Passive Magnetic Suspension by Diamagnetic Disk. *Proc. 25th International Conference on Electrical Machines and Systems,* pp. 1-6.

Pei, T. et al., 2022. Review of bearingless synchronous motors: Principle and topology. *IEEE Transactions on Transportation Electrification.*

Redemann, C., Meuter, P., Ramella, A. & Gempp, T., 2000. 30 kW Bearingless Canned Motor Pump on the Test Bed. *Proc. 7th International Symposium on Magnetic Bearings.*

Salazar, A., Chiba, A. & Fukao, T., 2000. A Review of Developments in Bearingless Motors. *Proc. 7th International Symposium on Magnetic Bearings.*

Schoeb, R. & Barletta, N., 1996. Principle and application of a bearingless slice motor. *Proc. 5th International Symposium on Magnetic Bearings.*

Sugimoto, H. et al., 2023. Novel Bearingless Motor Topology With Diamagnetic Salient-Pole Rotor. *IEEE Transactions on Industry Applications,* Volume 59, pp. 1639-1647.

Sun, X., Chen, L. & Yang, Z., 2013. Overview of Bearingless Permanent-Magnet Synchronous Motors. *IEEE Transactions on Industrial Electronics,* Volume 60, pp. 5528-5538.

Sun, X., Chen, L. & Yang, Z., 2014. Overview of bearingless induction motors. *Mathematical Problems in Engineering.*

Sun, X.-D., Shu, H.-Q. & Yang, Z.-B., 2012. An Overview and Development Trend of Bearingless Permanent Magnet Synchronous Motors. *China Mechanical Engineering,* Volume 23, pp. 2128-2135.

Xiang, Q., Ou, Y., Peng, Z. & Sun, Y., 2023. Review on Self-Decoupling Topology of Bearingless Switched Reluctance Motor. *MDPI Energies, Special Issue on Design and Control of Flywheel Energy Storage Systems,* Volume 16, p. 3492.

Applicability of flux switching permanent magnet bearingless motors in present-day and future manufacturing and transportation

Rafal P. JASTRZEBSKI [ab], Sadjad MADANZADEH [a], Atte PUTKONEN [a], Leonid CHECHURIN [a], Wolfgang GRUBER [c], Akira CHIBA [d]

[a] Lappeenranta-Lahti University of Technology LUT, 53850 Lappeenranta, Finland, rafal.jastrzebski@lut.fi; [b] University of Turku, 20014 Turku, Finland; [c] Johannes Kepler University Linz, 4040 Linz, Austria; [d] Tokyo Institute of Technology, 152-8550 Tokyo, Japan.

Abstract

Flux switching permanent magnet (FSPM) motors in both rotating and linear applications have been gaining popularity because of high tangential stress, low cost, low amount of permanent magnets required, and robust construction. The scope of this work is present-day and future manufacturing and transportation applications in industry 4.0 that take advantage of FSPM rotating and linear bearingless motors (BMs). FSMP BMs provide contactless oil-free operation, positioning accuracy, and magnet placement leading to robust and affordable construction. We review design challenges, trends, and opportunities. Tangential stress, normal stress, torque, and force density per motor volume are benchmarked based on available published data in recent years. The main challenges in design optimization, implementation in applications, and control are analyzed. The work concludes with discussion on scaling for higher powers and wider airgaps.

Keywords: *Flux switching permanent magnet motor, Bearingless motor, Linear levitated motor, Magnetic levitation.*

1. Introduction

Arguably the 1st industrial revolution (IR) has started in the late XVIII century with the introduction of steam power enabling large-scale manufacturing, mechanization, and textile industry. The development of these ideas resulted in rapid urbanization with many new cities emerging and rapidly growing in that era, for example, Detroit in USA, Manchester in UK, Lodz in Poland, Edo (modern-day Tokyo) in Japan or Tampere in Finland. In the late XIX century, the 2nd IR with railroads development, electricity generation, internal combustion engines, moving assembly lines, and mass production flourished. The 3rd IR started in the middle of the XX century with the digitization of manufacturing enabling semiconductors, computing, and digital communication networks. The 4th IR in early XXI century leverages and fuses technology breakthroughs in fields such as robotics, autonomous systems, 3D printing, and additive manufacturing.

Industry 4.0 needs adaptation of manufacturing space to new methods. Faster, more efficient, reliable, clean, safe production of higher-quality goods at reduced cost requires new solutions for assembly and packaging transportation lines. Emerging fully automated smart factories require cleanness, precision, integration, modularity, and cyber-physical system compliance. Environmental and technological needs will demand oil-free clean solutions.

In the last decade, perhaps the most intensely investigated applications for electrical machines have been electric vehicles (EVs) and hybrid electric vehicles (HEVs). In general, permanent magnet (PM) motors offer higher torque and energy densities than induction motors (IMs) and switched reluctance motors (SRMs). Interior PM synchronous motors (IPMSMs) when benchmarked for HEVs as presented by Kiyota et al. (2014) offered 71% higher maximum energy and maximum torque densities than SRMs. Short operation peak densities computed for current densities as high as 50 A/mm^2 and per volume measured by the iron core external dimensions without end windings reached 76 Nm/L and 22kW/L. Densities and efficiencies significantly are affected by operating conditions and design constraints. Yang et al. (2015) compared IPMSMs SRM, and IM drives for EV and HEV applications. In that study, peak power densities per kg for IPMSMs SRM, and IM have been 50 kW/30 kg, 50 kW/42 kg, and 50 kW/48 kg, and per volume 10.5 kW/L, 8.4 kW/L, and 8.1 kW/L.

The Flux switching permanent magnet (FSPM) motor offers similarly high energy density per volume as IPMSMs. The main advantages of FSPM over SPM and IPM technology are better mechanical integrity and high reliability for high-speed operation, better thermal dissipation, and smaller copper loss with concentrated windings. However, the main drawbacks include smaller PM utilization ratio, smaller output torque per kg of PMs (Cao et al., 2012). Performance metrics greatly can vary depending on the design and operating conditions. Cao et al., 2012 compared a FSPM with the IPM used in 2004 Prius HEV, which is nearly the same IPM as used by Yang et al. (2015) in his comparison. The peak-to-peak value of cogging torque was a little bigger than that of the Prius IPM. For FSPM, the average *d* and *q* axes inductances (0.343 mH and 0.48 mH) do not differ significantly and therefore, the reluctance torque of FSPM is small. The FSPM compared to IPM motor has been of slightly lower power density (9.2 kW/L vs. 10.1 kW/L). However, FSPM had lower weight (34 kg vs 39 kg). Hua et al. (2017) studied an outer rotor FSPM machine for in-wheel traction. For outer rotor configuration, FSPM has been reported to reach 40% (and 61.5% with wedge shaped magnets) more torque than SPM machine with the same rated current density (5 A/mm^2).

Manufacturing in Industry 4.0 and Smart Factory concepts develop at such a pace that the past conceptual solutions become the key drivers for success today. Magnetic levitation in rotating and linear motion systems is one of the examples. The basic solutions for factory logistics and manufacturing have been using mechanical chassis. The use of mechanical bearings and rollers creates harmful dust particles, which must be laboriously filtered out by centrifugal air compressors in the cleanrooms. Mechanical bearings and transmission also cause maintenance and reliability issues.

This paper discusses growing potential and technological advancement of levitated linear as well as rotating FSPM bearingless motors (BMs) especially in expanding industry 4.0 manufacturing sector. The FSPM BMs potential applications include efficient blowers, mixers, pumps and compressors with ultra-purity or vacuum requirements. The linear FSPM (LFSPM) BMs can be applied to transportation without mechanical conversion such as horizontal (production lines) and vertical (elevators) transporters, warehouses, and storage facilities. LFSPM motors require PMs only in the mover offering huge cost reduction per length of constructed track over SPM and IPM motor solutions. Installing contactless manufacturing space management and precise transportation as well as leveraging oil-free blowers, mixers, and pumps can affect how the factories are planned and designed to improve production efficiency, quality, time to market, and reduce cost and resources usage.

2. Potential markets for levitated FSPM machines

We generalize potential market estimates as of 2023 in this chapter. However, the practical implementation and commercial share and viability of levitated FSPM machines in various sectors is difficult to estimate and will vary depending on cost, scalability, regulatory considerations, and technological integration.

In terms of interest for levitated FSPM motors we can estimate the main global market values of the applications as

- semiconductors and silicon foundries $523 billion
- smart factories $97 billion
- elevators and transporters $82 billion
- warehouse automation $15.7 billion
- linear motion products $12.3 billion
- cleanroom technology $3.6 billion

In c. 2000, most of the World's chip production has been moved to Asia with a handful of producers. Twenty years later, sales of semiconductors reached $523.1 billion according to Semiconductor Industry Association (SIA) (2023), which was a 29.7% increase year to year. Electronic key component shortages brought discussion on the need to reestablish and diversify the semiconductor industry. Advanced manufacturing such as semiconductor manufacturing requires absolute cleanness.

Smart factories market size was valued at $79.41 billion in 2021 and is expected to grow at a CAGR of 10.46 % till 2023 and beyond according to Polaris Market Research (2022).

Elevators and transporters global market size was valued at $72.39 billion in 2022 and is expected to expand at CAGR of 6.4% from 2023 to 2030 (Grand View Research, 2021). Still, we should point out that entry markets might be much smaller, for example, the high-rise buildings elevator market size is only about $40 million.

Warehouse automation global market size was valued at $13.6 billion in 2021 and is projected to grow at a CAGR of 15.3% from 2022 (Allied Market Research, 2022).

The global linear motion products market is projected to grow from $11.5 billion in 2022 to $12.3 billion by 2023 (Fortune Business Insights, 2022). Examples of large companies in motion controller market include ABB Ltd., YASKAWA, Siemens, and Toshiba.

Similarly high cleanness levels are required by growing production segment of the pharmaceutical industry. For example, sterile drugs and biologics often require cleanroom production. New drug products, particularly biologics and cell therapies, require sterile manufacturing to ensure safety and efficacy. Additionally, the COVID pandemic has increased the demand for sterile drugs and vaccines. IQVIA Institute for Human Data Science (2022) forecasted spending on medication to reach $1.6 Trillion by 2025, which excludes spending on COVID-19 vaccines. The global cleanroom technology market size was estimated at $3.6 billion in 2022 and is expected to reach $3.8 billion in 2023 (Grand View Research, 2023).

Potential market sizes are large for levitated FSPM machines, but entry points are limited to special equipment primarily in silicon foundries and in cleanrooms.

3. Design considerations for bearingless operation

Applicability of BMs is broadening with performance improvements and rising realizable power levels (Chen et al. 2020). Rotors with active magnetic bearings (AMBs) and a separate motor unit have been easier to realize. They have been in industrial use with reported powers up to 29 MW. The constructed BMs reached 160 kW (Jastrzebski et al. 2021).

3.1 Rotating Applications

Motors with permanent magnets inserted in the stator core are rarer (Cheng et al., 2011). Those machines possess better rotor integrity and thermal dissipation advantage because of no PMs in the rotor. FSPM motors show torque densities on the same level as conventional PM motors. However, a standard FSPM machine suffers from high saturation. Evans and Zhu (2015) proposed the partitioned stator FSPM machine to relax the magnetic saturation. The stator partitioning separates the PM and copper and place PM within the rotor on a separate stationary body (inner stator). A second air gap is introduced between the salient rotor poles and the inner PM stator. Inherent saturation of FSPM technology contributes to cogging ripples. Pu et al. (2022) investigated and proposes a method to improve the electromagnetic torque of the FSPM motor considering the flux density harmonic reduction ratio.

Recent examples and technical evolution of rotating FSPM BMs have been introduced by Gruber and Radman (2017), Ding and Sarlioglu (2019) and Madanzadeh et al. (2022, 2023). Madanzadeh et al. (2022) introduced double stator concept to FSPM BMs. Fig. 1 shows example of high energy density per rotor volume geometry of such FSPM BM. The structure has discreate rotor poles and similarities to the magnetic geared motor (Kumashiro, 2022).

In the case of FSPM BMs in addition to torque also levitation force amplitude variations and foremost angle variations must be minimized. Good candidates for design optimization are maximizing mean torque, suspension force, and efficiency, when minimizing torque ripple, suspension force amplitude and angle variations.

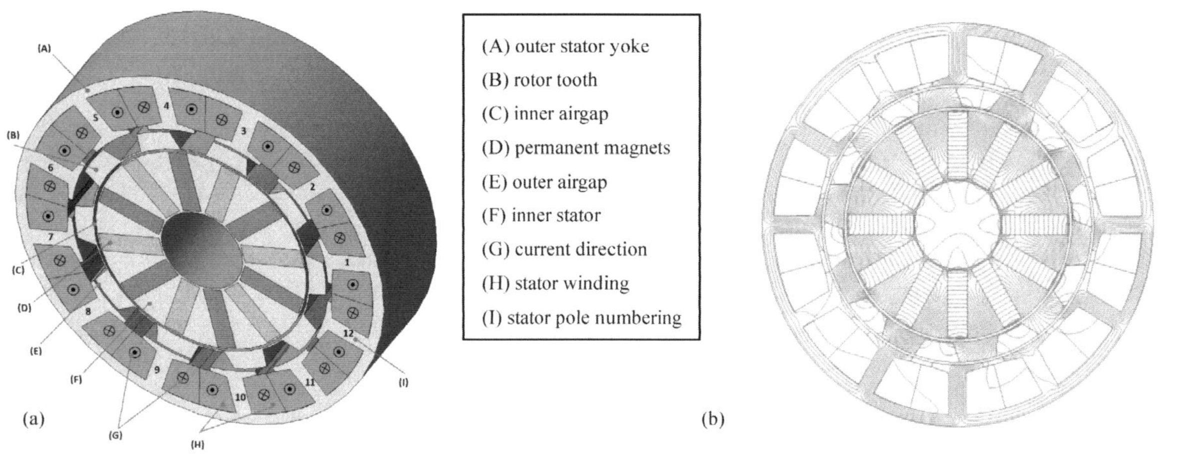

Figure 1 Double stator concept as in Madanzadeh et al. (2022): (a) 2D model of the FSPM BM with the magnets in the inner stator section. 12/10 (stator slot / rotor poles), configuration. (b) flux lines across 2D cross section when operating.

3.2 Linear Applications

The design structures from the rotating applications can be utilized also in the linear motion applications. Taking silicon foundries and smart factories as a potentially biggest market share for ultra-clean, precise, reliable but high-cost equipment demand for linear motion is rising. Substrate cassettes in the semiconductor manufacturing cleanroom or platform manipulators are traveling on wheels or rollers on the floor or as monorail transporter (with rollers) on the ceilings. This is also true for smaller range (within 1-2 meters) mechanical movement required for stationary manufacturing units (e.g., UV exposure process in mask-based etching, high-precision mechanical turning, high-speed packaging, and assembly stations), where the precision motion platforms can be used.

Technological advantages of linear LFSPM BMs over other functionally similar technologies are listed as follows. LFSPM BMs have PMs in the track making them significantly more affordable over linear levitated motors (LLMs) with PMs. At the same time, against reluctance linear motors, LFSPM BMs provide higher efficiency, higher energy density from the same volume, and smaller footprint. Against motors with mechanical rollers, they provide precision of movement, no maintenance needs, built-in monitoring and diagnostics, early warning, built-in weight estimate, and cleanness. Compared to rotating machines with transmission, no transmission, no gears, higher efficiency, no wires, no belts, no ropes are needed.

The well-studied configurations from the non-levitated machines include 12/10 and 12/14 (stator poles / rotor poles or mover slots / rail poles) (Cao et al., 2014). For rotating machine 12/14 configuration provides a higher torque with smaller ripple compared with the 12/10 configuration (Chen and Zhu, 2010). LFSPM motors produce thrust force without a need for conversion. Two additional teeth at the mover ends balance the magnetic circuit. For the levitated machine their width and distance can be varied to minimize the thrust and normal levitation force ripples (Fig. 2).

For FSPM machines significant number of geometry variants have been proposed in recent years to further enhance performance. For example, Tan et al. (2021) used discrete rail poles to increase thrust force density per volume. Metro lines have been one of possible applications where linear induction motors are in use in subways of, for example, Okamoto Green Line of Osaka Metro and Line 12 of Tokyo Metro in Japan. For railway transportation alternatives such as linear switched reluctance motors have been compared to FSPM motors when large 10 mm airgap is used (Cao et al. 2020). Specifically, the thrust, ripple, and efficiency of LFSPM motor was 2.13, 0.23, and 1.06 times that of switched reluctance counterpart, respectively.

FSPM levitated movers provide viable replacement for contact required motion platforms. For linear machines 12/14 configuration delivers commonly used thrust and normal force producing structure. For levitated platforms (Fig. 3a), with synchronous operated motors and at least 4DOF active control, eight three phase inverters are needed to supply opposite attractive force producing, e.g., 12/10 or 12/14 (Fig. 3b), configurations. Those can be replaced by joint or separated 6/5 (Fig.3b) or 6/7 configurations when short platforms are required. Side teeth of joint two modules can be eliminated but then normal force generation of each of joint modules is not magnetically independent. For example, when constructing simulation virtual prototype this means slightly more involved actuator modelling and more coupled control. Technology advancement has been presented in Jastrzebski et al. (2017), Sokolov et al. (2020), and Madanzadeh et al. (2023). Similarly, to the rotating BMs also linear BMs can have double stator (Madanzadeh and Jastrzebski, 2023).

Figure 4 illustrates 4DOF actively controlled (2DOF passively suspended) exemplary proof of concept. The commercial version would feature on board battery, and wireless energy transfer for self-powered untethered operation. Optimally, application integrated sensor technology that directly uses the rail poles as its target for gap and position detection should be installed (Jastrzebski and Tolsa, 2023). Similar platforms but using different levitation principles have been constructed in research labs (Meldrum et al., 2021, Zhou and Trumper, 2021). However, energy densities have been significantly lower limiting payload weight.

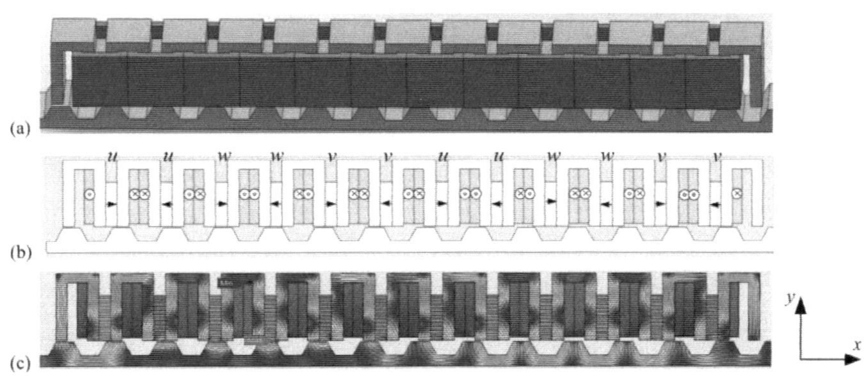

Figure 2 LFSPM BM as in Jastrzebski et al. (2017). (a) Isometric view of 12/14 LFSPM BM. (b) Coil and magnet arrangement. The laminations of the yoke are white, the 3-phase coils green, and the magnets yellow. (c) Flux lines and density distribution for the nominal normal load for i_d = 3.5 Arms and i_q = 0 A.

For BM operation, for example, in Jastrzebski et al. (2017) peak normal stress is 18.5 N/cm^2 and tangential stress (thrust force density) is 9.9 N/cm^2 with peak current densities reaching 13 A/mm^2 but for well saturated iron. The normal stress is about half of that corresponding to radial AMB. The normal force ripple is up to 7% at peak suspension current (and about 4% at lower currents) and nominal airgap. Mover to rail pole pitch ratio and structure optimization play significant roles in reducing normal and thrust force ripples without sacrificing normal and tangential stress values. In Madanzadeh et al. (2023) for 6/5 module at 5.2 A/mm^2 the peak normal stress is 11.8 N/cm^2 and the tangential stress is 5 N/cm^2. The stress values are slightly increased when two 6/5 modules are merged eliminating middle two side teeth, to 12.8 N/cm^2 and the tangential stress is 5.5 N/cm^2. Similar stress values should be possible for airgaps up to 1 cm with designs scaling based on results for non-levitated FSPM motors.

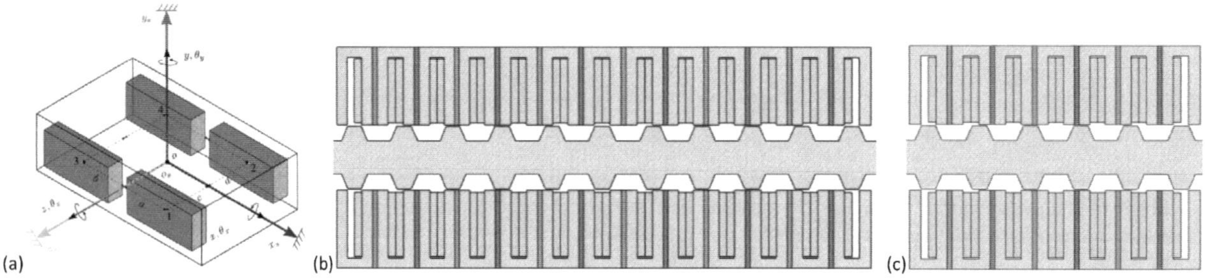

Figure 3 Levitated platform as in Madanzadeh et al. (2023): (a) Illustration of linear motion platform. (b) Exemplary cross section of 12/10 LFSPM BM module. 2DOF levitation is achieved using 2 3-phase inverters per armature. (c) Exemplary cross section of 6/5 LFSPM BM module. 1DOF levitation is achieved using 1 3-phase inverter per armature.

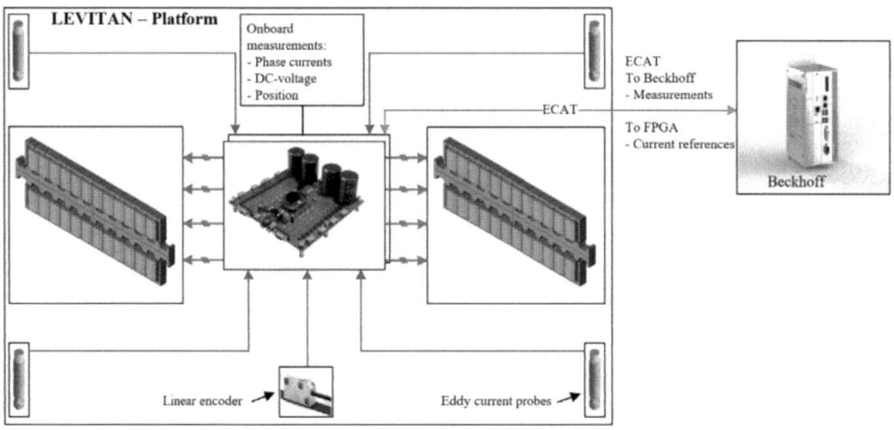

Figure 4 4DOF actively controlled (2DOF passively suspended) levitated linear motion platform with electronics similar to Mirić et al. (2021). The production version would feature integrated sensors, on board battery, and wireless energy transfer for self-powered untethered operation.

4. Conclusions

The FSPM BMs, levitated movers, and their virtual prototypes respond to demanding requirements in silicon foundries, smart factories, robotic assembly, additive manufacturing, pressing, forging, transporting, pharmaceuticals production, and other Industry 4.0 applications. High energy densities are achievable without need for PMs nor windings in the tracks. However, high number of inverters is required for active suspension.

The combination of magnetic levitation and vacuum technology offers improved performance, no friction, and enhanced stability, making it useful for high-speed or space exploration applications. Nevertheless, laboratory and industrial vacuum chambers for, for example, thin film deposition, vacuum drying, vacuum casting, vacuum packaging, composite material production are more likely early adopters.

Acknowledgements

This work was supported in part by Business Finland grant no. 1803/31/2022 (Linear Precision Motion Platform) and Academy of Finland grant no. 350880.

References

Allied Market Research (2022) Warehouse Automation Market by Component (Hardware, Software), by Application (Automotive, Food and beverage, E-Commerce, Pharmaceutical, Others), by End User Industry (Retailers, Manufacturers and distributors): Global Opportunity Analysis and Industry Forecast, 2021-2031. https://www.alliedmarketresearch.com/warehouse-automation-market-A17070 (accessed 01 of May 2023).

Cao R., Mi C., and Cheng M. (2012) Quantitative Comparison of Flux-Switching Permanent-Magnet Motors With Interior Permanent Magnet Motor for EV, HEV, and PHEV Applications, in IEEE Transactions on Magnetics, vol. 48, no. 8, pp. 2374–2384, Aug. 2012, doi: 10.1109/TMAG.2012.2190614.

Cao R., Cheng M., Mi C. C., and Hua W. (2014) Influence of Leading Design Parameters on the Force Performance of a Complementary and Modular Linear Flux-Switching Permanent-Magnet Motor, in IEEE Transactions on Industrial Electronics, vol. 61, no. 5, pp. 2165-2175, May 2014, doi: 10.1109/TIE.2013.2271603.

Cao R., Su E., and Lu M. (2020) Comparative Study of Permanent Magnet Assisted Linear Switched Reluctance Motor and Linear Flux Switching Permanent Magnet Motor for Railway Transportation, in IEEE Transactions on Applied Superconductivity, vol. 30, no. 4, pp. 1-5, June 2020, Art no. 3601205, doi: 10.1109/TASC.2020.2965874.

Chen, J. T. and Zhu, Z.Q. (2010) Comparison of All- and Alternate-Poles-Wound Flux-Switching PM Machines Having Different Stator and Rotor Pole Numbers, in IEEE Transactions on Industry Applications, vol. 46, no. 4, pp. 1406-1415, July-Aug. 2010, doi: 10.1109/TIA.2010.2049812.

Chen J., Zhu J., and Severson E. L. (2020) Review of Bearingless Motor Technology for Significant Power Applications, in IEEE Transactions on Industry Applications, vol. 56, no. 2, pp. 1377-1388, March-April 2020, doi: 10.1109/TIA.2019.2963381.

Cheng M., Hua W., Zhang J., Zhao W. (2011) Overview of stator-permanent magnet brushless machines, in IEEE Transactions on Industrial Electronics, vol. 58, no. 11, pp. 5087-5101, Nov. 2011, doi: 10.1109/TIE.2011.2123853.

Ding H. and Sarlioglu B. (2019) Design of a Novel Axial Flux-Switching PM Machine Integrated with Centrifugal Compressor with Radial Impellers. In IEEE Transportation Electrification Conference and Expo (ITEC), Detroit, MI, USA, pp. 1–6. doi: 10.1109/ITEC.2019.8790567.

Evans D. J. and Zhu Z. Q. (2015) Novel Partitioned Stator Switched Flux Permanent Magnet Machines, in IEEE Transactions on Magnetics, vol. 51, no. 1, pp. 1-14, Jan. 2015, Art no. 8100114, doi: 10.1109/TMAG.2014.2342196.

Fortune Business Insights (2022) Market research report, Nov. 2022, p. 120. https://www.fortunebusinessinsights.com/linear-motion-products-market-106917 (accessed 30 Jannuary 2023).

Grand View Research (2021) Elevators Market Size, Share & Trends Analysis Report By Type (Hydraulic, Traction), By Business (New Equipment, Maintenance, Modernization), By Application (Residential, Commercial), By Region, And Segment Forecasts, 2023 – 2030. https://www.grandviewresearch.com/industry-analysis/elevators-market-report (accessed 01 of May 2023).

Grand View Research (2023) Cleanroom Technology Market Size, Share & Trends Analysis Report By Product (Consumables, Equipment), By End-use, By Region, And Segment Forecast, 2023 – 2030, report ID: GVR-2-68038-216-7, p. 208. https://www.grandviewresearch.com/industry-analysis/cleanroom-technology-market (accessed 01 of May 2023).

Gruber W. and Radman K. (2017) Modeling and Realization of a Bearingless Flux-Switching Slice Motor. Actuators, 6(2): p. 12. doi: 10.3390/act6020012.

Hua W., Zhang H., Cheng M., Meng J., and Hou C. (2017) An outer-rotor flux-switching permanent-magnet-machine with wedge-shaped magnets for in-wheel light traction, in IEEE Transactions on Industrial Electronics, vol. 64, no. 1, pp. 69-80, Jan. 2017, doi: 10.1109/TIE.2016.2610940.

IQVIA (2022) Press release available at: https://www.iqvia.com/insights/the-iqvia-institute/reports/global-medicine-spending-and-usage-trends-outlook-to-2025 (accessed 01 of May 2023).

Jastrzebski R. P., Jaatinen P. and Pyrhonen O. (2017) Modeling and control design simulations of a linear flux-switching permanent-magnet-levitated motor. Mech. Eng. J., 4(5): pp. 1–12. doi: 10.1299/mej.17-00084.

Jastrzebski R. P., Kepsu D., Putkonen A., Martikainen I., Zhuravlev A., and Madanzadeh S. (2021) Competitive technology analysis of a double stage kinetic compressor for 0.5 MW heat pumps for industrial and residential heating, 2021 IEEE International Electric Machines & Drives Conference (IEMDC), Hartford, CT, USA, 2021, pp. 1-7, doi: 10.1109/IEMDC47953.2021.9449597.

Jastrzebski R. P., Tolsa K. (2020) Position sensor and a method for manufacturing the same, EP3980726A1(B1), FI130190B, WO2020245496A1.

Kiyota K., Sugimoto H., & Chiba, A. (2014). Comparing Electric Motors. IEEE Industry Applications Magazine, July-August, 12–20, doi: 10.1109/MIAS.2013.2288380.

Kumashiro A., Chiba A., Gruber W., Amrhein W., and Jungmayr G. (2022) Optimization of Stack Length in Magnetic-Geared Motor with Magnetically Suspended High-Speed Rotor, IEEE Energy Conversion Congress and Exposition (ECCE), Detroit, MI, USA, pp. 1–7, doi: 10.1109/ECCE50734.2022.9947687.

LUT and LAB Levitan project team (2023) Linear precision motion platform. Available at: https://levitan.fi/ (accessed 30 January 2023).

Madanzadeh S., Gruber W., Zhuravlev A. and Jastrzebski R. P. (2022) Self-Bearing Partitioned Stator Flux-Switching Permanent Magnet Motor. In 25th IEEE International Conference on Electrical Machines and Systems (ICEMS), Chiang Mai, Thailand, pp. 1–6. doi: 10.1109/ICEMS56177.2022.9983005.

Madanzadeh S., Gruber W. and Jastrzebski R. P. (2023) Self-Bearing Linear Flux-Switching Permanent Magnet Motor for Linear Motion Platform Application. In IEEE International Electric Machines and Drives Conference (IEMDC).

Madanzadeh S., Jastrzebski R. (2023) A bearingless electric drive, Finnish Patent Application no. 20235027.

Meldrum A. W., Robertson W., and Soong W. L. (2021) Planar Levitation and Propulsion of a Solid-State Craft over Conducting Surfaces, IEEE Energy Conversion Congress and Exposition (ECCE), Vancouver, BC, Canada, 2021, pp. 4382–4388, doi: 10.1109/ECCE47101.2021.9595372.

Mirić S., Giuffrida R., Bortis D. and Kolar J. W., (2021) Dynamic Electromechanical Model and Position Controller Design of a New High-Precision Self-Bearing Linear Actuator, in IEEE Transactions on Industrial Electronics, vol. 68, no. 1, pp. 744-755, Jan. 2021, doi: 10.1109/TIE.2020.2992943.

Polaris Market Research (2022) Smart Factory Market Share, Size, Trends, Industry Analysis Report, By Component; By Solution (SCADA, MES, Industrial Safety, PAM); By Industry (Process Industry, Discrete Industry); By Region; Segment Forecast, 2022 – 2030. https://www.polarismarketresearch.com/industry-analysis/smart-factory-market (accessed 01 of May 2023).

Pu Y., Xiang Z., Jiang M., and Zhu X. (2022) Investigation on Electromagnetic Torque of a Flux-Switching Permanent Magnet Motor From Perspective of Flux Density Harmonic Reduction Ratio, in IEEE Transactions on Magnetics, vol. 58, no. 2, pp. 1-6, Feb. 2022, Art no. 8101706, doi: 10.1109/TMAG.2021.3080500.

Semiconductor Industry Association's website (2021) https://www.semiconductors.org/global-sales-report/ (accessed 01 of May 2023).

Sokolov M., Saarakkala S. E., Hosseinzadeh R. and Hinkkanen M., (2020) A Dynamic Model for Bearingless Flux-Switching Permanent-Magnet Linear Machines. IEEE Transactions on Energy Conversion, 35 (3): pp. 1218–1227, doi: 10.1109/TEC.2020.2975082.

Tan Q., Wang M., and Li L., (2021) Analysis of a New Flux Switching Permanent Magnet Linear Motor, in IEEE Transactions on Magnetics, vol. 57, no. 2, pp. 1-5, Feb. 2021, Art no. 8201705, doi: 10.1109/TMAG.2020.3018457.

Yang Z., Shang F., Brown I. P., and Krishnamurthy M. (2015) Comparative Study of Interior Permanent Magnet, Induction, and Switched Reluctance Motor Drives for EV and HEV Applications, in IEEE Transactions on Transportation Electrification, vol. 1, no. 3, pp. 245–254, Oct. 2015, doi: 10.1109/TTE.2015.2470092.

Zhou L., and Trumper D. L. (2021) Magnetically Levitated Linear Stage with Linear Bearingless Slice Hysteresis Motors, in IEEE/ASME Trans. Mechatronics, vol. 26, no. 2, pp. 1084–1094, doi: 10.1109/TMECH.2020.3017435.

Design of an Active Magnetic Bearing for the Identification of Static and Dynamic Performances of a self-Aligning Hydrodynamic Journal Bearing

Jean-Yves Roger[(a)], Mohamed-Amine Hassini[(a)], Lucien Desprez[(a)], Jarir Mahfoud[(b)]

[(a)]*EDF R&D, 7 Gaspard Monge, Palaiseau, France*

[(b)] *University of Lyon, INSA-Lyon, CNRS UMR5259, LaMCoS, F-69621 Villeurbanne Cedex, France*

1 Abstract

Large Self-Aligning Hydrodynamic Journal Bearings (SAHJB) are used in many heavy-duty rotating machines such as steam turbines. In an EPR2 nuclear power plant, their diameter can reach up to 800mm. To test a representative small scale of such bearings (100mm diameter), a test rig is being designed where the SAHJB is mounted in the middle of a shaft supported by two active magnetic bearings (AMB) at its ends and operated up to 5000 rpm. The role of the AMBs is to apply enough static load to the SAHJB to obtain a Sommerfeld number on the tested SAHJB equivalent to the one used on the real turbine. The AMBs will also be used to impart small perturbations to the rotor movement in order to identify the stiffness and damping coefficient of the SAHJB for different bearing-rotor misalignment. To evaluate the forces in the magnetic bearing, a finite element FE approach was implemented using the inhouse software code_Carmel. Both a simplified structure of an electromagnet (a U shape core attracting and an I shape core) and a realistic magnetic bearing (8 magnetic poles) have been modelled. On the simplified structure, results of FE calculations with linear hypothesis on magnetic materials (7000 times vacuum permeability) are in very good agreement with the results of an equivalent analytical model. These theoretical results are also in good agreement with experimental data. The same FE model was utilized to evaluate the effect of the non-linearity of the magnetic core. When the saturation of the magnetic core occurs, the forces available within the airgap between the fix part and the mobile part of the electromagnet, decrease. For low saturation of the magnetic core, the obtained forces are in good agreement with the analytic model results. Considering the actual magnetic bearing, the same type of calculations was carried out, allowing evaluations of the effect of magnetic non-linearity and geometrical dimensions on bearing performances. These elements are used to evaluate both the actual forces available within the bearing and the optimal domain of use of the AMB in a pre-design stage.

Keywords: *AMB, design, Hydrodynamic bearing*

1. Introduction

The objective of the paper is to give details on the design of a magnetic bearing which will be used to analyze the static and dynamic behavior of Self-Aligning Hydrodynamic Journal Bearings. These bearings are heavily loaded to prevent any hydrodynamic instability (whirl instability). They are also capable to accommodate large amount of rotor to bearing misalignment. Such feature is important to ensure bearing reliability and reduce alignment requirements which in turn facilitate maintenance operations.

A Thermo-Elasto-Hydrodynamic numerical model implemented within LEGOS, a Python library developed by EDF R&D to predict the static and dynamic behavior of hydrodynamic and hydrostatic bearings. The pressure field within the oil film gap is predicted using a finite volumes model that solves the generalized Reynolds equation while considering film rupture and reformation. The Reynolds equation is coupled with a 3D Energy equation to predict the temperature field within the oil film. This hydrodynamic model (Reynolds and energy equation) is also coupled with a finite element thermo-mechanical model of the bearing solid parts using a Neumann-Dirichlet strategy to solve the conjugate heat transfer problem between the oil film and the bearing and predict its elastic deformation that impact the oil film gap. The details of the model can be found in .

Like any theoretical model, it requires experimental data are necessary to validate the developed model. The validation is performed with respect to static performances, i.e., pressure field, inlet and outlet temperature, oil flow rate, etc.., and dynamic behavior, i.e., the stiffness and damping coefficients. The latter are usually identified by imparting small perturbation to the bearing using a hydraulic or electrical shaker and measuring the generated force. It is also possible to impart the perturbation directly to the rotor usually through ball bearing [2] or using a highly stiff hydrostatic bearing [3] or AMBs [4]. Because of the SAHJB unique feature to self-align, AMBs were selected to apply the necessary vertical load, generate a controlled misalignment while imparting small perturbation to the rotor. This will lead to geometrical variations of the airgap within the magnetic bearing and of the magnetic behavior of the actuator.

To achieve these goals, numerical studies were carried out to optimize the design of the AMBs and to evaluate the effect of misalignment on their performances. The design of the magnetic bearing followed two steps: a first step was to design a simple electromagnet actuator to evaluate the methods and tools that will be used. The second step was to design the full-scale magnetic bearing.

2. Test rig description

Figure 1 present the designed test rig that will be used to carry out the validation of the SAHJB theoretical model. A rotor operated up to 5000 rpm is supported by two AMBs at its ends. It is driven by an electric motor using a flexible coupling. Both the electric motor and the coupling are not represented. The SAHJB is a 100 mm diameter and is placed between the AMBs. The bearing is lubricated using an ISOVG 68 mineral oil. The active surface of the hydrodynamic bearing is coated using white metal. Small holes are machined within the lower lobe to measure the hydrodynamic pressure that build up with the thin film fluid. The bearing is also equipped with temperature probes that measure the temperature of the solid part just underneath the coating. The bearing has two pockets, one on the leading edge and one on the trailing edge of the lower lobe. The oil is supplied to the bearing through the leading-edge pocket. The lower lobe is supported using a lower plate that allow the bearing to tilt in the two directions perpendicular to the bearing axial axis. The other plates are not in contact with the bearing casing. The flexible couple will ensure the rotor axial constraint since the axial forces are negligible.

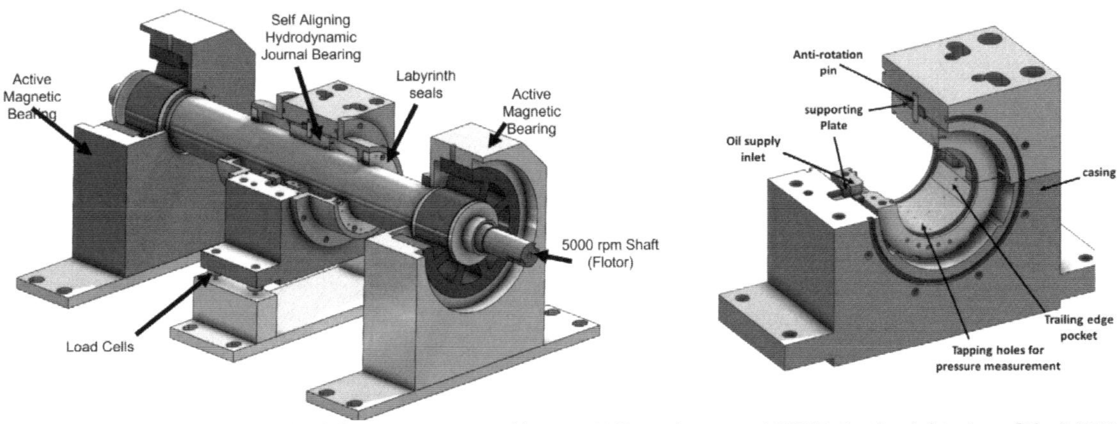

Figure 1 View of the test rig. On the left a rotor supported by two AMBs and a central SAHJB. On the right, view of the SAHJB

3. Test case

In order to evaluate the design process a first modelling has been carried out on a simple structure. Theorical and experimental approaches have been done.

3.1. Numerical approach

The electromagnet actuator design is depicted in Figure 2. The geometry and the mesh are generated using the Salome platform using the shaper and smesh modules (website Salome-platform [5]). The shaper module allows to draw a parametric model which will be used especially for optimizing the design. The generated meshes were both 2D and 3D models. The 2D modelling is sufficient to evaluate the forces but 3D modeling were also used to validate the computation in that case also. The method used to evaluate the forces is the well-known virtual work method.

Figure 2 Design of the first electromagnet U-shape and I-shape magnetic core and winding

To compare the forces obtained using Finites Elements computation, we use the simplified analytical forces from [6]:

$$F = \mu_0 . A_0 . \left(\frac{N.I}{2.\delta_0} \right)^2 \quad (1)$$

Where, F is the total forces available in the airgap, μ_0 the vacuum permeability, A_0 the area of magnetic pole, N the number of turns of windings, I the applied current and δ_0 is the nominal radial clearance.

3D modeling has been made, one with linear magnetic properties and two with the non-linear properties of M270-35A electrical steel material, with and without the holes in the I and U shape. With linear properties values are very close to the theorical forces. With the non-linearity considered, we see a large reduction of the available forces for high current values due to saturation and small effects of holes in the magnetic circuit on the force value. As shown in Figure 3, the finite element model is in good agreement with the analytical solution in the simple case of a single electro-magnet. The effect of holes is also negligeable.

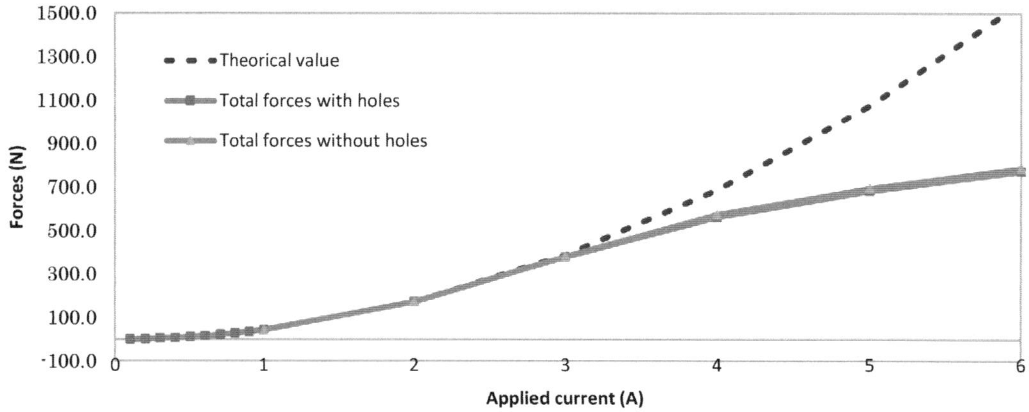

Figure 3 Forces values with theoretical approach and EF approach with and without holes in the magnetic circuit

4. Magnetic bearing

For this study, we choose to use a heteropolar-type radial AMB with an 8 poles structure (Figure 4). The aim of the study is to design the magnetic bearing according to the required performances. As a base design, we used some fixed dimensions as: an airgap of 0.5 mm and an outer diameter of the rotor shaft of 55mm. In addition, an analytical approach set the number of turns of each winding to 191 turns of a 1.1mm^2 wire. Thus, the minimum size of the winding can be estimated. Other dimensions, as the inner and outer diameter of the stator, the outer diameter of the rotor and the height of the teeth, are parameters of the study. The size of the slot has also been considered a possible parameter of design.

Figure 4 Base design of the first magnetic bearing

4.1 Effect of teeth spanning on available forces

A study of the effects of the teeth form have been done, considering all other dimensions fixed, and especially the surface used to compute the forces in each case. We have modeled: case (a): teeth are rectangular (0.0° between radial direction and the border of tooth, case (c): slots are rectangular (22.5° between radial direction and the border of tooth) and a case (c) which is in between (11.25°). The computed forces are presented on Figure 6.

The opening of each tooth increases the value of the available forces. The larger is the teeth at his root, the higher is the force. This can be explained both by the reduction of the saturation in the teeth and also by the reduction of the reluctance value due to the length reduction of the magnetic iso-value lines.

For the next optimization, the larger opening will be used, corresponding to a rectangular slot, case (c).

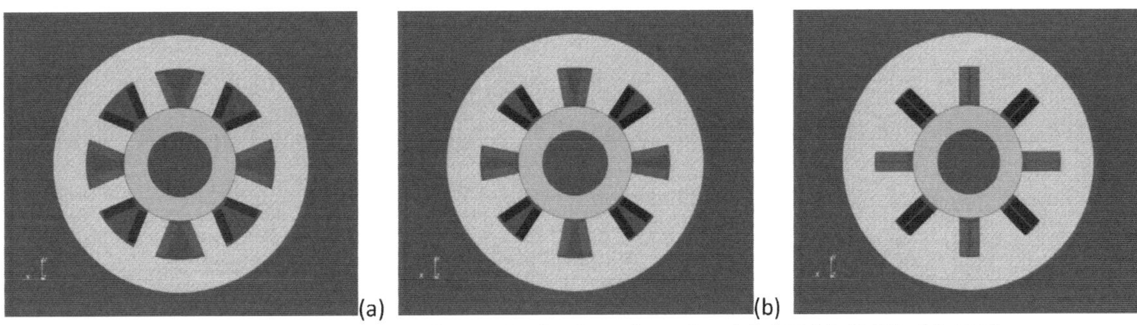

(a) (b) (c)

Figure 5 Study of the teeth spanning for three dimensions (a): 0.0° (b): 11.25° - (c): 22.5°

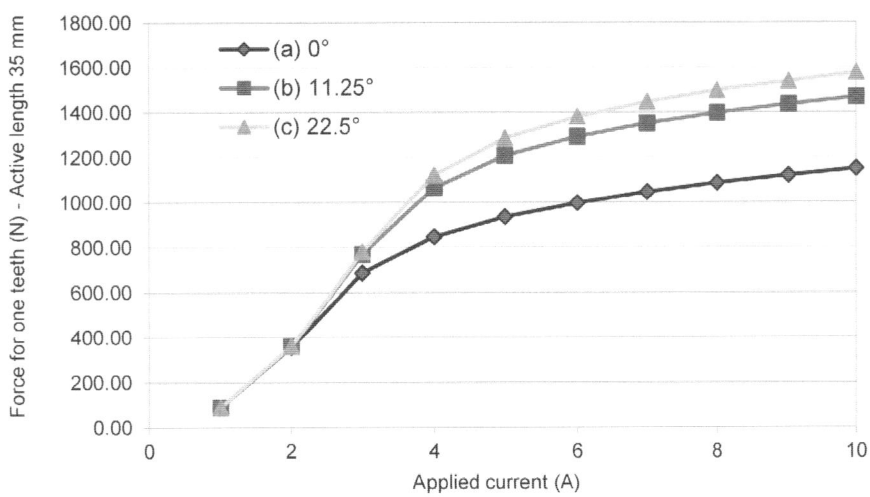

Figure 6 Forces obtained with three teeth spanning (a): 0.0° - (b): 11.25° - (c): 22.5°

4.2 Design optimization

In this case, we consider the slots rectangular. A design optimization has been carried out to maximize the value of the available force in the airgap of each tooth. The size of the shaft and the external diameter of the stator are fixed. The aim is to evaluate the best ratio between the three heights: stator yoke height, h_{yoke}, tooth height, h_{tooth}, and rotor lamination height, h_{rot_iron}. h_{rot_iron} can be expressed as a function of h_{yoke} and h_{tooth}. Thus, only two variables are tested. In addition, we consider that the surface of the winding window, $S_{winding}$, is fixed by the product $h_{winding}$. $w_{winding}$, with $h_{winding} = h_{tooth}$ - 4mm, the height of the winding and $w_{winding}$ the width of the winding.

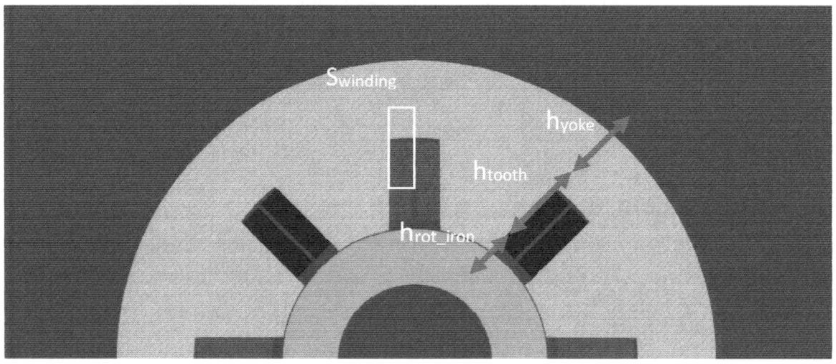

Figure 7 Optimized dimensions of the magnetic bearing

To achieve the optimization, we used the 3DVar technic available in the module ADAO of the Salome platform [5]. This classical and robust algorithm in static data assimilation, performs a state estimation by variational minimization of the objective function J [8]:

$$J(\mathbf{x}) = (\mathbf{x} - \mathbf{x}^b)^T \mathbf{B}^{-1}(\mathbf{x} - \mathbf{x}^b) + (\mathbf{y} - \mathbf{y}^{obs})^T \mathbf{R}^{-1}(\mathbf{y} - \mathbf{y}^{obs}) \tag{2}$$

The objective function is the sum of two terms, one measuring the distance to the background x^b, the other measuring the distance to the additional observation vector y^{obs}. These two terms are weighted by the inverse covariance matrices of the corresponding error B and R [8].

For the chosen external diameter of the stator, the optimized dimensions are given in **Table 1**:

Table 1 Obtained dimensions for the magnetic bearing parts in two cases

	Dimensions	*Case 1*	*Case 2*
D_{ext_stator}	External stator diameter	240.0mm	300.0mm
D_{int_stator}	Internal stator diameter	122.0mm	141.8mm
D_{ext_rotor}	External rotor diameter	121.0mm	140.8mm
D_{int_rotor}	Internal rotor diameter	55.0mm	55.0mm
h_{teeth}	Slot height	32.3mm	39.5mm
w_{slot} encoche	Slots width	24.5mm	19.4mm

If we compute the ratio of the stator and rotor iron part of the magnetic bearing, according to the external stator diameter excluding casing, we obtain:

Table 2 ratio of stator and rotor part over the external diameter

	Case 1	*Case2*
stator iron height ratio	49,2%	52,7%
rotor iron height ratio	27,5%	28,6%
rotor shaft diameter+airgap ratio	23,3%	18,7%
Height tooth ratio	13,5%	13,2%

The tooth height ratio remains the same in the two cases. As the external diameter value increases, the ratio on the stator side becomes greater than in the rotor part of the bearing. The forces obtained are given in Figure 8. As the external stator diameter increases, the value of the forces increases. This can be linked to the fact that both the surface in the airgap where forces are developed increases and the saturation in the magnetic material decreases.

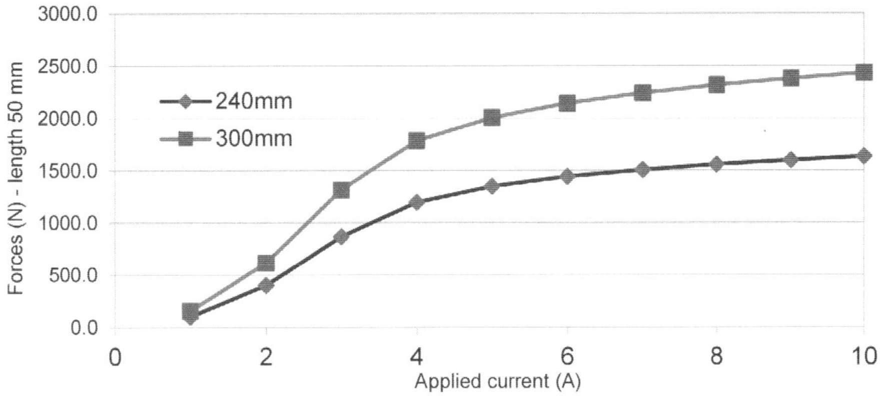

Figure 8 Calculated forces on one tooth for two external stator diameters with optimized dimensions

4.3 Misalignment force

In order to control rotor misalignment within the SAHJB, we have to apply a torque of 86 N.m. The value was predicted by the TEHD model and the detail are out of the scope of the present paper. Considering the dimensions of the rotor, the static forces to be applied by the AMBs is 235.6N.

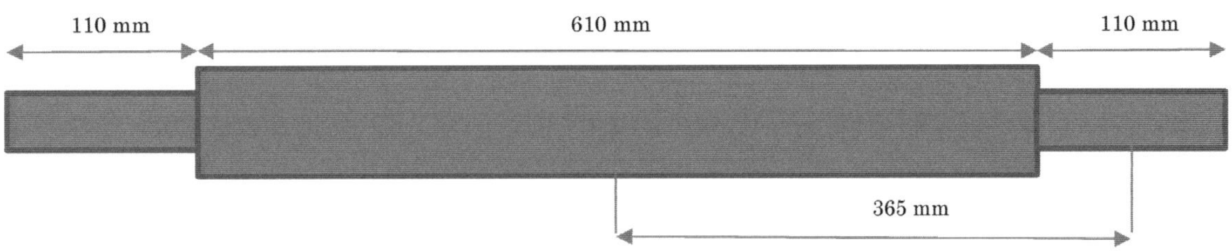

Figure 9 Rotor shaft dimensions used

The value of this force is quite small compares to the static force needed to load the SAHJB and don't appear to be dimensioning. The variation of the airgap is of bigger concerns.

4.4 Evaluation of the effect of misalignment of the rotor shaft

In the case of self-alignment of the SAHJB, we consider that the rotor tilts of 0.04°. Thus, we can estimate the variation of the airgap on each magnetic bearing considering the dimensions of the bench.

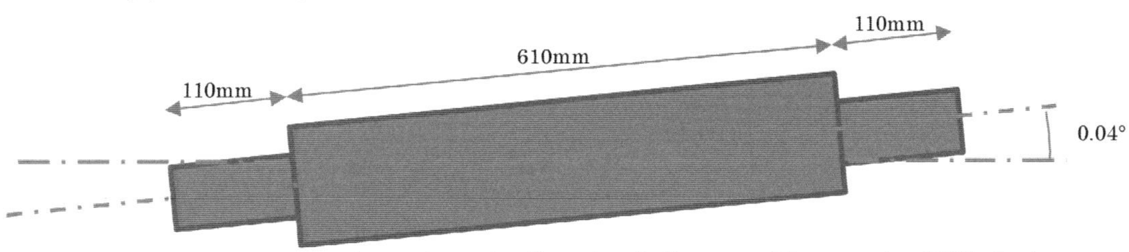

Figure 10 Dimensions used to evaluate the effect of a misalignment of the rotor after SAHJB tilted

The value of the airgap reduction in the middle of the bearing is around 0.256mm and we will use this value to evaluate the effect of the airgap variation on the available forces on each tooth. A 3D study must be done in order to fully estimated the effect but as a first approach, we used a 2D approach considering a mean reduction of 0.256 mm along the full length of the bearing. We consider that the variation of the airgap is anti-symmetrical on the two bearings.

Results obtained are depicted in Figure 11. We can observe, for a given applied current value, the effect of the variation of the airgap size on the y direction with both an increase of the forces where the airgap is reduced and a decrease of the forces where the airgap value is increased. Hence, we will have to consider the variation of the forces into the control part of the bearing.

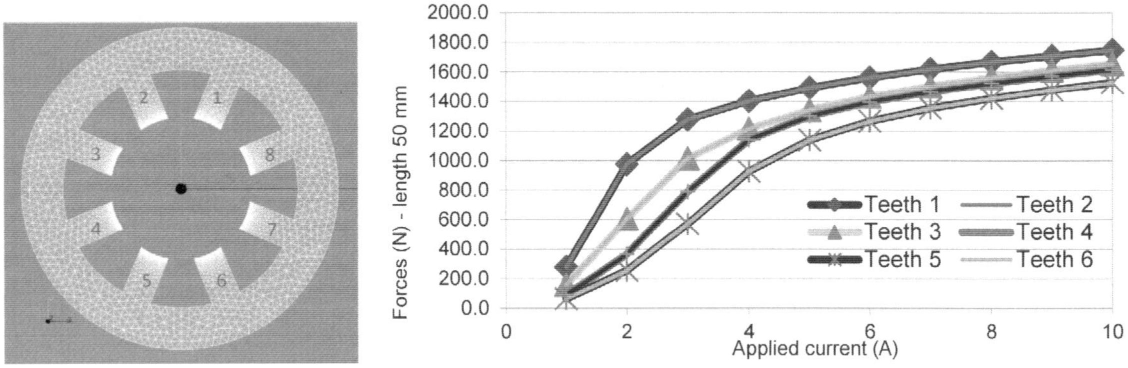

Figure 11 Effect of an airgap reduction of 200 µm on the y direction

5 Conclusions and perspectives

A first modelling has been carried out to evaluate the size of the magnetic bearing needed to be added on both sides of the SAHJB. An optimization technic associated with a parametric CAD drawing and meshing are available to test other dimensions and material in order to achieve the target forces required both by the stability of the rotor and the test of the SAHJB in align or misalign configuration. Further investigations are ongoing to obtain the final design of the AMBs.

A first prototype of the simple electromagnet actuator has been made to validate the modeling done. At this stage, static testing was performed to validate the geometrical design. Testing in real operating conditions are planned where power amplifiers, controllers and sensors design will be considered.

6 References

[1] Hassini, M., Zhang, S., Chatterton, S., and Pennacchi, P. (November 28, 2022). "Theoretical and Experimental Comparisons for Rotordynamic Coefficients of a Multiscratched Tilting Pad Journal Bearing." ASME. J. Eng. Gas Turbines Power. February 2023; 145(2): 021008. https://doi.org/10.1115/1.4055482

[2] Chatterton, S, Pennacchi, P, Vania, A, Hassini, MA, & Kuczkowiak, A. "Effect of Scratches on a Tilting-Pad Journal Bearing." *Proceedings of the ASME Turbo Expo 2020: Turbomachinery Technical Conference and Exposition. Volume 10B: Structures and Dynamics*. Virtual, Online. September 21–25, 2020. V10BT29A006. ASME. https://doi.org/10.1115/GT2020-14700

[3] Jolly P, Hassini MA, Arghir M, Bonneau O. Identification of stiffness and damping coefficients of hydrostatic bearing with angled injection. Proceedings of the Institution of Mechanical Engineers, Part J: Journal of Engineering Tribology. 2013;227(8):905-911. doi:10.1177/1350650113482613

[4] Voigt, A. J., Mandrup-Poulsen, C., Nielsen, K. K., and Santos, I. F. (January 4, 2017). "Design and Calibration of a Full Scale Active Magnetic Bearing Based Test Facility for Investigating Rotordynamic Properties of Turbomachinery Seals in Multiphase Flow." ASME. *J. Eng. Gas Turbines Power*. May 2017; 139(5): 052505. https://doi.org/10.1115/1.4035176

[5] Salome platform - https://www.salome-platform.org/

[6] Gerhard Schweitzer and Eric H. *Maslen Magnetic Bearings: Theory, Design, and application to rotating* machinery Springer (2009).

[7] Der Hagopian, J. and Mahfoud, J., "Electromagnetic Actuator Design for the Control of Light Structures", Smart Structures and Systems, Vol. 6, No. 1 (2010), 29-38.

[8] Talagrand O., Assimilation of Observations, an Introduction, Journal of the Meteorological Society of Japan, 75(1B), pp.191-209, 1997

Comparative analysis of the Performances and reliability of a Switching Power Amplifier (SPA) for Active Magnetic Bearing Systems: SiC MOSFET Vs Si IGBT Version

André De Andrade [a], Lakdar Sadi-Haddad [a], Ramdane Lateb [a], Joaquim Da Silva [a]

[a] SKF Magnetic Mechatronics, 2 rue des champs, 27950 Saint Marcel, France

Abstract

This paper explores the potential advantages of using wide bandgap (WBG) semiconductors, specifically Silicon-Carbide (SiC), over traditional Silicon (Si) in power electronic systems. The authors present a comprehensive comparison between the SiC MOSFET switching power amplifier (SPA) and the Si IGBT SPA, assessing their efficiency, weight, and footprint, as well as the behavior of the SiC SPA in presence of long power cables. Additionally, the durability and reliability of the SiC SPA are analyzed through environmental aging tests such as thermal cycling, vibration, shock tests, and EMC tests. The results show that the SiC SPA is a promising technology with significant advantages over traditional Si SPAs, such as reduced losses, increased efficiency, and decreased weight. However, further qualification tests are needed to ensure its reliability and durability under more demanding conditions. The SiC SPA has the potential to become a valuable solution for a wide range of applications, including oil and gas platforms and sub-marine applications.

Keywords: *Silicon Carbide, Switching Power Amplifier, Power density, Active Magnetic Bearing, efficiency, carbon footprint*

1. Introduction

Wide bandgap (WBG) semi-conductors, such as Silicon-Carbide (SiC), are widely recognized in power electronic circles for their significant advantages over traditional Silicon (Si), such as high-power density, lower losses, and smaller carbon footprint due to their lower energy consumption during operation [1]-[3]. This make them an attractive option for applications where energy efficiency and environmental impact are a concern. Nevertheless, the substantial cost of SiC components remains a major hurdle for their broader implementation in non-embedded industrial applications [4], and despite recent progress in wafer manufacturing [5]-[6].

SiC components were first used in IT infrastructures, renewables, and telecommunication. In automotive application, SiC MOSFET device was applied first by Tesla and followed by others especially in Asia [7]. Some comparative studies between Si IGBT and SiC MOSFET, analyzing three phase inverters [8]-[9] or a power factor controller [10], have shown a significant improvement in efficiency, weight, and size even by increasing the switching frequency. Additional benefits have been highlighted as well, such as a reduction of the heat sink, the output filter with a conclusion of a better cost-effective solution [8]-[9]. Despite the interest in the more electrical aircraft development [11], SiC semi-conductors have limited track record in high reliability applications.

Efforts have been made to comprehensively address the reliability aspects of WBG power devices, which warrant careful consideration [12]. Reliability studies have shown a time drift of the threshold voltage, gate oxide reliability concern and some short-circuit failures [13]-[17], attributed to the relative novelty of the product [18]-[19]. Besides the component itself, its operation within the circuit and the system is equally of importance.

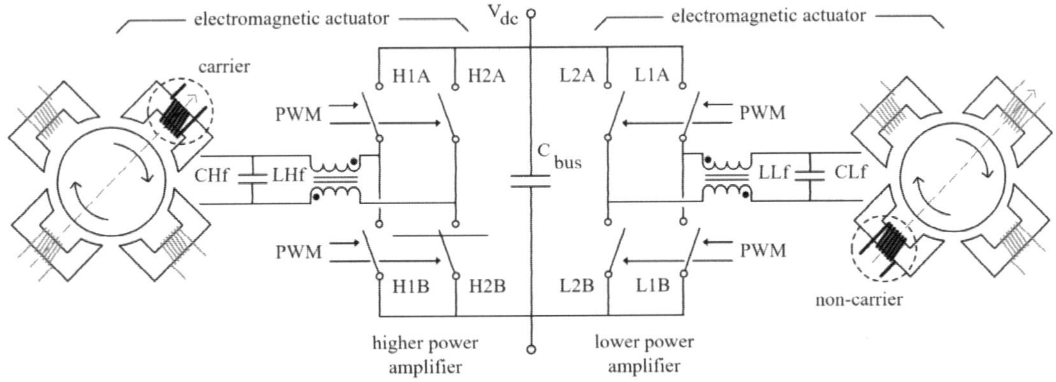

Figure 1 One axis electromagnetic actuator.

A higher dv/dt of SiC MOSFET can increase ringing and adversely affect the EMC/EMI performance, and higher switching frequency causes increased level of common-mode voltage leading to leakage/stray currents such as electric discharge machining type current and capacitive type current [20].

Controllable dv/dt and di/dt behavior of SiC MOSFET has been studied for telecom applications [21], for DC/DC boost converters [22] and for a three-phase inverter using long cables [23]. The use of SiC also holds potential value in the Oil & Gas industry, particularly in applications involving active magnetic bearings controlled by switching power amplifiers (SPA), such as compressors and turbo-expanders. SiC's benefits make it an attractive choice for embedded applications, such as platforms, floating production, storage, and offloading vessels (FPSO), floating storage and offloading units (FSU), and sub-marine applications requiring reliable operation at significant depths. The paper aims to present a comprehensive comparison between two power electronic systems: the SiC MOSFET SPA and the Si IGBT SPA supplying active magnetic bearings (AMBs).

Figure 1 shows a typical representation of two SPA bridges supplying one axis electromagnetic actuator. To achieve this, the authors will assess their efficiency (i.e., losses), weight, and footprint. Additionally, they will investigate the behavior of the SiC SPA in presence of long power cables.

Furthermore, the authors analyze the durability and reliability of the SiC SPA by subjecting it to various environmental aging tests such as thermal cycling, vibration, shock tests, and EMC tests. Through these tests, the authors aim to provide a better understanding of the SiC SPA's performance in real-world conditions.

2. Switching Power Amplifier

2.1. Half bridge Vs full bridge power amplifier

The asymmetric half-bridge (Figure 2) converter is widely recognized as the most used topology in Active Magnetic Bearings (AMBs) due to its simplicity of implementation. It only requires two controllable power components (IGBT or MOSFET) and two diodes. This topology is preferred for its ease of use and practicality in AMB applications. it is considered the most mature and suitable option for AMB's differential control. In AMBs, the coil currents responsible for controlling one Degree of Freedom (DOF) of motion are typically equal to the bias current, with the addition or subtraction of the control current. The coil current primarily undergoes changes near the bias current. Given these characteristics, the unidirectional current supplied by the half-bridge SPA proves to be sufficient for AMBs.

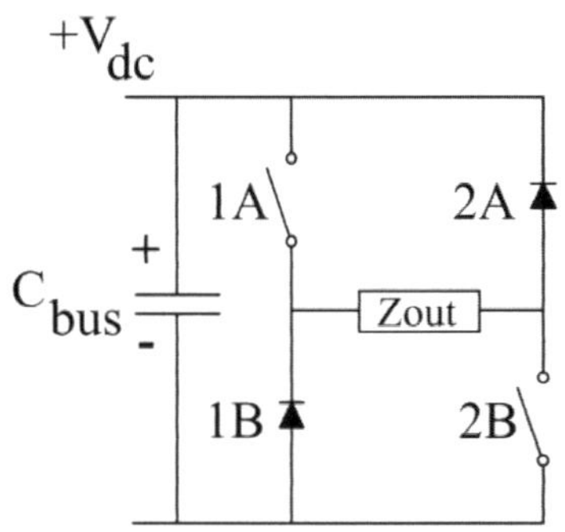

Figure 2 Half bridge topology.

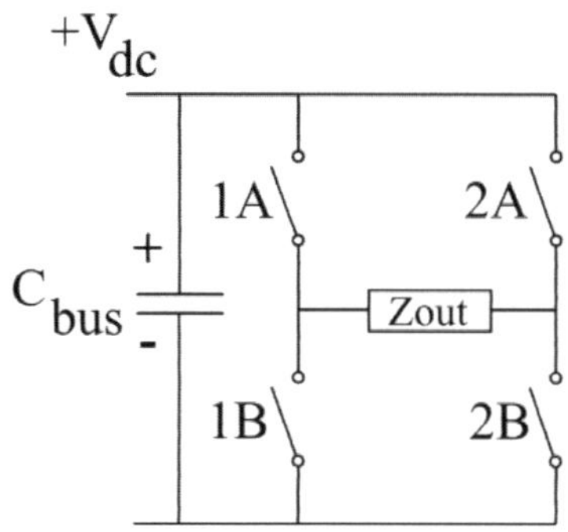

Figure 3 Full bridge topology.

However, it is important to note that the asymmetric half-bridge converter has two potential drawbacks when the design criteria prioritize overall efficiency and dynamic performance of the system. In such cases, the full bridge topology (Figure 3) may be worth considering for two reasons.

Firstly, it allows for the minimization of losses by employing MOSFETs (Figure 4) instead of diodes, resulting in improved efficiency.

Secondly, the full bridge configuration enables the injection of a negative current, which can enhance the dynamic behavior of the magnetic bearing, especially in the presence of eddy currents. In the following paragrah, a comparison will be made between a Si IGBT half-bridge SPA and a SiC MOSFET full-bridge SPA, exploring their respective characteristics and performance in AMB applications for $V_{dc} = 300\ V$ and $i_{amp} = 30\ A$.

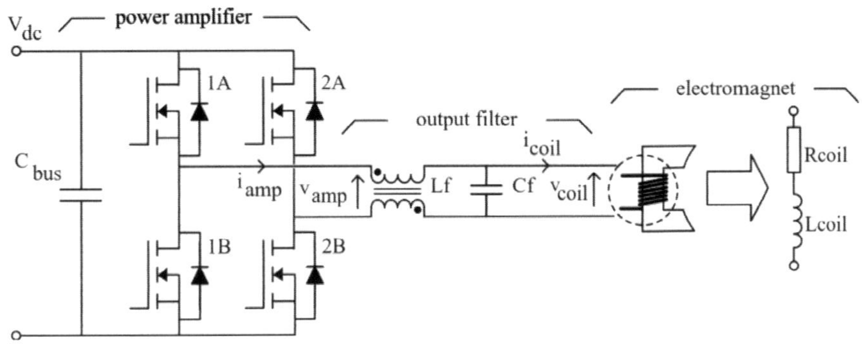

Figure 4 Full bridge SPA supplying one coil of the AMB actuator.

2.2. Integration

Figure 5 shows a visual comparison between Si and SiC SPAs and Figure 6 exhibits the SiC-SPA currently in qualification process. In the following, this paper brings performances on its design. The benefits of SiC are listed:

i) load Frequency: 8x faster.

ii) power Losses: 50 % lesser.

iii) heatSink volume: minimized.

iv) output cabinet filter: integrated within SiC-SPA.

v) system footprint: 65 % saved space.

vi) weight: 48.5 % lighter.

Figure 5 Layout Si (top) vs SiC (bottom).

Figure 6 SiC SPA: currently in qualification process.

2.3. Losses

As observed in previous publications, it is not surprising that the losses are lower in SiC-based configurations compared to those in Si-based configurations (Figure 7). This trend holds even when the switching frequency is multiplied by a factor of 10 or 14. The SiC version of the SPA consistently exhibits lower losses. Additionally, increasing the switching frequency has the added benefit of reducing the output filter of the SiC-based device, while maintaining low losses, as compared to the Si-based version of the device.

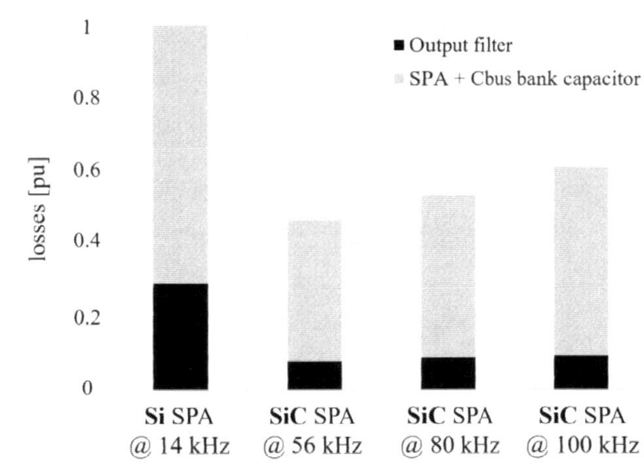

Figure 7 Comparison of losses between Si and SiC configuration..

2.4. Switching Mode

The IMPA operates in 3-level PWM method (cf. Figure 8). Benefiting from the increase of the switching frequency f_{sw} (SiC technology) the output load frequency $f_{ap}$is twice that applied to the command of each SiC-MOSFET ($f_{ap} = 2/T_{sw}$). This is a double impact: bandwidth increase and filtering optimization. The current ripple in the load and the voltage ripple in the capacitor bank terminals are reduced. Figure 9 shows the operation in the first quadrant. In STEPS 1 and 3 the energy is transferred from the capacitor $\boldsymbol{C_{bus}}$ to the bearing. Figure 10 shows the operation in the fourth quadrant. In STEPS 1 and 3 energy flows from the bearing to the capacitor $\boldsymbol{C_{bus}}$. STEPS 2 and 4: freewheeling steps.

When operating with any constant current, the duty cycle $D \approx 0.5$. The ripple Δi_{amp} tends to zero. The Table 1 illustrates some of the equations impacting the design.

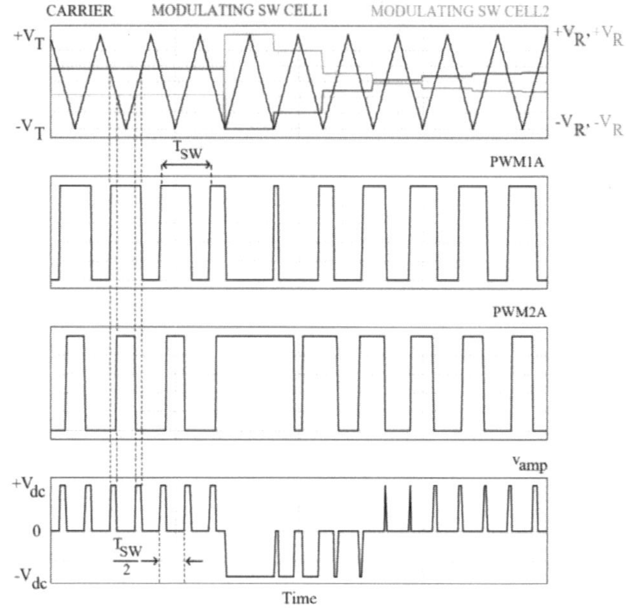

Figure 8 3-level PWM waveforms.

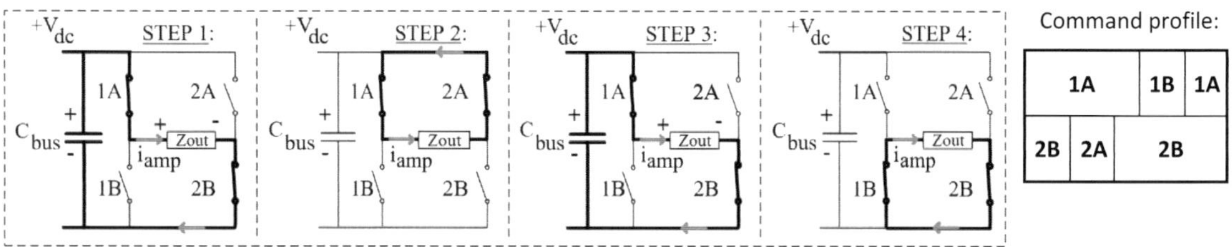

*Figure 9 Switching sequency in steady state configuration for $i_{coil} > 0$ and 2-levels v_{amp} **v** $[0, +V_{dc}]$.*

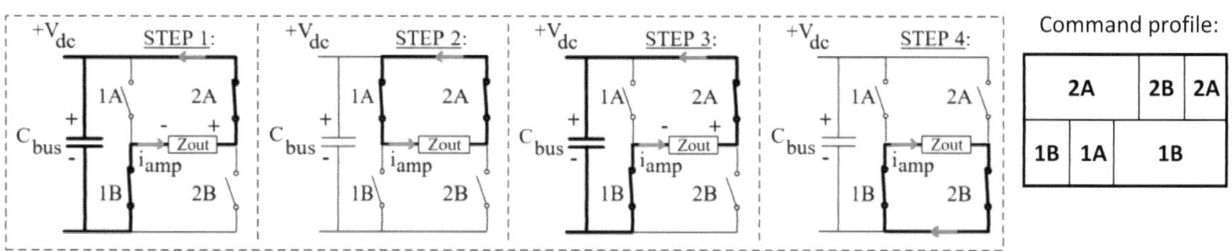

*Figure 10 Switching sequency in steady state configuration for $i_{coil} > 0$ and 2-levels v_{amp} **v** $[0, -V_{dc}]$.*

Table 1 Summary of 3-level PWM method equations

Duty cycle for leg 1:	Duty cycle for leg 2:	Modulation index:
$D_1(t) = 0.5\,[m\,sin(\omega t) + 1]$	$D_2(t) = 0.5\,[m\,sin(\omega t - \pi) + 1]$	$m = V_R/V_T$
Duty cycle cases:		
$0 \leq D \leq 1/2$	$1/2 \leq D \leq 1$	$D = 1/4 \; or \; D = 3/4$
i_{amp} current ripple:		
$\Delta i_{amp} = \frac{V_{dc}}{L_f f_{sw}}(1-2D)D$	$\Delta i_{amp} = \frac{V_{dc}}{L_f f_{sw}}(2D-1)(1-D)$	$\Delta i_{amp}^{MAX} = \frac{1}{8}\frac{V_{dc}}{L_f f_{sw}}$
v_{coil} volage ripple:		
$\Delta v_{coil} \approx \frac{1}{4}\frac{V_{dc}}{L_f C_f f_{ap}^2}(2D-1)D$	$\Delta v_{coil} \approx \frac{1}{4}\frac{V_{dc}}{L_f C_f f_{ap}^2}(2D-1)(1-D)$	$\Delta v_{coil}^{MAX} \approx \frac{1}{32}\frac{V_{dc}}{L_f C_f f_{ap}^2}$

2.5. Passive Components

2.5.1. DC-link film Capacitor Bank C_{bus}

The dc bus capacitor needs to deal with the i) ripple current due to power amplifiers switching; ii) voltage fluctuation due to the source lead inductance; iii) voltage transient due to leakage inductance and SiC power device switching and iv) overvoltage due to regeneration the energy stored in the electromagnetics to the C_{bus} capacitor.

Inside the cabinet, all power amplifiers are connected to the fixed and non-regenerative common DC bus. In practice, the case (iv) is the most restrictive event. Improving the reliability, film technology is very suitable. Looking a trade-off between high RMS current values and reducing PCB (printed circuit board) footprint, the criterion considers the highest capacitive density per pcb area. For this, references from different manufactures (Kemet, TDK, Vishay, etc.) with different shapes could be evaluated. Figure 11 and Figure 12 show some examples.

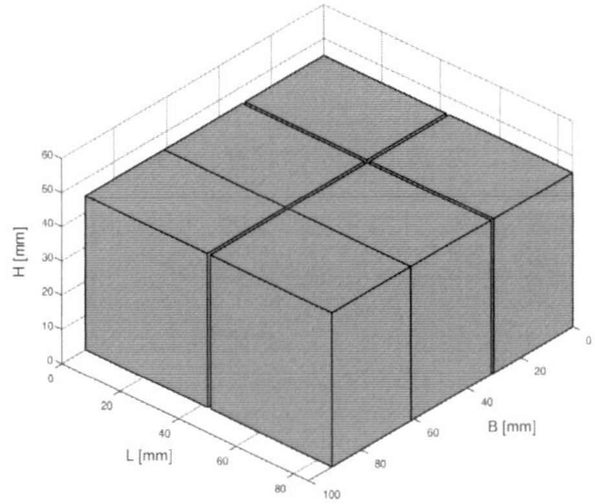

Figure 11 quantity = 6; total density: 5.6 µF/cm².

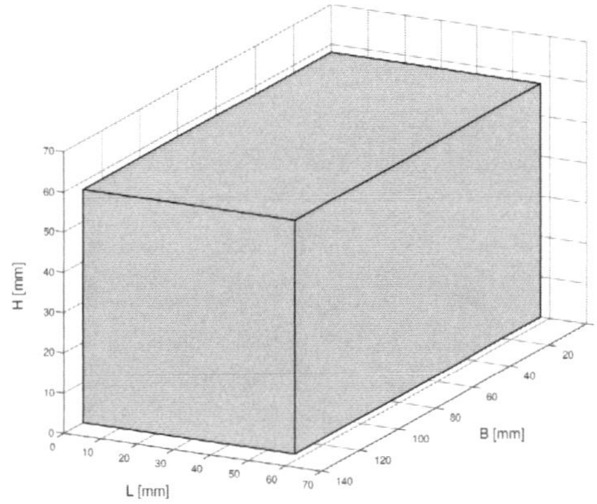

Figure 12 quantity = 1; total density: 6.4 µF/cm².

2.5.2. Output filter

In differential mode, many roles are assigned to this function (protecting the coil against the high dV/dt from the SiC, master the output HF current ripple reducing the cable losses, symmetrize the switching nodes at the output of the power amplifier. Different parameters are involved: i) power inductor ripple current; ii) output capacitor voltage ripple; iii) switching frequency value; iv) cutoff frequency; and v) mechanical volume for integration.

2.5.2.1. Output capacitor C_f

Figure 13 shows the relationship between C_f and its voltage ripple ΔVC_f and the output load frequency f_{ap}. A constant ripple current on the power inductor is considered and fully absorbed by the capacitor. For this analysis, due to SiC power device, the frequency range is large and the cutoff frequency cf. (1).

It is interesting to note that the largest C_f variation occurs in the first tens of kilohertz.

$$f_{cutoff} = \frac{1}{2\pi\sqrt{L_f C_f}} < \frac{f_{ap}}{2} \quad (1)$$

For the design, meeting the balance between power amplifier losses, thermal and EMC (dV/dt increase) behavior, and the compactness of passive elements, a trade-off to be kept in mind.

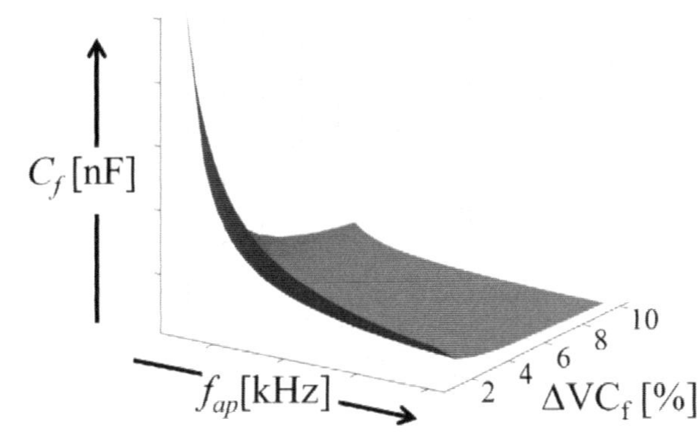

Figure 13 Output capacitor dependency.

2.5.2.2. Output Power Inductor L_f

The power inductor must handle a specified DC current within the electromagnetic actuator. Looking for the compactness, the design integrates the benefits of SiC-MOSFETs high switching frequency. Regarding the 3-level PWM modulation, two benefits integrate the design: i) frequency of the current ripple is increased twice: $f_{ap} = 2 f_{sw}$; ii) current ripple amplitude minimized: $\Delta i_{amp}^{D=1/2} \approx 0$ (in steady state the required output voltage for the magnet is low: $D \approx 1/2$).

The approach to design considers the equation (2). The peak inductive energy storage (left side) cannot exceed the maximum magnetic energy storage capability of the core material (right side).

$$\frac{1}{2}L_f\left(i_{amp}^{peak}\right)^2 = \frac{1}{2\mu}\left(B_{core}^{MAX}\right)^2 v_{core}^{MIN} \quad (2)$$

v_{core}^{MIN} is the minimum core volume and $B_{core}^{MAX} < B_{SAT}$ (magnetic non-saturation condition). This criterion (2) impacts the choice of the magnetic powder core: the v_{core}^{MIN} is proportional to the permeability. A toroidal core geometry and single-layer wound (better to heat extraction at high temperatures) are considered. Figure 14 shows the industrial packaging integrating both inductors: L_{Hf} and L_{Lf}.

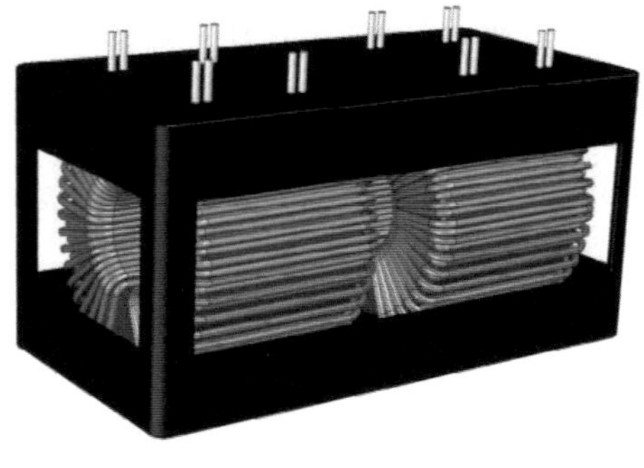

Figure 14 Industrial manufacturing packaging.

2.6. EMC Input Filter

For industrial applications, electromagnetic constraints are concerned, and requirements are set in a regulatory context. The switching power amplifiers generate significant electromagnetic conducted (EMC) interference in a broad spectrum and special attention must be considered related to the high switching frequency and fast time transitions of voltages and currents using SiC-MOSFETs. To minimize the global effort during the qualification of the whole active magnetic system, the approach to design is suppressed the disturbances per axis to an acceptable level.

Figure 15 shows (V_S: input voltage supply) a simplified EMC input filter structure and Figure 16 an equivalent common mode (CM) circuit. In common mode, the attenuation is achieved by the effect of the choke L_{MC} and the two C_Y capacitors. In differential mode (DM), the attenuation is carried out by the C_X and the dispersion inductance on L_{MC}.

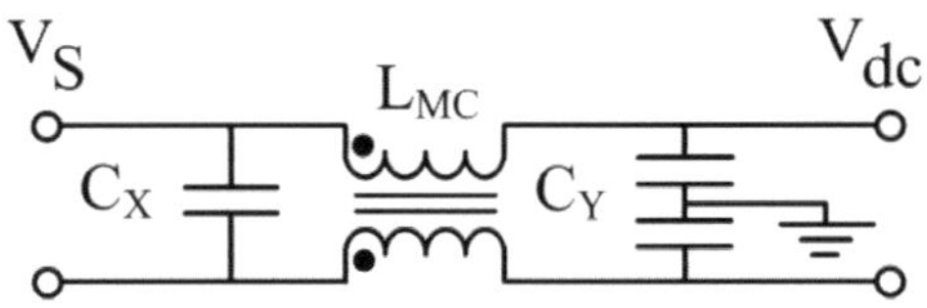

Figure 15 EMC Input filter topology.

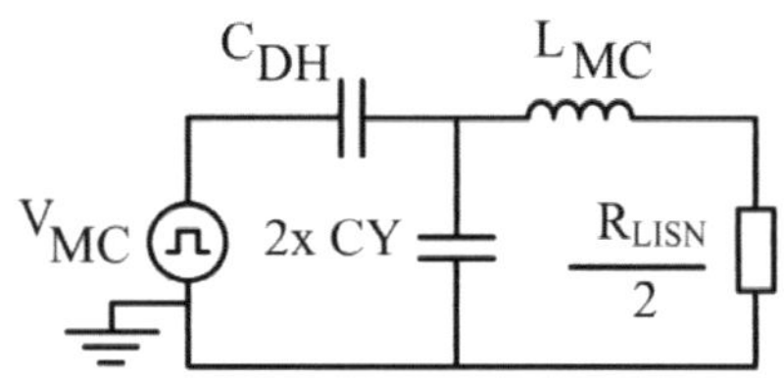

Figure 16 CM equivalent circuit (C_{DH}: parasitic capacitance).

For EMC tests for measurements, Figure 17 and Figure 18 summarize the equivalent circuits setup using the line impedance stabilization network (LISN).

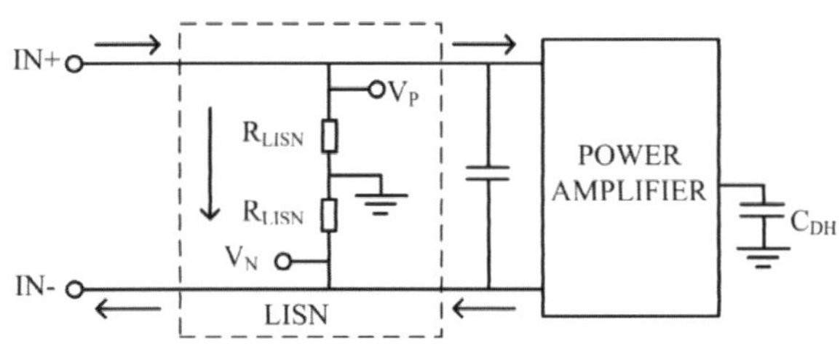

Figure 17 Equivalent circuit for differential mode conduction.

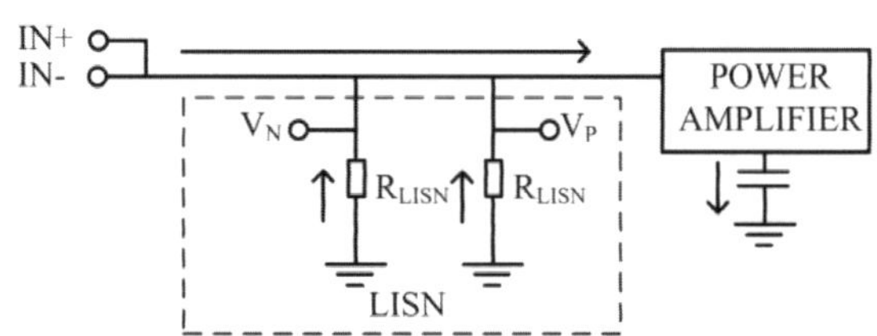

Figure 18 Equivalent circuit for common mode conduction.

L_{MC}, C_X and C_Y values result from an experimental and simulation approach. The EMC signature pf the power amplifier performing from 150 kHz to 30 MHz according to the IEC 61000-6-4.

2.7. Long cables influence

Figure 19 shows a simplified grounding diagram with a shielded cable. In the field, the cable length can reach up to 700 meters! R_C(resistance), L_C (inductance), C_P (capacitance) and G_P (conductance), are the parameters modeling the cable. Each parameter is a frequency-dependent function.

In addition to EMC impact, two potential issues are identified when connecting the magnetic bearings to a power amplifier with long cables: overvoltage and leakage currents. For predict and mitigate these different issues, a model with a series of elementary cells and (n+1) conductors is shown in Figure 20.

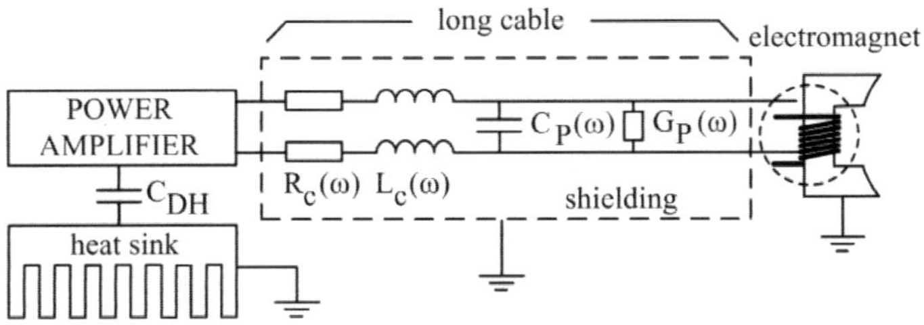

Figure 19 Simplified grounding diagram.

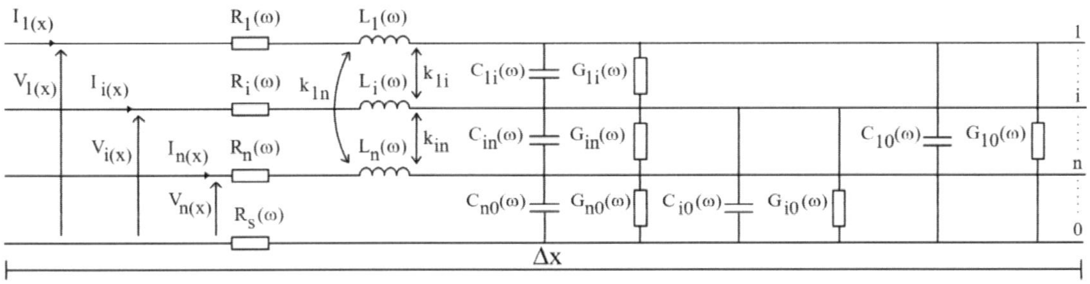

Figure 20 Elementary cell of multiconductor transmission

Based on the MTL (Multiconductor Transmission Line) theory, the cable can be modeled with a series of elementary cells. The work in [24] presents a practical solution for power cable characterization with the impedance analyzer. A cable with 12 conductors from Nexans is considered and the frequency band from 10 kHz to 30 MHz.

A proposal of a specific adapter, with spring connections, allows the power cables characterization for electromagnetic simulation. Considering the cable geometry and the uncertainty of measurement, with 76 impedance measurements, the development provides a protocol extracting the cable parameters in function of the frequency.

3. Environmental testing of the SiC Switching Power Amplifier

It is crucial to rigorously evaluate the reliability of SiC under extreme environmental conditions. Robust testing under challenging conditions is essential to ascertain the true performance and durability of SiC.

SiC-SPA qualification is currently in progress. In this chapter some qualification tests are presented. To carry out environmental tests, established standards set by the International Electrotechnical Commission (IEC) and the American Petroleum Institute (API) are considered: i) API-17F for thermal and vibration testing; ii) IEC 61000-6-4 for measurement of conducted disturbance and radiated field.

3.1. Thermal tests

Thermal testing is a crucial process used to assess the thermal fatigue resistance of solder joints, interconnections, and top contacts in electronic devices. Inside the climate chamber, this testing involves subjecting the Device Under Test (DUT) to hot and cold temperatures, while monitoring its performance during and after the test. After completion, an external visual inspection of the marking is also performed. Figure 21 and Figure 22 show the profiles carried out.

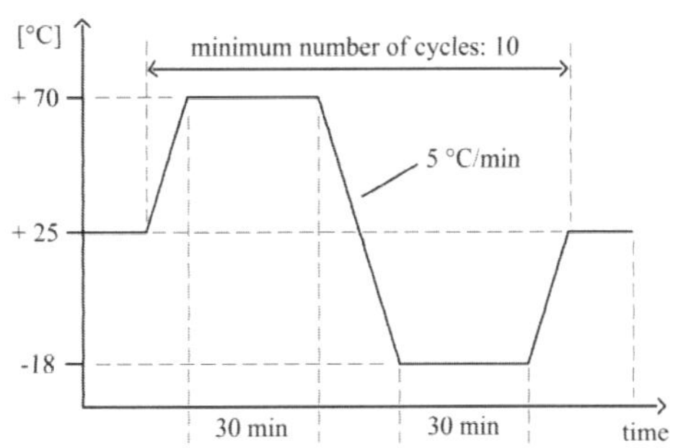

Figure 21 Cycling testing profile.

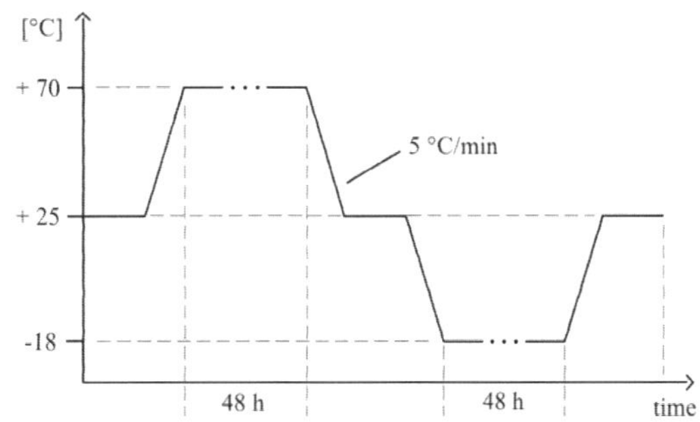

Figure 22 Soaking testing profile.

Several thermocouples are located inside (glued on the different electronic boards) and outside the casing. To evaluate the safety margin, Figure 23 summarizes it for 50 different electronic components. This is the difference between the maximum values from each manufacturer (extracted from datasheets) and the maximum temperature measured during the tests when the power amplifier delivers 100 % of its capacity (full load in operating point).

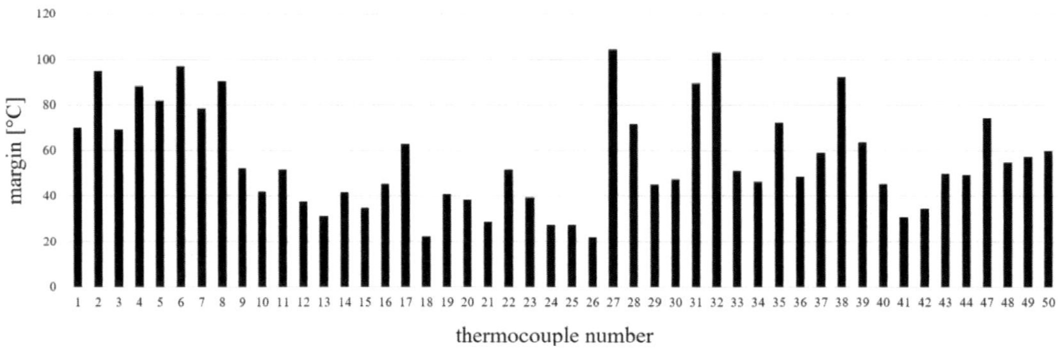

Figure 23 Temperature margins for 50 components equipped by thermocouples.

3.2. EMC/EMI

Figure 24 shows the measurement of conducted emissions generated by SPA. It the figure compares the signature without filtering to the solution with EMC filtering and improvement of the earth connections to the casing. In view of the standard (frequency $\in [150\ kHz;\ 30\ MHz]$) it is interesting to notice the net improvement impact.

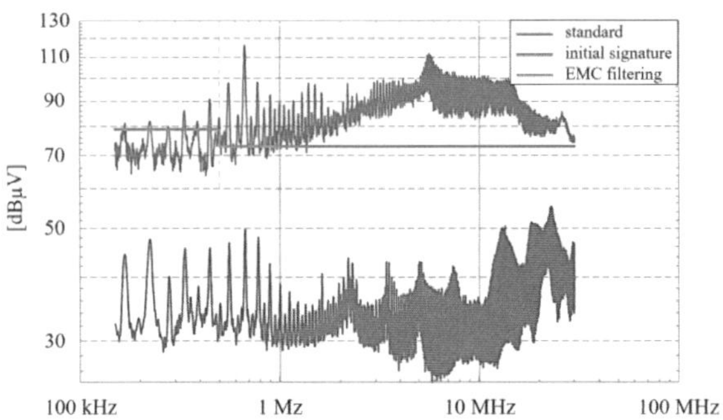

Figure 24 Measurement of conducted emissions.

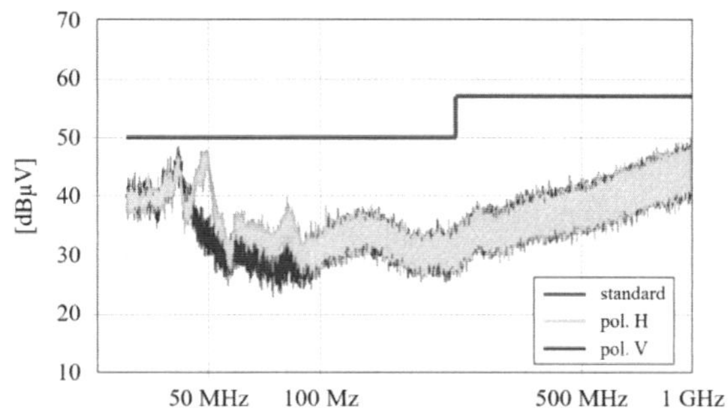

Figure 25 Measurement of radiated field.

Figure 25 shows the measurement of radiated emissions generated by SPA, results obtained inside an anechoic chamber. The horizontal (pol. H) and vertical (pol. V) polarized plans of oscillation were catch up. According to the standard (frequency $\in [30\ MHz;\ 1\ GHz]$) the results are very satisfactory.

Through careful design modifications and improvements, the measured results demonstrate unequivocal compliance with the standard, with emissions levels below the specified limits.

3.3. Vibration

During vibration and shock testing, it's important to ensure that the testing surface, such as the table surface, remains flat to accurately transmit the vibrations from the shaker to the DUT. The DUT is fastened to the platform (Figure 27). The SPA shall be undertaken with the DUT powered and with continuous monitoring of functions.

Before manufacturing the casing for the SPA, a finite element model predicting the mechanical modes was developed (Figure 26). The vibration test is a confirmation of the concept carried out for this design.

Figure 26 Modeling of mechanical modes.

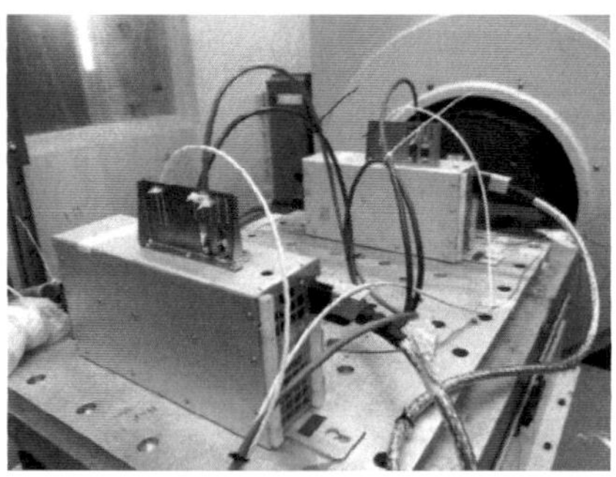

Figure 27 Assembly on vibration machine.

3.3.1. Sine sweep fatigue analysis

Test to find resonant frequencies that can excite the natural frequencies of the DUT. Test profile:

Table 2 Sine sweep profile

Parameter	*Value*
sweep rate	1 oct/min
frequency band	5 – 1000 Hz
number of sweeps	2 (1 up + 1 down)
level	5 – 25 Hz: 2 mm
	25 – 1000 5 g

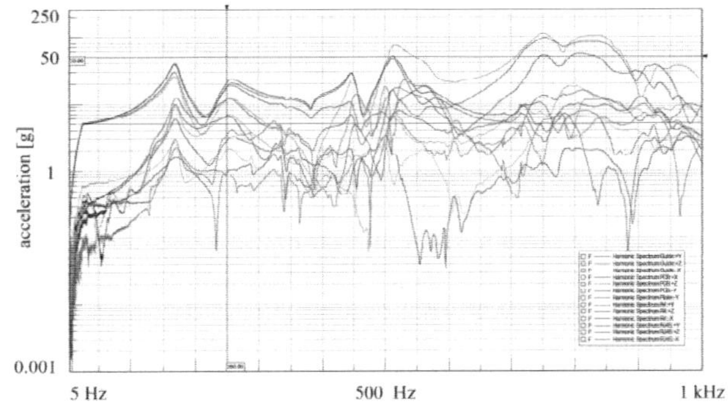

Figure 28 Frequency research following Y axis.

Figure 28 shows the experimental results for Y-axis. Each signature corresponds an accelerometer placed for acceleration measurement inside the DUT.

3.3.2. Shock test

Test to assesses the ability to survive a transient event by running a sharp transfer of energy. Test profile:

Table 3 Shock profile

Amplitude	*Duration*	*Number of shocks*	*Duration between pulses*
30 g	11 ms	4 positives and 4 negatives	1 s

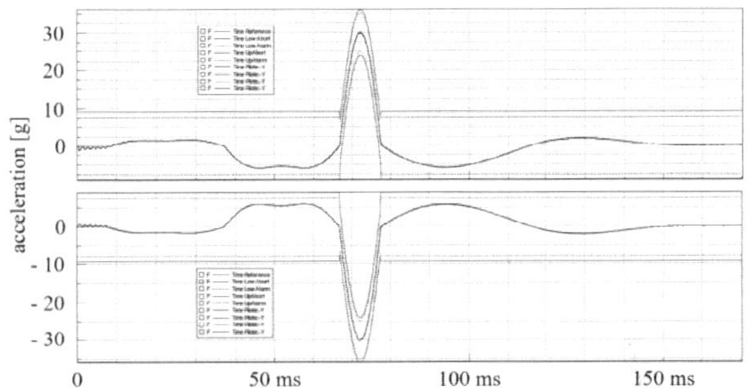

Figure 29 Driving curves during shock tests following the Y axis.

Figure 29 shows the experimental results for Y-axis.

3.3.3. Random vibration fatigue

This test includes all the forcing frequencies which a shaker vibrates and excites all DUT resonances. Test profile:

Table 4 Random profile

Frequency [Hz]	*Slope [dB/octave]*	*Amplitude [g^2/Hz]*
20 – 80	+ 3	-
80 – 350	-	0.04
350 – 2000	- 3	-
RMS	-	6

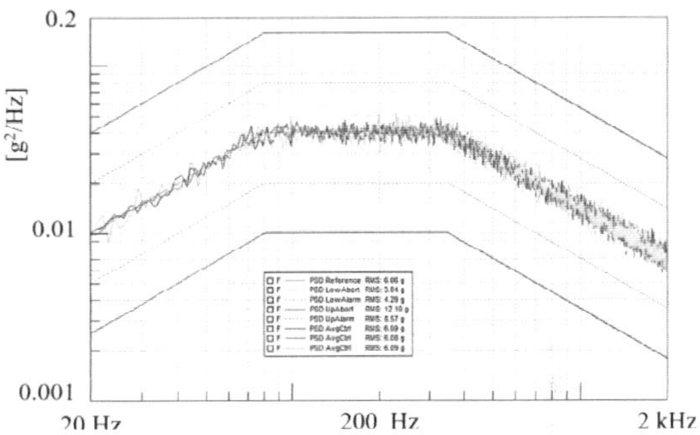

Figure 30 Driving curves during random following Y axis.

Figure 30 shows the experimental results for Y-axis.

The experimental tests showed the robustness behavior regard to all profiles: no risk of plastification identified.

4. Conclusion

The SiC SPA has been analyzed and compared to the Si SPA, revealing its potential as a more efficient and lightweight solution with reduced losses even at very high switching frequency. Moreover, the SiC SPA has undergone several environmental cycling tests and has successfully met IEC and API standards. However, further qualification tests need to be conducted to ensure its reliability and durability under more demanding conditions, such as when mounted on a cabinet subject to vibrations, shocks, and EMC.

Overall, the SiC SPA is a promising technology that offers significant advantages over traditional Si SPAs and has the potential to become a valuable solution for a wide range of applications.

References

[1] A. Elasser, T.P. Chow, "Silicon Carbide Benefits and Advantages for Power Electronics Circuits and Systems". IEEE proc, Vol. 90, pp. 969-986, 2002, [CrossRef]

[2] J. Bielaall, et all, "SiC vs. Si-Evaluation of potentials for performance improvement of Inverter and DC-DC Converter Systems by SiC Power Semiconductors", IEEE Trans. on Industrial Electronics, Vol. 58, No. 7, July 2011, [CrossRef]

[3] J. Millán, et all, "A Survey of Wide Band Gap Power Semiconductor Devices," IEEE Trans. on Power Electronics., Vol. 29, No. 5, pp. 2155-2163, May 2014, [CrossRef]

[4] T. Neyer, "What's the difference between Silicon Carbide and Silicon", website electronicdesign.com, March 2, 2022, [CrossRef]

[5] S. Daryanani, "The road to 200-mm SiC Production", Website Power Electonic news, June 14, 2022, [CrossRef]

[6] M. Di Paolo Emilio, "SiC Power Devices: Lowering costs to Drive Adoption", Website Power Electonic news, January 18,2022, [CrossRef]

[7] W. Cai, et all, " Review and Development of Electric Motor Systems and Electric Powertrains for New Energy Vehicles", Automotive innovation , vol 4, Feb 2021, [CrossRef]

[8] STmicroelectronic, " E-mobility SiC Traction Inverter", APEC 2020, March 2020, [CrossRef]

[9] M. Nitzsche, et all, " Comprehensive comparison of a SiC MOSFET and Si IGBT based Inverter", PCIM 2019, May 2019, [CrossRef]

[10] E. Ayerbe, J.Shao, "SiC Mosfet Based Bi-Directional 3 phase AC/DC converter", internal presentation CREE 2017

[11] P. W. Wheeler, "The more electric aircraft: Why aerospace needs power electronics?" Power Electronics and Applications, 2009. EPE '09. 13th European Conference on, pp. 1-30, Sept. 2009, [CrossRef]

[12] A.J. Wileman, et all, "A roadmap for reliable power electronics for more electric aircraft", Progress in Aerospace Sciences, Vol 127, Nov 2021, [CrossRef]

[13] B.J. Nel, S. Perinpanayagam, " A brief overview of SiC MOSFET failure modes and design reliability", 5th Int Conf TesConf 2016, [CrossRef]

[14] C. Chen, et all, " Study of Short-circuit Robustness of SiC MOSFETs, analysis of the failure modes and comparison with BJTs", Microelectronics Reliability,Vol 55, pp 1708-1713, August 2015, [CrossRef]

[15] A. Castellazzi , et all, "Sic Power MOSFETs performance, robustness and technology maturity", Microelectronics reliability Vol 58, pp 164-176, March 2016, [CrossRef]

[16] S. Jagdale, "Reliability and robustness tests for Next-generation high voltage SiC Mosfets", Power electronic news, June 2022, [CrossRef]

[17] S. Daryanani, "The SiC MOSFET reliability studies at the Ohio State University: Threshold Voltage Stability and Gate Oxide Screening Challenges", Power electronic news, Sept 2022, [CrossRef]

[18] M. Furukawa, et all, " Improved reliability of 1.2 kV SiC MOSFET by preventing the intrinsic body diode operation", PCIM 2020, [CrossRef]

[19] L. C. Yu, et all, "Reliability issues of SiC MOSFETs: A technology for high temperature Environments", IEEE Trans. on Device and Materials Reliability, Vol.10, NO.4, pp 418-426, Dec 2010, [CrossRef]

[20] Y.Xu, et all, " Experimental assessment of high frequency bearing currents in an induction motor driven by a SiC Inverter", IEEE Access, vol 9, pp 40540-40549, March 2021, [CrossRef]

[21] D. Aggeler, J. Biela, J.W. Kolar, "Controllable dV/dt behavior of the SiC MOSFET/JFET Cascode, an alternative hard commutated Switch for telecom applications", APEC 25th conference, Feb 2010, [CrossRef]

[22] S. Hazra, et all, "High switching performance of 1700V, 50A SiC power MOSFET over Si IGBT/BiMOSFET for advanced power conversion applications", IEEE transaction on power electronics, Vol. 31, Issue 7, pp 4742-4754, July 2016, [CrossRef]

[23] M. di Benedetto, et all, "Efficiency Comparison of 2-level SiC inverter and soft switching Snubber SiC inverter for electric motor drives", Energies 2021, Vol .14, issue 16, pp 1-17, 2021, [CrossRef]

[24] T. G. Bade, J. Roudet, A. Derbey, A. de Andrade, L. Sadi-Haddad, "Adapter for the Impedance Measurement of Power Cable with an Impedance Analyzer" IEEE Transactions on Electromagnetic Compatibility, 14 March 2023 [CrossRef]

AUTHOR INDEX

Linz Center of Mechatronics GmbH
Altenberger Straße 69
4040 Linz
Osterreich

ISBN 978-1-7138-9516-9

CURRAN ASSOCIATES INC.
proceedings
.com